PHYSIOLOGY

by

Charles J. Grossman, Ph.D.

Associate Professor of Physiology and Biophysics,
University of Cincinnati College of Medicine, and
Professor of Biology, Xavier University

Sulzburger & Graham Publishing, Ltd.
New York

Dedication:
For Elaine. Without her support and help I could never have completed this project.

Editor-in-Chief, Publisher:
Neil C. Blond, Esq.

Graphic Illustration:
Jeremy Borenstein
Daniel Borenstein

Interior Design:
David Gregory

P.O. Box 20058
Park West Station
New York, NY 10025

ISBN 0-945819-42-0
Printed in The United States of America

Contents

1. Levels of Organization and Homeostasis 5
The Cell as a Living System 5

2. Membrane Physiology 15
Structure and Composition 15

3. Nervous System Neurophysiology and the Action Potential 31
Function of the Nervous System 31
Events That Lead to an Action Potential 43
Generation of an Action Potential 46

4. Central Nervous System 61
Introduction 61
Organization of the Central Nervous System (CNS) 63

5. Special Senses 89
Sensory receptors and transduction 89
The Sense of Smell 95
The Sense of Taste 98
Vision 100
Hearing and Balance 109
Nerve Pathways that Conduct Auditory Stimuli 113
Balance or Equilibrium 114

6. Muscle Physiology 115
Muscle Mechanics 131
Smooth Muscle V. Cardiac Muscle 132

7. Cardiac Physiology 139
The Cardiovascular System: An Overview 139
The Cardiac Conduction System 144

8. The Vascular System and Control of Blood Pressure 161
Pathways of Circulation 161
The Involvement of the Lymphatic System 174

9. The Blood 187
Introduction 187
Blood Cells 188

Blood Typing 195
Hemostasis 199
The Blood Clotting Systems 201

10. Defense Mechanisms 205
Lymphatic System 205
Development of Immune Effector Cells 207
The Specific Immune Response 213

11. Respiratory System 231
Internal and External Respiration 231

12. The Urinary System and Renal Physiology 263
Functions of the Urinary System 263
The Kidneys 263
The Counter-Current Mechanism 279
Elimination of Urine 285

13. Fluid, Electrolite and Acid-Base Balance 289
The Intra-cellular and Extra-cellular Compartments 289

14. The Digestive System 309
Functions of the Digestive System 309

15. The Endocrine System 339
Endocrine Glands 348
Hormones of the Endocrine System 354

16. The Reproductive System 385
Development of the Reproductive System 385
Primary v. Accessory Reproductive Organs 389
Male and Female Sexual Response Cycle Is Similar 401
Female Reproductive System 403
Menstrual Cycle 411
Puberty 419
Menopause 419
Lifestyle of the Sperm and Ovum 420

Chapter 1
LEVELS OF ORGANIZATION AND HOMEOSTASIS

THE CELL AS A LIVING SYSTEM

The cell is the smallest structurally integrated unit that is considered alive. Any smaller structures, such as viruses or rickettsia, are not truly alive unless they become associated with a cell. Once within the confines of the cell or plasma membrane, they can proceed to hijack the cell machinery and duplicate their own genetic material.

Whether a cell exists singularly or as a unit in a multicelled organism, it can perform certain basic functions. These functions denote a living system and are as follows:

Absorption — the ability to obtain O_2, water and nutrients from the environment that surrounds the cell.

Metabolism— the ability to perform chemical reactions that utilize the nutrients (food substances)and combine them with O_2 to generate energy. Other chemical reactions involve the synthesis of substances such as proteins, carbohydrates and lipids that are utilized by the cell for growth or for the performance of a particular cellular process.

Excretion — the removal or elimination of waste produced from the cell. These waste products are excreted into the surrounding environment and include CO_2 and other substances generated as byproducts of metabolism.

Secretion— the ability to release into the surrounding environment substances that are not necessarily waste products, but instead fulfill certain extra-cellular purposes. These secretory substances are manufactured through cellular metabolism and may act extra-cellularly as communication agents (hormones), enzymes, protective agents (mucous), adhesive agents, antibacterial substances, and so on.

Growth— increases in size usually without any major change in shape within a multicelled organism. However, if growth takes place while cellular differentiation is in progress, the resulting new cell may be quite different in structure and function from the originating cell.

Responsiveness— the ability to receive information from the environment surrounding the cell. This allows the cell to respond to environmental stimuli.

Movement — the ability to move materials from one intra-cellular location to another thus allowing the cell to carry out necessary inter-cellular processes. In addition, many cells also undergo changes in cellular shape that may lead to changes in the surrounding extra-cellular environment. Examples include pseudopodia generation in amoeba, and muscle contraction in multicelled organisms.

Reproduction — the ability of cells, as well as multicelled organisms, to generate additional exact copies of themselves.

LEVELS OF ORGANIZATION

As pointed out, the basic unit of life within the human organism is the cell. However, smaller structural elements comprise the cell. Accordingly, the following list by size, from smaller to larger, describes the levels of organization from atoms to the entire organism.

Atom — for purposes here, consider atoms as the smallest structural unit, although smaller subatomic particles also exist. Atoms such as hydrogen and oxygen form molecules such as water.

Molecules — formed from atoms; can be built up to form larger structures called macromolecules.

Macromolecules — composed of molecules and atoms and include such substances as proteins, lipids, carbohydrates or nucleic acids.

Organelles — formed from macromolecules and include such structures as the plasma membrane, mitochondria, endoplasmic reticulum and the nucleus.

Cells — are the smallest form of life and function to maintain homeostasis and energy balance. There are many different types of cells and although they all possess many common structures, through the process of differentiation they become specialized to provide a particular function within the multicelled organism. Examples of specific types of cells include nerve cells, muscle cells, epithelial cells and connective tissue cells.

Tissues — are formed from similarly differentiated cells (e.g., nerve cells form nervous tissue, muscle cells form muscle tissue; epithelial cells form epithelial tissue and connective tissue cells form connective tissues) and as a result of their complex organization they provide particularly important functions within the organism.

Organs — built up from various types of tissue, they act to provide a particular function within the multicelled organism. For example, the heart is a pump for blood, the liver is a complex chemical processing plant and the adrenal gland is an important source of endocrine hormones. Individual organs supply specialized functions for the organism.

Systems — combining groups of organs together creates systems. For example, the liver, intestine, stomach, and pancreas compose the digestive system that functions to process food. The heart and blood vessels compose the circulatory system that bathes all cells with fluid, supplies cells with nutrients and O_2 and removes waste products and CO_2. Systems provide specific functions that are required to maintain homeostasis and thus the life processes.

Organism — is a fully functional living individual who can fulfill all life functions.

HOMEOSTASIS

Homeostasis is the ability of an individual cell or an entire organism to maintain a controlled internal environment. Millions of years ago in the primordial oceans, the plasma membrane developed.

This was probably the greatest single achievement towards life because it allowed cells to be separated from the outside environment. Since the plasma membrane is semipermeable, it controls the movement of substances into and out of the cell. With this development, the cell was able to maintain a controlled internal environment (i.e., homeostasis) even though the outside environment was

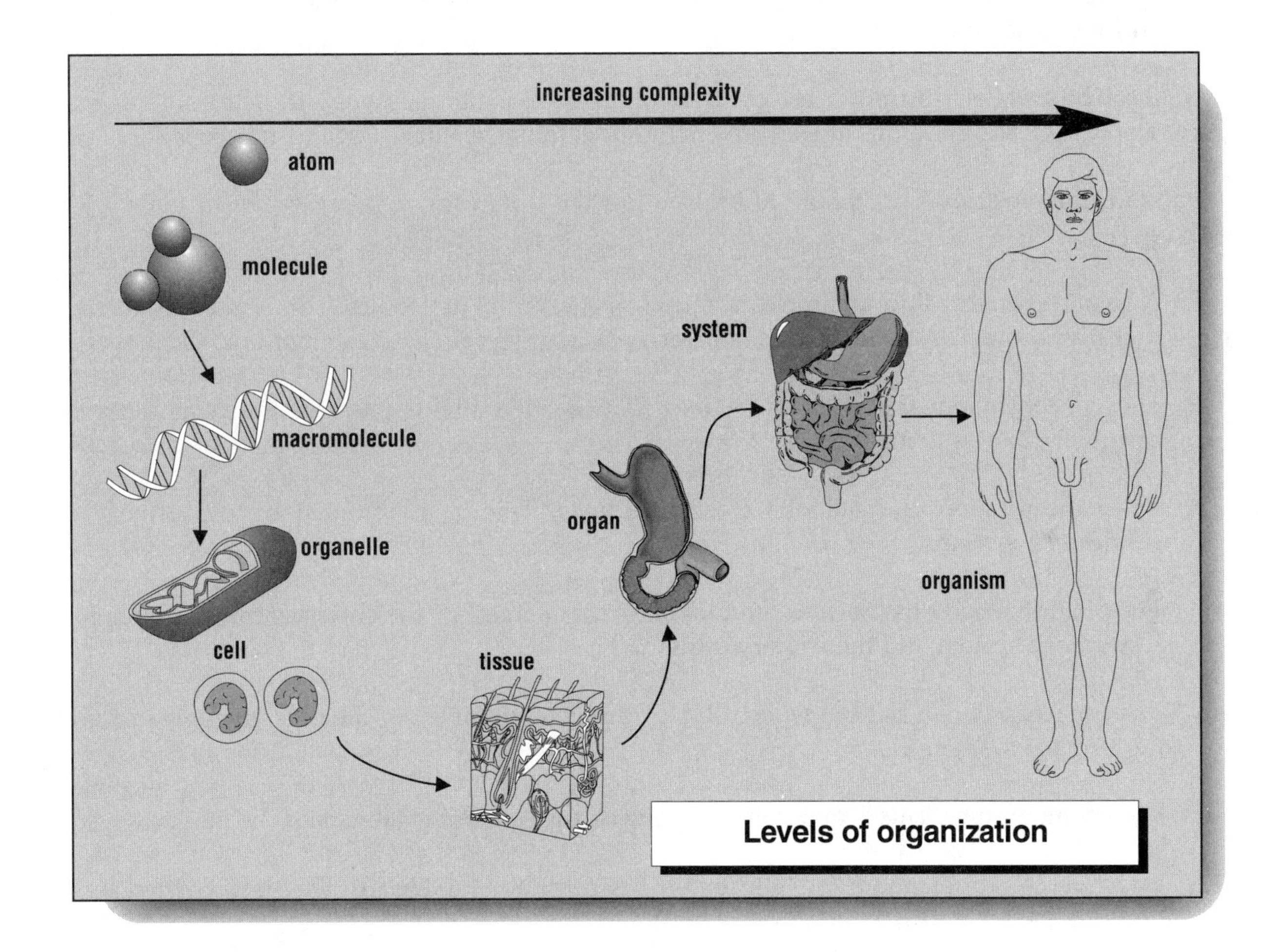

FIGURE 1.1

constantly changing. Therefore, cells could now concentrate on other important processes such as energy generation, growth and reproduction.

As multicelled organisms developed, it was still necessary to maintain homeostasis within the entire

group of cells. However, additional problems arose with multicelled organisms because the cells on the inside of the mass were now unable to obtain nutrients and O_2, or to discharge waste products into the outside ocean. To overcome this problem, multicelled organisms were forced to develop specific systems that could maintain the necessary requirements for the entire organism. For example, the respiratory system regulated gas requirements for the entire organism, while the cardiovascular system transported the gases to all cells of the organism. However, if any one of these systems fails to fulfill its particular function, the entire organism will be placed at risk because homeostasis is no longer maintained.

1. Each cell must maintain intra-cellular homeostatic processes to survive. Specialized cells form specialized tissues, organs and systems that will then, in turn, maintain homeostasis for the entire multicelled organism.

2. It can be said that during disease processes alterations in homeostasis within the organism take place. If these alterations are severe enough, death of the entire organism occurs.

To maintain homeostasis (and therefore life) within the organism, it is necessary to maintain the following:

1. A balanced energy flow and the concentration of nutrient molecules. This is maintained by the digestive system, by cellular metabolic processes and by the nervous and endocrine systems.

2. The concentration of the gases O_2 and CO_2. This is maintained by the respiratory and cardiovascular systems.

3. The concentration of metabolic waste products. This is maintained by the urinary and respiratory systems.

4. The pH of the body tissues and fluids. This is maintained by the chemical buffer systems, the respiratory system and the urinary system.

5. The concentration of salts and electrolytes. This is maintained by the digestive system and by the urinary system.

6. Body temperature. This is maintained by the cardiovascular system, and the nervous system.

7. Effective communication and regulation between body systems. This is provided by the nervous system and the endocrine system.

8. Ability to control the environment, gather food for nutrition and protect the organism. This is provided by the muscular system, the nervous system and the endocrine system.

NEGATIVE FEEDBACK

Homeostasis is maintained by self-regulating, control mechanisms that can sense a change away from a normal set point and can activate processes to correct the fluctuation. The most common method for

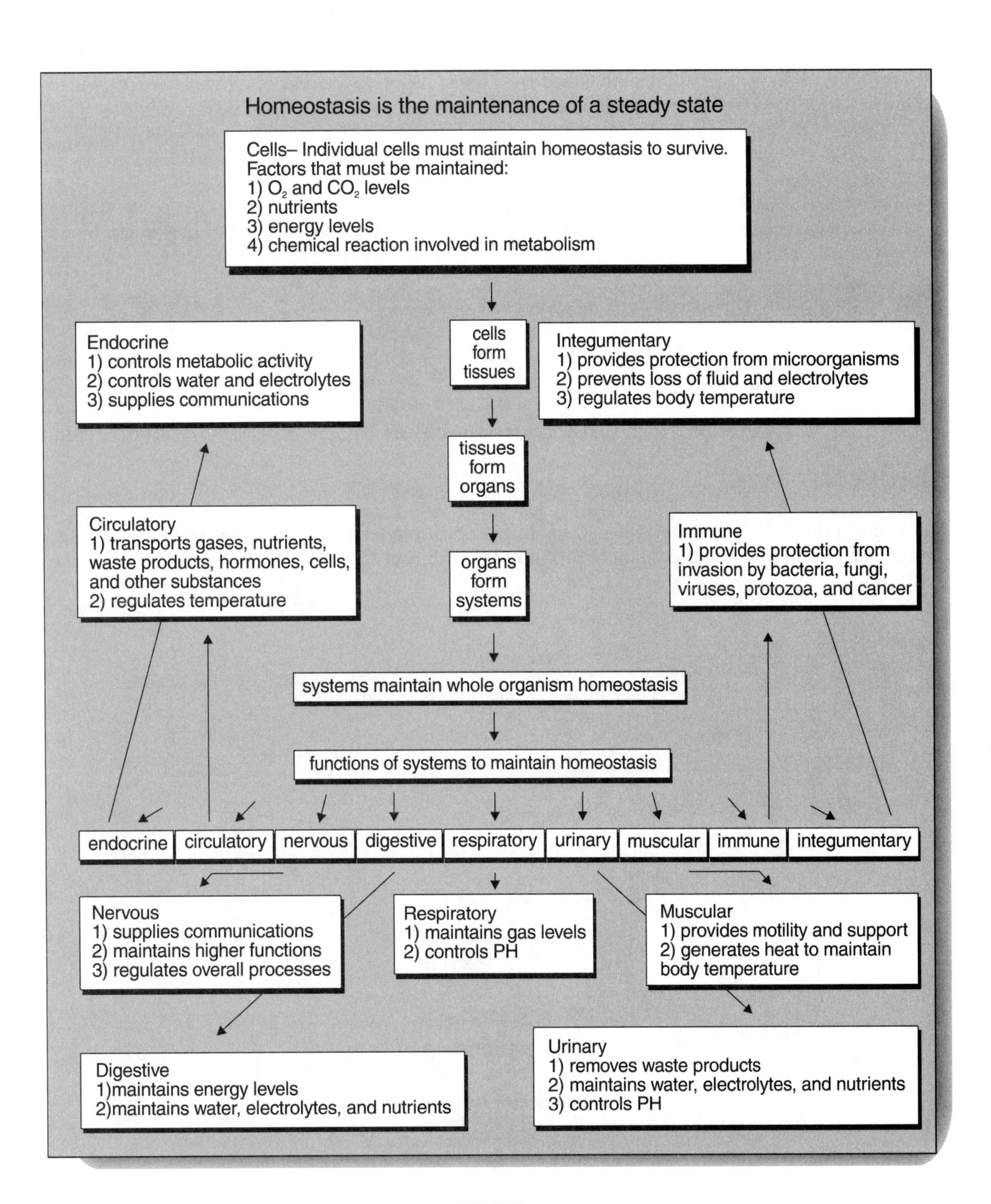

Homeostasis is the maintenance of a steady state
Cells– Individual cells must maintain homeostasis to survive.
Factors that must be maintained:
1) O_2 and CO_2 levels
2) nutrients
3) energy levels
4) chemical reaction involved in metabolism
cells form tissues
tissues form organs
organs form systems
systems maintain whole organism homeostasis
functions of systems to maintain homeostasis
Endocrine
1) controls metabolic activity
2) controls water and electrolytes
3) supplies communications
Integumentary
1) provides protection from microorganisms
2) prevents loss of fluid and electrolytes
3) regulates body temperature
Circulatory
1) transports gases, nutrients, waste products, hormones, cells, and other substances
2) regulates temperature
Immune
1) provides protection from invasion by bacteria, fungi, viruses, protozoa, and cancer
endocrine
circulatory
nervous
digestive
respiratory
urinary
muscular
immune
integumentary
Nervous
1) supplies communications
2) maintains higher functions
3) regulates overall processes
Respiratory
1) maintains gas levels
2) controls PH
Muscular
1) provides motility and support
2) generates heat to maintain body temperature
Digestive
1)maintains energy levels
2)maintains water, electrolytes, and nutrients
Urinary
1) removes waste products
2) maintains water, electrolytes, and nutrients
3) controls PH

FIGURE 1.2

self-regulation is the process of negative feedback. Some negative feedback pathways are presented in **Figure 1.4** for blood pressure and **Figure 1.5** for maintenance of blood glucose levels.

1. As blood pressure drops, sensors in the vascular system monitor this drop and stimulate nerve pathways.

2. This increases heart rate and stroke volume and also causes constriction of peripheral blood vessels.

3. The outcome is an increase in blood pressure.

4. As the pressure rises, the vascular sensors note this increase and stop stimulating the nerves. (This is the negative feedback process.)

Negative feedback regulation is found in all body systems. These regulatory processes will be covered.

POSITIVE FEEDBACK

This process does not maintain homeostasis. In fact, positive feedback promotes certain necessary changes that build up and become more pronounced with time. For example, for parturition to take

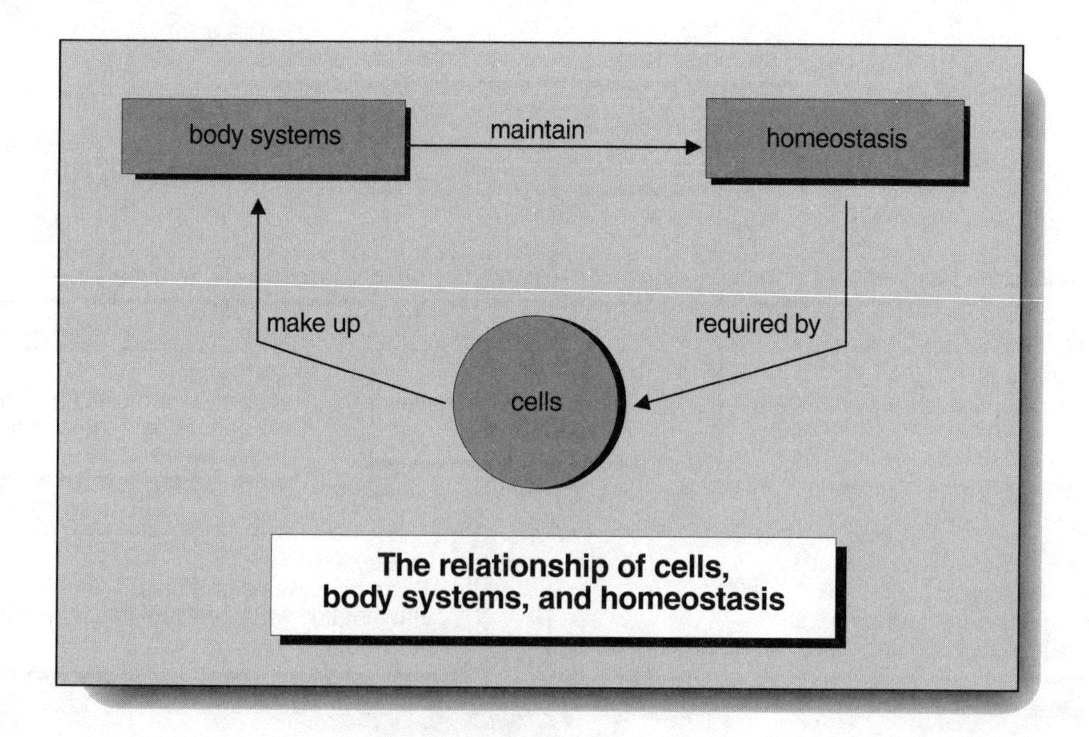

FIGURE 1.3

place, uterine contractions must be initiated and increase during labor. The release of oxytocin, one hormone that promotes these contractions, is an excellent example of positive feedback.

1. A variety of factors at the end of gestation promote initial uterine contractions.

2. Stimulus of the cervix by the baby's head activates a neurological reflex that reaches the brain and eventually the hypothalamus.

3. As a result of the hypothalamic stimulation, oxytocin is released from the posterior pituitary into the blood.

4. The oxytocin circulates to the uterus and stimulates increased uterine smooth muscle contractions.

5. These contractions increase stimulus on the cervix and eventually result in increased release of oxytocin from the pituitary.

6. Through this positive feedback, uterine contractions increase until the baby is delivered.

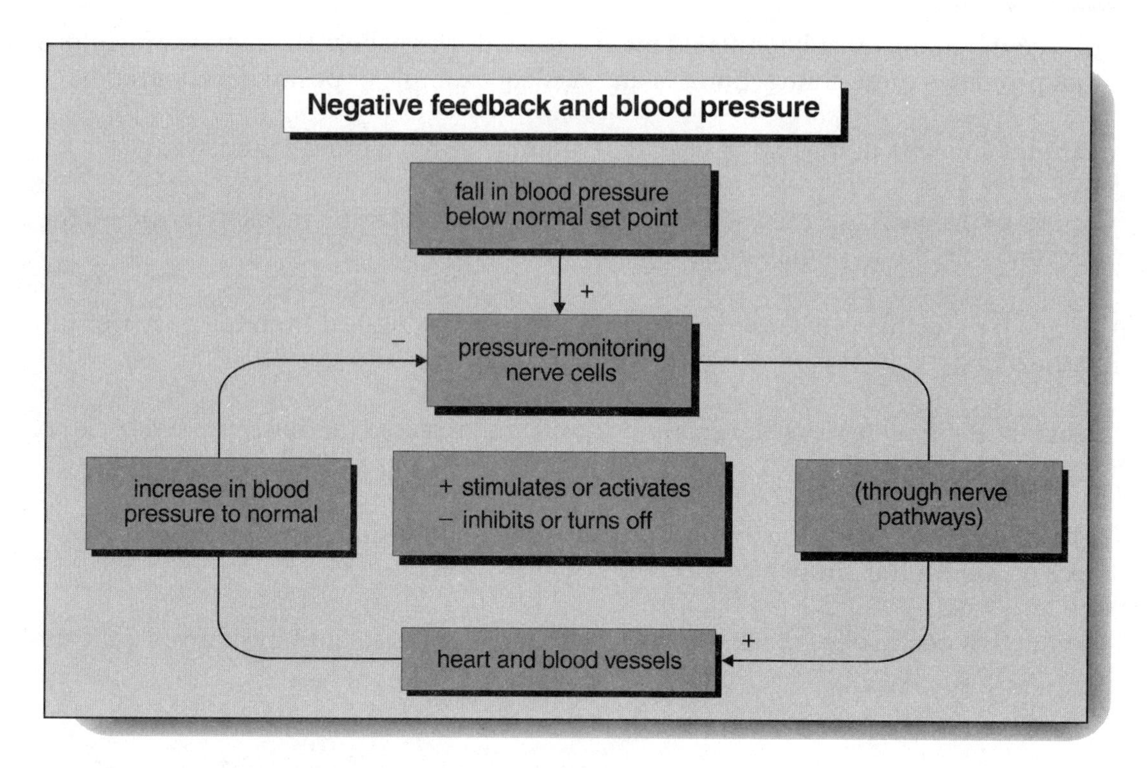

FIGURE 1.4

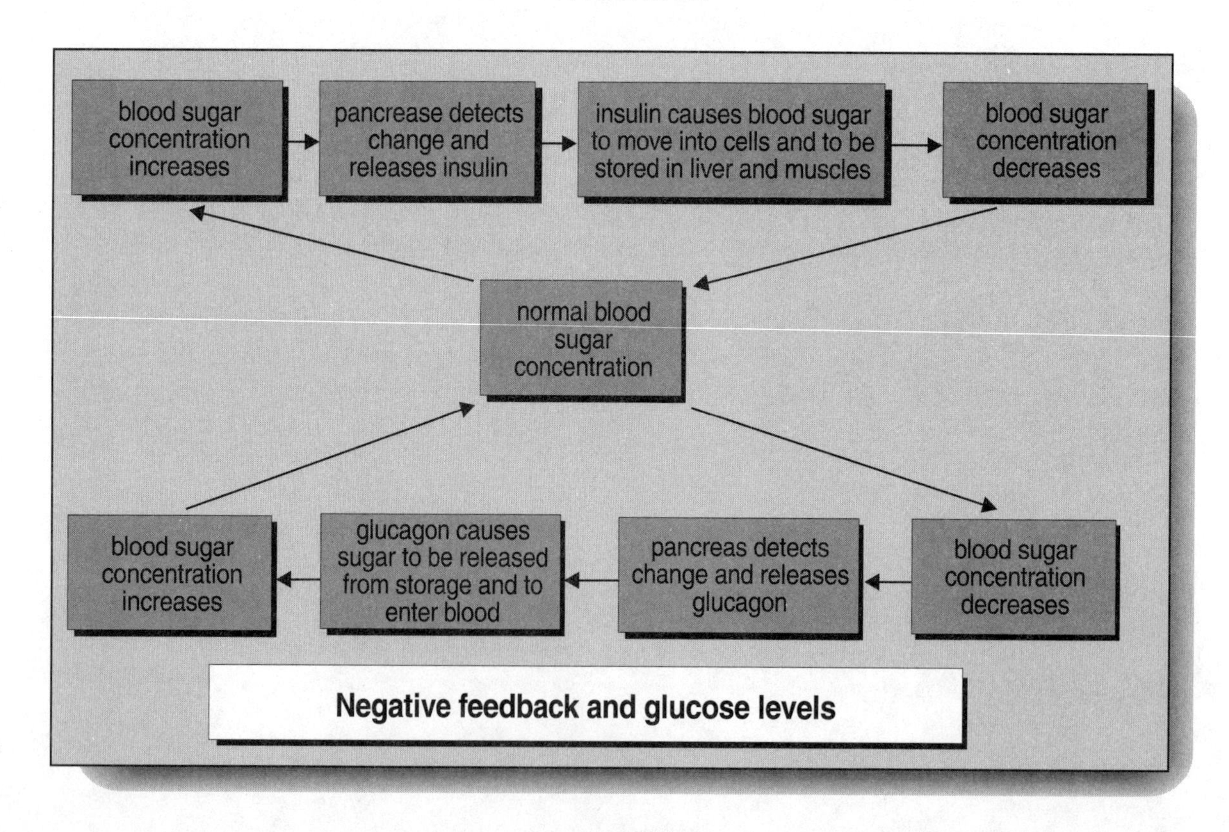

FIGURE 1.5

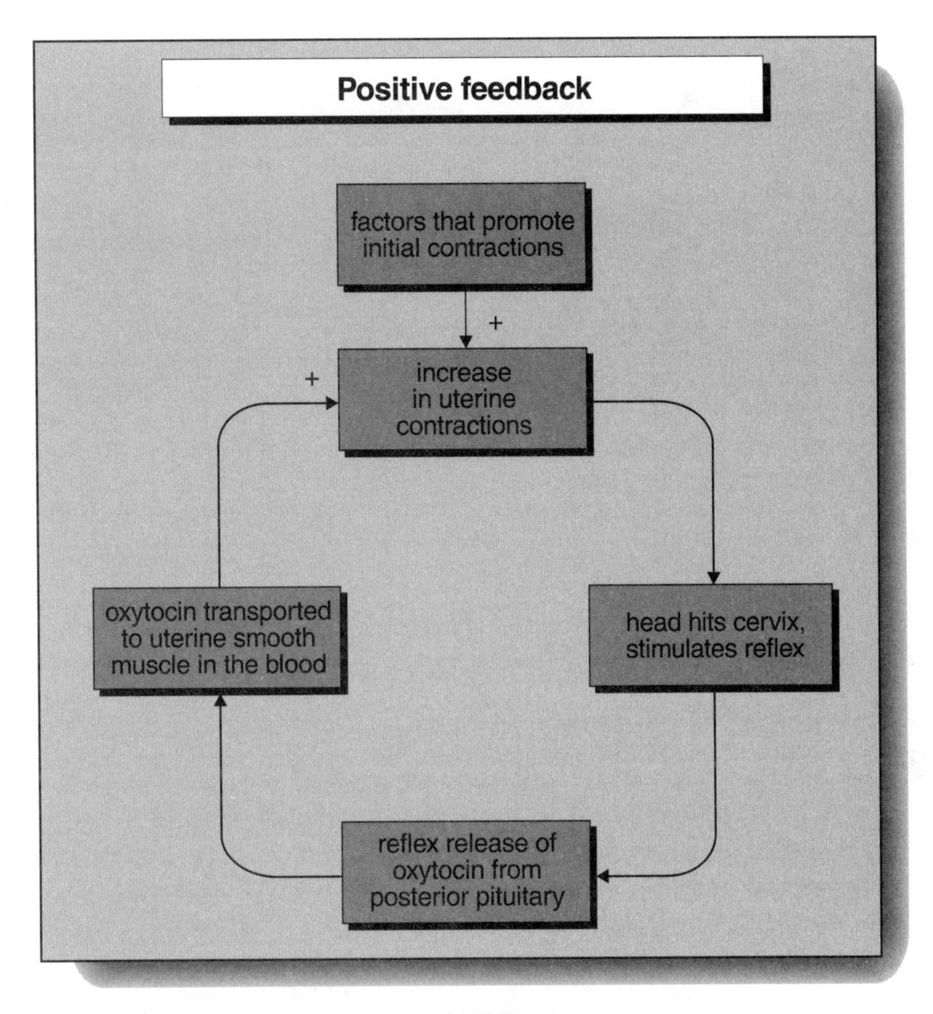
Positive feedback
factors that promote initial contractions
+
increase in uterine contractions
+
head hits cervix, stimulates reflex
reflex release of oxytocin from posterior pituitary
oxytocin transported to uterine smooth muscle in the blood

FIGURE 1.6

CHAPTER 2
MEMBRANE PHYSIOLOGY

STRUCTURE AND COMPOSITION

The cell (plasma) membrane is the outermost limit of the cell. As mentioned in the last chapter, development of the plasma membrane probably was the single most important event leading to the formation of a cell that could maintain homeostasis. The survival of every cell depends on the integrity of the plasma membrane because it maintains the controlled, internal environment necessary for cell life—even though the external environment may be seriously chaotic. In addition, the cell membrane is intimately involved in a variety of metabolic events. Finally, it acts as a communication link allowing cells to remain in contact, and to function appropriately with other cells.

The plasma membrane is too thin to see, even at the resolving power of the best light microscopes. It was first observed when cells were stained and viewed under the electron microscope. In these early electron micrographs, the plasma membrane appeared as two dark lines separated by a clear zone (a trilaminar structure). Studies suggested that the two outer lines were composed of protein, while the middle of the "sandwich" was lipid. Later studies demonstrated a much more complex structure, but the trilaminar arrangement is still partially correct.

More recent studies have shown that the plasma membrane is composed of lipids mostly phospholipids. Since individual phospholipids are formed with a polar, hydrophilic head (Choline + Phosphate + Glycerol) and two nonpolar, hydrophobic tails (Fatty Acids) **(Figure 2.1)**, they arrange themselves with their heads outward and their tails inward in a double layer **(Figure 2.3)** in what is called a lipid bilayer. In this arrangement the water molecules can interact on both the outer and inner surfaces, but are inhibited from passing through the membrane. Thus, the membrane acts as a barrier to water and substances that dissolve in water. In addition, dispersed among the phospholipid molecules are molecules of cholesterol that prevent the fatty acid chains from packing together and forming crystals, which would disorganize the membrane and reduce fluidity.

FLUID MOSAIC MODEL

The phospholipid and cholesterol molecules that form the membrane allow a wide range of fluid-like motion to take place between the components. The membrane can change shape as the cell is compressed. If a sharp object is inserted into the membrane and then withdrawn, the membrane may heal the hole. Imbedded in the phospholipid ocean are various protein molecules **(Figure 2.3)**. Some of the proteins may extend all the way through the membrane (from the outer to the inner side), while others may only project on the outer or inner sides. Since the membrane is like a fluid, the proteins are able to float in the phospholipid ocean like boats, and this protein mobility allows them to perform specialized functions such as capping (see below) during receptor interactions. In addition, small numbers of carbohydrate molecules are found on the outer surface of the membrane, either attached to the proteins (glycoprotein) or to the lipid (glycolipid).

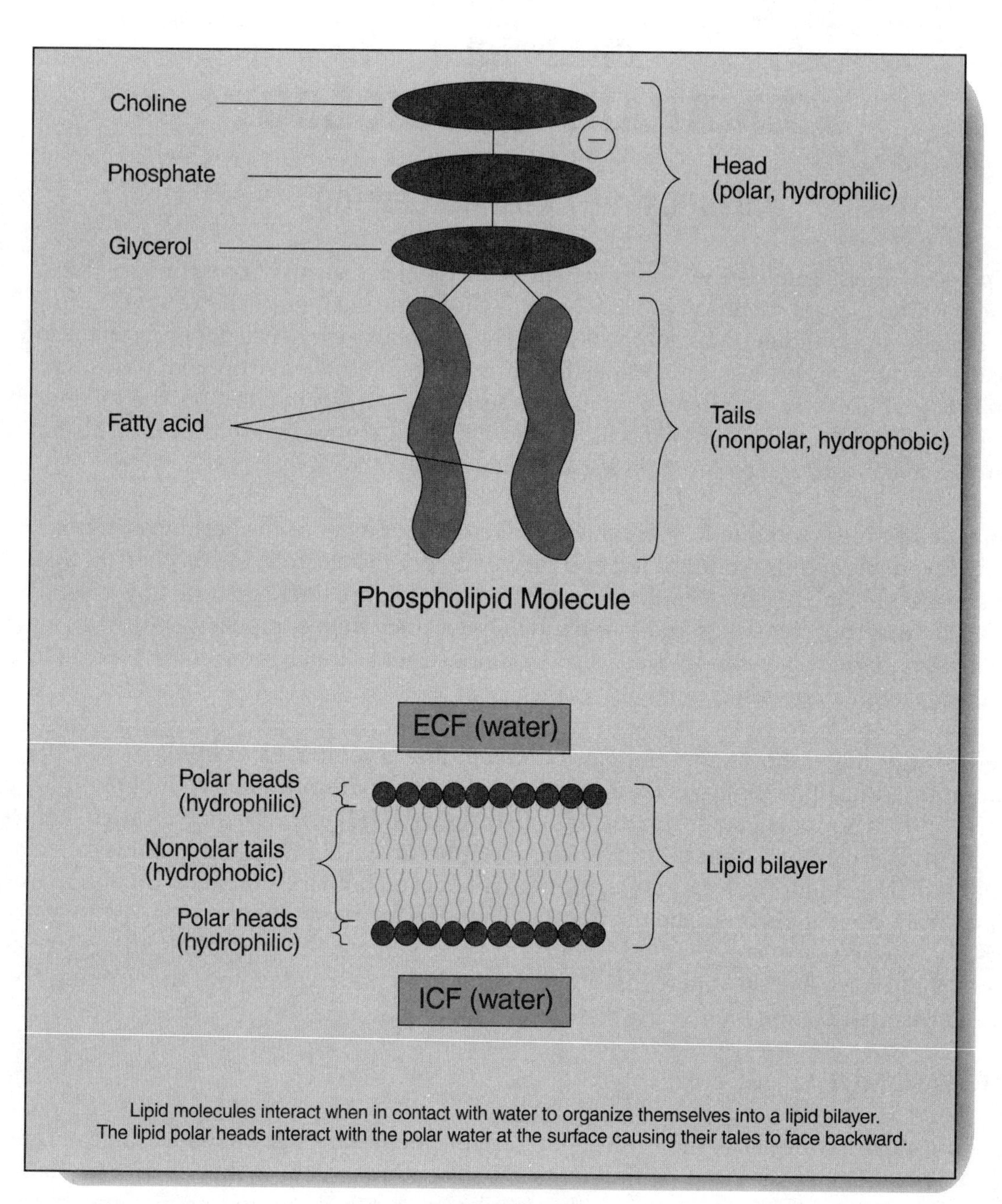
Choline
Phosphate
Glycerol
Head
(polar, hydrophilic)
Fatty acid
Tails
(nonpolar, hydrophobic)
Phospholipid Molecule
ECF (water)
Polar heads
(hydrophilic)
Nonpolar tails
(hydrophobic)
Polar heads
(hydrophilic)
Lipid bilayer
ICF (water)
Lipid molecules interact when in contact with water to organize themselves into a lipid bilayer.
The lipid polar heads interact with the polar water at the surface causing their tales to face backward.

FIGURE 2.1

These carbohydrate moieties are thought to be involved in receptor interactions. This structural arrangement of the plasma membrane is sometimes called the *fluid mosaic model*.

MEMBRANE PROTEINS

Since the phospholipid layer is impermeable to water and substances dissolved in water, certain membrane proteins that span the phospholipid layer act as membrane channels. The channels are highly selective and admit only those substances designed to fit or pass through the particular channel. Thus we speak of Na^+ or K^+ channels, for example. Arrangement of the charges within the interior of the protein channel is believed to control the selectivity of the channel. In addition, many of these channels can be selectively opened or closed to control the movement of the substances. These channels are said to be *gatted* because the protein configuration can alter to open or close the channel.

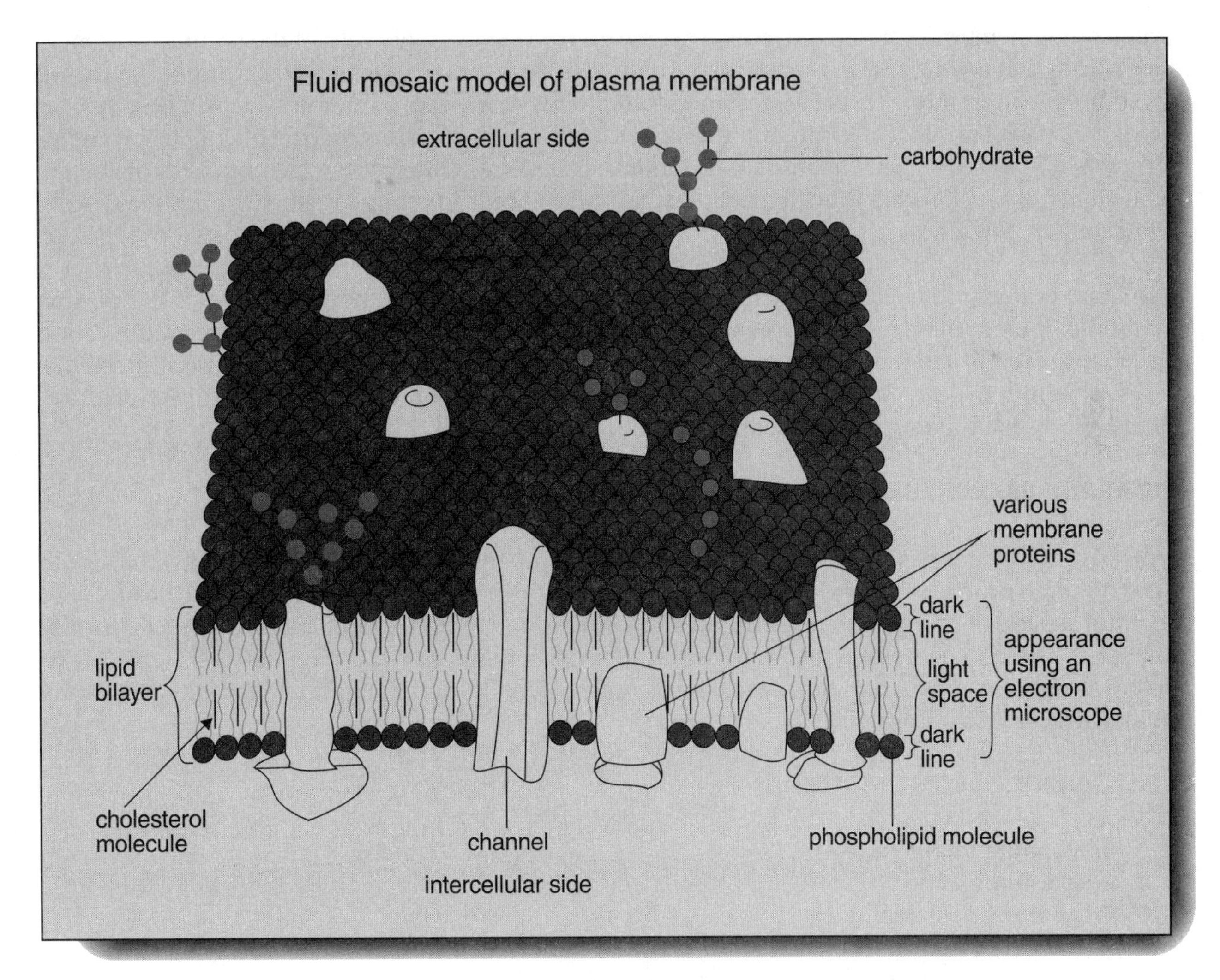

FIGURE 2.2

Along with the proteins that act as specific membrane channels, there are additional proteins that can act as carrier molecules. These proteins may or may not extend all the way through the membrane but are capable of attaching to a particular substance (e.g., an amino acid) on one side of the membrane and then transporting it through the membrane to the other side where the substance is then released. Movement of substances by carrier proteins may or may not require the expenditure of energy and the carrier proteins are relatively specific for the substance that each can transport.

Proteins or glycoproteins, located on the outer side of the membrane, can also act as specific receptors that bind hormones or other messenger-type molecules. Binding of hormones specifically to a particular membrane-bound receptor may then affect additional enzyme proteins located on the inner side of the membrane and thus alter intra-cellular events.

Binding of a hormone to receptors on the extra-cellular side of the membrane may open or close specific gatting proteins and alter transport of a particular substance through a membrane channel (e.g., insulin binding to insulin receptors opens glucose channels and glucose enters the target cell).

Certain outer membrane proteins, especially some glycoproteins, allow cells of the immune system to identify cells that belong or don't belong to the body. A foreign cell or other substance (termed antigen) will stimulate an immune response designed to remove the invader from the system. These surface identification proteins that allow immune system identification are known as the Major Histocompatibility (MHC) antigens (see chapter 10, the immune system). Other outer membrane proteins are involved in cell to cell recognition. Such recognition allows cells to group together to form the specific architecture of tissues.

Membrane-bound proteins that function as enzymes are located on the inner side of the plasma membrane. Some of these enzymes are involved in receptor events that lead to cellular communication (e.g., adenyl cyclase and the second messenger hypothesis). Others are involved in various metabolic processes within the cell. In addition, the filamentous meshwork on the inner side of the membrane is involved in maintaining cellular shape and cellular movements.

MEMBRANE CARBOHYDRATES

Membrane carbohydrates (including the glycoproteins and glycolipids) are clearly involved in cell to cell recognition processes. Carbohydrate surface markers on the outer membrane may also be involved in certain aspects of cell growth regulation, or the inhibition of uncontrolled cell growth (i.e., surface inhibition). It has been suggested that when surface inhibition ceases, cancer cells grow with uncontrolled vigor and disrupt other tissues and organs. Finally, outer membrane carbohydrates may stabilize the membrane proteins.

COMPONENTS OF THE MEMBRANE

A. Phospholipid Bilayer
B. Cholesterol
C. Membrane Proteins
    1. Channels
    2. Receptors

3. Enzymes
4. Membrane Transport Agents
5. Cell Identification Molecules

D. Membrane Carbohydrates
1. Glycoproteins
2. Glycolipids
3. Cell Identification Molecules
4. Cell Growth Regulators

FUNCTIONS OF CELL MEMBRANE

1. Maintains cellular homeostasis — promotes a stable internal environment within the cell.
2. Semipermeable — specifically regulates the kind, rate and amount of specific substances that can cross the membrane in either direction.
3. Acts as a communication medium (via external membrane receptors) allowing the cell to respond to external stimuli (chemical and physical factors).
4. Acts as a recognition system (via external membrane components) to distinguish itself from others. Allows the cell to be recognized by cells of the immune system as part of the same organism. Allows the cell to function as a unit with other cells of the same organism.
5. Involved in generation of electro-chemical potentials (action potentials, generator potentials, receptor potentials). This is necessary for muscle and nerve functions.

IONIC CHANNELS

Movement across membranes is controlled by specific channel proteins. Structural changes in the proteins can act to open or close the channels. The most important ions that cross the membrane through such channels are as follows:

1) Na^+ and K^+ cross the membrane through specific channels that may open for a short time and then close again. Alterations in the flow of these ions across the membrane result in the generation of electrochemical potentials (action potentials). In nerve and muscle cells, Na^+ movement through such channels is usually from outside to inside by means of passive diffusion. In nerve and muscle cells some K^+ normally moves out of the cells, continuously through channels, by passive diffusion. However, at certain times during the end stage of an action potential, K^+ moves through yet another channel from inside to outside.

2) Ca^{++} generally moves from outside to inside through a channel by passive diffusion. These channels open to admit the Ca^{++} when the membrane proteins are stimulated by various mechanisms such as hormone-receptor binding, electrical stimulation, or physical stretch of the membrane. Entrance of the Ca^{++} into the cell can trigger structural changes in certain intracellular proteins. This will lead to alterations in cellular response.

MEMBRANE TRANSPORT

A variety of mechanisms promote movement of substances through cell membranes. We can divide these transport processes into physical mechanisms and physiological mechanisms.

PHYSICAL MECHANISMS OF TRANSPORT

Physical mechanisms are those transport processes that can take place in either a living or a non-living system. In fact, such physical mechanisms of transport can be duplicated in a chemistry laboratory with the right equipment. Here are four physical mechanisms of transport.

Diffusion
Dialysis
Osmosis
Filtration

Diffusion. It takes place when molecules or ions migrate from an area of higher concentration to lower concentration (down a concentration gradient). The force of movement is simply due to molecular motion (kinetics) and does not require input of energy from another source. For example, place a drop of red food coloring into a beaker of water and with time the molecules will diffuse throughout the entire beaker, turning the water pink. The majority of substances that cross the cell membrane move by diffusion.

O_2 crosses the cell membrane because the concentration of this gas is higher in the tissue fluids than inside the cell. Na^+ moves from outside the cell to inside the cell (through a Na^+ channel) because the concentration of this ion is higher outside the cell than inside the cell.

Various factors can alter the rate of diffusion of a substance.

1) The concentration of the substances on the two sides of the membrane determine the concentration gradient. The greater the difference in the concentration on the two sides, the larger will be the concentration gradient, and the greater will be the rate of diffusion.

2) Permeability of the membrane to the substance. The more permeable, the greater will be the rate of diffusion. For example, if more Na^+ channels open in the membrane, the permeability will increase, and the rate of diffusion will also increase.

3) Physical or electrical properties of the substance can affect the rate of diffusion.

4) The shorter the distance that the substance moves, the greater is the rate of diffusion.

5) In addition to the movement of substances due to the presence of a concentration gradient, movement can also result from an electrochemical gradient. For example, the movement of Cl^- ions across cell membranes is frequently the result of an electrical attraction due to the active transport of Na^+ (see following material). As the Na^+ ions are transported, their positive charges build up and an electrochemical gradient is generated. This will attract the negatively charged Cl^-, which will then move to follow the Na^+.

Dialysis. This is the process of separating small molecules from larger molecules in a liquid. If a semipermeable dialysis membrane is used as a barrier, only those molecules of the appropriate

size (usually smaller molecules) will be able to pass through the holes in the membrane, but larger molecules will not penetrate. To separate the smaller molecules from the larger ones, therefore, the dialysis membrane is placed in a buffer solution. Smaller molecules inside the dialysis membrane will then move by diffusion into the outer buffer solution leaving the larger molecules inside. Molecular movement takes place because the concentration of the molecules (e.g., salts) are greater inside the dialysis membrane than on the outside, and thus a concentration gradient is present. This principle is utilized in hemodialysis where smaller molecular waste products, such as urea, are separated from other substances by passing the patient's blood through a dialysis machine. As urea concentration increases in the buffer solution, the concentration on both sides of the dialysis membrane will begin to reach equilibrium. Under these conditions, the gradient will decrease and dialysis will stop. Therefore, to maintain the concentration gradient, the old buffer solution must be continuously removed and fresh buffer solution must be continuously added.

Osmosis. Another special form of diffusion is known as osmosis. Here the substance that is diffusing across the cell membrane is water. Like other substances that move by diffusion, water also moves as a result of a concentration gradient. Thus, for water to move from the outside of the cell to the inside of the cell, the water concentration must be greater on the outside than on the inside. To understand how water concentration can vary, it will be helpful to discuss the process of osmosis by using the red blood cell (RBC) as a model.

A solution that contains the same concentration of salt and water as the cytoplasmic contents of a RBC is termed an *isotonic solution*. Since the saltwater concentration can be measured in units of milliosmolarity (mOsmo), an isotonic solution is said to be 300 mOsmo. Placing a RBC in such an isotonic solution will not alter the net movement of water across the membrane in either direction since the concentration of water on the two sides of the membrane is equal. However, if the RBC is placed in a solution containing more salt and less water than an isotonic solution (i.e., greater than 300 mOsmo) there will be more water inside the RBC than outside and water will move out of the cell. A solution that has an osmolarity greater than 300 mOsmo is a hypertonic solution. A RBC placed in a hypertonic solution will lose water and will therefore shrink. If, on the other hand, the RBC is placed in a solution that contains more water and less salt than the RBC (ie: less than 300 mOsmo) water will move from outside—where it is more concentrated—to the inside. Such a solution is said to be hypotonic and a RBC placed in such a solution will swell and rupture.

Type of Solution	mOsmo	Effect
Isotonic	300 mOsmo Same salt-water concentration as RBC.	Cell normal
Hypertonic	Greater than 300 mOsmo More salt - less water than RBC.	Cell shrinks
Hypotonic	Less than 300 mOsmo, Less salt - more water than RBC.	Cell swells and ruptures

Filtration. If molecules are forced to pass through a membrane, but the driving force is neither that of diffusion nor osmosis, this is termed *filtration*. In filtration, different sized molecules can be separated from each other depending on the pore size of the filtration membrane. In the chemistry laboratory, a mixture of larger and smaller molecules can be separated by passing them through a paper filter. In this case, the force causing the filtration is due to gravity acting on the liquid. As most students know, the rate of filtration can be increased if a vacuum is applied to the collection reservoir. In the body, filtration takes place when substances pass between the endothelial cells that compose the capillary walls. Again, smaller substances such as water, salts, amino acids, glucose and so on can penetrate through the pores between the endothelial cells. However, larger substances such as proteins and blood cells do not normally penetrate. Here, the force driving this filtration is blood pressure (or hydrostatic pressure) generated by the pumping action of the heart **(Figure 2.3)**.

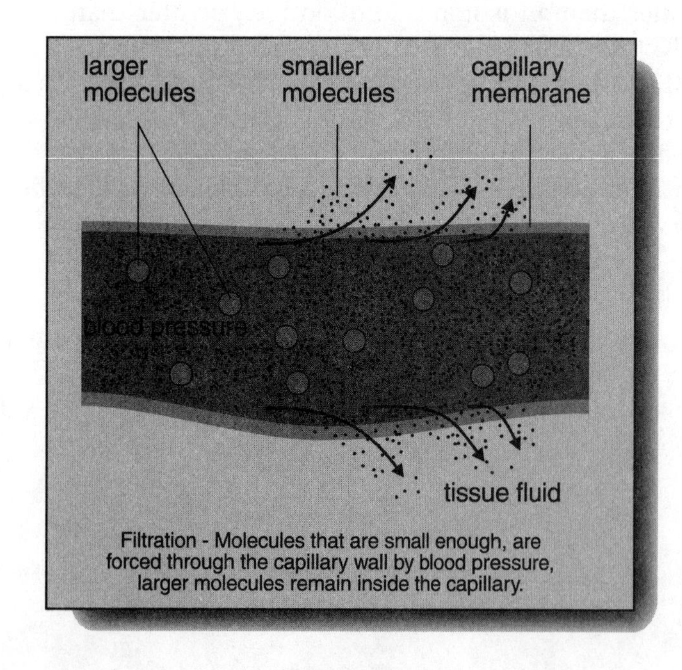

Filtration - Molecules that are small enough, are forced through the capillary wall by blood pressure, larger molecules remain inside the capillary.

FIGURE 2.3

PHYSIOLOGICAL MECHANISMS OF TRANSPORT

Physiological transport mechanisms can only function in a living cell because they require the expenditure of energy. In addition, various protein molecules are usually required for these transport processes to function. Such proteins can only be assembled and activated in the environment of a living cell.

PHYSIOLOGICAL TRANSPORT MECHANISMS

A. Carrier Mediated Transport
   1. Facilitated Diffusion
   2. Active Transport
B. Endocytosis and Exocytosis
   1. Pinocytosis
   2. Phagocytosis
   3. Receptor Mediated Endocytosis

Carrier Mediated Transport. Certain membrane transport processes require the presence of carrier proteins that span the lipid bilayer. These proteins are designed to bind with substances that require transport across the membrane. According to theory, the substances specifically bind to the carrier protein. This binding triggers a conformational change in the protein which may or may not be energy dependent. The structural alteration in the protein then transports the substance across the membrane and releases the substance on the other side. The amount and rate of substances transported by carrier mediated transport systems can be affected by the following parameters:

Saturation of the Carriers — there are only a limited number of carrier protein binding sites capable of binding the substance to be transported in the plasma membrane. If all the binding sites are used up (i.e., saturated), then maximum transport of the substance has been reached and no additional transport is possible. This limit is defined as the transport maximum (T_m). Units of T_m are usually in amount/time; (e.g., 100 mg/min).

Specificity of the Carriers — each carrier protein is structurally formed with a specific site that binds with high specificity to a specific substance to be transported. Thus, a glucose carrier is specific for glucose and a Na^+ carrier is specific for Na^+. Chemicals with closely associated chemical structures may also bind to the same carrier molecule as the substance it was designed to transport.

Affinity of the Carriers — affinity is a measure of the strength of binding between the carrier and the substance to be transported. Substances that have a better fit in the binding site of the carrier (i.e., are more specific for the carrier) usually have a higher affinity as well.

Competition for the Carriers — since closely related compounds may bind to the same carrier with about the same specificity, they may block each other as they both try to bind to the carrier. Substances with higher affinities tend to bind to the carrier better than substances with lower affinities. Thus the higher affinity substances may compete for the carrier binding and prevent the lower affinity substances from binding. In addition, the concentration of the substance can affect its ability to compete for binding sites on the carrier. The greater the concentration of the substance, the more effectively will it compete for binding sites. Thus, if a carrier is capable of transporting two closely related substances, the presence of both will reduce the rate of transport of either across the membrane since they compete for binding.

FACILITATED DIFFUSION AND ACTIVE TRANSPORT

Facilitated diffusion is one type of carrier mediated transport and it is a specialized form of diffusion. Many sugars (e.g., glucose) and certain other insoluble lipids are too large to pass through membrane

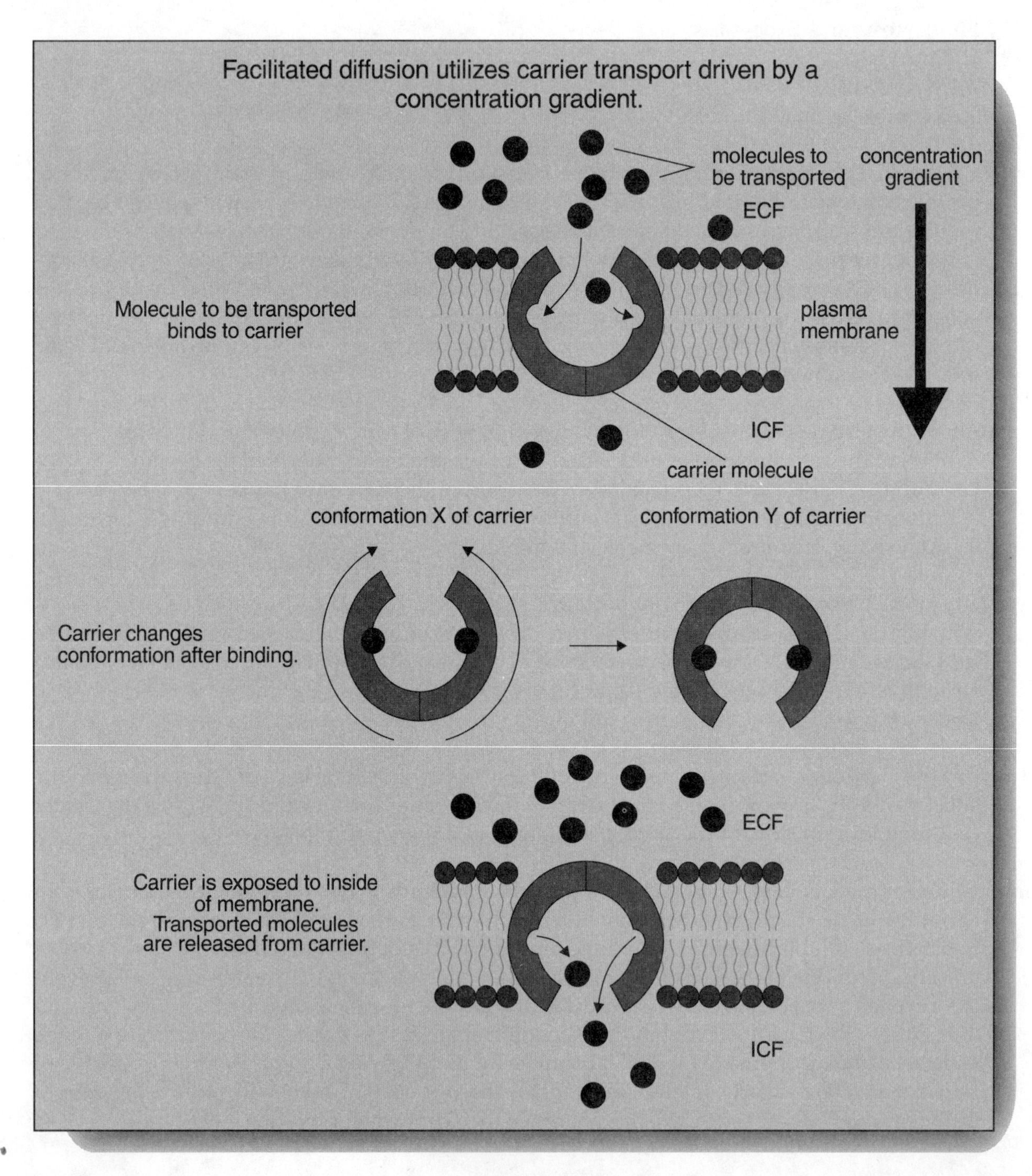

FIGURE 2.4

protein channels. To transport these substances into the cell, the substance combines with a specific carrier protein and then passes through the lipid bilayer to be released on the inner side of the membrane. Since the driving force is actually the concentration gradient of the substance, no expenditure of cellular energy is required. Although this process is sometimes classified as a physical method of transport, it is carrier-mediated and is in truth a physiological method **(Figure 2.4)**.

Active transport is quite similar to facilitated diffusion but for one very important difference—cellular energy is required. Since the substances to be transported are moving against a concentration gradient (i.e., up hill, not down hill), the force of diffusion is working against the transport. To counteract the concentration gradient, the transport pumps must be supplied with energy. Thus, the carrier protein attaches to the substance to be transported on one side of the membrane, and through a conformational change transports and releases the substance on the other side of the membrane. However, for the conformational change to take place, energy must be supplied. Usually this energy is from the hydrolysis of ATP (Adenosine Triphosphate)—the universal chemical energy compound of the cell **(Figure 2.5)**.

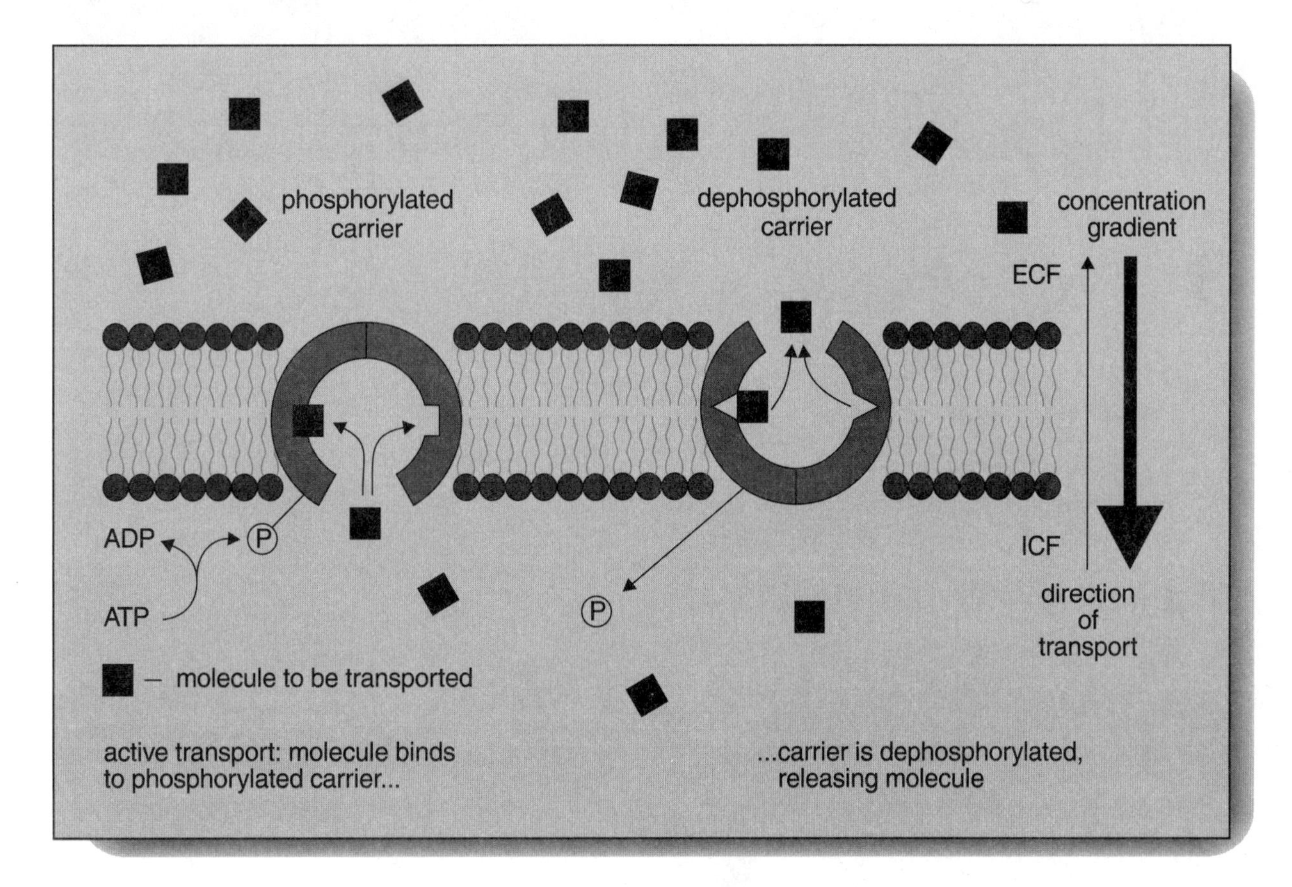

FIGURE 2.5

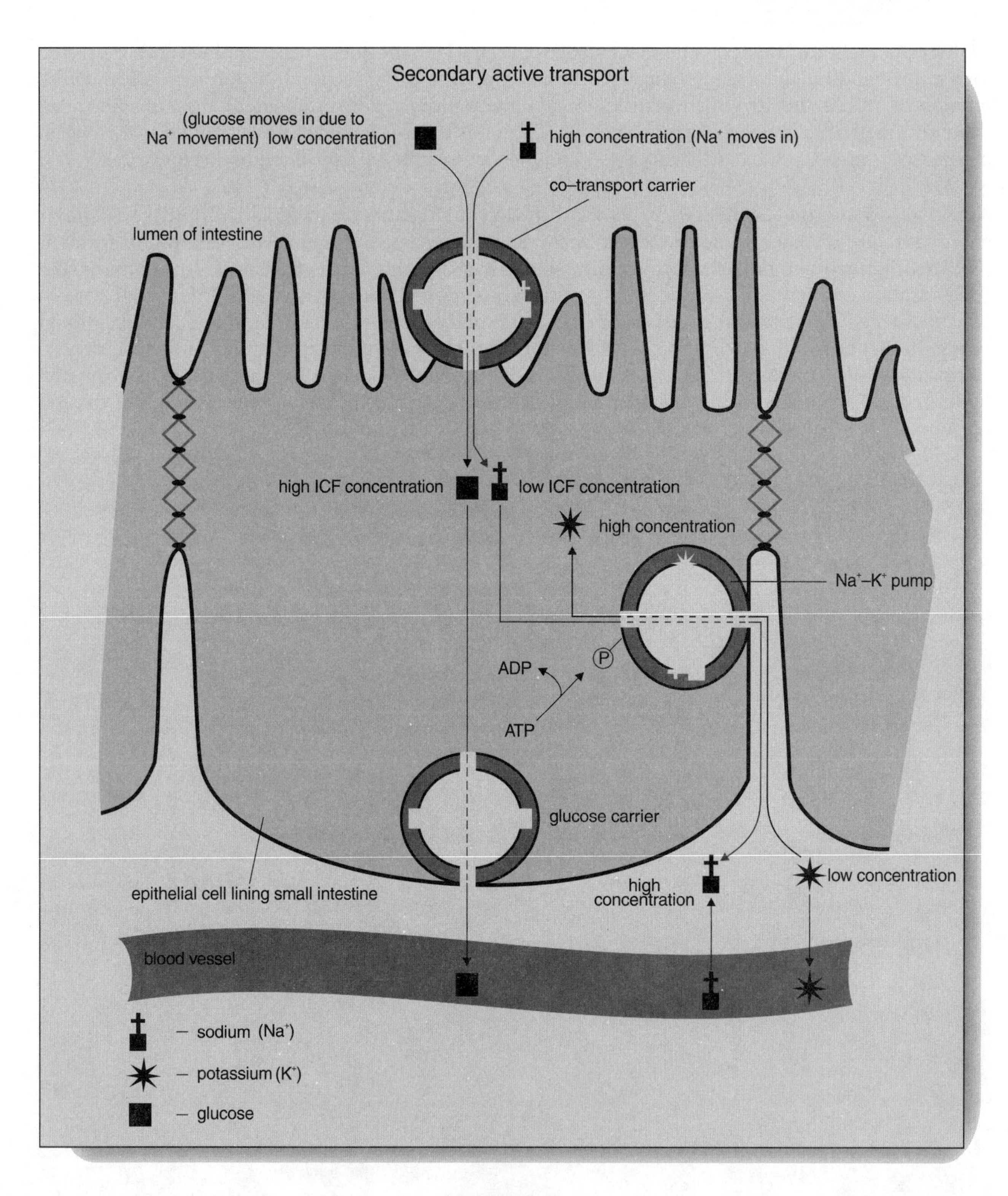
Secondary active transport
(glucose moves in due to
Na+ movement) low concentration
high concentration (Na+ moves in)
co–transport carrier
lumen of intestine
high ICF concentration
low ICF concentration
high concentration
Na+–K+ pump
P
ADP
ATP
glucose carrier
epithelial cell lining small intestine
high
concentration
low concentration
blood vessel
– sodium (Na+)
– potassium (K+)
– glucose

FIGURE 2.6

Active transport pumps may function to transport a single substance or may be linked to transport two molecules together. A commonly linked active transport pump is that which functions to transport Na^+ out of cells, but at the same time transports K^+ back in (**Figure 2.6**).

If the pumps directly use the energy from ATP for transport of a particular substance, this is termed *primary active transport*. However, if energy is required for the process of transport but is not directly needed to operate the pump, this is termed *secondary active transport*. Thus, the transport of glucose from the lumen of the intestine into the blood is by means of secondary active transport. The actual driving force is the primary active transport provided by the linked Na^+/K^+ ATP pump that generates a Na^+ gradient. Since this Na^+/K^+ ATP pump reduces the level of intra-cellular Na^+, the movement of Na^+ back into the cell then operates the linked Na^+/glucose pump, bringing glucose into the cells (**Figure 2.6**).

ENDOCYTOSIS AND EXOCYTOSIS

Endocytosis and exocytosis are forms of vesicular transport which move large substances such as macromolecules into (endocytosis) or out of (exocytosis) a cell. Such large particles are enclosed in a plasma membrane-bound vesicle. In endocytosis, the particle or substance is trapped in the vesicle, which encloses it. The vesicle then moves to the cytoplasm of the cell where it may fuse with a lysosome to form a digestive vesicle. Enzymes in the lysosome degrade the contents of the vesicle and then the digested material is released into the cytoplasm. In some cases the degraded contents may be transported back out of the cells by exocytosis **(Figure 2.7)**.

In the case of exocytosis, the vesicle moves to the inner side of the outer plasma membrane and fuses with it. The contents of the vesicle are then released to the outside of the cell. Formation of the vesicles and fusion and release are all energy-dependent processes requiring hydrolysis of ATP **(Figure 2.7)**.

Pinocytosis — or cell drinking—is endocytosis of dissolved substances (small molecules) transported into cells from outside. Most cells are capable of pinocytosis (**Figure 2.8**).

Phagocytosis — or cell eating—is endocytosis of larger substances such as particles or bacteria. White blood cells (leukocytes) are able to phagocytosize bacteria, which are then usually killed when lysosomal fusion releases digestive enzymes into the vesicles. Certain leukocytes are capable of phagocytosis, and other cells of the reticuloendothelial system are also phagocytic **(Figure 2.8)**

RECEPTOR MEDIATED ENDOCYTOSIS

When receptor molecules on the outer membrane bind a ligand (such as a hormone in an endocrine target cell, or an antigen in an immune effector cell), the binding triggers a conformational change. Initially, the receptor molecules may be equally spaced around the entire circumference of the cell. However, after binding with the particular ligand, the complex then undergoes capping. During capping all the receptor-ligand complexes move to one pole of the cell forming a CAP. After this is completed, the receptor-ligand complexes are enclosed in an endocytic vesicle and engulfed into the cell. In the cytoplasm, the receptor-ligand complexes then trigger various cellular responses.

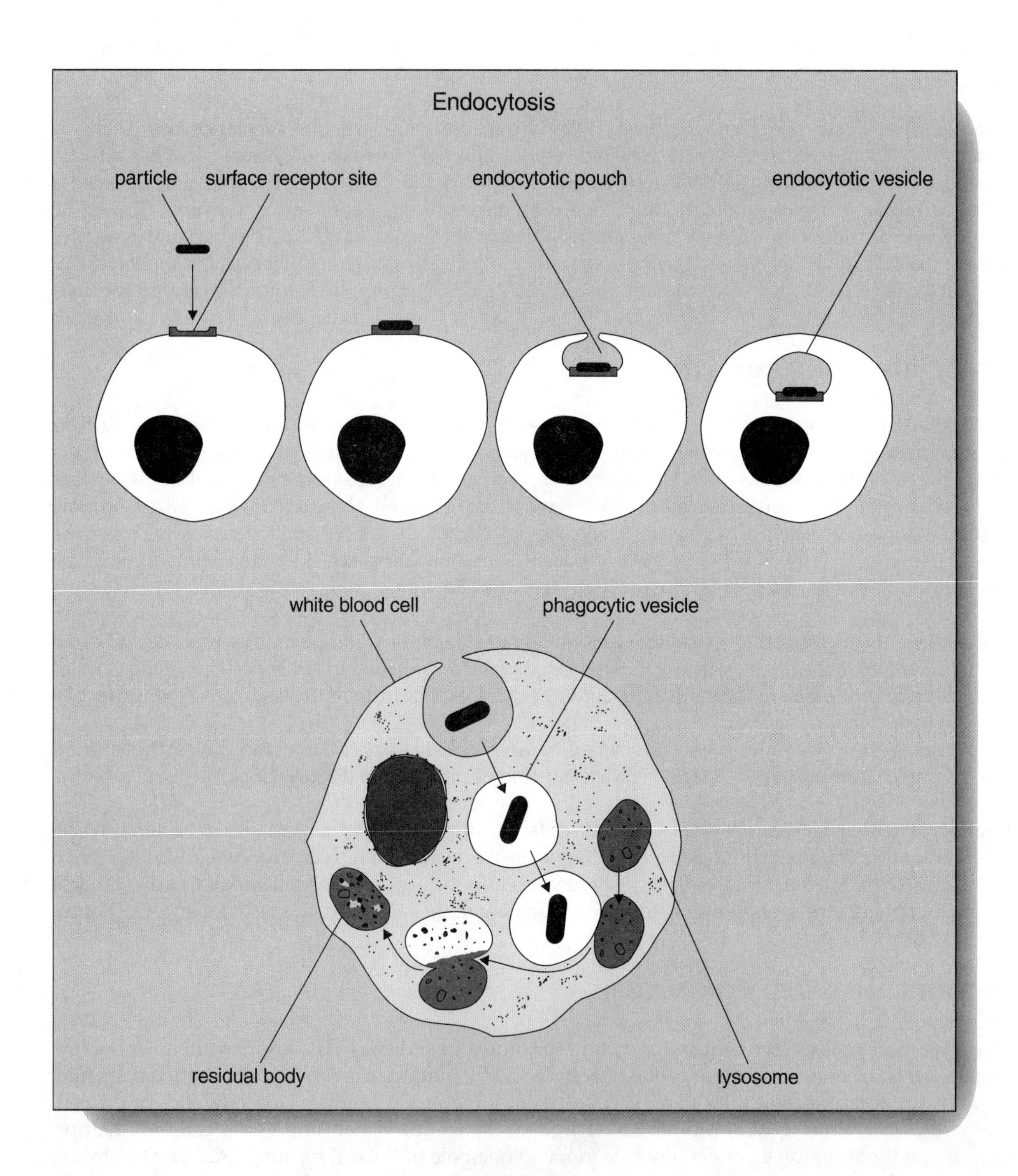
Endocytosis
particle
surface receptor site
endocytotic pouch
endocytotic vesicle
white blood cell
phagocytic vesicle
residual body
lysosome

FIGURE 2.7

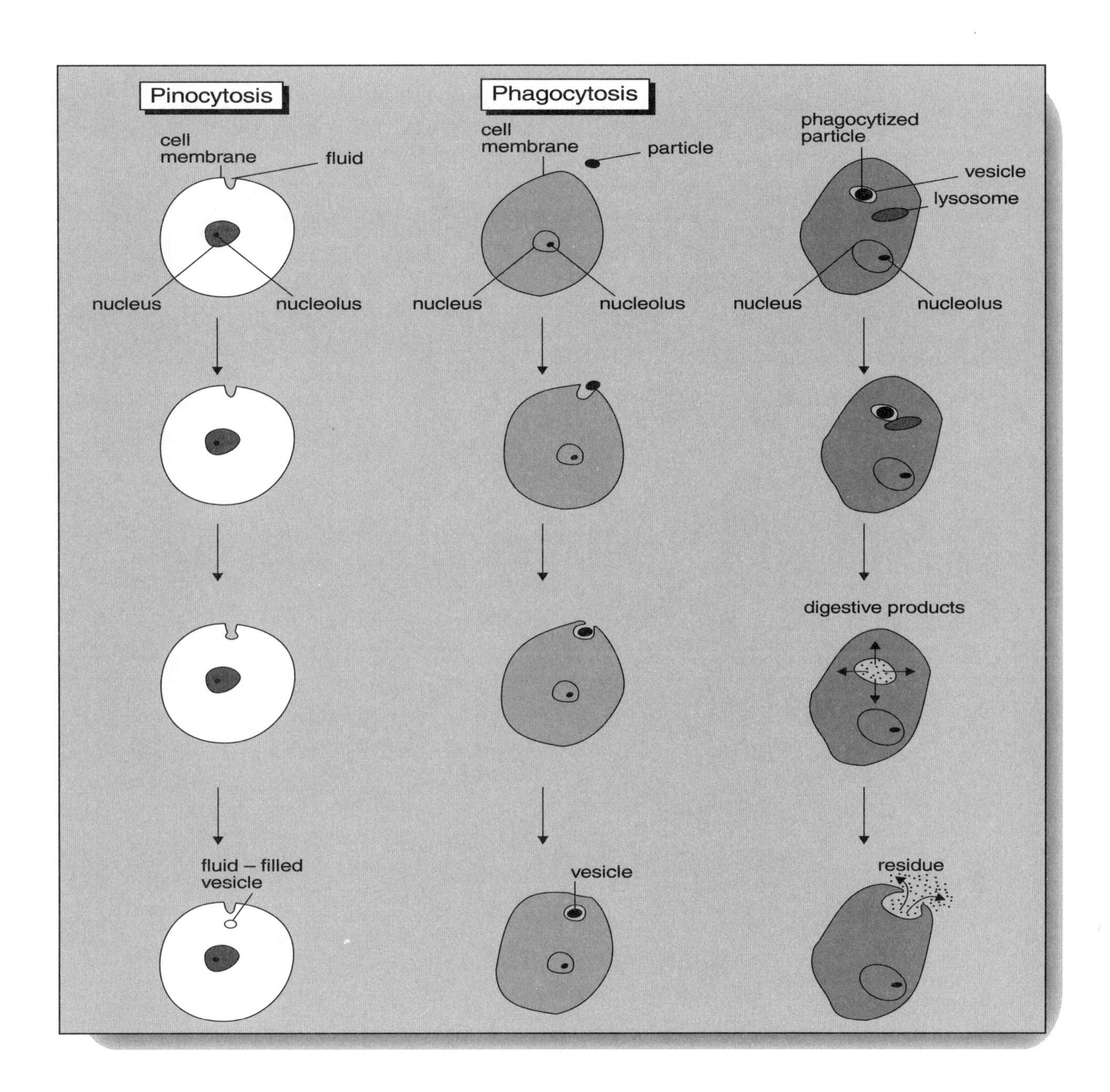
Pinocytosis
Phagocytosis
cell
membrane
fluid
nucleus
nucleolus
cell
membrane
particle
nucleus
nucleolus
phagocytized
particle
vesicle
lysosome
nucleus
nucleolus
digestive products
fluid – filled
vesicle
vesicle
residue

FIGURE 2.8

CHAPTER 3
NERVOUS SYSTEM NEUROPHYSIOLOGY AND THE ACTION POTENTIAL

FUNCTION OF THE NERVOUS SYSTEM

The nervous system is a communication system. Neurons, the cells that compose the nervous system, perform this communication. Accessory or neuroglial cells are also found in the nervous system and perform functions that support and protect the neurons themselves. The nervous system can be subdivided in the peripheral and central divisions. Sensory information relating to changes in the external or internal environment is gathered by specialized sensory receptors present on receptor cells or neurons. This information is then transmitted by peripheral sensory neurons to the central nervous system. In the central nervous system, highly complex (and poorly understood) interactions between nerve cells process the information and eventually generate a response (output). This is then transmitted back to motor neurons of the peripheral nervous system. These cells control other effector organs: skeletal, smooth and cardiac muscles and glands. Thus, the nervous system can detect changes occurring either externally or internally, make decisions on a course of action to follow and promote corrective measures. Such corrective measures might include changing body motility, effecting circulatory or gastric function, or activating glandular secretions **(Figure 3.1)**.

STRUCTURE OF NEURONS

Like other cells, a neuron is constructed with a cell body containing the usual cytoplasmic organelles: nucleus, Golgi apparatus, microtubules, lysosomes, mitochondria, cytoplasmic inclusions, plus Nissl bodies, similar to the rough endoplasmic reticulum of other cells. The neuron is covered by a limiting plasma membrane called a neurolemma. Extending from the cell body are two kinds of nerve fibers—dendrites and axons. Although most neurons usually have many dendrites, only one axon is present in any neuron. While the dendrites are relatively short in most neurons, the axon is frequently quite long. Dendrites possess many chemical receptors for neurotransmitters along their lengths, although such receptors can also be found on the cell bodies and even on the axon. Dendritic spines are also frequently present on dendrites that appear to act as contact points at synaptic connections.

Long slender neurofibrils, located within the axon, help maintain the axon structure. These neurofibrils also transport chemical transmitter from its site of synthesis in the cell body to the other end of the axon (the axon terminal) where the transmitter is stored in vesicles until released. Axons arise from the axon hillock. While only a single axon is present in any neuron, the axon may branch into a number of collaterals after leaving the cell body. At the terminal end of each collateral are fine branches: the axon terminals or the presynaptic terminals (or the boutons). Vesicles containing neurotransmitter are present in the axon terminal as well as many mitochondria. **(Figures 3.2)**.

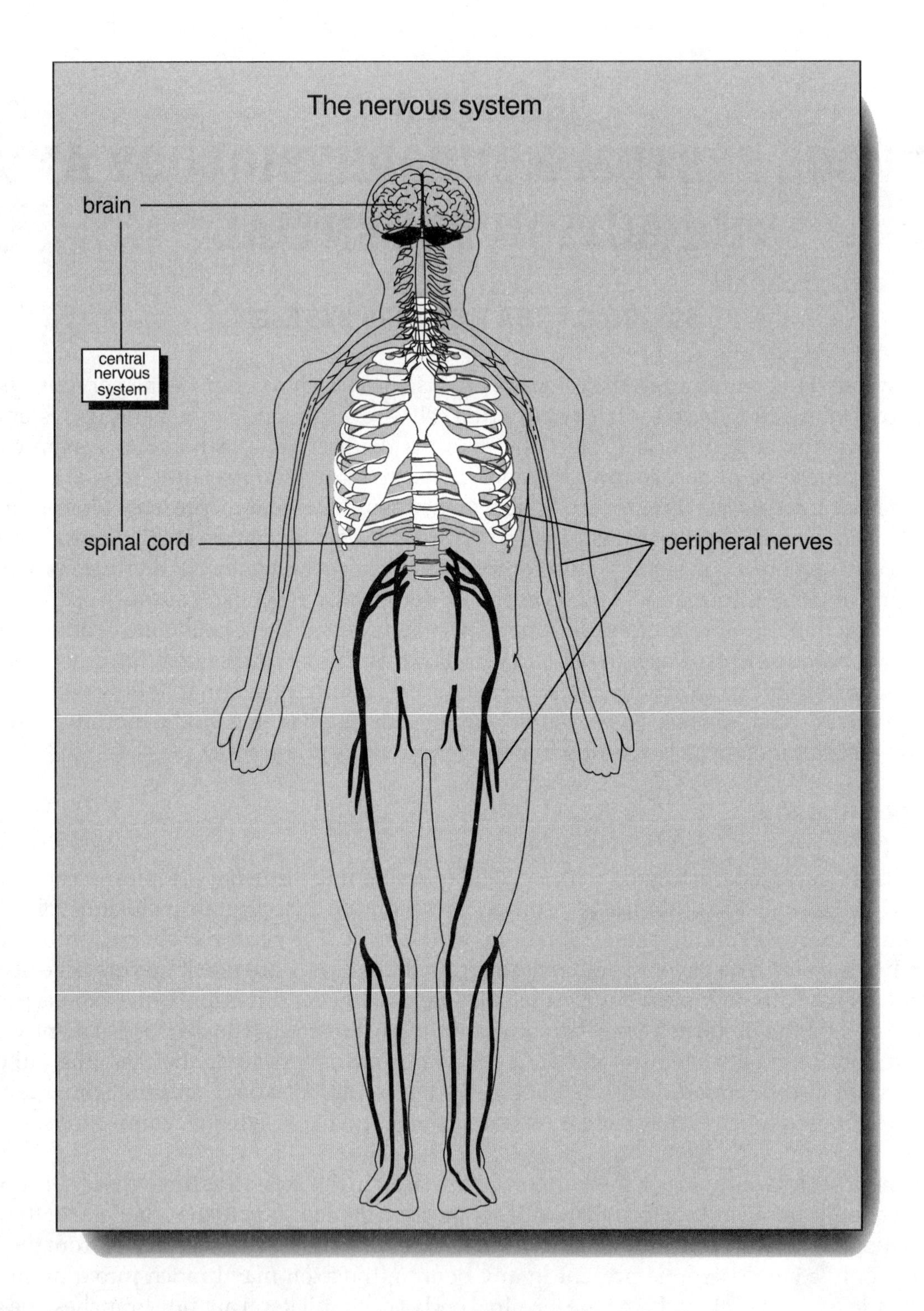

FIGURE 3.1

MYELINATION OF NERVE FIBERS

Large axons of the peripheral nerves are commonly encased, either individually or in groups, by a covering formed by a neuroglial cell called a *Schwann Cell*. Axons may be encased in groups by

Schwann cells containing cytoplasm and nuclei. Such a covering is called a neurolemmal sheath and these axons are then said to be unmyelinated. On the other hand, single axons may be covered by multiple layers of the lipid-protein called myelin formed by the individual Schwann cell wrapping its cell membranes around the axon many times. The outermost portion of this myelin sheath is still called the neurolemma because it contains the Schwann cell cytoplasm and nucleus. Groups of myelinated fibers appear white upon gross anatomical examination, while areas that contain neuron cell bodies appear gray. Since a single myelinated axon is usually quite long, many individual Schwann cells are required to form the myelin covering along the entire axon length. The junction between the myelin coverings formed by two adjacent Schwann cells is called a *Node of Ranvier*. Such nodes play very important roles in the conduction of action potentials (see following material).

OTHER NEUROGLIAL CELLS

Neuroglial cells act as accessory cells in the nervous system. The only neuroglial cells present in the peripheral nervous system are the Schwann cells, however, there are four types of neuroglial cells found in the central nervous system.

ASTROCYTES

In the central nervous system (CNS), astrocytes are closely associated with blood vessels through foot-like processes. Extensions from the astrocyte cell body also contact neurons. Such an arrangement may act to support the neurons. In addition, astrocytes are believed to regulate the movement of nutrients from the blood vessel into neurons. This ability may account for the presence of the so-called blood/brain barrier. Such a blood/brain barrier accounts for the observation that many substances and drugs present in the general circulation are unable to enter the CNS. Such a protective function of the blood/brain barrier is useful because it acts to isolate the CNS from the effects of an ingested substance. On the other hand, it can become a problem if it inhibits the entrance of antibiotics to the CNS during an infection.

OLIGODENDROCYTE

This cell is the source of the myelin covering on nerve fibers (axons) in the CNS. However, while the oligodendrocyte is capable of forming myelin, it does not form a neurolemmal sheath. This may be one reason why fibers of the CNS (unlike fibers of the Peripheral NS) are unable to regenerate (see following material).

MICROGLIAL CELLS

These small cells present in the CNS are phagocytic and perform the functions attributed to fixed macrophages. Their function is to phagocytize cellular debris and bacteria that may be present during an infection. (Thus, they are classified as members of the Reticuloendothelial system.) Usually such cells increase during an infection in the CNS.

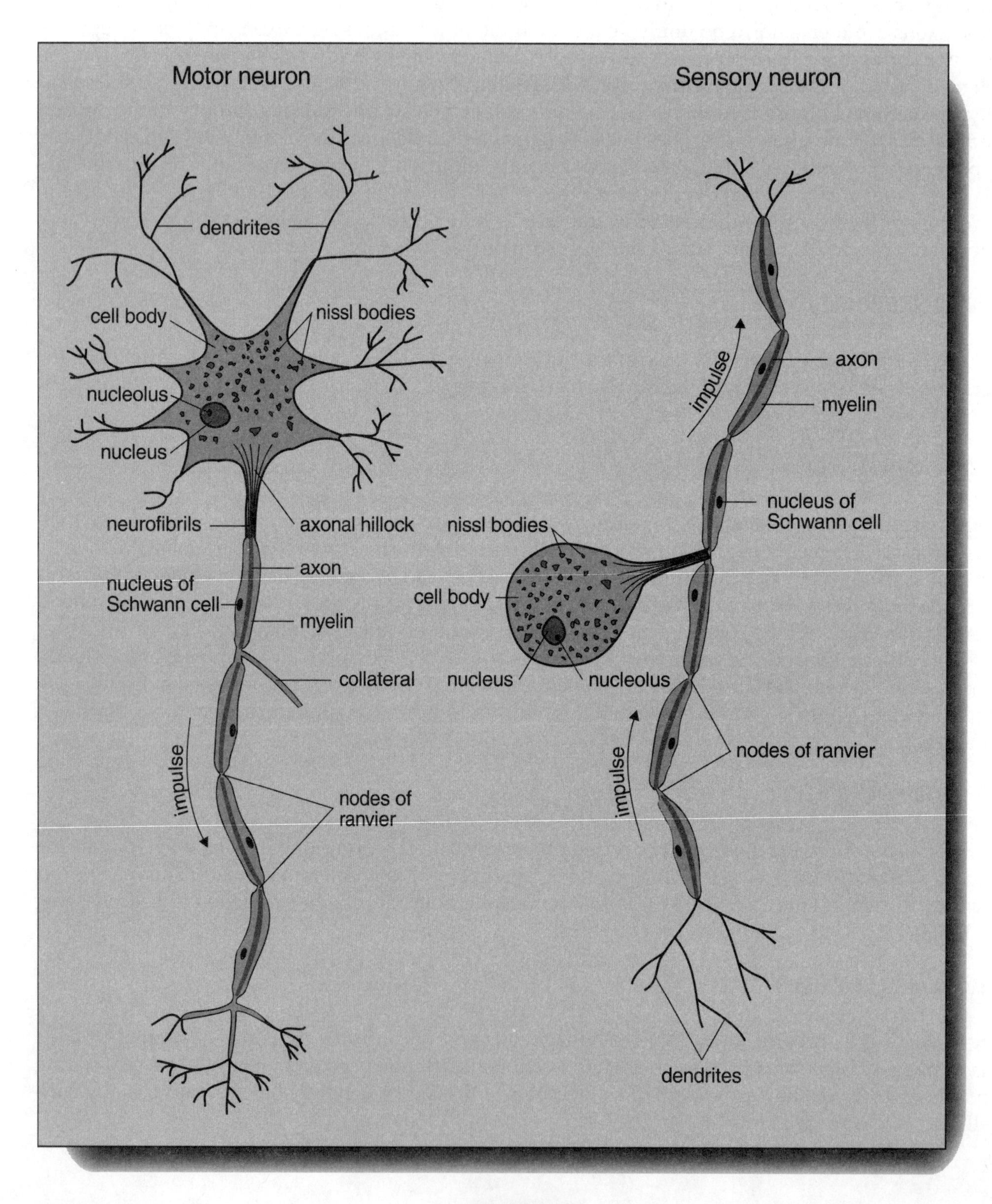

Motor neuron
Sensory neuron
dendrites
cell body
nissl bodies
nucleolus
nucleus
neurofibrils
axonal hillock
axon
nucleus of Schwann cell
myelin
collateral
impulse
nodes of ranvier
axon
myelin
impulse
nucleus of Schwann cell
nissl bodies
cell body
nucleus
nucleolus
impulse
nodes of ranvier
dendrites

FIGURE 3.2

EPENDYMA

These cells are cuboidal or columnar in shape and may be ciliated. They form epithelial-like membranes that line the hollow spaces inside the ventricles of the brain. In addition, they compose the structures called *choroid plexuses* found in all four ventricles. The choroid plexuses are the source of the cerebral spinal fluid (CSF) that flows throughout the ventricular system.

Neuroglial cells (Table 3.1)

Type	Location	Function
Schwann cell	Peripheral NS	Forms Neurolemmal sheaths. Is responsible for myelination of axons.
Astrocytes	Central NS	Blood - brain barrier. Regulates movement of substances from the circulatory system to neurons.
Oligodendrocytes	Central NS	Responsible for myelination of nerve fibers in central NS. Does not form neurolemmalsheaths.
Microglial cells	Central NS	Act as fixed macrophages in CNS. Function as phagocytic cells. May be classified as part of Reticuloendothelial system.
Ependyma	Central NS	Cuboidal or columnar cells that may have cilia. Form membrane covering inside of the ventricles. Form choroid plexuses that make cerebral spinal fluid.

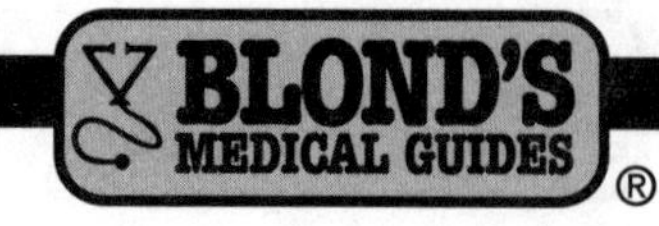

REGENERATION OF PERIPHERAL NERVES

When a nerve cell body is injured, the usual outcome is the death of the cell. However, if the axon of a peripheral nerve is cut, but the cell body remains intact, the axon may be able to regenerate. Initially, the cut segment distal to the cell body will degenerate all the way back to the skeletal muscle fiber. In addition, the Schwann cells that formed the neurolemmal sheath lining the degenerated axon will also degenerate, but will not completely disappear. These partially degenerated Schwann cells will act as a pathway to guide the growing tip of the newly regenerating axon. Growth from the proximal end of the axon will be able to follow the tube of sheath cells and may eventually reinnervate the skeletal muscle fiber and restore function. As the axon regrows, the Schwann cells also regenerate. Therefore, as a result of the presence of Schwann cells in the peripheral nervous system, peripheral nerves are able to regenerate. However, within the CNS, oligodendrocytes do not form a neurolemmal sheath (although they do form myelin) and therefore nerves of the CNS rarely regenerate if the fibers are cut.

MEMBRANE POTENTIALS

The term *potential difference* refers to an unequal number of electrically charged (+ and -) particles at two locations. In the case of a cell, if there is an unequal distribution of charges on two sides of a plasma membrane, the potential difference of charges across the membrane is called the *membrane potential*. Since opposite charges (+ charge and - charge) attract each other, but like charges (two + charges, or two - charges) repel each other, work is required in order to separate unlike charges on two sides of a membrane. On the other hand, if opposite charges are allowed to move towards each other through the force of attraction, the resulting flow of *current* will perform work. This is how electricity in your home can be harnessed to operate electric motors, appliances and so on. Further, this is how nerve and muscle cells are able to use electrical energy to send messages (called *action potentials*) across the cell membranes (see following material). However, before discussion of action potentials, one must consider the underlying mechanisms responsible for generation of membrane potentials.

THE CONCEPT OF MEMBRANE POTENTIAL

To understand how membrane potentials are generated, consider a plasma membrane surrounded on both sides by an electrolyte solution composed of both + and - charged ions **(Figure 3.3)**. If the distribution of + and - charges is equal on both sides of the membrane, there will be no net difference in charges on the two sides of the membrane and thus we will have electrical neutrality. However, if charge separation takes place in such a way that more + charges are present on one side of the membrane and more - charges are present on the other, this will produce a net difference in the charges on the two sides of the membrane. If we measure this charge difference on the two sides of the membrane with a *voltage* meter, we will be able to read a potential difference between the two sides (or two points). In other words, the charge separation across the membrane results in a membrane potential. Electrical potential is always measured in units called volts (V). However, in a cell, the electrical potential is so small that the units are in millivolts (1000 mV = 1 V). The greater the concentration of charges separated by the membrane, the greater the membrane potential. In our model, since the only area of charge separation is across the membrane, the volume of electrolyte solution not directly in conjunction with the plasma membrane will still have equal charge distribution and will therefore be electrically neutral.

GENERATION OF A MEMBRANE POTENTIAL

To generate a membrane potential, charge separation must be promoted across the membrane, but how is such separation of charges accomplishable? This separation of charged particles (ions) is primarily due to differences in the membrane permeability to Na^+, K^+ and the large intra-cellular proteins (A^-). According to the model, Na^+ and K^+ ions are able to penetrate the membrane only if they can pass through specific Na^+ and K^+ channels. As can be seen in **Table 3.2**, there is a significantly greater concentration of Na^+ ions in the extra-cellular fluid, and a greater concentration of K^+ ions present in the intracellular fluid. These differences in concentration are due primarily to the action of the linked Na^+/K^+ ATP-driven active transport pump capable of transporting intracellular Na^+ from the inside to the outside, while at the same time transporting extracellular K^+ from the outside to the inside. Note that, because only a few Na^+ channels are open only a small amount of Na^+ will be able to cross back into the cell from the outside to the inside by passive diffusion. (Remember that substances move from an area of higher to lower concentration by passive diffusion.) On the other hand, many more K^+ channels are open in the membrane and thus, a larger amount of K^+ is able to move by diffusion from inside to outside of the cell.

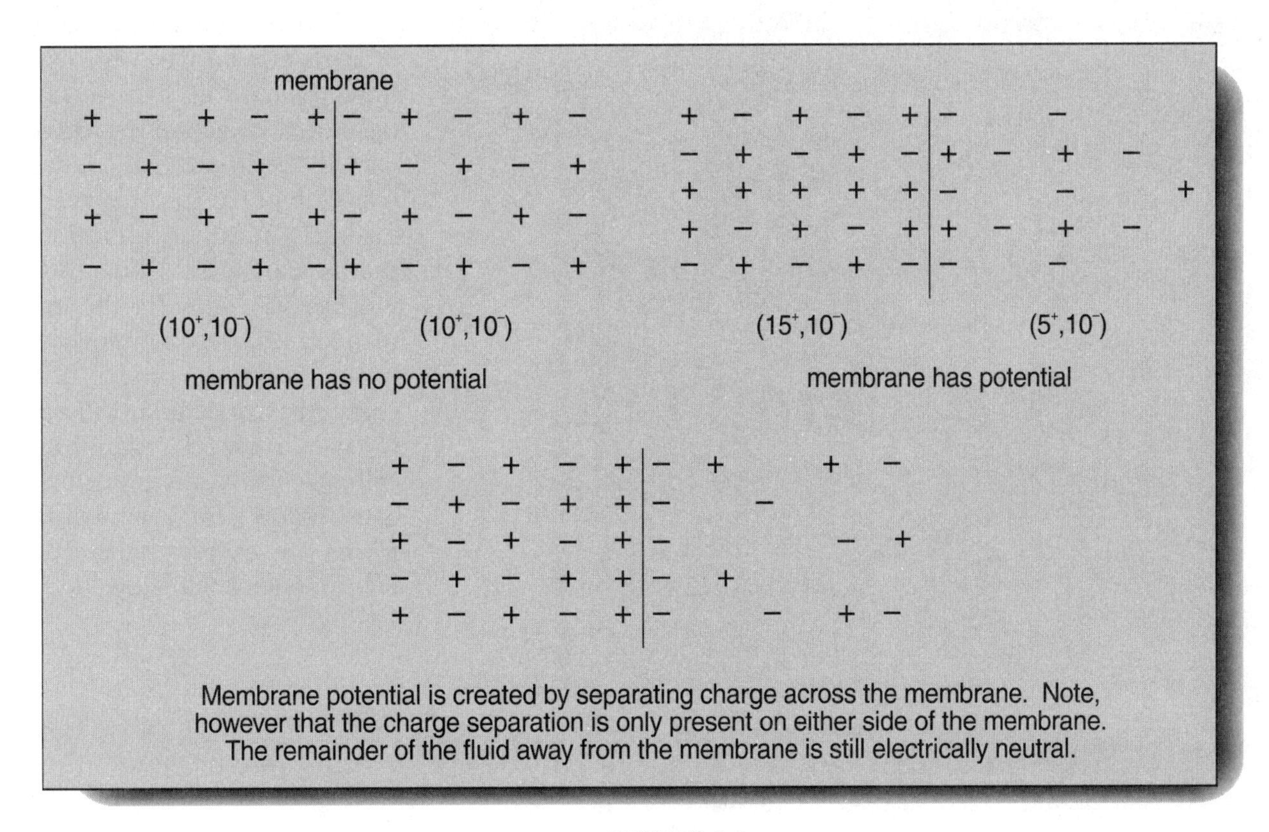

FIGURE 3.3

K^+ EQUILIBRIUM POTENTIAL

Consider a hypothetical situation where only K^+ and A^- are present and Na^+ is not involved: If we initially assume that the concentration of the intracellular K^+ is equal to the concentration of the intracellular A^-, the net charge inside the cell would be neutral (for each K^+ there is an A^- to balance it, thus forming a neutral pair.) However, while the K^+ can move down its concentration gradient and leave the inside of the cell, the A^- cannot cross the membrane (these proteins are too large to penetrate). Thus, we now find that as the positively charged K^+ leaves the cell, the remaining negatively charged A^- are no longer balanced by these positive charges. Under these conditions, the inside of the cell begins to build up negative charges resulting from the presence of unpaired A^-, while the outside begins to build up positive charges as a result of the presence of the unpaired K^+ **(Figure 3.4)**. These charges congregate directly on the membrane surface because unlike charges attract.

The question now must be asked: What is to prevent the K+ ions from continuously leaving the cell by diffusion until the intracellular K^+ concentration becomes equal to the extracellular concentration? Two mechanisms prevent this from happening. One (which we have already mentioned) is the linked Na^+/K^+ pump, and the other is described in the force diagram **(Figure 3.4A)** for K^+. Since K^+ is a positively charged ion, it will be attracted to the negatively charged A^- remaining inside the cell. This results in the build-up of charges on the inner and outer membrane surfaces. However, as more and more K^+ leaves the cell by diffusion, more and more A^- will build up inside. Thus the force of attraction between the extra-cellular K^+ and the intra-cellular A^- will continue to increase. Eventually, the outward force of diffusion driven by the chemical concentration gradient of intra-cellular K^+ will become equal to the inward electrical attraction force between the positively charged extra-cellular K^+ and the negatively charged intra-cellular A^-. At this point because the outward and inward forces are exactly balanced, K^+ movement will stop.

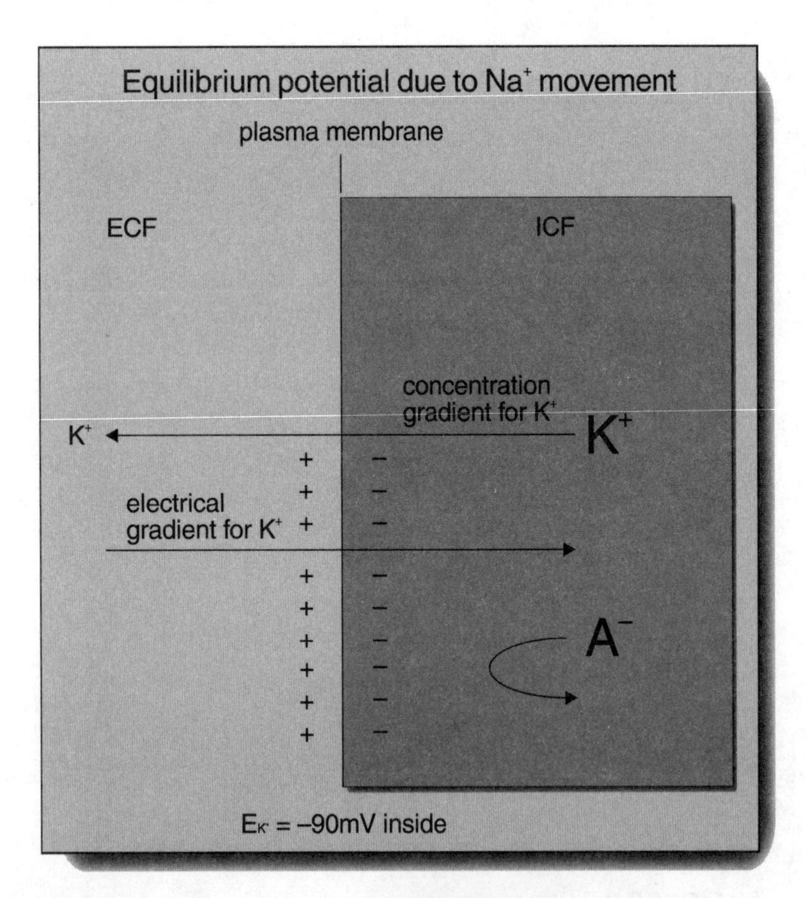

FIGURE 3.4

Extra and Intra-cellular Ion Concentration (Table 3.2)			
Ion	**Extra-cellular**	**Intra-cellular**	**Relative/ Permeability**
Na+	150 Mmoles/lt	15 Mmoles/lt	1
K+	5 Mmoles/lt	150 Mmoles/lt	50—75
A-	0 Mmoles/lt	65 Mmoles/lt	0

If we now measure the voltage across the membrane by placing one electrode of the millivolt meter on the inside of the membrane and one on the outside, we will find that there is a reading of -90 mV, termed the *equilibrium potential* for K^+ or E_K. (Note that the sign defines the polarity of the charge on the inside of the membrane.)

N_A^+ EQUILIBRIUM POTENTIAL

In addition to the presence of an E_K we must also consider the effect of Na^+ ions on the membrane potential. Under hypothetical conditions where there is no K^+ (or A^- for that matter), Na+ movement will elicit an intra-cellular positive charge. Since Na^+ ions are at a higher concentration on the outside of the cell than on the inside, a gradient exists that tries to push Na^+ into the cell. However, since there are fewer open Na^+ channels available, only limited amounts of Na^+ will be able to penetrate into the cell. As the positive Na^+ enters, it leaves behind negative Cl^- on the outside of the membrane. Eventually, the force of the chemical gradient that causes Na^+ to enter the cell will be balanced by the electrical force of attraction between the Na^+ on the inside of the membrane and the Cl^- on the outside **(Figure 3.5)**. The hypothetical outcome would thus be the development of an *equilibrium potential* for Na^+ or an E_{Na} which can be measured as +60 mV (the + charge is on the inside of the membrane).

THE N_A^+/K^+ EQUILIBRIUM RESTING MEMBRANE POTENTIAL (THE TRUE SCENARIO)

However, under normal resting conditions, in a real cell, K^+, Na^+, Cl^- and A^- are all interacting together. Here the membrane permeability of K^+ is two to three fold greater than for Na^+. Thus

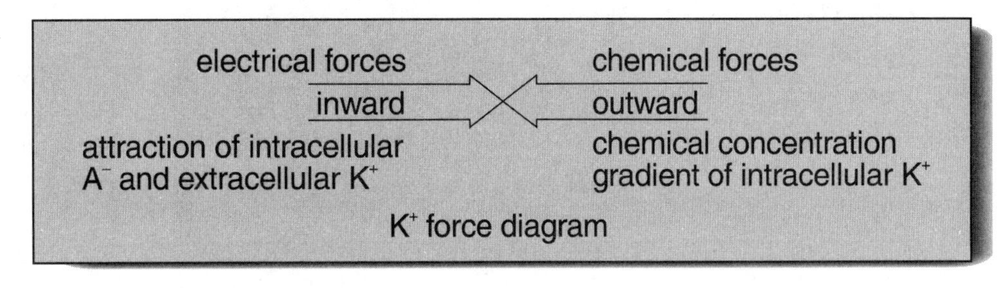

FIGURE 3.4A

only a small amount of Na^+ enters the cell, but a large amount of K^+ leaves. There are two forces acting together that account for the movement of Na^+ into the cell. First, the Na^+ concentration gradient pushes the Na^+ from the higher concentration extra-cellularly to the lower concentration intra-cellularly. Second, the intra-cellular negative charge attracts the positively charged Na^+ ions into the cell. When the Na^+ enters the cell it neutralizes some of the negative charges inside. This is sometimes termed the "short circuit Na^+ current." The effect is to reduce the internal negative charge from -90 mV (due to E_K alone) to a -70 mV (due to the combination of E_K and E_{Na}). Thus, in a normal cell the *equilibrium resting membrane potential* is in the neighborhood of -70 mV **(Figure 3.6)**.

ACTIONS OF THE Na^+/K^+ PUMP

Because the equilibrium resting membrane potential is -70 mV, it doesn't counterbalance the flow of K^+ outward or the flow of Na^+ inward. Recall that to balance the K^+ concentration gradient requires a value of -90 mV (the E_K) be maintained. In addition, since Na^+ movement into the cell is being driven both by the chemical concentration gradient, and by the electrical gradient, there is nothing to oppose this inward movement. As a result, Na^+ ions continuously leak in and K^+ ions continuously leak out. If this situation remains unchecked, the intra-cellular concentration of the K^+ will continue to drop while that of Na^+ will rise. To maintain these concentrations constant, (and thereby also maintain a constant equilibrium resting membrane potential), the cell has developed the linked Na+/K+ pump **(Figure 3.7)**. As Na^+ leaks in it is removed by the pump which replaces K^+ that leaks out. Obviously, since both Na^+ and K^+ are being pumped against their individual concentration gradients (Na^+ outward and K^+ inward), this pumping "uphill" requires the input of energy which is supplied by ATP.

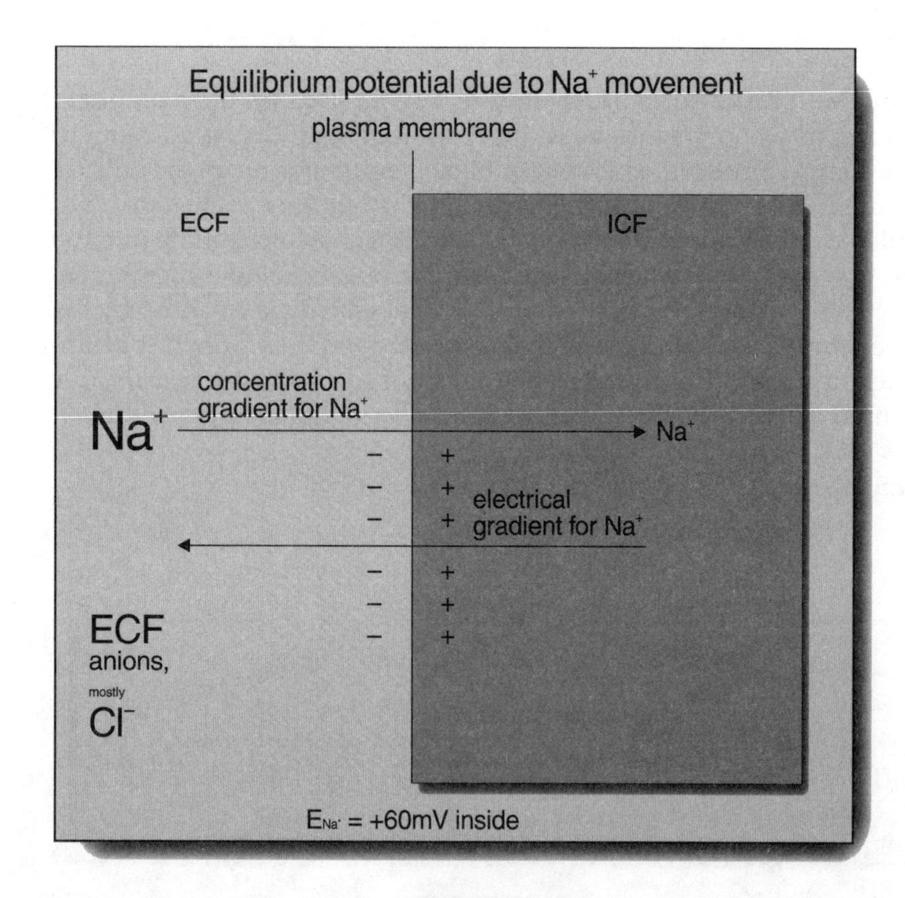

FIGURE 3.5

Action of the Na^+/K^+ Pump

1. Pumps Na^+ out of the cell and K^+ into the cell.

2. Moves against the ionic chemical concentration gradients, and is thus active transport.

3. Requires ATP as the energy source.

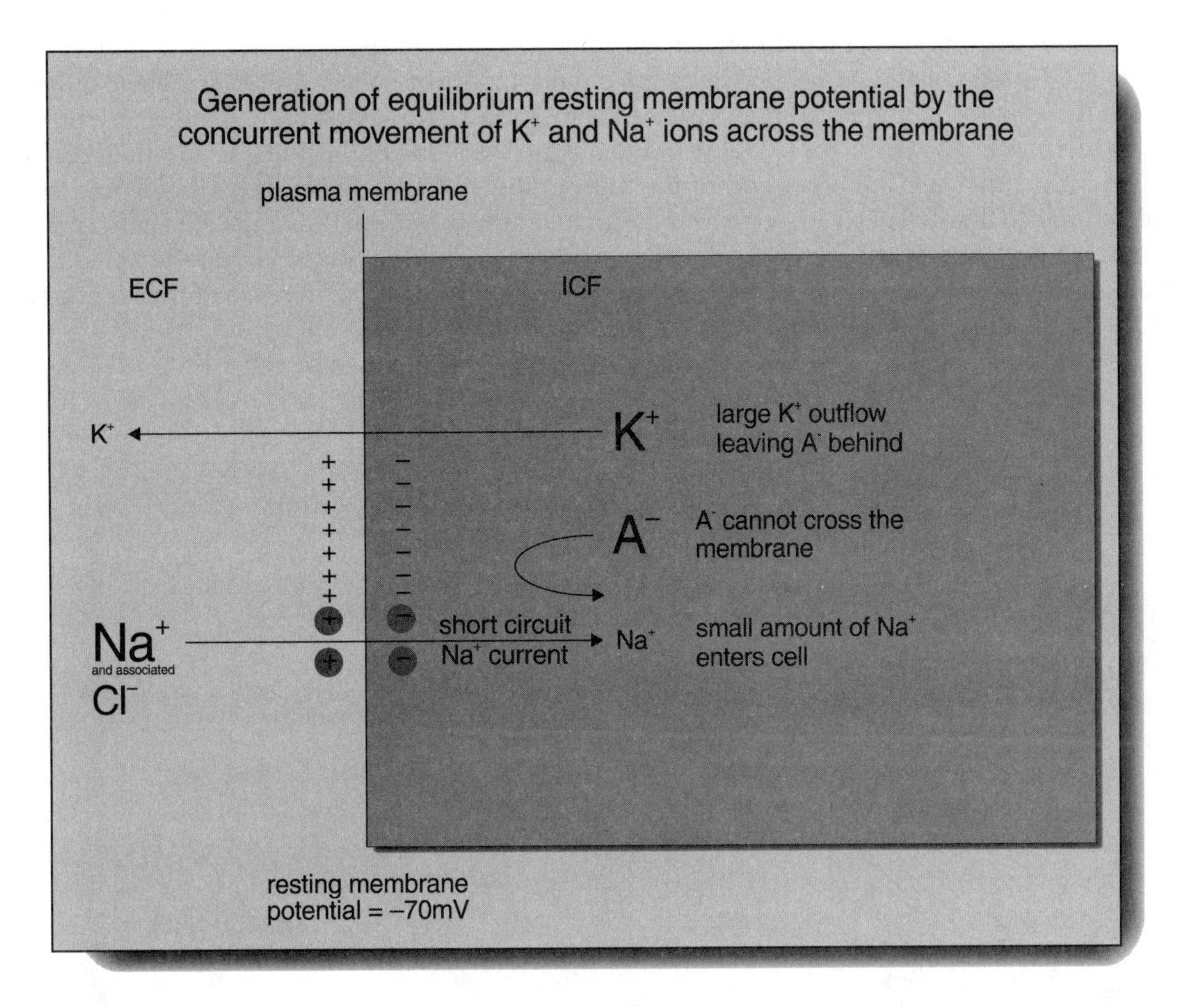

FIGURE 3.6

4. Is responsible for the initial Na^+ and K^+ concentration differences across the membrane.

5. Is responsible for maintaining the Na^+ and K^+ concentration differences across the membrane in the face of leakage of these ions into and out of the cell.

THE EQUILIBRIUM RESTING MEMBRANE POTENTIAL AND CL^- IONS

Cl^- is an extra-cellular ion which is able to passively move across the membrane in most cells. Under hypothetical conditions, the equilibrium potential of Cl^- is equal to -70mV. Movement of Cl^- across the cell membrane is entirely due to the presence of the electrical gradient resulting from K^+ and Na^+, and since this is equal to -70 mV (with the negative charges inside the cell) Cl^- is pushed out by repulsion.

THE CELL AS A BATTERY

A battery is a source of potential chemical energy. The chemical reactions in a battery generate charged particles that have the potential to do work. The usual garden variety flashlight battery is rated at 1.5

volts. Since voltage is the potential difference measured between the positive (+) side of the battery and the negative (-) side of the battery, voltage is actually a sort of concentration gradient of charged particles. If leads of a voltage meter were placed across the + and - ends of the battery it would read 1.5V. Furthermore, if a conductor (wire) was attached to the - side of the battery, and then connected to a light bulb, attaching the other end of the bulb to the + side of the battery, this would create a complete circuit. Charged particles (electrons, [e^-]) could now flow from the - end of the battery through the light bulb and back to the + end. As this flow of charged particles (a *current* flow) passed through the bulb, it would be able to do work, in this case to light up the bulb with some of the kinetic energy supplied by the moving e^- **(Figure 3.9)**.

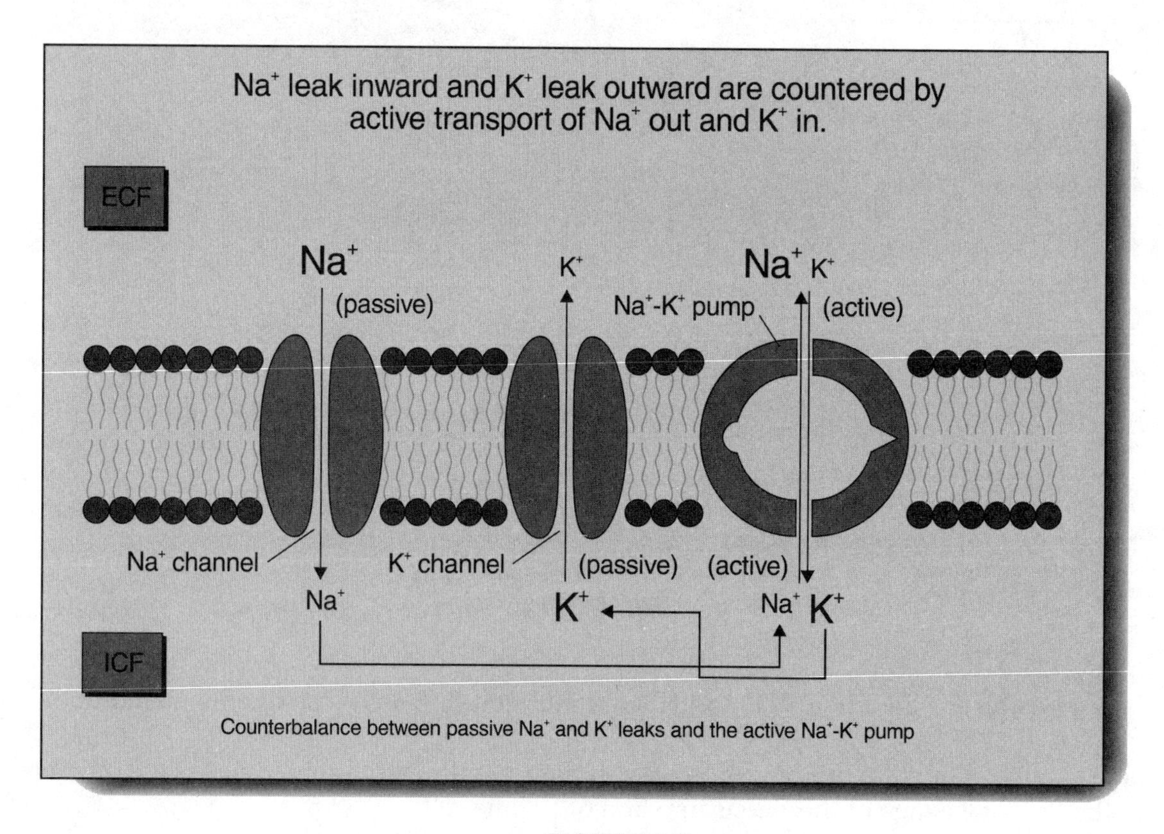

FIGURE 3.7

Let us now look at a cell as if it too were a battery. Like the flashlight battery, one side (the inside) is negative and the other side (the outside) is positive. If we place the leads of a millivolt meter across the membrane—one lead inside on the - side of the battery—and the other outside, on the + side of the battery, we can read the voltage (-70mV). If we were to attach conductors across the membrane to a very tiny light bulb, hypothetically at least, if the bulb were small enough, we could generate a current flow to light the bulb **(Figures 3.8 and 3.9)**. Obviously, we will not use this cellular battery for such a purpose, but current flow can send messages from one part of a nerve or muscle cell to another. Such messages are called *action potentials*.

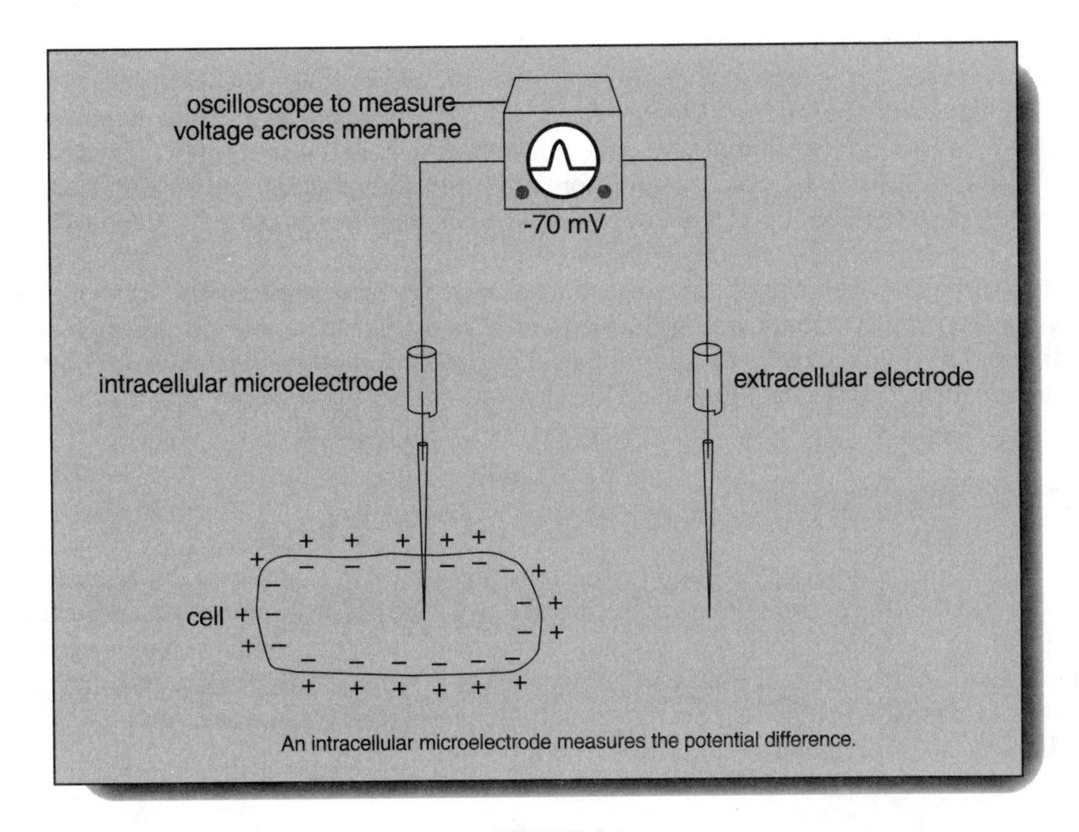

An intracellular microelectrode measures the potential difference.

FIGURE 3.8

EVENTS THAT LEAD TO AN ACTION POTENTIAL

LOCAL CURRENT FLOW

Cell membranes in all living cells produce an equilibrium resting membrane potential. Not all cells utilize this resting membrane potential as effectively as do nerves and muscle cells. Since cell membranes separate charges, they create a potential difference (voltage) that is capable of doing work (if these charged particles are able to move). However, movement of particles down the membrane can only be accomplished if a limited and discrete area of membrane undergoes charge reversal. For the moment, do not worry about how such a local area of charge reversal is produced. Assume that at this one active spot, and at no other, the outside of the membrane becomes negative while the inside becomes positive. The surrounding adjacent inactive areas of

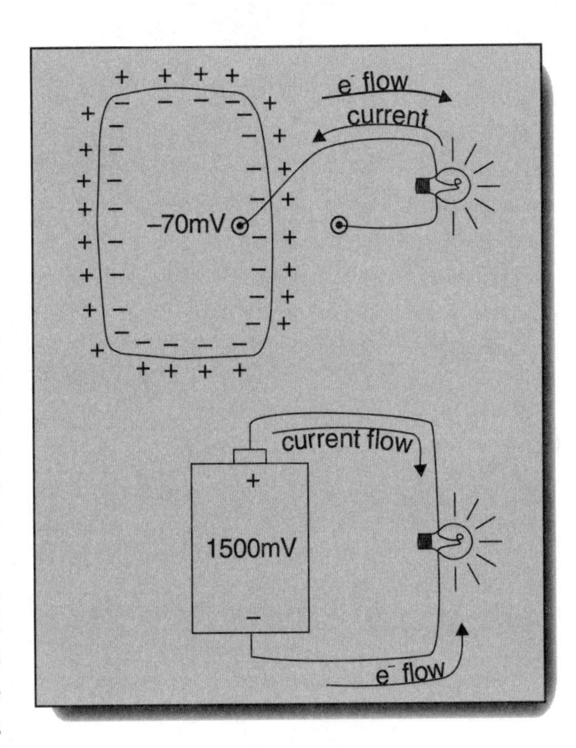

FIGURE 3.9

membrane are not affected, and here the outside remains positive while the inside remains negative. Because of this reversal of charge at a single local site, charged particles are now able to flow. If we assume that the movable charged particles are electrons (e^-) then we must conclude that these e^- should flow from the areas of negative charge on the membrane surfaces to the areas of positive charge on the membrane surfaces. This is a correct assumption, although nonetheless, convention states that the direction of *current flow* is from positive to negative (opposite to the direction of the moving e^-).

The reason for the confusion regarding direction of current flow is probably attributable to Ben Franklin who first proposed that moving charged particles were responsible for electricity. However, since he did not know which of his two hypothetical particles were movable, he incorrectly chose the positive particle instead of the negative one. Thus, by convention, the direction of current flow is from positive to negative, although actually e^- move from negative to positive.

DECREMENTAL CONDUCTION

Now for a look at these moving charges (i.e., this current flow). The charges will migrate away from the active site and thus the current flow will be directed from the positive regions towards the negative regions. However, the kinetic energy of the moving particles will rapidly dissipate due to resistance in the membrane, and the current flow will get weaker as it moves farther away from the site of the original charge reversal. Current flow decreases, getting weaker with distance. Eventually, the current flow will become so weak that it will disappear. **(Figure 3.10)**.

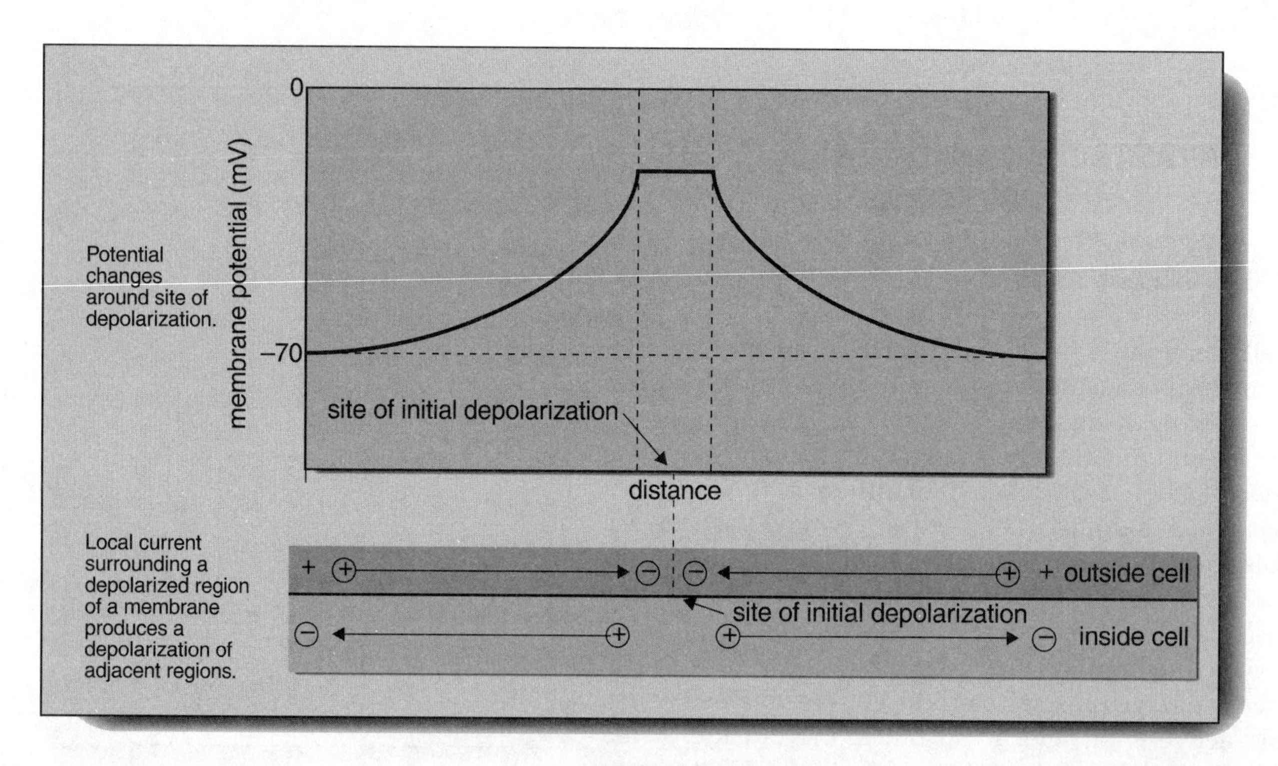

FIGURE 3.10

GENERATION OF CHARGE REVERSAL IN THE RESTING MEMBRANE

Any time charge reversal takes place in a membrane, a certain amount of current will flow. Various stimuli (chemical, electrical or physical processes) at the membrane can trigger local charge reversal at one active spot. If the intensity of the local charge reversal is proportional to the intensity of the stimulus, the charge reversal is defined as a *graded potential*. Since the magnitude of the graded potential is directly proportional to the intensity of the triggering stimulus, a bigger stimulus will promote a bigger graded potential. In other words with a larger triggering event the amount of local charge reversal will be greater than with a smaller triggering event. Taking this one step further, with a bigger stimulus there will be a larger current flow generated. **(Figure 3.11)**.

SENSORY RECEPTORS AND TRANSDUCTION OF INFORMATION

Communication between the body and the outside world is provided by the nervous system. Stimuli generated by the outside environment are conducted into the central nervous system (CNS) to be processed. These stimuli provide information regarding conditions in the outside world. For example, information in the form of electromagnetic energy (heat, light), mechanical energy, or chemical changes are transmitted by the peripheral nervous system to the CNS to be processed. This input of information allows the CNS to respond to changing external conditions and alter the function of the various body systems designed to maintain total body homeostasis. Naturally, for the nervous system to do the job, the external stimuli (i.e., light, heat, chemical, mechanical) must be converted from their original form into the common language that the nervous system understands. This common language is in the form of *action potentials*. The process of converting diverse stimuli into action potentials is known as *transduction*.

For example, as light energy strikes specialized receptor cells of the retina (rods and cones) the energy contained in the light particle/waves (i.e., photons) is transduced into action potentials conducted to the CNS by the optic nerve.

Conversion of the stimulus energy into action potentials is mediated by specialized sensory receptors that actually perform the process of transduction. In order to convert the original stimulus energy (say, light energy) into action potentials, the sensory receptor actually transduces the stimulus energy into a graded potential. Thus, with a bigger stimulus energy, a larger graded potential will be produced, which will result in a larger flow of current on the sensory receptor membrane.

Taking the above example one step further, when light energy strikes the rods and cones, it is transduced into a graded potential. This results in a flow of current in the receptor cell and eventually generates action potentials that are conducted into the CNS.

RECEPTOR POTENTIALS AND GENERATOR POTENTIALS

Sensory receptors that transduce external stimulus energy can be divided into two basic types: (1) receptors connected directly to the sensory nerve, and (2) receptors in the form of separate cells. If the receptor membrane is part of the dendrite of the sensory nerve (called the *afferent sensory nerve*), this type of receptor membrane will produce a graded potential that is called a *generator potential*. If the sensory receptor is on a separate cell that is in close contact with the afferent sensory nerve, the graded

potential that it produces here will be called a *receptor potential*. In either event, the local current flow that is created as a byproduct of these graded potentials will decrease with distance (becoming weaker as it moves into the neurolemma). However, if the flow of current is large enough, it may be able to stimulate the neurolemma to produce an action potential. A stimulus that induces a local current flow in the nerve that in turn causes an action potential to be generated, is a *threshold stimulus*.

GENERATION OF AN ACTION POTENTIAL

As just mentioned, although all cells of the body are capable of producing equilibrium resting membrane potentials, only nerve and muscle cells have learned to utilize this membrane potential to send messages (in the form of action potentials) across the cell membrane. An action potential is actually a disturbance in the resting membrane potential. Such a disturbance takes the form of a rapid transient change in the polarity of the membrane potential at a single delineated site on the membrane. At this one active spot, and at no other, the outside of the membrane becomes negative while the inside becomes positive. However, the surrounding inactive areas of membrane are not affected and here the outside remains positive while the inside remains negative.

Action potentials are produced when a local flow of current causes the depolarization (a reversal of charge) at a small segment of the membrane. If this charge reversal results in a threshold stimulus, it will initiate the opening of limited numbers of Na^+ channels in the nerve membrane. As these channels open, a combination of the Na^+ chemical gradient and the Na^+ electrical gradient causes Na^+ ions to pass from the outside to the inside of the nerve cell. This depolarizes the inside of the cell (it becomes more positive). As the inside becomes more positive, the structures of the membrane proteins change

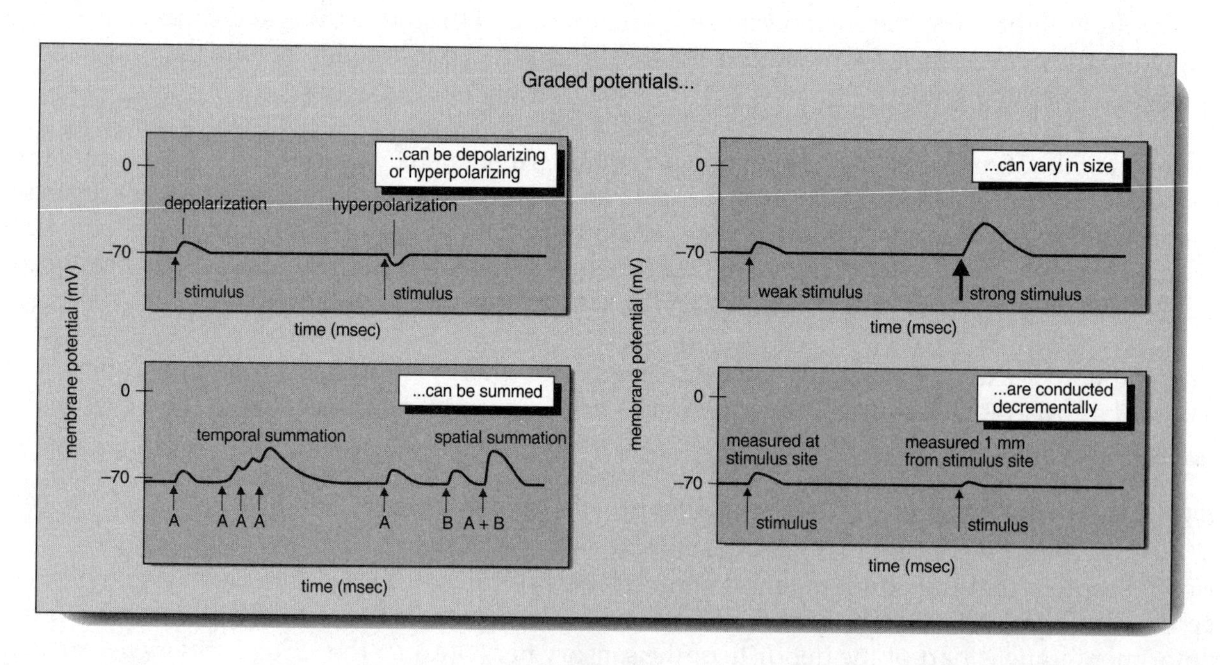

FIGURE 3.11

leading to the opening of more Na^+ channels (the membrane permeability to Na^+ increases). In turn, this increases the inflow of Na^+ which makes the inside more positive. **(Figure 3.12)**. The resulting positive feedback generates an in-rushing flow of Na^+ that depolarizes the membrane leading to +30 mV (a value close to the Na^+ equilibrium potential). This increase in Na^+ permeability leads to the so-called rising phase, or the depolarization phase of the action potential.

Depolarization — when the resting membrane potential that is normally -70 mV moves towards positive.

Repolerization — when the resting membrane potential moves back towards negative (i.e., towards -70 mV).

Hyperpolarization — when the resting membrane potential becomes more negative then -70 mV, and usually reaches approximately -90 mV **(Figure 3.13)**.

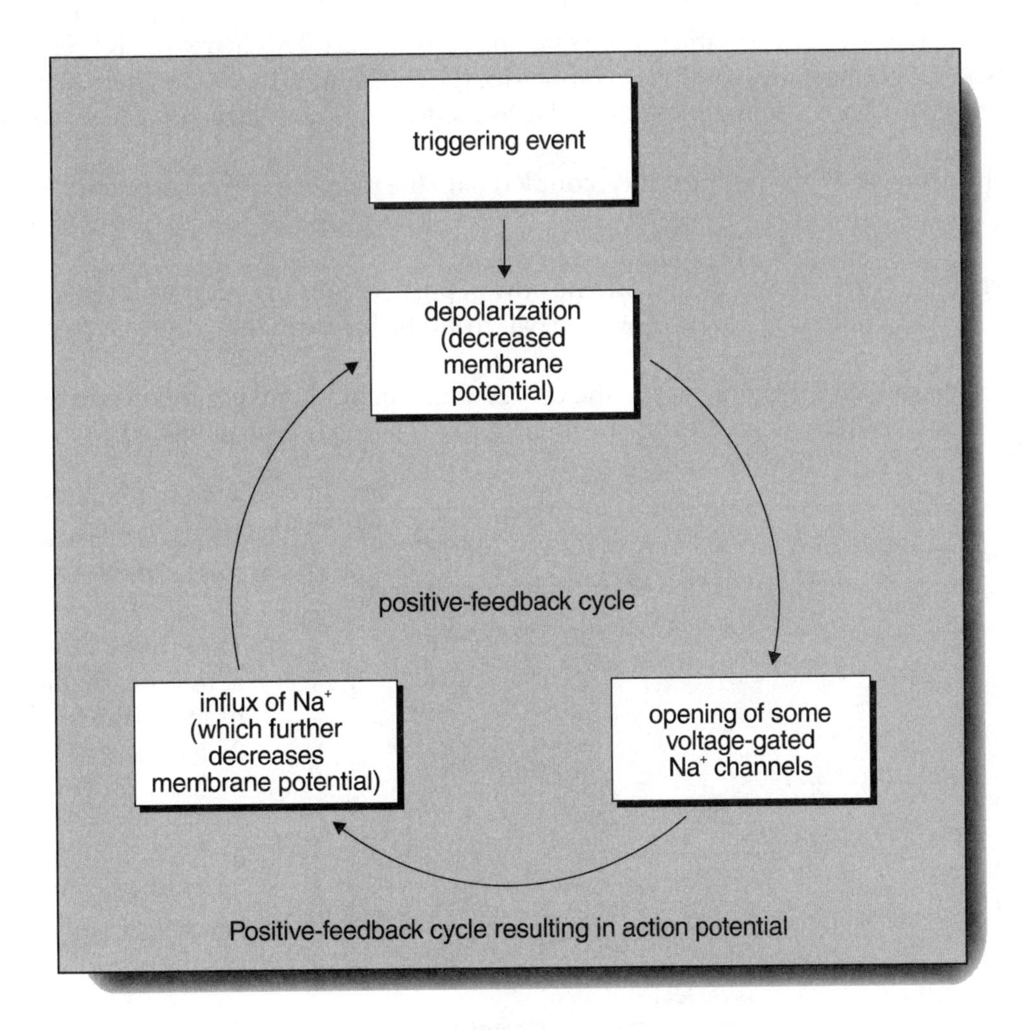

FIGURE 3.12

Voltage — a measure of potential energy. It is due to the potential difference in the number of charged particles (usually e^-) when one point is compared to another. It results from differences in the number (or concentration) of charged particles present at these two locations.

Current — the flow of charged particles between two points through a conductor. This movement denotes kinetic energy that can be harnessed to do work. In equivalent chemical terms it could be described as the movement of a substance down a concentration gradient from a point of higher concentration to one of lower concentration. If there is a greater voltage between the two points as a result of a greater potential difference (a larger concentration gradient of charged particles), the resulting flow of current will be greater and more work can be done.

At the peak of the Na^+ inflow, the Na^+ channels now rapidly close (Na^+ permeability decreases) cutting off the ionic flow. Simultaneously, K^+ channels open in the membrane (K^+ permeability increases) and additional K^+ flows out of the cell. Remember that throughout this process there is some outflow of K^+ going on, since this is the mechanism that leads to the resting membrane potential. Thus, the membrane is initially permeable to some K^+ but towards the end of the action potential this permeability increases about 300 times more than the value at rest. This increased K^+ permeability and the decreased Na^+ permeability combine to repolarize the membrane (it becomes more negative again). This repolarization phase is sometimes referred to as the *falling phase of the action potential* **(Figure 3.16)**.

Why does the decrease of Na^+ permeability, coupled with the increased K^+ permeability, repolarize the nerve?

1) First, when Na^+ inflow is turned off, the flow of positive ions into the nerve ceases. In addition, the Na^+/K^+ pump will now remove the Na^+ from the inside making the inside more negative.

2) As increased amounts of K^+ leave the cell, A^- is left behind making the inside more negative. Thus, as K^+ outflow increases the inside negative charge is re-established.

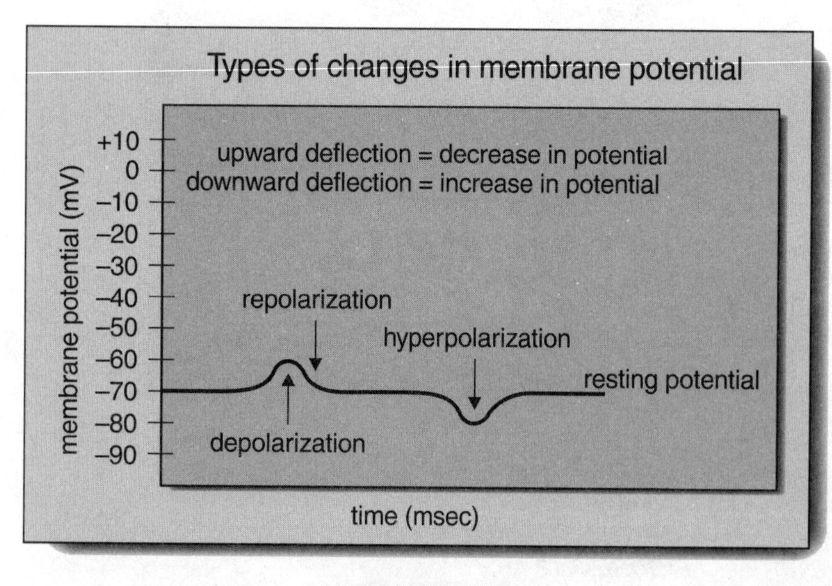

FIGURE 3.13

STEPS LEADING TO THE RISING PHASE OF THE ACTION POTENTIAL

1. Current flow depolarizes a minute segment of membrane to threshold.
2. Na^+ channels open.
3. Na^+ permeability increases.
4. Na^+ enters nerve.
5. The inside of the nerve cell becomes more positive; it depolarizes.
6. Repeat # 1 through 5 until the inside of the nerve reaches +30 mV.

STEPS LEADING TO THE FALLING PHASE OF THE ACTION POTENTIAL

1. Na^+ channels close at the peak of the Na+ inflow.
2. Na^+ permeability decreases.
3. Na^+ flow into the nerve is significantly reduced.
4. K^+ channels open.
5. K^+ permeability increases.
6. K^+ outflow from the nerve increases (300 fold).
7. Resting membrane potential becomes more negative (it repolarizes).
8. Due to the excessive outflow of K^+, the resting membrane potential undershoots. It becomes more negative than -70 mV and will probably reach -90 mV or -100 mV (hyperpolarization phase).
9. The majority of the K^+ channels now close again.
10. The Na^+/K^+ pump now re-establishes the normal levels of ions inside the cell.
11. Resting membrane potential returns to -70 mV.

THE ALL-OR-NONE LAW OF ACTION POTENTIALS

To generate an action potential, a depolarizing stimulus is required to alter the protein structure of the Na^+ gates on the channels in the membrane. If this stimulus is sufficiently large (termed a *threshold stimulus*), the Na^+ channels will open wide enough to admit a self-sustaining flow of Na+ ions into the cell. Under these conditions, the inflow of the Na^+ continues to open the Na^+ channels and results in the rising phase of the action potential. Thus, a threshold stimulus will result in the generation of a full-blown action potential. However, if the depolarizing stimulus is not strong enough, the Na^+ channels may open only slightly causing a small, and short-lived, depolarization in the resting membrane potential (sometimes called an *excitatory post synaptic potential*, or EPSP; see following material). This weak stimulus cannot generate an action potential because the Na^+ channels do not stay open, but close immediately. This weak stimulus is thus called a *subthreshold stimulus.*

On the other hand, if the depolarizing stimulus is larger than a threshold stimulus (called a *suprathreshold stimulus*), the Na^+ channels will indeed open and an action potential will be generated. However, the number of Na^+ channels that open due to a suprathreshold stimulus will be exactly the same as the number that open with a threshold stimulus. Thus, the action potential generated after a suprathreshold stimulus will be exactly the same size, and last the same length of time, as that generated with a threshold stimulus! This effect is termed the all-or-none law or effect of the action potential. Either you get the entire action potential or you don't get any action potential, there is no in-between effect. (**Figure 3.14**).

WHAT PURPOSE IS SERVED BY THE GENERATION OF THE ACTION POTENTIAL?

Initiation of a single action potential at a single delineated site on the membrane will generate current flow that will move away from the initiation site. This current flow will weaken with distance and will eventually disappear completely due to resistance in the membrane. Thus, if we were to initiate an action potential at the cell body, the message (in the form of the current flow) would not reach the axon terminal, or even if it did, it would be too weak to stimulate the release of transmitter at the synapse.

To correct this problem, the current flow generated by our first action potential will cause the next adjacent segment of membrane to reach threshold. The result will be the generation of a new action potential in this next segment (call it segment B). Again, this new action potential will produce a current flow that will produce a threshold stimulus in the next adjacent segment (call it segment C) and this process will continue all the way down the nerve axon (for perhaps hundreds of adjacent segments). Thus, at the axon terminal the final action potential to be generated will provide adequate current flow to allow the release of transmitter at the synapse. Clearly, the repeated generation of action potentials, one after the other, all the way down the axon, acts as a current amplifier and keeps the signal strong.

MYELINATION AND VELOCITY OF CONDUCTION

To send a stimulus down a nerve, action potentials must be repeatedly generated one after the other. Since each action potential takes a finite amount of time to be generated (approximately 8 msec), transmission down an axon will be affected by the number of action potentials to be generated. To speed up the velocity of conduction, we need to make the current propagate farther before a new action potential is produced. If the nerve axon is covered with a myelin sheath, the current flow will be forced to jump between nodes of Ranvier. At each node a new action potential can be generated, but action potentials cannot be generated where the axon is covered by myelin. If we compare the velocity of conduction between a myelinated and an unmyelinated nerve, we find that in the myelinated fiber action potentials are generated at 1 mm intervals on the nerve (the distance of adjacent nodes of Ranvier) whereas in the unmyelinated nerve, the adjacent action potentials are generated much closer together. Thus, in the myelinated fiber it takes less action potentials to send the message to the end of the nerve; or to put it another way, the velocity is 50 x faster in a myelinated nerve as compared to an unmyelinated fiber. This jumping of the action potential between nodes is termed *saltatory conduction*. In the body, the myelinated nerve tracks carry more important information than do the unmyelinated tracks **(Figure 3.15)**.

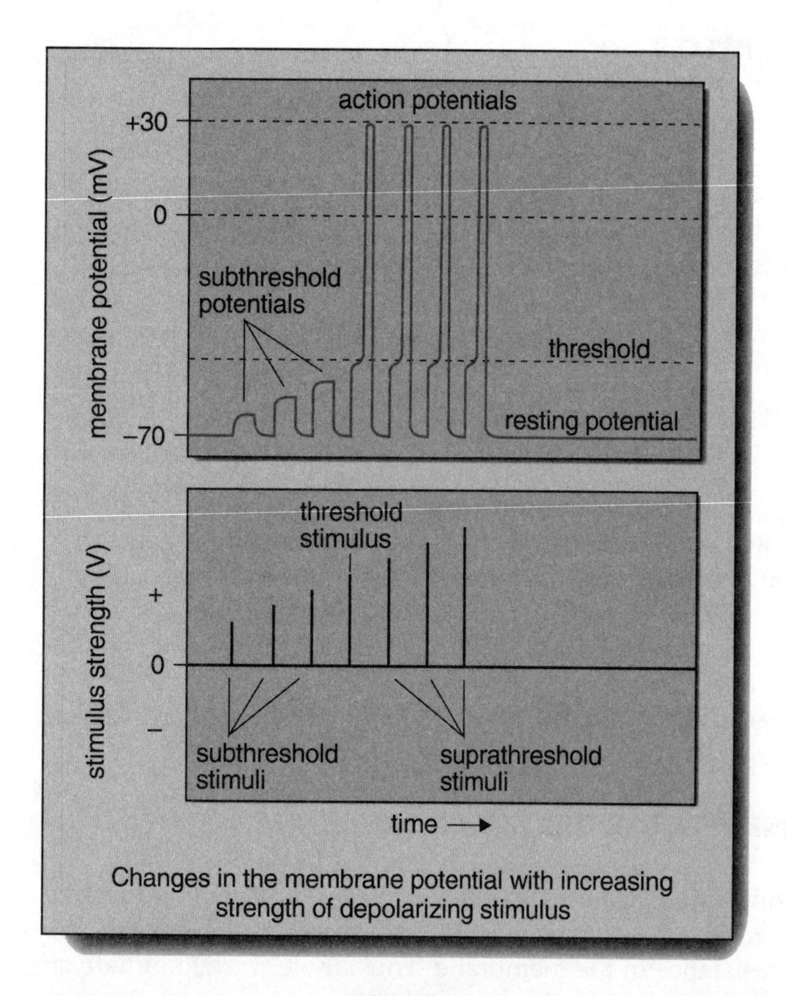

Changes in the membrane potential with increasing strength of depolarizing stimulus

FIGURE 3.14

FIBER DIAMETER AND VELOCITY OF CONDUCTION

Fiber diameter also affects velocity. Simply stated, the larger the diameter, the faster the velocity. Thus, large myelinated fibers, such as motor neurons that control skeletal muscle cell contraction, can conduct action potentials at the phenomenal rate of 120 meters/sec (360 miles/hr). On the other hand, velocity in the unmyelinated small diameter nerves is only 0.7 meters/sec (2 miles/hr). Clearly, myelination and diameter play a very important role in the speed of action potential conduction.

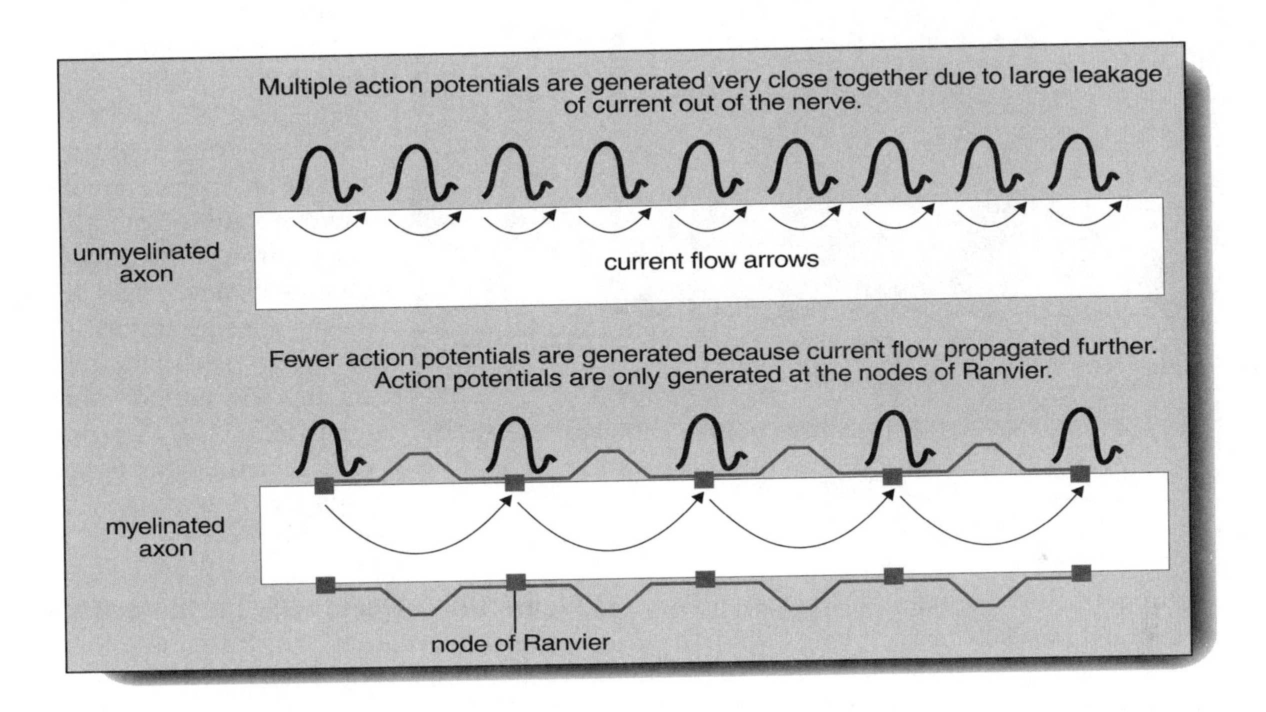

FIGURE 3.15

SYNAPSES AND VELOCITY OF CONDUCTION

For stimuli to cross a synapse, a chemical transmitter must be released (see below). This process is time consuming and therefore will slow down impulse conduction through a nerve track (i.e., neurons linked together by synapses.) It therefore follows that nerve tracks with few synapses will conduct information faster than nerve tracks with many synapses.

REFRACTORY PERIOD

The nervous system is unidirectional. This means that it sends action potentials down the nerve in only one direction (from cell body to axon terminal) and not the other way (from axon terminal to cell body), although under certain conditions conduction in the reverse direction can take place. In addition,

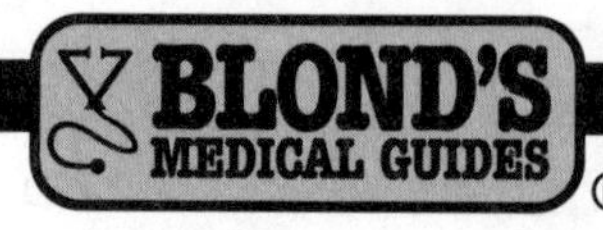

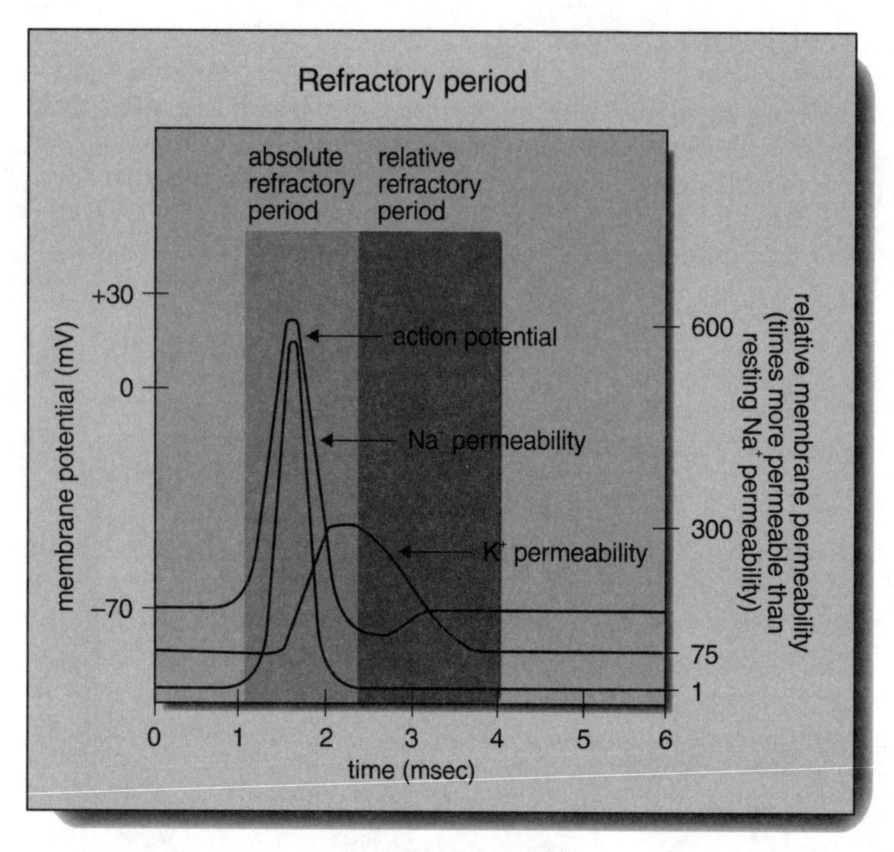

FIGURE 3.16

stimuli can only cross the synapses in one direction (from the presynaptic side to the post synaptic side; see following). To account for the unidirectional nature of action potential conduction down individual axons, we must consider the process termed *refractory period*. The generation of an action potential in a local segment of membrane requires the opening and closing of, first, Na^+ and then K^+ channels. After the completion of an action potential in this one active segment (call it segment A), these protein channels must reset themselves before they are capable of being reactivated. Thus, during the period immediately following the completion of the action potential, segment A is said to be in refractory period.

The entire refractory period probably lasts for 4 msec **(Figure 3.16, 3.17 and 3.18)**, but this period can be subdivided into two parts, as follows: (1) The absolute refractory period which begins at the time the Na^+ channels open (in the rising phase of the action potential), and ends when the Na^+ gates return to their resting configuration (lasting for about 1 msec). (2) The relative refractory period which corresponds to the period from the falling phase of the action potential to the end of the hyperpolarization phase, when the K^+ channels are resetting to their closed condition (lasting for about 2 additional msec). During the absolute refractory period, another action potential cannot be activated in segment A, no matter how large the applied stimulus (because the Na^+ gates are not ready to open again). On the other hand, during the relative refractory period, a suprathreshold stimulus may be able to generate a new action potential in segment A, but a regular threshold stimulus will not be able to do so. Obviously, after the end of the relative refractory period, a regular threshold stimulus will again be able to elicit another action potential in segment A.

To account for the unidirectional conduction of action potentials in an axon, again consider segment A and the additional axon segments B, C and D **(Figure 3.17)**. If we apply a threshold stimulus to segment A, an action potential is now generated here, and here only. Current will flow from segment A to adjacent segment B; the current will act as a threshold stimulus in segment B, and a new action potential will be generated in segment B. Meanwhile, segment A will become refractory. The action potential generated in segment B will cause current to flow forward to adjacent segment C and in

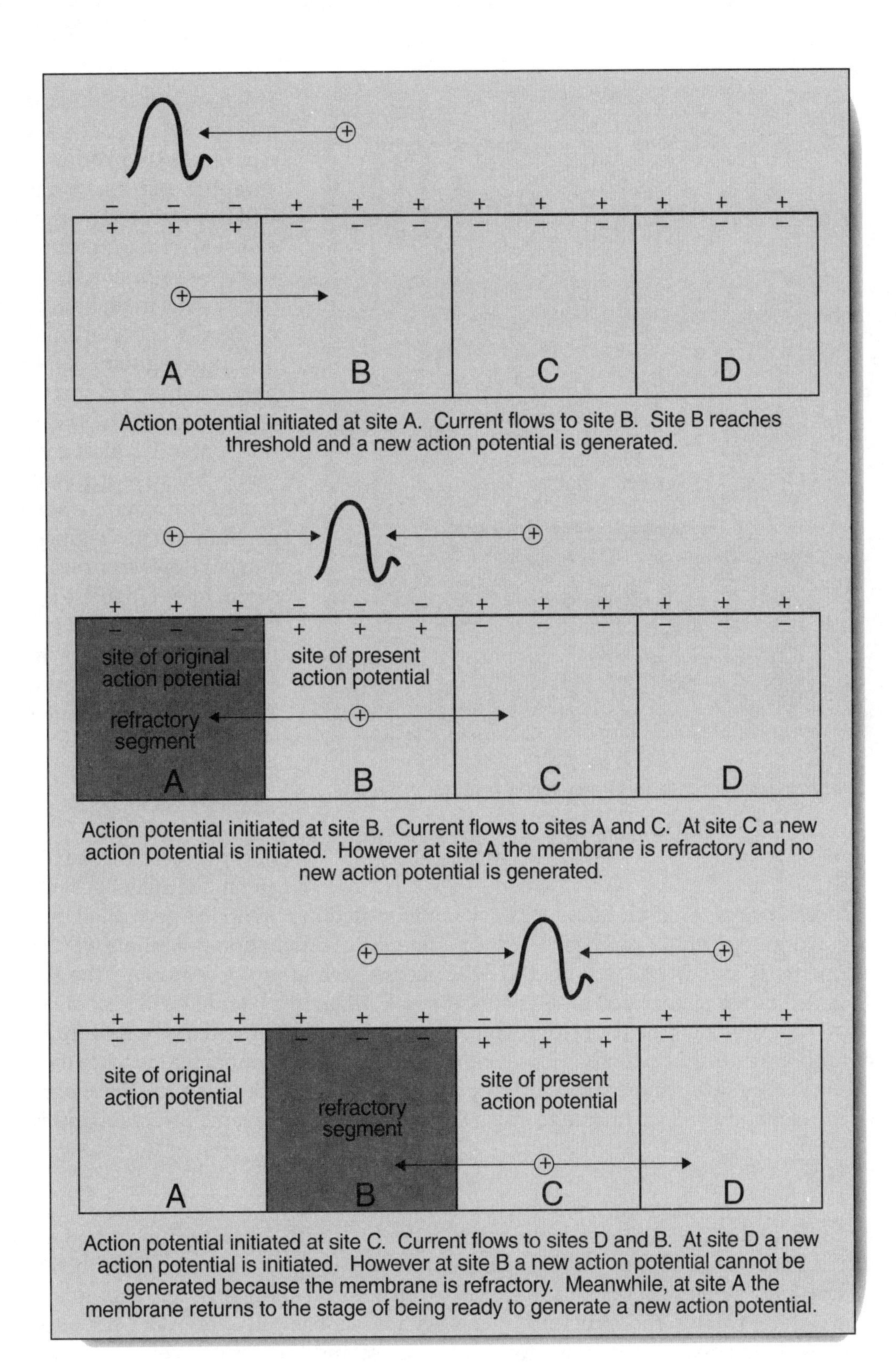
A
B
C
D
Action potential initiated at site A. Current flows to site B. Site B reaches threshold and a new action potential is generated.
site of original action potential
refractory segment
site of present action potential
A
B
C
D
Action potential initiated at site B. Current flows to sites A and C. At site C a new action potential is initiated. However at site A the membrane is refractory and no new action potential is generated.
site of original action potential
refractory segment
site of present action potential
A
B
C
D
Action potential initiated at site C. Current flows to sites D and B. At site D a new action potential is initiated. However at site B a new action potential cannot be generated because the membrane is refractory. Meanwhile, at site A the membrane returns to the stage of being ready to generate a new action potential.

FIGURE 3.17

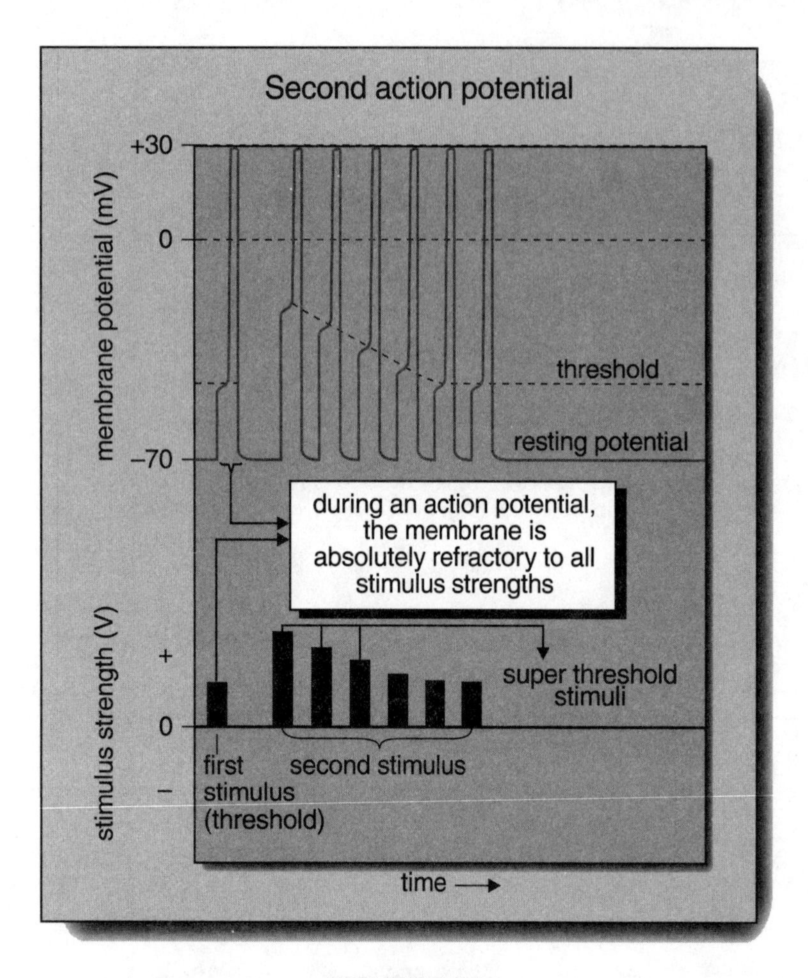

FIGURE 3.18

addition current from segment B will also flow in the reverse direction back into segment A. In segment C the current will produce a threshold stimulus and a new action potential will be produced in segment C. However, even though current flows backward to segment A, since this segment is still in the refractory period—no new action potential will be produced in segment A. Now the action potential based at segment C will generate current flow to stimulate an action potential in segment D. Meanwhile, because segment B is now refractory, no action potential will be generated in this segment even though current flows from segment C back to segment B. Finally, the refractory period will have ended in segment A, so that if this segment is stimulated with a new threshold stimulus it can again generate an action potential.

CONDUCTION IN THE REVERSE DIRECTION

This process can take place as follows: given segments A,B,C,D,E (none of which are refractory), if we stimulate segment C and an action potential is generated here, current will flow forward to segment D and backward to segment B. Thus, action potentials will be generated in both segments B and D and segment C will become refractory. Continuing the scenario, action potentials will now be generated in segments A and E, segments B and D will become refractory, and segment C will no longer be refractory. Here the action potential will move both towards the axon terminal and towards the cell body. However, while the action potential that reaches the axon terminal will be able to release transmitter and thus the stimulus will cross the synapse, the action potential that reaches the cell body will be unable to cross the synapse in the reverse direction and will dissipate.

SYNAPSES

As previously mentioned, the distal end of the nerve fiber is the axon terminal. At the axon terminal (also called the *synaptic knob or bouton*), the nerve is linked to another structure through a synapse. Nerve axons may synapse at one of three types of structures: another nerve, a muscle cell, or a gland cell. Generally speaking, the structure of the synapse is basically the same for all three types of terminations. First consider the nerve to nerve synapse and the various components involved in its construction. Later in the chapter on muscle contraction we will discuss the arrangement of various

nerve to muscle synapses. Recall that normal action potential conduction is directed from the cell body to the axon terminal. When describing the structures of the synapse, that nerve on the axon terminal side of the synapse is described as the *presynaptic nerve or fiber*. The nerve on the other side of the synapse (the dendritic side) is described as the *postsynaptic nerve or fiber*. Since the location of the synapse determines which fiber is the presynaptic and which the postsynaptic, a single fiber can act as both **(Figure 3.19 and 3.20)**. With respect to the structures in the presynaptic nerve terminal, within this region there are many membrane-bounded vesicles that contain neurotransmitters. In addition, mitochondria are present here obviously to supply ATP. The ATP is needed to energize the events during the synaptic transmission.

A space exists between the presynaptic and the postsynaptic side known as the *synaptic cleft*. Located on the postsynaptic side of the axon is the subsynaptic membrane. Within the structure are receptors that are designed to specifically bind the neurotransmitter released on the presynaptic side. These receptors are highly specific for individual transmitters, but it is quite possible to find that the subsynaptic nerve membrane may possess receptors for many different transmitters. On the other hand, a single synaptic knob is only capable of releasing one type of neurotransmitter. Finally the last major components of the synapse are the degrading enzymes that are designed to inactivate and remove transmitters that have served their purpose (see following). A large variety of synaptic

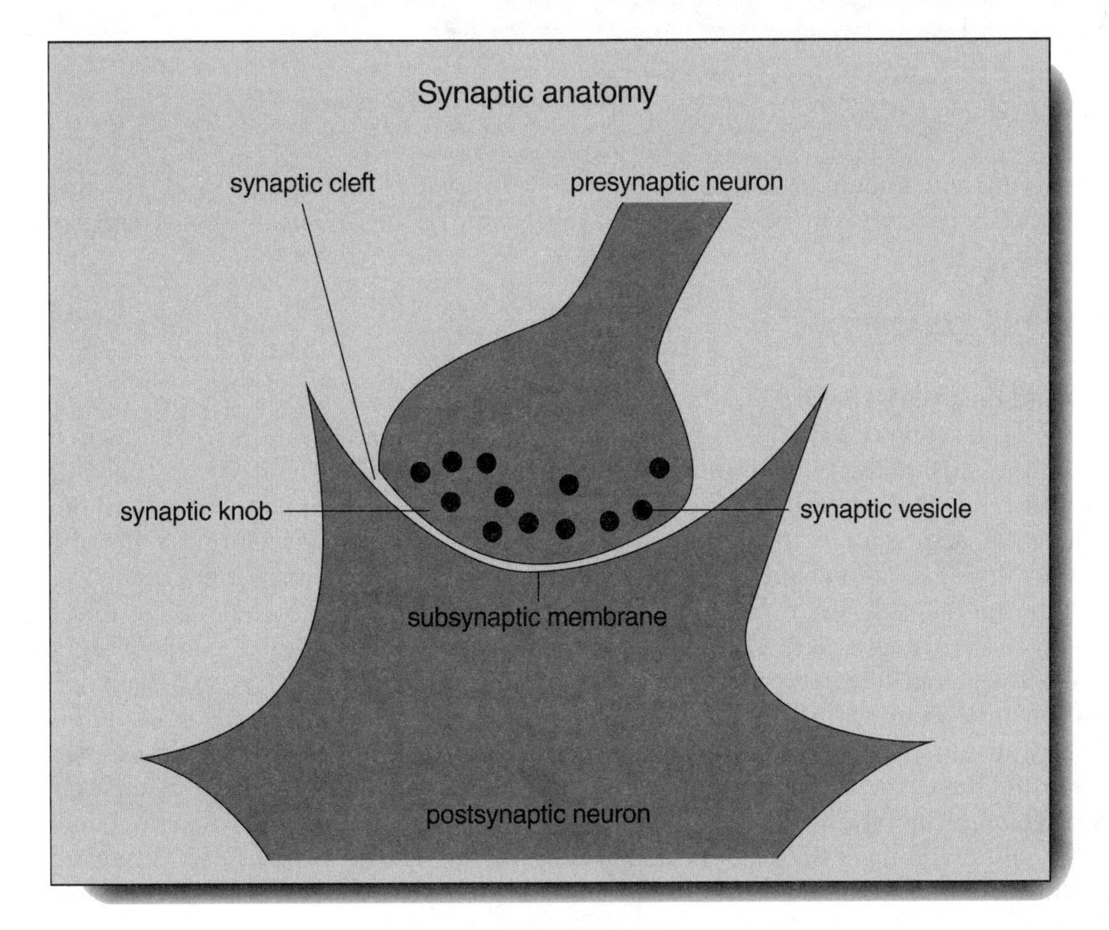

FIGURE 3.19

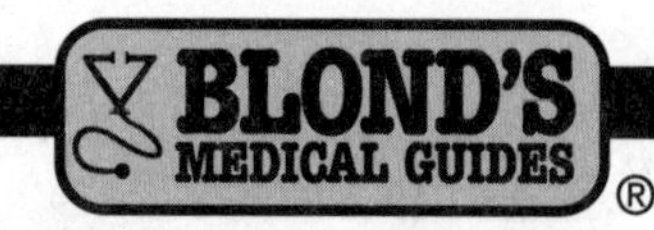

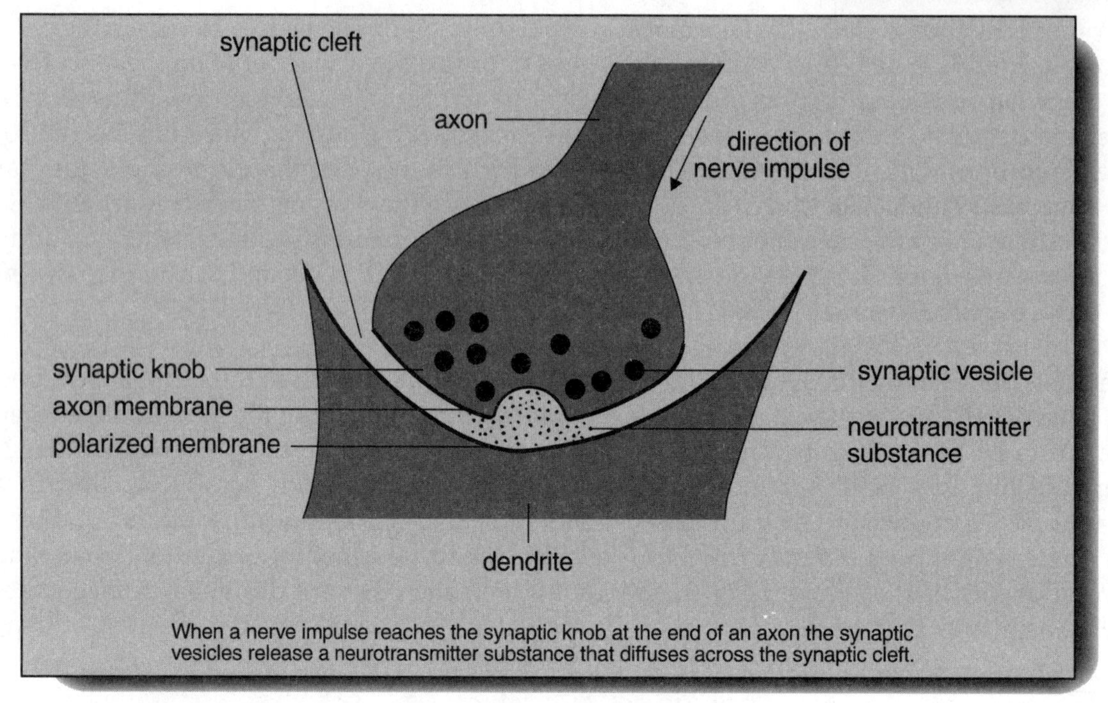

When a nerve impulse reaches the synaptic knob at the end of an axon the synaptic vesicles release a neurotransmitter substance that diffuses across the synaptic cleft.

FIGURE 3.20

arrangements are now known to be present in the nervous system and while the simple axodendritic synapses are the most common, others such as axon-axonic or the more complex synaptic triads are also possible.

EVENTS AT THE SYNAPSE

To understand the events that lead to the release of neurotransmitter at the synapse, remember that the purpose of the action potential is to generate and maintain an adequate current flow in the nerve cell membrane. Thus, at the axon terminal the depolarization within the nerve cell membrane will create a current flow that enters the synaptic knob. This current flow will depolarize the membrane of the synaptic knob and result in an increase in Ca^{++} permeability of the axon terminal membrane. The resultant Ca^{++} flow into the synaptic knob induces movement of the transmitter-containing vesicles and they now merge with the membrane of the synaptic knob. This results in the release of the transmitter into the synaptic cleft. This exocytic process is probably energy dependent and probably depends on the changes in microtubule or filament structure within the axon terminal. Since the concentration of neurotransmitter is greater on the presynaptic side than on the postsynaptic side of the cleft, the transmitter now diffuses down its concentration gradient towards the postsynaptic side. The transmitter then binds to the receptors located on the postsynaptic side in the subsynaptic membrane. This specific binding of transmitter to receptors alters the structure of ionic channels in the subsynaptic membrane, increasing the permeability of this membrane to ionic flow. This resultant flow will affect the resting membrane potential and will lead to either a depolarization (an EPSP) or in some cases a hyperpolarization (an IPSP).

EPSP AND IPSP

At a single synapse, the release of neurotransmitter can either result in a depolarization of the resting membrane potential, or a hyperpolarization of the resting membrane potential. If the effect is to depolarize the resting membrane potential then this is termed an *excitatory postsynaptic potential* (or EPSP). If the effect is hyperpolarization, this is termed an *inhibitory postsynaptic potential* (or IPSP). In the case of an EPSP, the depolarization brings the resting membrane potential closer to the threshold level within that neuron. **(Figure 3.21 and 3.22).**

For example, assume that to generate an action potential in a particular neuron we must depolarize it to -50 mV (thus -50 mV is the threshold level). Binding of transmitter at this synapse raises the resting membrane potential from -70 mV to -60 mV. Although this depolarization does not quite reach -50 mV, still the resting membrane potential is now closer to threshold than it was before. It is for this reason that this is termed an *excitatory* post synaptic potential. If a combination of EPSPs were all to stimulate this neuron at the same time, even though each alone was not capable of raising the nerve to threshold, the combination together may reach threshold. This is a type of summation of stimuli termed *spatial summation.*

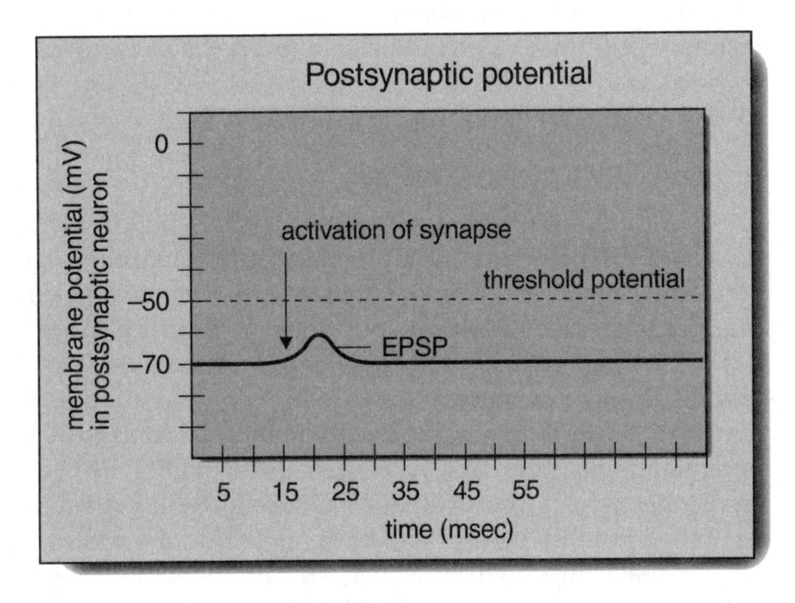

FIGURE 3.21

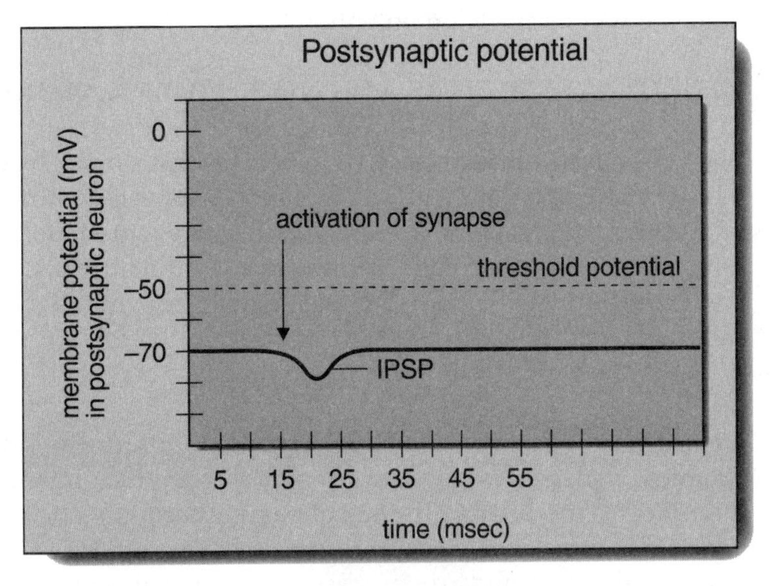

FIGURE 3.22

IPSPs are conceptually the exact reverse of EPSPs; here the resting membrane is hyperpolarized. This makes it harder for the nerve to reach threshold, and thus harder to generate an action potential. Returning to the model, recall that the combination of two subthreshold EPSPs generated by two *excitatory* presynaptic neurons is able to raise the postsynaptic neuron to threshold. However, if a separate *inhibitory* presynaptic neuron is also stimulated to release its neurotransmitter on the same postsynaptic neuron, it will generate an IPSP. Combining the two EPSPs

with the IPSP will prevent the postsynaptic cell from reaching threshold. Thus, the *inhibitory* postsynaptic potential (IPSP) from this inhibitory neuron prevents an action potential in the postsynaptic cell.

A neurotransmitter that results in EPSPs probably acts by opening Na^+ channels in the subsynaptic membrane of the postsynaptic cell. The increased Na^+ entering into the postsynaptic nerve results in depolarization and the EPSP. Neurotransmitter that results in IPSPs probably acts by opening Cl^- channels in the subsynaptic membrane of the postsynaptic cell. The resultant inflow of Cl^- thus makes the interior of the postsynaptic cell more negative. It has also been suggested that in some nerves IPSPs are generated when K^+ channels open in the subsynaptic membrane of the postsynaptic cell. The resultant outflow of K^+ would also hyperpolarize the neuron. **(Figure 3.23).**

SPATIAL AND TEMPORAL SUMMATION

In the case of *spatial summation*, subthreshold EPSPs generated from two or more separate presynaptic fibers (inputs) simultaneously applied to the same postsynaptic nerve cell can be combined together (summated).This results in raising the resting membrane in the postsynaptic cell to threshold and thus generating an action potential. On the other hand, it is sometimes possible to summate EPSPs that originate from a single presynaptic fiber and produce an action potential in the postsynaptic nerve cells if the frequency of the postsynaptic input is increased. This is termed *temporal tummation*. This means that the number of action potentials/time propagating down the axon of the presynaptic cell will be increased from, for example, 1 action potential/min to 100 action potentials/min. Since each action potential in the presynaptic fiber generates an EPSP on the postsynaptic side of the synapse, it can be envisioned that 1 action potential/min will result in the release of one burst of transmitter and, thus, 1 subthreshold EPSP/min will be generated in the postsynaptic cell. Under these conditions the cell will not reach threshold. However, if the frequency is increased to 100 action potentials/min, 100 bursts of transmitter will be released at the synapse. Under these conditions the concentration of transmitter will build up and may result in raising the resting membrane potential to the threshold level. Thus the outcome will be an action potential.

NEUROTRANSMITTERS AND NEUROMODULATORS

The type of transmitter released by a nerve is determined by the type of neuron involved. For example, motor neurons contain acetylcholine in their vesicles while terminal sympathetic fibers (the postganglionic adrenergic fibers) possess vesicles that contain norepinephrine. A wide range of neurotransmitters as well as neuromodulators can be found in both the peripheral and central nervous systems. Neurotransmitters found in the vesicles may be synthesized in the cell bodies of the neurons or in the axon terminals themselves. If synthesized in the cell body, the transmitter is transported down the axon via the neurofibrilar system within the axon.

Neurotransmitters function at the synapse by producing EPSPs and IPSPs; however, some types of chemical messengers released by neurons produce more complex responses. Such effects can be described as modulating the signal passing through a particular synapse, and thus these substances are called *neuromodulators*. It is not always easy to distinguish between neurotransmitters and neuromodulators, and in many cases one chemical may produce both effects. However, one distinction is that while neurotransmitters function mainly by controlling the opening and closing of ion channels in the membrane, neuromodulators promote biochemical changes. It is believed that neuromodulators may function through second messenger systems (i.e., via cyclic AMP).

Although neuromodulators (like neurotransmitters) may affect the membrane potentials, their action is slower. For example, while neurotransmitters may act within milliseconds, neuromodulators take longer (minutes, hours or even days) and are associated with such events as learning, motivation or sensory motor activities. In the example of **Figure 3.24**, axon terminal A may release a neuromodulator that will, in turn, regulate neurotransmitter released by axon terminal B. Such an effect has been reported for *pain gates* where *endorphins* and *enkephalins* act to modulate the transmission of pain stimuli across synapses by the neuropeptide, *substance P*, in a pain pathway.

REMOVAL OF NEUROTRANSMITTERS FROM THE SYNAPSE

If the neurotransmitter remains bound to its receptor in the subsynaptic membrane, the ionic channels will remain open and action potentials will continue to be generated in the postsynaptic fiber. Obviously it is necessary to turn off action potential generation in the postsynaptic fiber once the

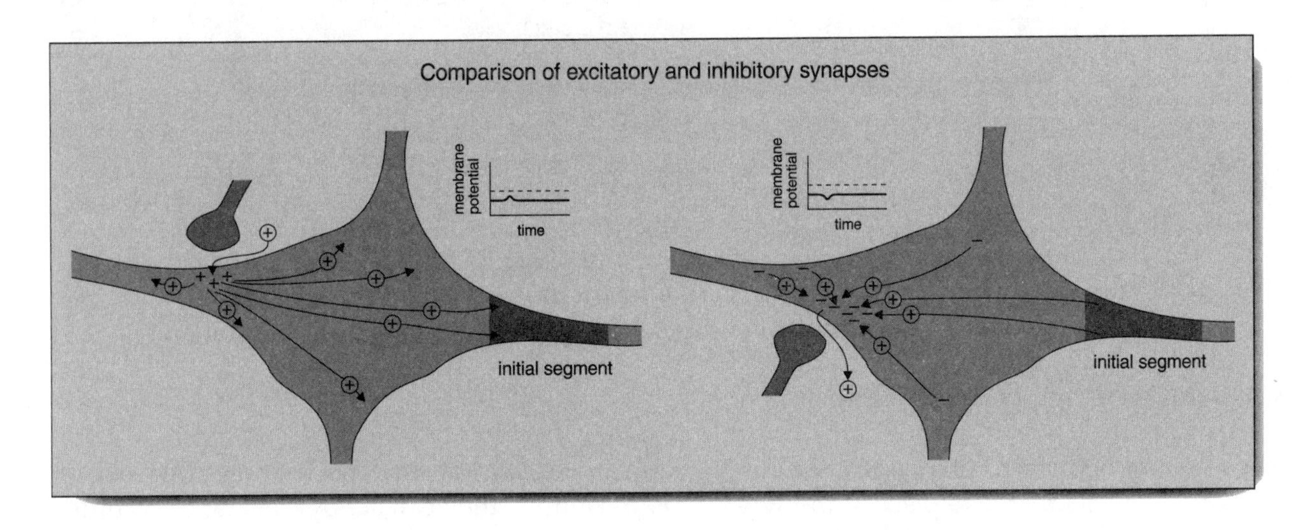

FIGURE 3.23

stimuli in the presynaptic fiber are terminated. To do this the neurotransmitter must be removed from its binding sites on the receptors.

Removal is accomplished as follows:

1. Specific enzymes inactivate and degrade the transmitter substance within the subsynaptic membrane itself.

2. The transmitter simply diffuses away from its binding site and dissipates.

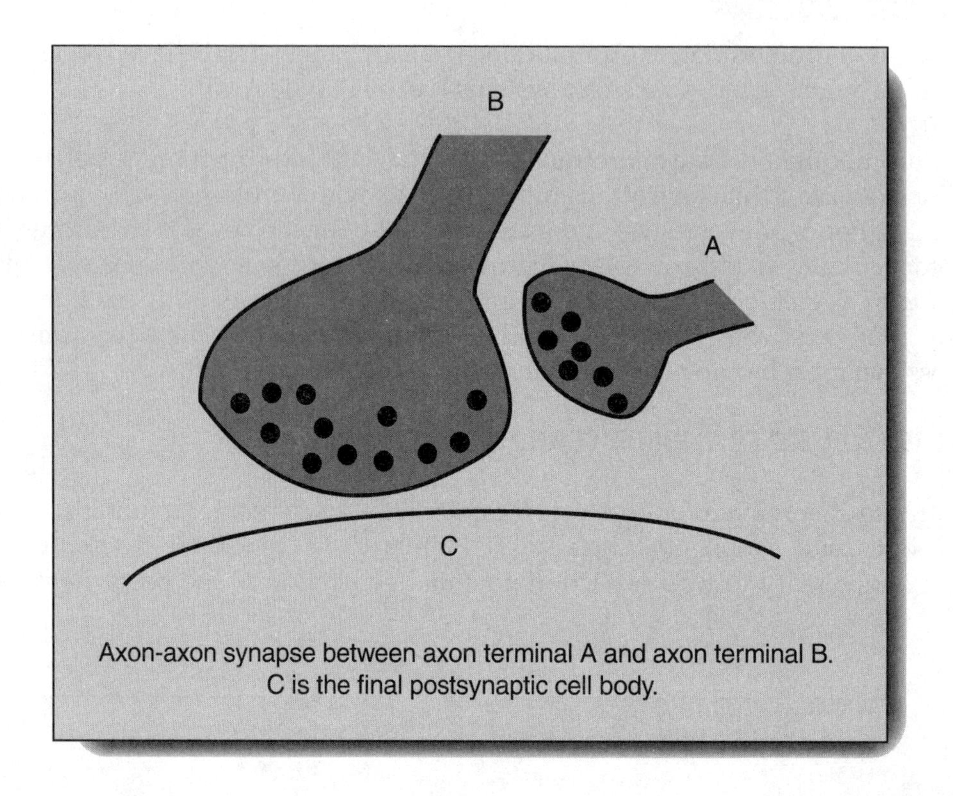

Axon-axon synapse between axon terminal A and axon terminal B.
C is the final postsynaptic cell body.

FIGURE 3.24

3. The transmitter substance, or its degraded components, is reabsorbed into the presynaptic terminal by active transport. After entering the axon terminal the transmitter can be resynthesized or stored for reuse, or it can be further destroyed by enzymes in the axon terminal.

DRUG ACTION AT THE SYNAPSE

Drugs that act on the nervous system usually function at the synapse itself. Such drugs affect one or more of the events that take place at the synapse. For example, they can act by causing transmitter release, receptor binding, degradation, reuptake, or synthesis.

Chapter 4
CENTRAL NERVOUS SYSTEM

INTRODUCTION

The nervous system is organized into two main divisions: the central and the peripheral nervous system. The central nervous system is composed of the brain and spinal cord while the peripheral nervous system consists of nerves that carry information to or from the central nervous system and innervates the various structures located in the periphery.

THE CENTRAL NERVOUS SYSTEM

PROTECTION AND NOURISHMENT

Protection of the central nervous system is provided by the skeletal system. The brain is enclosed in the bony covering of the skull (cranium) while the spinal cord is protected because it passes through the vertebral canal within the vertebrae that compose the backbone. In addition, the brain and spinal cord are also protected by three layers of membrane located inside the skull and vertebral canal. The membranes that surround the brain and spinal cord are referred to as the meninges. To further cushion the elements of the central nervous system, cerebral spinal fluid (CSF) surrounds the brain and spinal cord. This CSF not only allows the central nervous system to float in liquid, but it also provides some nourishment to the tissues and is also involved in certain aspects of respiratory control (see chapter 11, on respiration).

MENINGES

The three layers of meninges are as follows:

1) **Dura mater** — the outermost layer that is in contact with the bones of the skull or vertebral column is the Dura mater (tough mother). This is an inelastic membrane inside of which are small regions called dural sinuses, filled with blood. In addition, larger cavities in the dura called venous sinuses are also filled with blood. In the venous sinuses, blood drains from the brain to return to the heart. In the dural sinuses, CSF passes out of the subarachnoid spaces (see following material) and re-enters the blood.

2) **Arachnoid mater** — the middle layer of the meninges is the Arachnoid mater (spider mother). This membrane is highly vascularized, almost like a spider web in appearance, with blood vessels running within the spider web region called the subarachnoid space. This subarachnoid space overlies the pia mater and is filled with CSF that has circulated out of the ventricles of the brain. Within the subarachnoid space this CSF is reabsorbed into the arachnoid villi and then passes into the dural sinuses and into the blood.

3) **Pia mater** — the innermost membrane of the meninges is called the Pia mater (soft or gentle mother) and is in direct contact with the nerve tissue of the brain and cord. Since it is in close

contact with the nervous tissue it follows the contours of the brain and in certain areas dips deeply into the brain to supply the tissues with a blood supply.

CEREBRAL SPINAL FLUID (CSF)

The CSF is manufactured by the structures called the *choroid plexuses* that are found in the ventricles of the brain. The choroid plexuses are formed from vascularized regions of the Pia mater that pass through the ependymal layer of cells lining the ventricles. Movement of CSF is from the two lateral ventricles into the third and then the fourth ventricle, down the central canal of the spinal cord and then returning up the outer layers of the spinal cord within the subarachnoid space. CSF also diffuses out of the fourth ventricle and into the subarachnoid space. The CSF then flows through the subarachnoid space surrounding the brain and finally passes out through the arachnoid villi (also called the arachnoid granulations) into the dural sinuses and back into the blood circulation. This flow of CSF is maintained at a pressure of approximately 10 mm Hg. Reduction in the pressure after a spinal tap (even by 1 or 2 mm Hg) can result in severe spinal headaches that may take days to correct. In some instances, the only way to correct this problem is to perform a blood patch.

Approximately 150 ml of CSF fills the CSF spaces and this volume is generated and reabsorbed about three times a day. In the event that the production is greater than the removal, the increased pressure that builds up will lead to Hydrocephalus, resulting in brain damage if untreated. The major functions of the CSF are:

1) to act as a cushion for the brain and spinal cord against shock and physical trauma, and

2) to allow exchange of substances between the blood and the nerve tissue.

The composition of the CNS is slightly different from the plasma (Na^+ concentration is greater and K^+ concentration is less than that in the plasma compartment); however, the concentration of other substances such as glucose and dissolved gases will be similar to the plasma. Additionally, dissolved CO_2 present in the plasma freely diffuses into the CSF and here, in the form of H^+ ions, it will affect the central respiratory chemoreceptors and increase the rate and depth of respiration.

THE BLOOD/BRAIN BARRIER

The blood/brain barrier is composed of a specialized arrangement of endothelial cells that make up the capillaries in the brain. Unlike endothelial cells of capillaries in other body locations, the endothelial cells in capillaries of the brain do not have pores that open where the cells contact each other. Instead, the cell contacts are closed with tight junctions, and thus movement of substances from the blood to the brain tissue must be across the endothelial cells (not through pores). This significantly limits what can cross from the blood into the brain. It should be mentioned, however, that while aqueous substances are significantly limited in passing through this barrier, lipid soluble substances cross the cell membranes quite easily. In addition to the endothelial cell barrier, what surrounds the brain capillaries are specially designed glial cells called *astrocytes*. The "foot processes" of the astrocytes contact the capillaries and supply an additional barrier to the movement of substances out of the capillaries. Specifically astrocytes are believed to:

1) Play a role in signalling the capillary endothelial cells to maintain intact tight junctions,
2) Assist in the transport of substances across the cell membranes, and
3) Act to structurally support the capillaries and the surrounding tissue.

ORGANIZATION OF THE CENTRAL NERVOUS SYSTEM (CNS)

The structural elements composing the CNS can be organized both structurally and functionally. The most common organizational hierarchy is as follows:

A. Forebrain
1. Cerebrum
   (a) Cerebral Cortex
   (b) Basal nuclei
2. Diencephalon
   (a) Hypothalamus
   (b) Thalamus

B. Cerebellum
C. Brain stem

THE CEREBRAL CORTEX

This is the largest part of the human brain and is organized into the left and right hemispheres. The two hemispheres are interconnected by means of a number of commissure tracts. The largest of these commissure tracts is the Corpus Callosum. Since the two hemispheres are connected through such tracts, they function as a single unit (much like one large computer). However, if the Corpus Callosum is cut, each hemisphere functions individually and the resulting behavior is sometimes quite strange. For example, the left and right hands both fight to pick up the same object. The outer covering of each hemisphere is composed of gray matter (neuron cell bodies) while the inner layers are made of white matter (myelinated tracts). In addition, within the white matter are embedded other gray areas composed of cell bodies called ganglia (e.g., the Basal Ganglia). At the simplest level, the cerebral cortex functions much like a computer, and thus input (sensory) information flows into the cerebral cortex, and information processing takes place in many areas; but output (motor) information flows back to the body. Within the cortex, complex fiber tracts transmit information from one processing area to another. Additionally the cortex supplies subtle interpretations to the information being processed that are expressed as human emotional behaviors. The most complex thought processes based within the cerebrum lead to an individual's sense of self, and the ability to perceive that present actions will result in certain future outcomes. It is not clear if the brains of "lower" animals are similarly capable of this type of complex conceptualization, although animal rights advocates believe that they have similar abilities.

THE CORTICAL LOBES AND THEIR FUNCTIONS

The cortex (and thus the individual hemispheres) are organized into both functionally and structurally distinct areas. Particular areas are designated as lobes and the cell bodies located at the surface extend down to the area of underlying white matter. Neurons within discrete lobar regions function in groups

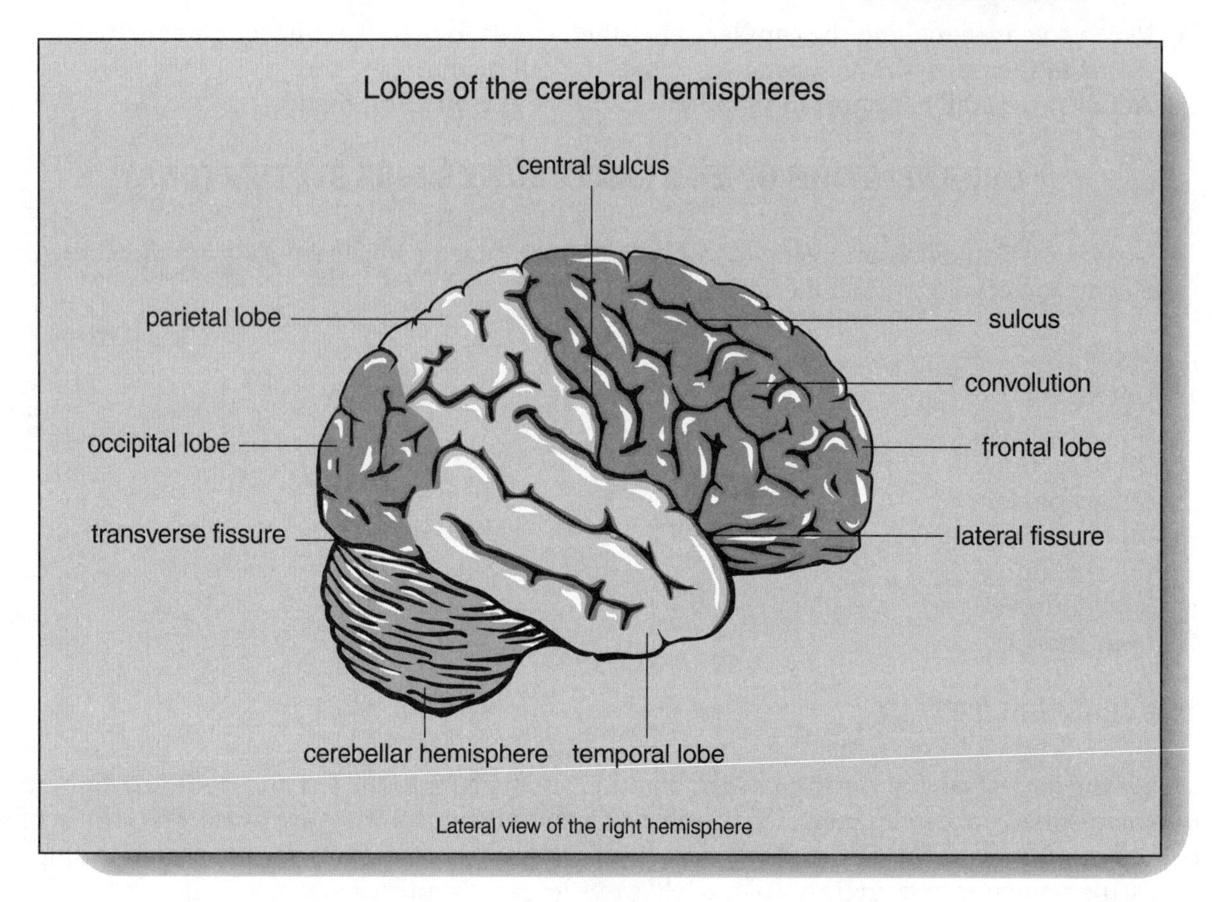

FIGURE 4.1

involved in various aspects of information processing or interpretation. There are four anatomically defined major lobes and one minor lobe located in each hemispere, and the hemispheres are anatomically similar to each other (i.e., there is a right and left temporal lobe, a right and left parietal lobe, and so on). However, while some of their functions are similarly represented in both hemispheres, other functions—frequently the higher functions—are discretely located in one or the other hemispere alone. A variety of anatomical landmarks, such as shallow or deep folds, are used to define the various areas. Specifically, the four major lobes in each hemispere are the **(Figure 4.3)**:

Occipital
Temporal
Parietal
Frontal

The minor lobe, only visible if the temporal lobe is deflected, is the insula.

The left and right hemispheres are separated by the longitudinal fissure. The parietal and frontal lobes within each hemisphere are separated by the central sulcus, while the frontal and temporal lobes are separated by the lateral fissure. Between the parietal and occipital lobe is the parietooccipital notch **(Figure 4.1)**.

Much is known about the functions of each of the lobes, much remains yet to be explained. It should be noted that for these discrete functions to be expressed correctly, all areas of the brain must interact together as a single complex unit.

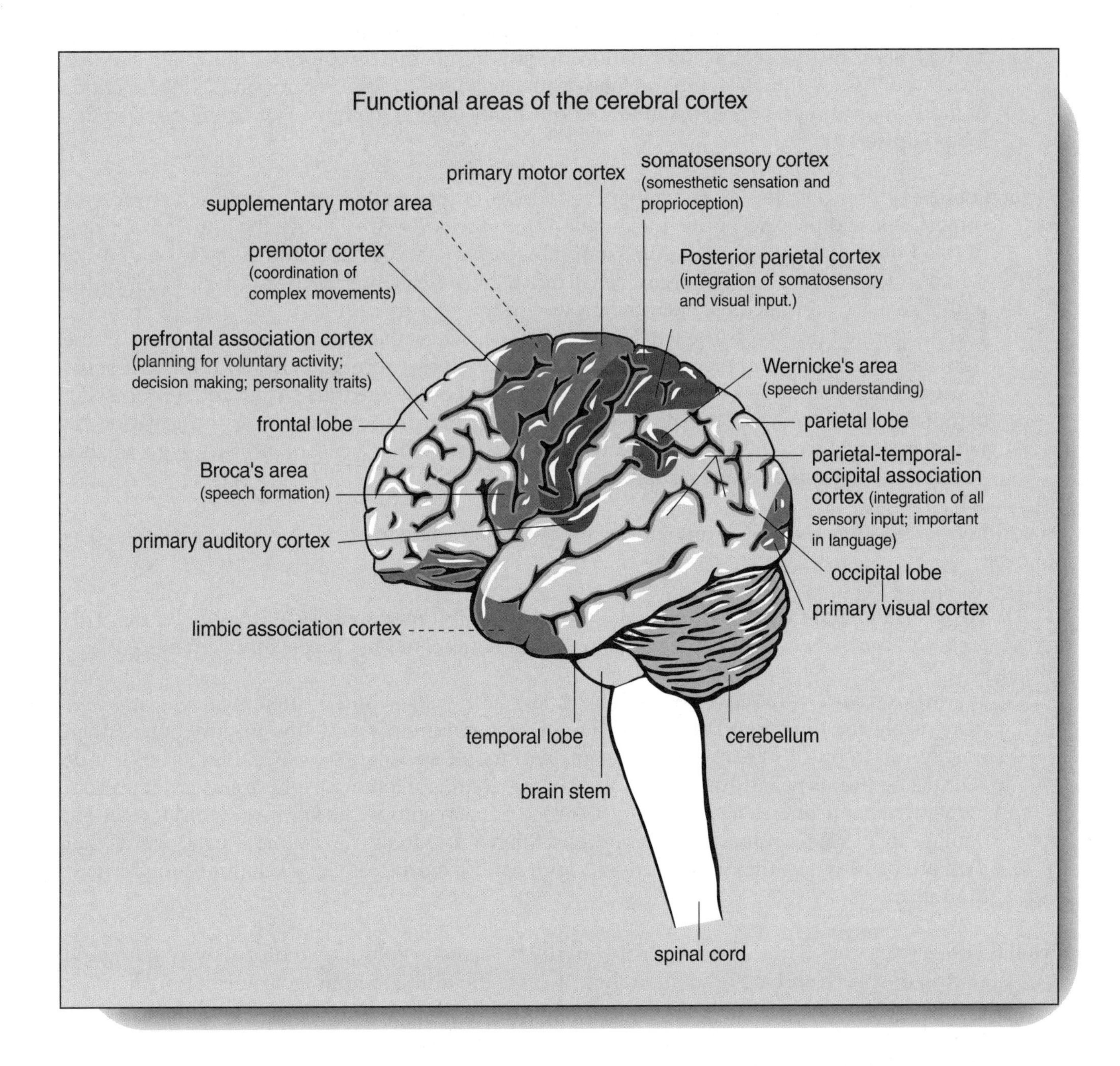

FIGURE 4.2

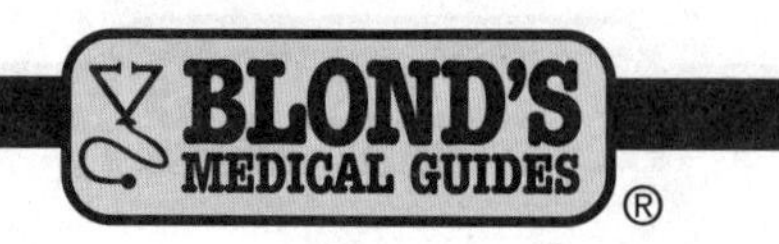

The major anatomical regions and functions **(Figures 4.2)** attributed to the various lobes, and regions within the lobes, are as follows:

Occipital Lobes — located within the occipital lobes is the primary visual cortex surrounded by the higher order visual cortex. Here visual information originating from the retinas is initially processed. In addition, the memory of visual patterns is contained in this region.

Temporal Lobes — the primary auditory cortex surrounded by the higher order auditory cortex is located here. In this region information originating in the receptors of the inner ears are processed. Also within the temporal lobes is the *limbic association cortex* involving the functions of motivation and emotional behavior. Additionally certain memory areas are located in this lobe **(Figure 4.2)**.

Parietal Lobe — within both the left and right parietal lobes is the somatosensory cortex which receives somesthetic and proprioception information from the body. The site for the initial processing of this information is located at the front of each parietal lobe and immediately behind the central sulcus **(Figure 4.2)**. It is organized in much the same manner as the primary motor cortex with a sensory homunculus. Each region within the sensory cortex receives inputs from the various parts of the body. The left sensory cortex receives the majority of its inputs from the right half of the body and the right sensory cortex receives information from the left half of the body and most of the sensory tracts cross at the level of the medulla. After initial processing of the sensory information, the somatosensory cortex then projects this sensory information via white matter fibers (myelinated axons) to adjacent higher sensory areas where further processing, analysis and integration takes place.

In addition to the somatosensory cortex there are other important areas within the parietal lobe, as follows:

1) **Posterior Parietal Cortex** — involved in integration of somatosensory and visual inputs. This area is important because it integrates information involved in complex movements.

2) **Wernicke's Area** — required for understanding both spoken and written messages. It is also responsible for the formulation of coherent speech patterns and this information is then transferred to Broca's area where the control of motor speech takes place. Like Broca's area, Wernicke's area is usually only developed in the dominant left hemisphere and is not present in the right hemisphere. While damage to Broca's area results in the failure of word formation, damage in Wernicke's area produces *aphasia*. Such individuals cannot understand words that they see or hear and they cannot choose appropriate words to convey the meaning of their thoughts.

Frontal Lobes — the cells within this area are primarily involved in voluntary motor activity, in muscle control of speech and in higher thought processes including temporal organization, planning and problem solving. Within the frontal lobe and located close to the central sulcus in each hemisphere is the primary motor cortex. This region is responsible for voluntary motor control of skeletal muscles throughout the body. Neurons within this region send motor projection tracts down through the brain stem. The majority of these axons then cross at the level of the medulla and continue down the spinal cord to synapse on motor neurons located in the gray matter ventral horns (sometimes called *ventral horn cells*) at each level of the cord. Axons from these motor neurons then exit the cord via the ventral roots and synapse on skeletal muscles at neuromuscular junctions. Since the axon tracts originating from the neurons in the left motor cortex cross to the right side at the medulla, the right motor cortex controls the left half of the body and vice versa.

It is known that stimulation of the discrete regions in the right or left motor cortex results in movement of specific regions of body. A map of the organization of these discrete regions of the primary motor cortex can be seen in **Figure 4.4**.

Stimulation of the neurons located towards the bottom (near the temporal lobe) results in control of muscles located in the tongue; stimulation of the neurons above the region results in control of muscles of the face; above this, the neurons control muscles in the hands; towards the top of the motor cortex, the neurons control muscles of the elbow, shoulder, and trunk; and the neurons located around the back of the motor cortex (within the longitudinal fissure) control muscles of the knee, ankle and toes, etc.

In addition to the primary motor cortex located in each frontal lobe, there are other important regions within the frontal lobes (**Figure 4.2**).

1) **Supplementary Motor Area** — located deep within the frontal lobe, and involved in programming of complex movements.

2) **Premotor Cortex** — involved in the coordination of complex sequential movements.

3) **Prefrontal Association Cortex** — involved in planning for voluntary activity, decision making and personality traits.

4) **Broca's Area** — involved in coordination of complex motor actions of the mouth, tongue and larynx which make speech possible. A person with an injury to this area may be able to understand spoken words but cannot speak. This area is usually only present in the left cerebral hemisphere.

5) **Limbic Association Cortex** — is primarily located on the inner and bottom surface of the temporal lobe. However, a smaller portion is present on the bottom surface of the frontal lobes (see earlier material on temporal lobe, for details).

HEMISPHERIC DOMINANCE

Although many basic functions are shared in both hemispheres, in most individuals one hemisphere usually possesses certain dominant functions. In more than 90% of the population, language-related activities (speaking, writing and reading) are localized to the left hemisphere. This hemisphere is also dominant for various complex intellectual functions that require verbal, analytical and computational skills. Thus, mathematical ability, organizing and planning of events, time sense, and so on are located in the left hemisphere.

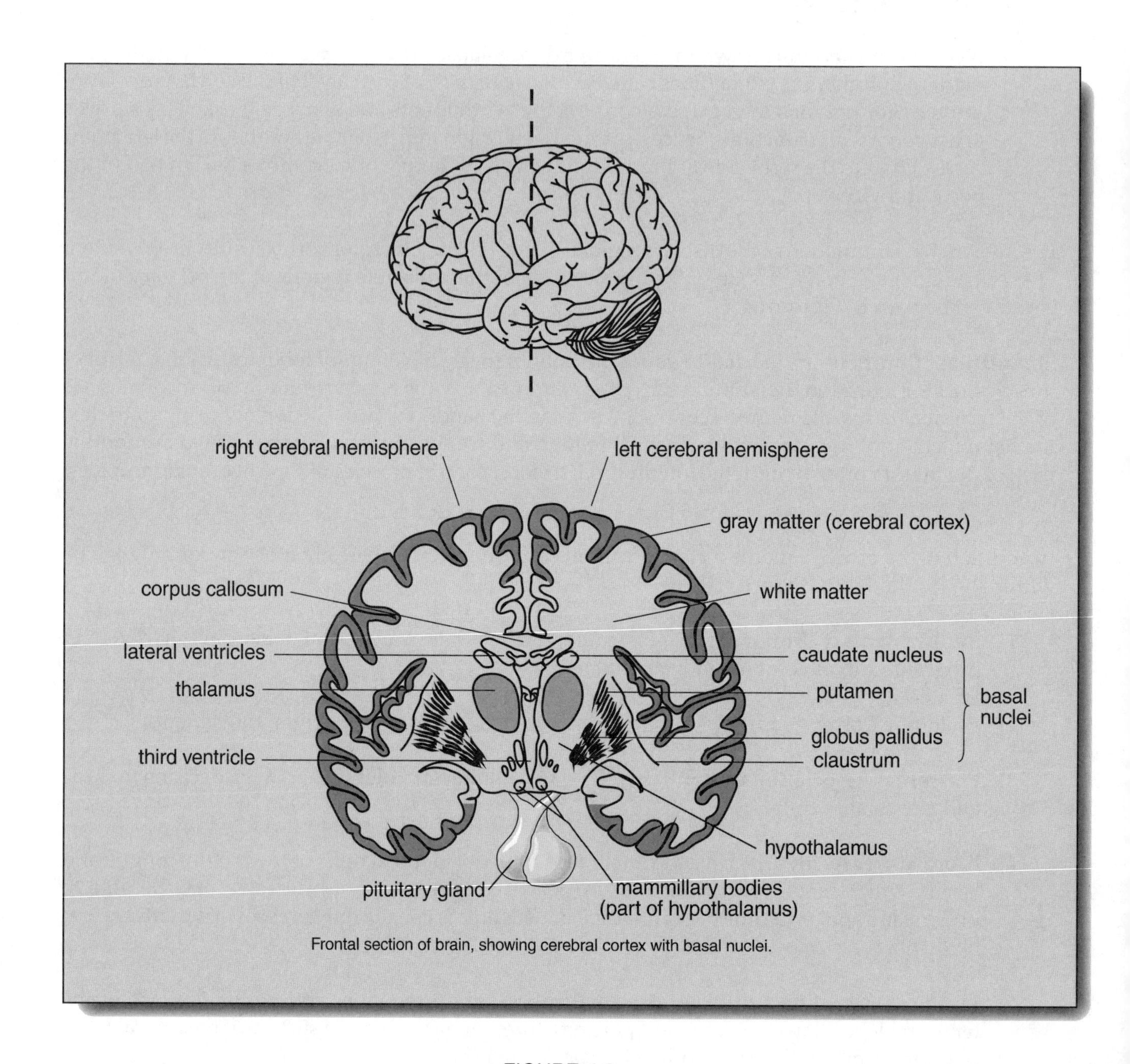

Frontal section of brain, showing cerebral cortex with basal nuclei.

FIGURE 4.3

While the left hemisphere is dominant in the majority of the population, the right hemispheric dominance is present in some individuals. In still others, the hemispheres are co-dominant. Although right-handedness is also more common in the populations, all right-handed individuals are not necessarily left hemisphere dominant.

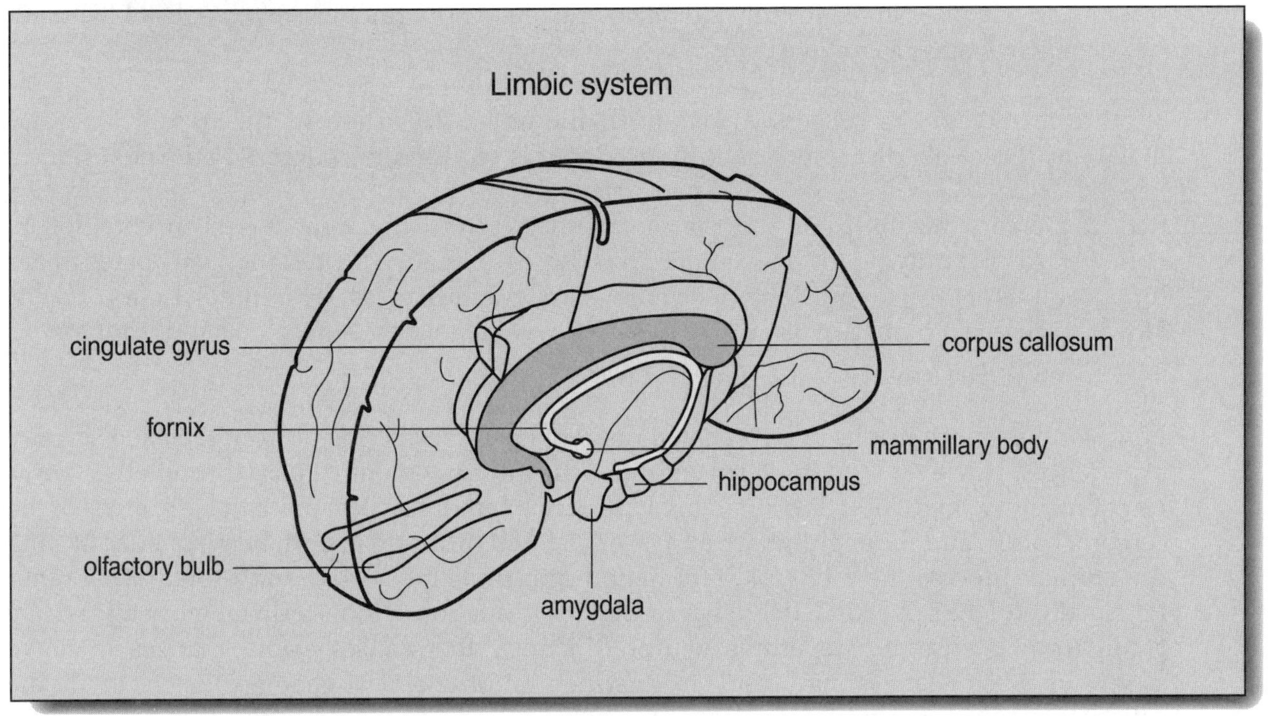

FIGURE 4.4

Right v. Left Handedness and Hemispheric Dominance
90% of right-handed adults are left hemisphere dominant.
10% of right-handed adult are right hemisphere dominant.
64% of left-handed adults are left hemisphere dominant.
20% of left-handed adults are right hemisphere dominant.
16% of left-handed adults have co-dominant hemispheres.

The non-dominant hemisphere (usually on the right) specializes in non-verbal functions including: motor tasks that require orientation of the body in three dimensional space; understanding and interpreting musical patterns; understanding and interpreting visual experiences; emotional and intuitive thought processes; emotional aspects of language in speech expression.

STORAGE OF MEMORY

Information storage in the brain is a complex process that is only partially understood. Most memory storage takes place in the cerebral cortex. However, it is now known that certain other parts of the

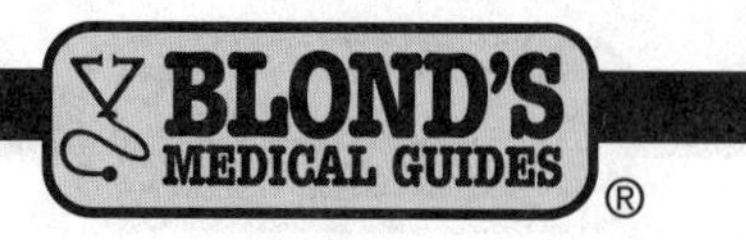

nervous system are involved in the sorting and retrieving of memory traces or *engrams*. Establishment of memory occurs in stages, as follows:

1) Short-term memory is processed mainly in the anterior portion of the frontal lobe and thalamus. It is easily disrupted and displaced, but is required to progress to the next stage.

2) Recent memory results from short-term memory information that has been consciously reviewed (as in studying for an examination). Recent memory processing takes place in the hippocampus of the temporal lobe. Such memory engrams may last for minutes, hours or days. However these engrams are usually only stored weakly and are subject to significant loss as a result of disuse.

3) Long-term memory results from repeated recall of new information or from repeated sensory experiences. However, for long-term engrams, the information must pass through the short/ recent memory stages first. Processing of long-term memories takes place in large areas of the cerebral cortex, and the engrams are apparently stored diffusely in both hemispheres. During the storage process, new bits of information appear to be sorted and stored with other information of the same kind. It is believed that the formation of long-term memory traces may result from intense and repetitive neuronal activity that causes stable changes in nerve pathways. Such changes may result from structural or chemical changes that facilitate the action of certain synapses.

ELECTROENCEPHALOGRAPHY

Electrodes placed on the scalp can be used to detect extra-cellular current flow arising from electrical activity within the cerebral cortex. The electroencephalogram (EEG) records this extra-cellular current generated by the collective postsynaptic potential activity in the neuronal cell bodies and dendrites located in cortical layers close to the recording electrodes.

EEG patterns can be used as a clinical tool to diagnose certain cerebral dysfunctions. EEGs can also be used to distinguish various stages of sleep, as well as in the legal determination of brain death. Common wave patterns seen in the EEG **(Figure 4.5)** are as follows:

1) **Alpha Waves** — which are recorded most easily from the posterior regions of the head have a frequency of 8 to 13 cycles/sec. These waves are present when a person is awake but resting with the eyes closed, and disappear when the eyes are open **(Figure 4.6)**. These waves also disappear during sleep.

2) **Beta Waves** — are present if a wakeful person has his or her eyes open **(Figure 4.6)**, and beta waves have a frequency of more than 13 cycles/sec. They are usually recorded in the anterior region of the head and occur when a person is actively engaged in mental activity.

3) **Theta Waves** — have a frequency of 4 to 7 cycles/sec and can be recorded mainly in the regions of the parietal and temporal lobes. They are normally present only in children, although some adults may produce these wave patterns in the early stages of sleep or during emotional stress.

4) **Delta Waves** — have a frequency of less than 4 cycles/sec and occur during sleep. They may originate from the inactive cerebral cortex during the time when it is not being stimulated by the reticular formation.

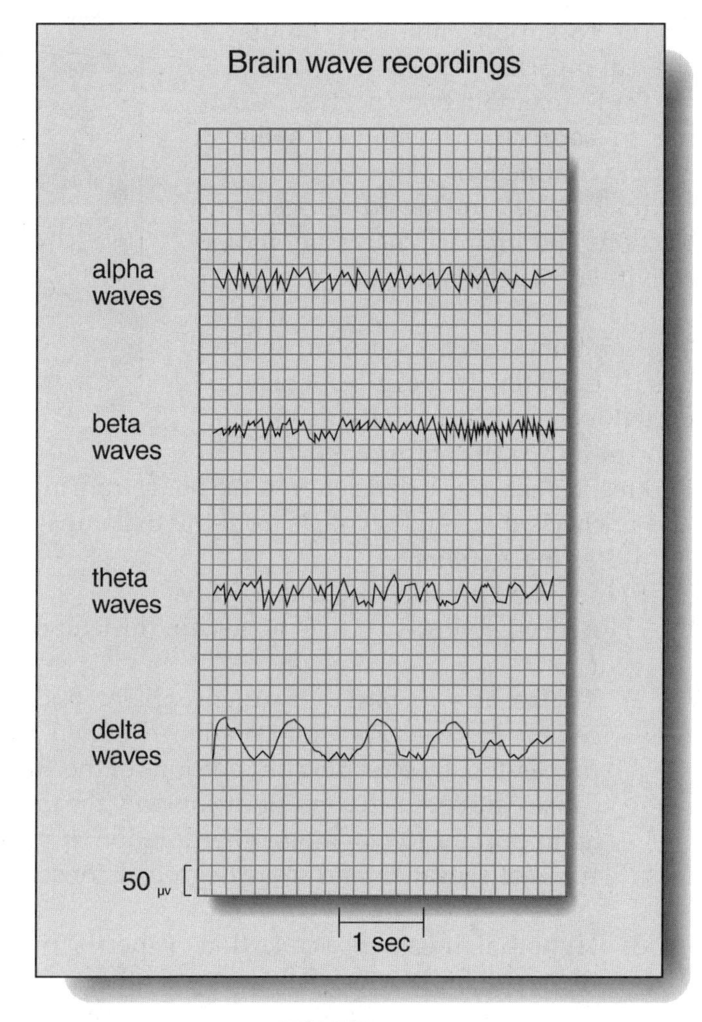

FIGURE 4.5

PATTERNS OF SLEEP

The more advanced organisms undergo some form of sleeping behavior, however, the actual function provided by sleep is far from certain. Two forms of sleep have been identified, as follows:

1) **Normal Sleep** — also known as slow wave sleep, occurs when a person is tired. It takes place when the cortex does not receive stimuli from the reticular formation. This form of sleep is restful and dreamless and is accompanied by a reduction in blood pressure, body temperature and respiration rate.

2) **Paradoxical Sleep** — also sometimes known as REM sleep, it occurs when some areas of the brain are activated while others remain asleep. This form of sleep is episodic—the episodes occur about every 90 minutes, and last from 5 to 20 minutes. The incidence of these episodes also increase as the person becomes rested. During this type of sleep, the person is experiencing dreams; in addition, the respiration rate increases, heart rate becomes irregular and eye muscles are stimulated producing the "Rapid Eye Movements" or REMs.

SUBCORTICAL STRUCTURES

Buried deep within the white matter of the myelinated axon tracts in the cortex are groups of nerve cell bodies that act as important relay centers for various functions. These subcortical structures and some of their important functions are as follows **(see Figure 4.3)**:

1) **Basal Nuclei (basal ganglia)** — consist of the Caudate nucleus, Putamen, Globus pallidus and Claustrum.
   The major functions of these structures are:

(a) To supply inhibitory inputs to skeletal muscles.

(b) To maintain body posture by acting to coordinate sustained skeletal muscle contractions.

(c) To block unwanted muscle contractions and maintain necessary motor activity in skeletal muscles.

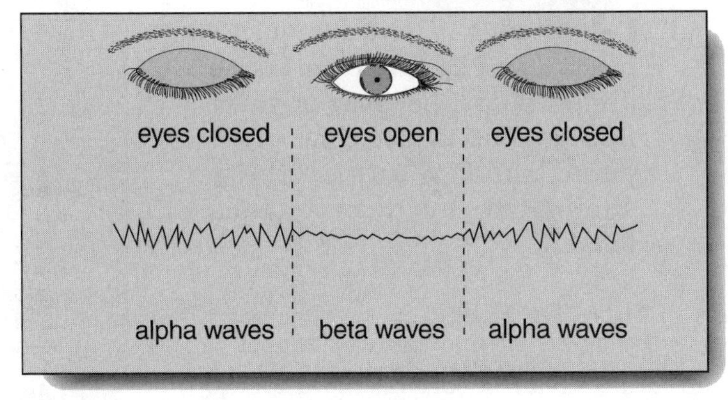

FIGURE 4.6

The inhibitory action supplied by the basal ganglia is counteracted by excitatory stimuli from the thalamus that originate in the motor cortex. In Parkinson's disease, a deficiency in the neurotransmitter, dopamine, reduces these inhibitory stimuli and unchecked action potentials from the thalamus and cortex then result in:

a) Resting tremors.
b) Increased muscle tone and rigidity.
c) Reduced ability to initiate or sustain difficult or complex motor behavior.

2) **Thalamus** — is part of the diencephalon and it serves as an important relay and integration center. Here sensory information destined for the sensory cortex undergoes preliminary processing. By screening out portions of the "unimportant" sensory information, the thalamus causes the cortex to concentrate on only the important stimuli. In addition, the thalamus plays an important role in motor coordination. It provides an inexact level of awareness of certain types of sensation and some degree of consciousness.

3) **Hypothalamus** — is a collection of specific nuclei and associated fibers and is interconnected to the cerebral cortex, thalamus and other parts of the brain stem. It is therefore able to receive and send information to and from these regions. The hypothalamus plays a central role in maintaining homeostasis by regulating various visceral activities and by serving as an important link between the nervous and endocrine systems. Among its important functions are the following:
(a) Regulation of heart rate and arterial blood pressure.
(b) Regulation of body temperature.
(c) Regulation of water and electrolyte balance. Control of thirst and urine output.
(d) Control of hunger and regulation of body weight.
(e) Control of smooth muscles of the gastrointestinal tract and of glandular secretions of the stomach and intestines.
(f) Production of neurosecretory substances that stimulate the pituitary gland to release various hormones. These hormones are then involved in the regulation of growth, control of other endocrine glands, and influence reproductive physiology.
(g) Regulation of sleep and wakefulness.
(h) This region also plays a role in emotional and behavioral patterns, including psychosomatic events.

4) **Limbic System** — involved in emotional behavior; **(Figure 4.4)** it is not a separate structure, but involves various forebrain structures that surround the brain stem and interconnect by complex neural pathways. The structures of the limbic system include portions of the lobes of the cerebral cortex, basal nuclei, thalamus and hypothalamus. Behavioral patterns that are regulated by limbic system function include laughing, crying, blushing, and preparation for attack or defense when angered. These emotional behaviors are probably very ancient in their origins and were initially designed to allow the organism to survive and interact with its environment and with outside influences.

MIDBRAIN

The midbrain, or mesencephalon, is a short section of the brain stem located between the diencephalon and the pons. Composed of many bundles of myelinated nerve tracts, it acts as a connection between higher centers and the spinal cord. In addition, within the midbrain are areas of gray matter composed of nerve cell bodies that act as relay centers. Within the midbrain are the following structures:

1) Cerebral aqueduct that connects the third and fourth ventricles.
2) Cerebral peduncles — two prominent bundles of nerve fibers that include the corticospinal motor tracts.
3) Corpora quadrigemina containing visual reflex centers (superior colliculi) and auditory reflex centers (inferior colliculi).
4) Red nucleus that communicates with the cerebellum and with reflex centers in the spinal cord and is involved in body posture.

PONS

This area appears as a rounded bulge on the underside of the brain stem. This structure contains nerve fibers that relay impulses to and from the medulla and cerebellum, and from the cerebrum to the cerebellum. It also contains various nuclei that relay impulses from the peripheral nerves to the higher brain centers, while other nuclei function as regulatory centers involved in rate and depth of respiration.

MEDULLA OBLONGATA

The medulla is an enlargement at the upper termination of the spinal cord and extends from the level of the foramen magnum to the pons. Located on either side of the medulla are large bundles of nerve fibers called the *olive* that connect to the cerebellum. Ascending tracts within the medulla link the spinal cord to the brain, and descending pathways within the medulla connect the brain to the spinal cord. Within the medulla, white matter myelinated tracts surround a central gray core composed of the neuron cell bodies. However, the gray matter is divided into nuclei that act as relay centers. Major relay centers here include:

1) **The Nucleus Gracilis and Nucleus Cuneatus** — that receive sensory impulses from the fasciculus gracilis and fasciculus cuneatus tracts within the spinal cord. Pathways from these nuclei then synapse in the thalamus and eventually end in the sensory cortex of the cerebral hemispheres.

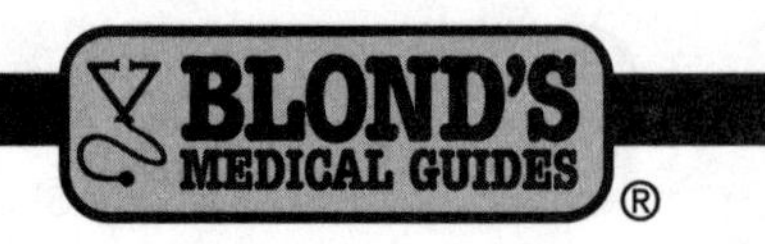

2) **Cardiac Center** — impulses from here are transmitted to the SA, AV and ventricles of the heart and regulate heart rate and contractility.

3) **Vasomotor Center** — impulses from here travel to smooth muscles in the walls of blood vessels. The resulting vasoconstriction causes an increase in total peripheral resistance and a rise in blood pressure. Other cells of the vasomotor center can produce the opposite effect resulting in dilation of the blood vessels and a fall in blood pressure.

4) **Respiratory Center** — functioning in conjunction with the center in the Pons, this area regulates the rate, rhythm and depth of respiration.

5) **Nonvital Reflex Centers** — groups of nerve cells in the medulla regulate so-called nonvital reflexes such as coughing, sneezing, swallowing and vomiting.

RETICULAR FORMATION

The reticular formation consists of widely scattered areas of gray matter and a complex network of nerve fibers located in the medulla, pons and midbrain. They extend from the upper portion of the spinal cord into the diencephalon. The reticular formation connects with centers in the hypothalamus, basal ganglia, cerebellum and cerebrum, and is associated with fibers of the major ascending and descending pathways. It appears to act as the primary arousal center for the cerebral cortex. In the absence of this arousal, the cortex remains inactive and unable to interpret sensory information or carry out thought processing. Thus, decreased activity in the reticular formation results in sleep. In individuals where the reticular formation is damaged, this may lead to unconsciousness and a comatose state.

In addition, this area seems to act as a filter for incoming sensory information blocking certain types of information from reaching the higher centers. It may, for example, inhibit or reduce pain stimuli from reaching the cortex and it may function to allow the cortex to concentrate on "important" as opposed to "unimportant" information.

CEREBELLUM

The cerebellum is attached to the upper part of the brain stem. It functions as a major regulator of motor activity by modifying the output of the major motor pathways that originate in the motor cortex of the hemispheres. It is composed of these three functionally distinct segments:

1) **Vestibulocerebellum** — involved in the maintenance of balance and control of eye movement.

2) **Spinocerebellum** — involved in the regulation of muscle tone and coordination of skilled, voluntary movements. Information originating within the corticomotor centers are adjusted by the spinocerebellum. In addition, the spinocerebellum compares the information from the higher centers with the performance by the muscle groups and makes corrections as required. It also makes predictions of future movements and corrects for any potential deviations from the proposed and expected outcome of the movement.

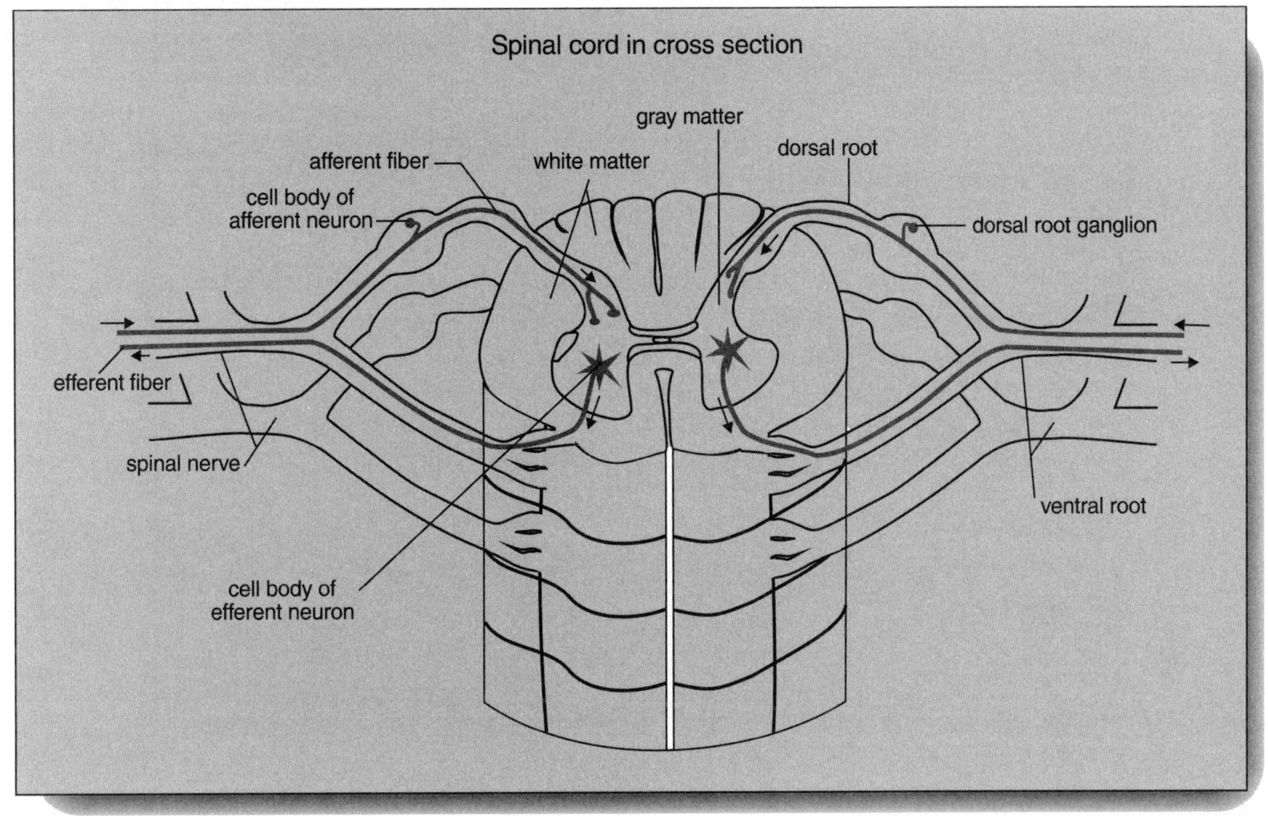

FIGURE 4.7

3) **Cerebrocerebellum** — involved in the planning and initiation of voluntary activity. Neurons from this region provide input to the cortical motor areas.

SPINAL CORD

The spinal cord extends from the brain stem and is about 45 cm long and 2 cm in diameter. It passes through the vertebral canal within the vertebrae of the spine, and paired spinal nerves emerge from the spinal cord between adjacent vertebrae. Spinal nerves are named according to the regions of the spinal cord, and thus there are 8 pairs of cervical nerves, 12 pairs of thoracic nerves, 5 pairs of lumbar nerves, 5 pairs of sacral nerves, and 1 pair of coccygeal nerves. White matter myelinated nerve tracts are found in the outer layers of the cord, while the central portions of the cord are composed of gray matter (neuron cell bodies that are not myelinated). Sensory information enters the cord on the dorsal side via afferent sensory fibers. Cell bodies of these fibers are located outside of the central nervous system in the dorsal root ganglia. Motor information exits the cord from the ventral side via efferent fibers **(Figure 4.7)**.

Tracts that ascend within the cord carry sensory information from the peripheral nervous system to the higher centers of the brain. Tracts that descend carry motor information from the higher centers of the brain to peripheral motor neurons that control various effector organs (skeletal muscle, smooth muscle, cardiac muscle or glands). Tracts that begin or end in the cerebral hemispheres in the motor

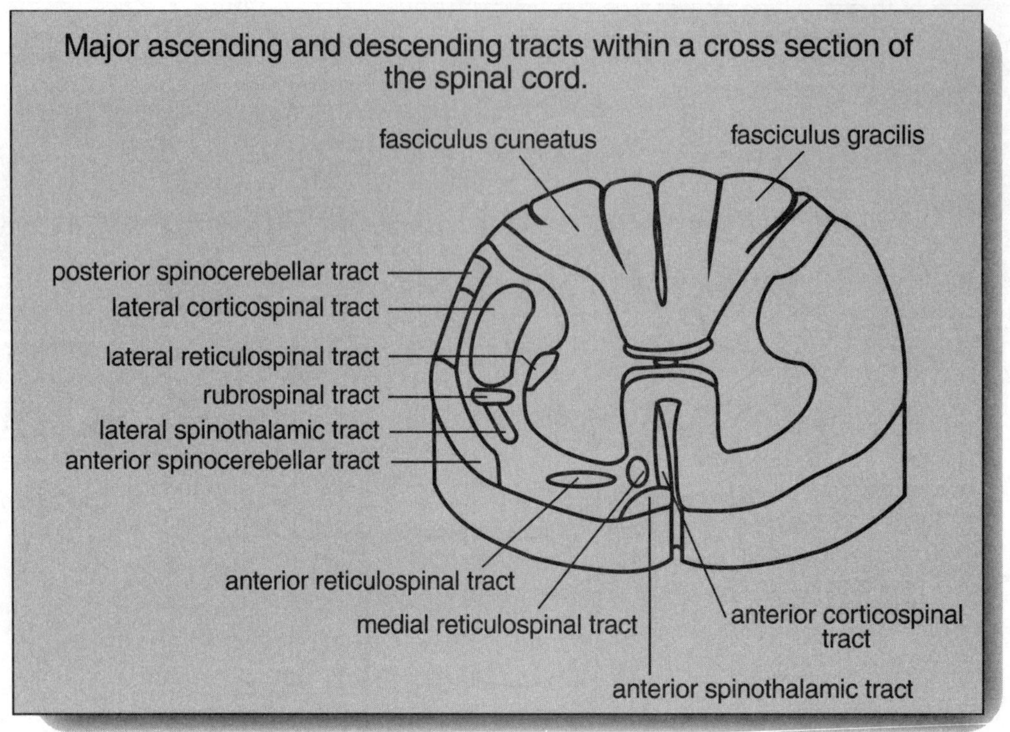

FIGURE 4.8

or sensory cortex, or in areas that attach to the motor or sensory cortex such as the thalamus, are considered conscious tracts. Tracts that begin or end in the cerebellum or in structures that are associated with the cerebellum, such as the basal ganglia, are unconscious tracts. Examples of the major conscious and unconscious motor and sensory tracts are presented in **Figure 4.8**. The majority of these tracts cross at the level of the medulla so that a particular hemisphere receives information from, and controls structures on, the opposite side of the body. An example of such a conscious sensory tract is the Fasciculus cuneatus **(Figures 4.8 and 4.9)**, while an example of such a conscious motor tract is the cortical spinal **(Figures 4.8 and 4.10)**. However, some tracts may cross at each level of the cord (spinothalamic tracts) while a few remain uncrossed.

REFLEXES

Reflexes are automatic, unconscious responses to various stimuli that originate in receptors located either inside or outside of the body. The automatic responses generated to these stimuli then assist in the maintenance of homeostasis by initiating effects in skeletal, smooth or cardiac muscles or in glandular secretions. There are two types of reflexes:

1) **Simple Reflexes** — are "wired" into the nervous system and consist of unlearned responses (withdrawal of the hand after the finger is burned).

2) **Conditioned Reflexes** — which result from learning a particular behavior (e.g., stopping at a red light).

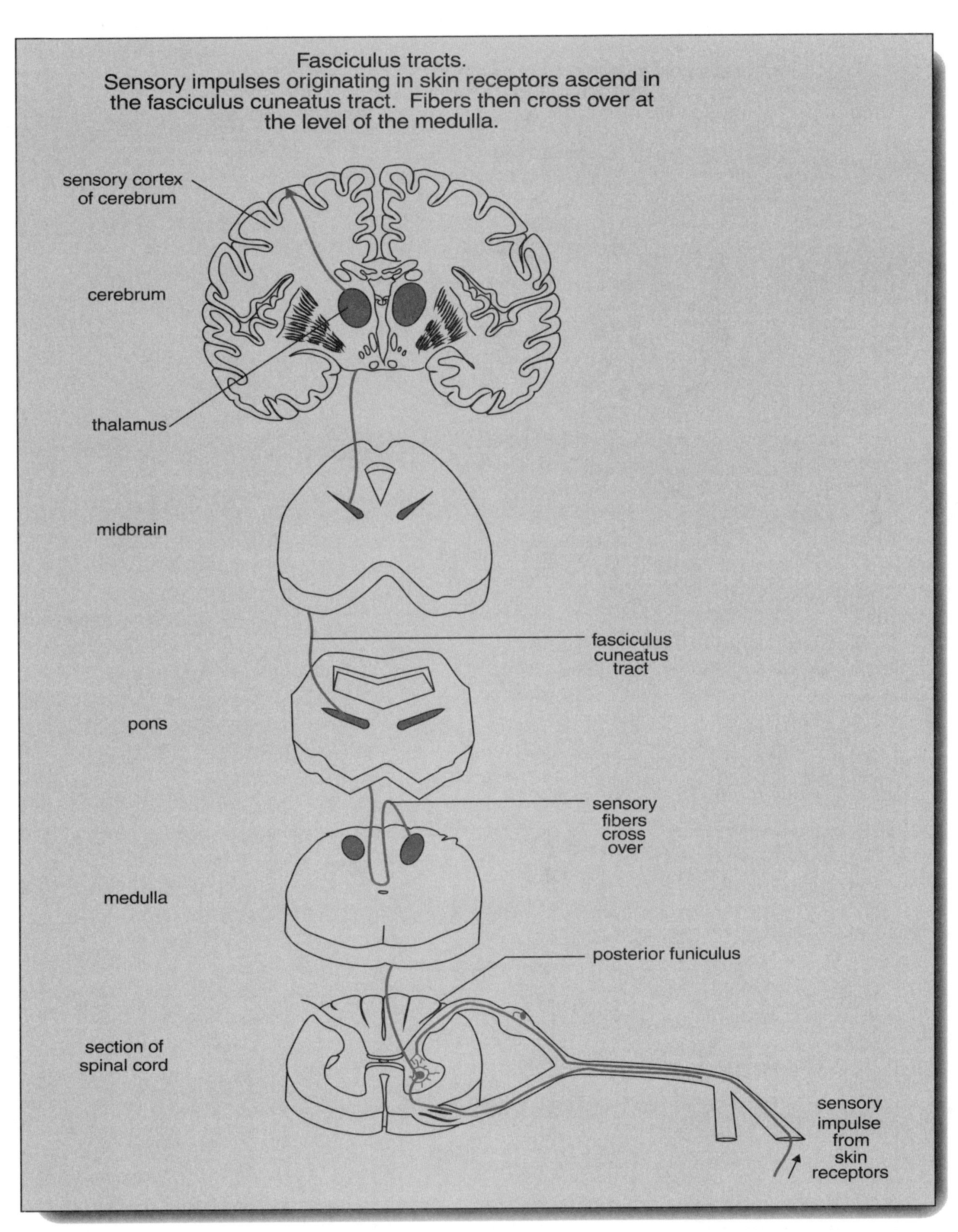
Fasciculus tracts.
Sensory impulses originating in skin receptors ascend in the fasciculus cuneatus tract. Fibers then cross over at the level of the medulla.
sensory cortex of cerebrum
cerebrum
thalamus
midbrain
fasciculus cuneatus tract
pons
sensory fibers cross over
medulla
posterior funiculus
section of spinal cord
sensory impulse from skin receptors

FIGURE 4.9

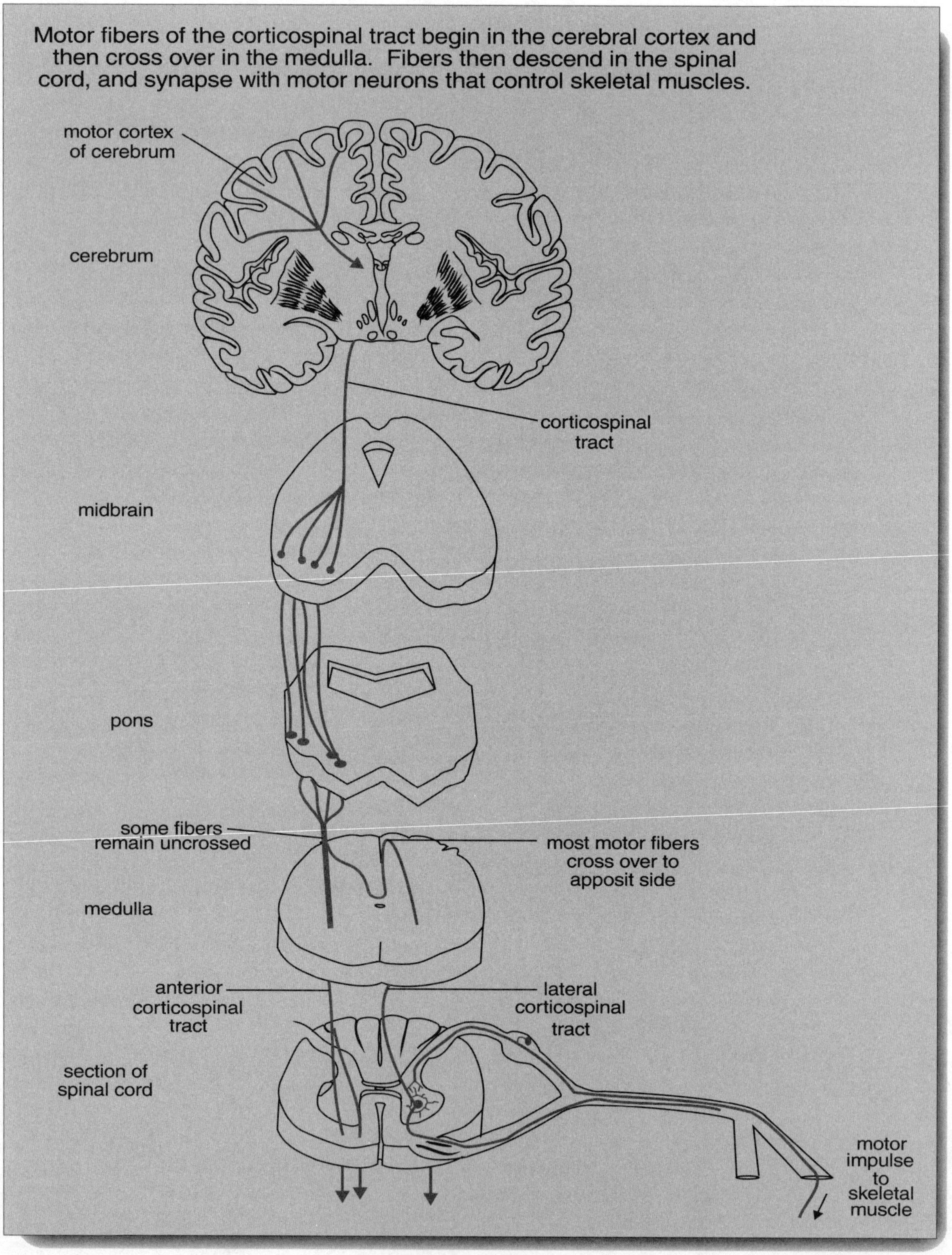

Motor fibers of the corticospinal tract begin in the cerebral cortex and then cross over in the medulla. Fibers then descend in the spinal cord, and synapse with motor neurons that control skeletal muscles.

motor cortex of cerebrum

cerebrum

corticospinal tract

midbrain

pons

some fibers remain uncrossed

most motor fibers cross over to apposit side

medulla

anterior corticospinal tract

lateral corticospinal tract

section of spinal cord

motor impulse to skeletal muscle

FIGURE 4.10

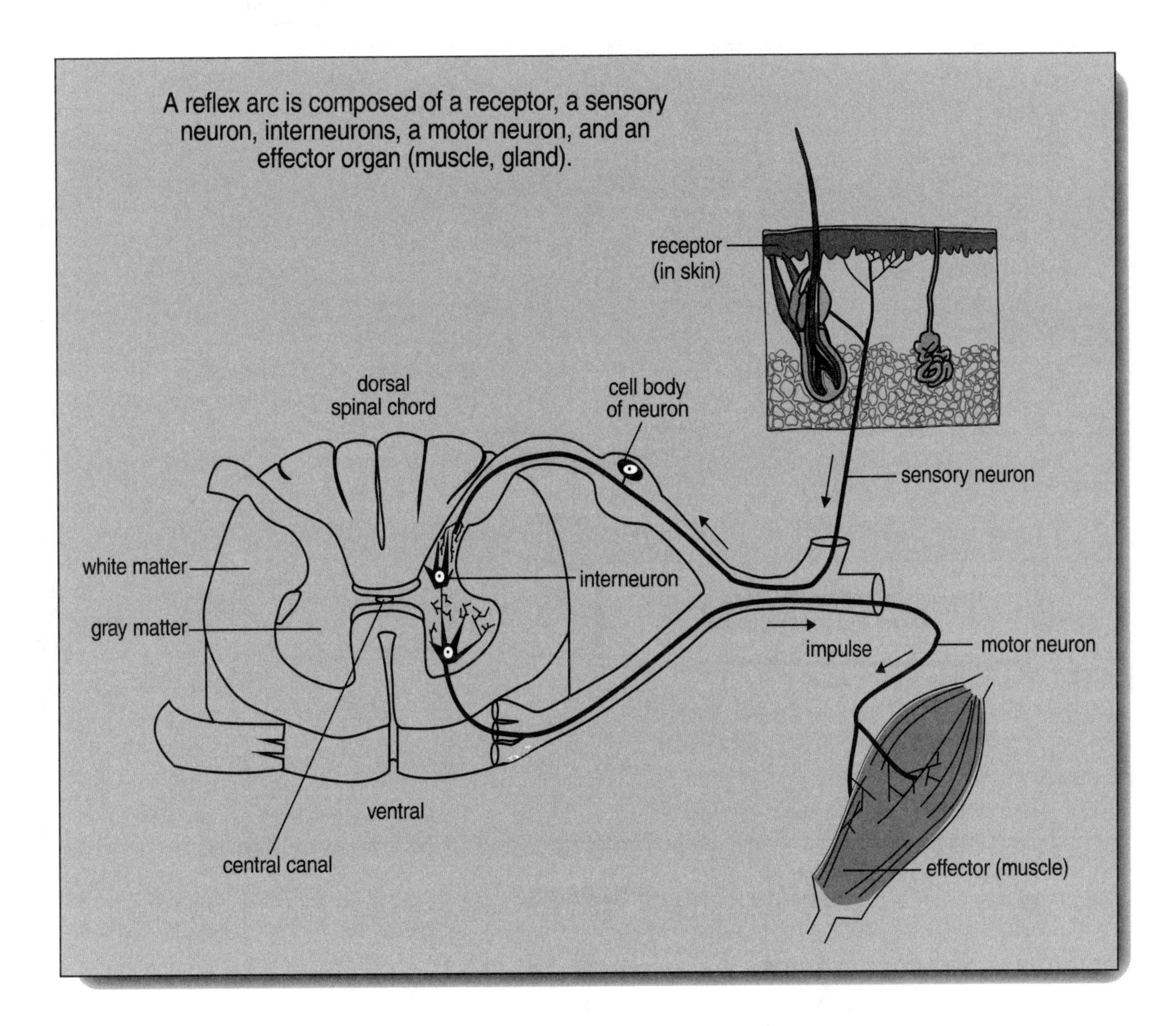

FIGURE 4.11

Reflexes function via reflex arcs **(Figure 4.11)** which are composed of the following basic elements:

1) Sensory receptor
2) Afferent pathway (or afferent sensory nerve)
3) Integrating center (or interneurons)
4) Efferent pathway (or efferent motor nerve)
5) Effector organ (skeletal muscle, smooth muscle, cardiac muscle or gland)

An example of the knee jerk reflex is presented in **Figure 4.12**. A comparison between a somatic reflex arc and an automic reflex arc is found in **Figure 4.13**.

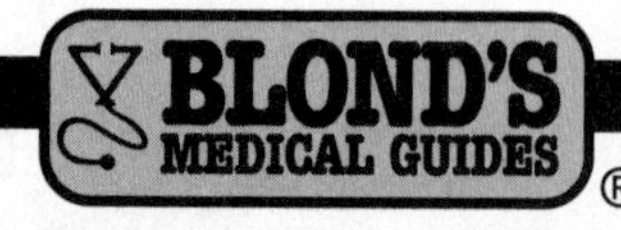

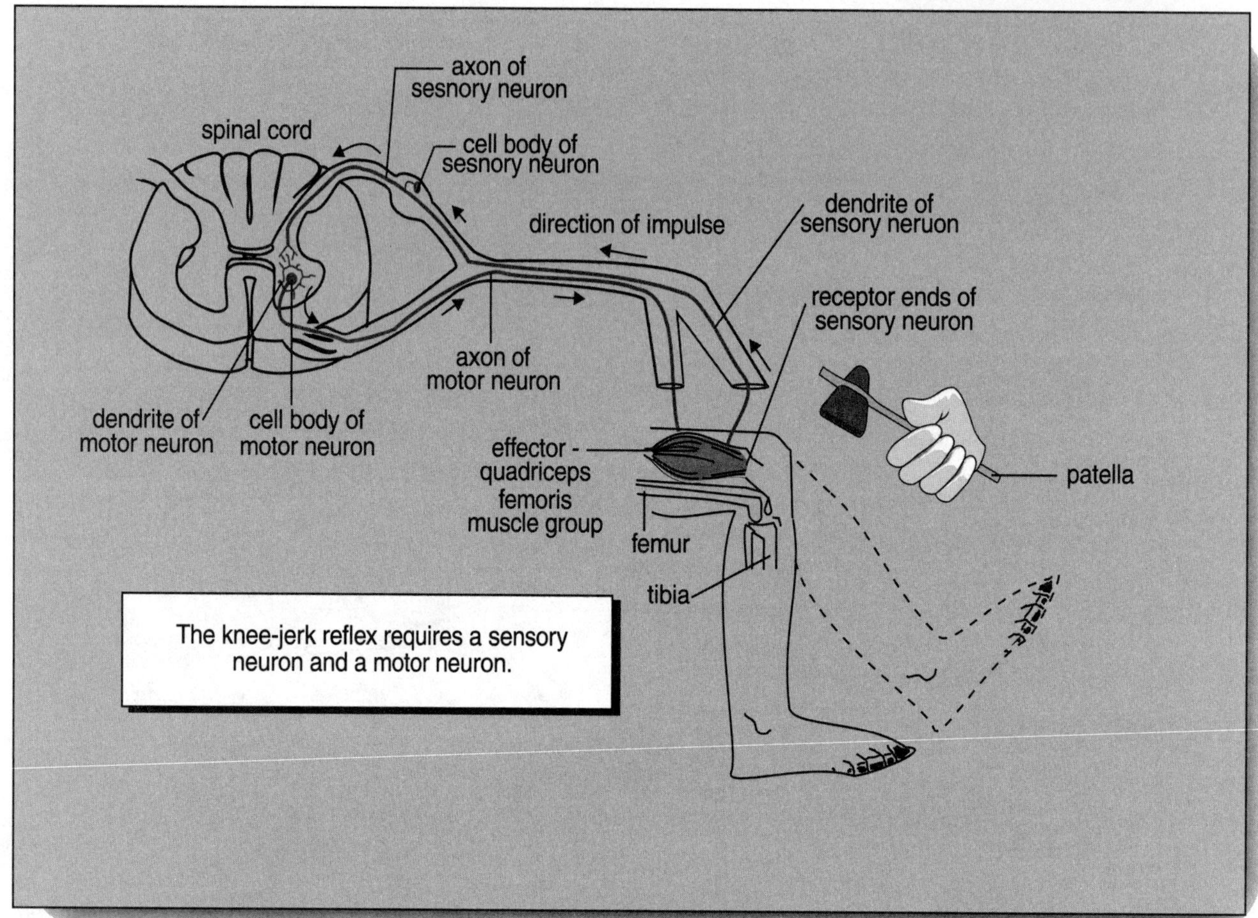

FIGURE 4.12

PERIPHERAL NERVOUS SYSTEM

The peripheral nervous system (PNS) consists of nerves that originate from the central nervous system (CNS) and connect to other structures of the body. Elements of the PNS include the cranial nerves that arise from the brain, and spinal nerves that arise from the spinal cord. The PNS can be further subdivided into the somatic and autonomic nervous systems.

1) **The Somatic Nervous System** — is involved with conscious activities and consists of cranial and spinal nerves that connect the CNS to the skin and skeletal muscles.

2) **The Autonomic Nervous System** — is involved with unconscious activities and can be further subdivided into:

   a) The Sympathetic division, and
   b) The Parasympathetic division.

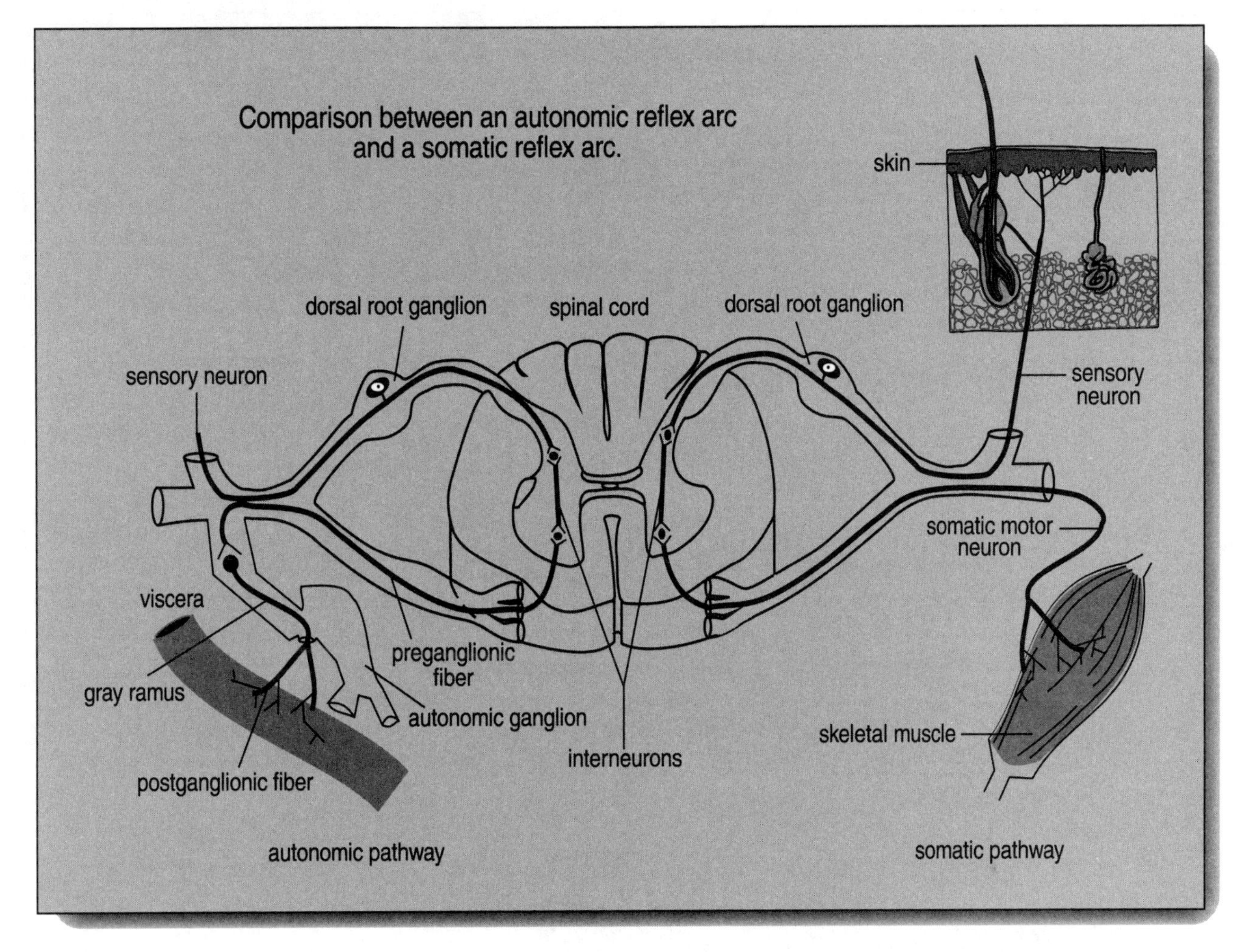

FIGURE 4.13

It consists of fibers that connect the CNS to the visceral organs (heart, GI tract, glands, and so on).

CLASSIFICATION OF NERVES

Nerves can be classified on the basis of their functional differences:

1) **Sensory Neurons (afferent neurons)** —carry impulses from the periphery of the body into the brain and spinal cord. These neurons have specialized receptors at the dendritic ends capable of initiating action potentials. Such neurons are responsible for sensing changes that take place inside or outside of the body and conveying this information into the CNS for further processing. The majority of sensory neurons are unipolar in structure.

2) **Interneurons (association neurons)** — are present within the brain and spinal cord. They are multipolar and form links between other neurons. Interneurons transmit information from one part of the brain or spinal cord to another.

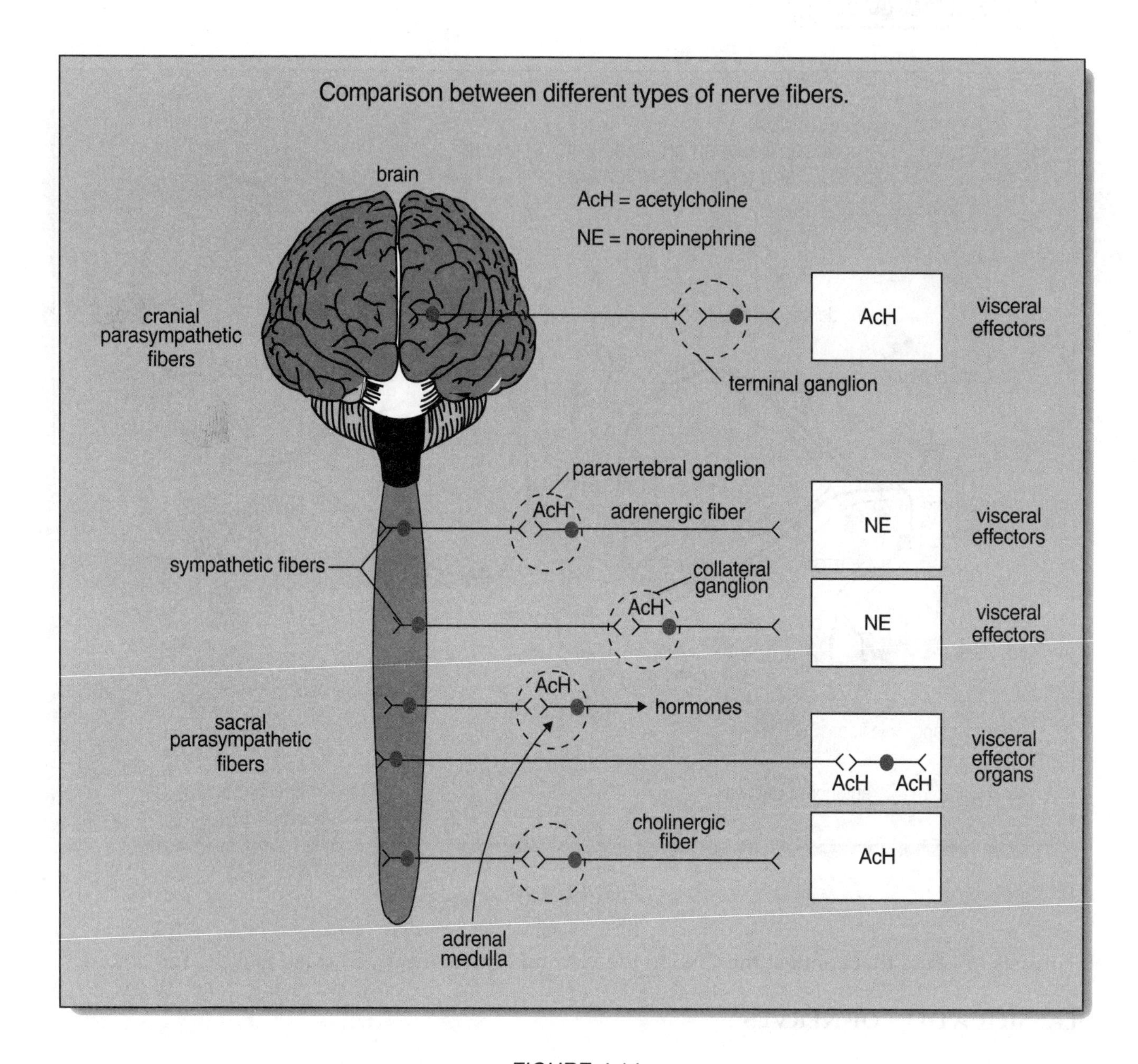

FIGURE 4.14

3) **Motor neurons (efferent neurons)** — carry impulses out of the CNS to effector organs in the periphery. Motor neurons are multipolar in structure.

Nerves and nerve fibers are different. Although a nerve fiber or axon is an extension from the neuron cell body, bundles of such axons surrounded by myelin and connective tissue are termed *nerves*. Such nerves appear anatomically as white bands in the tissue. If the axons within such a nerve all originate from sensory neurons, then the nerve is said to be a *sensory nerve*. If the axons are all from motor

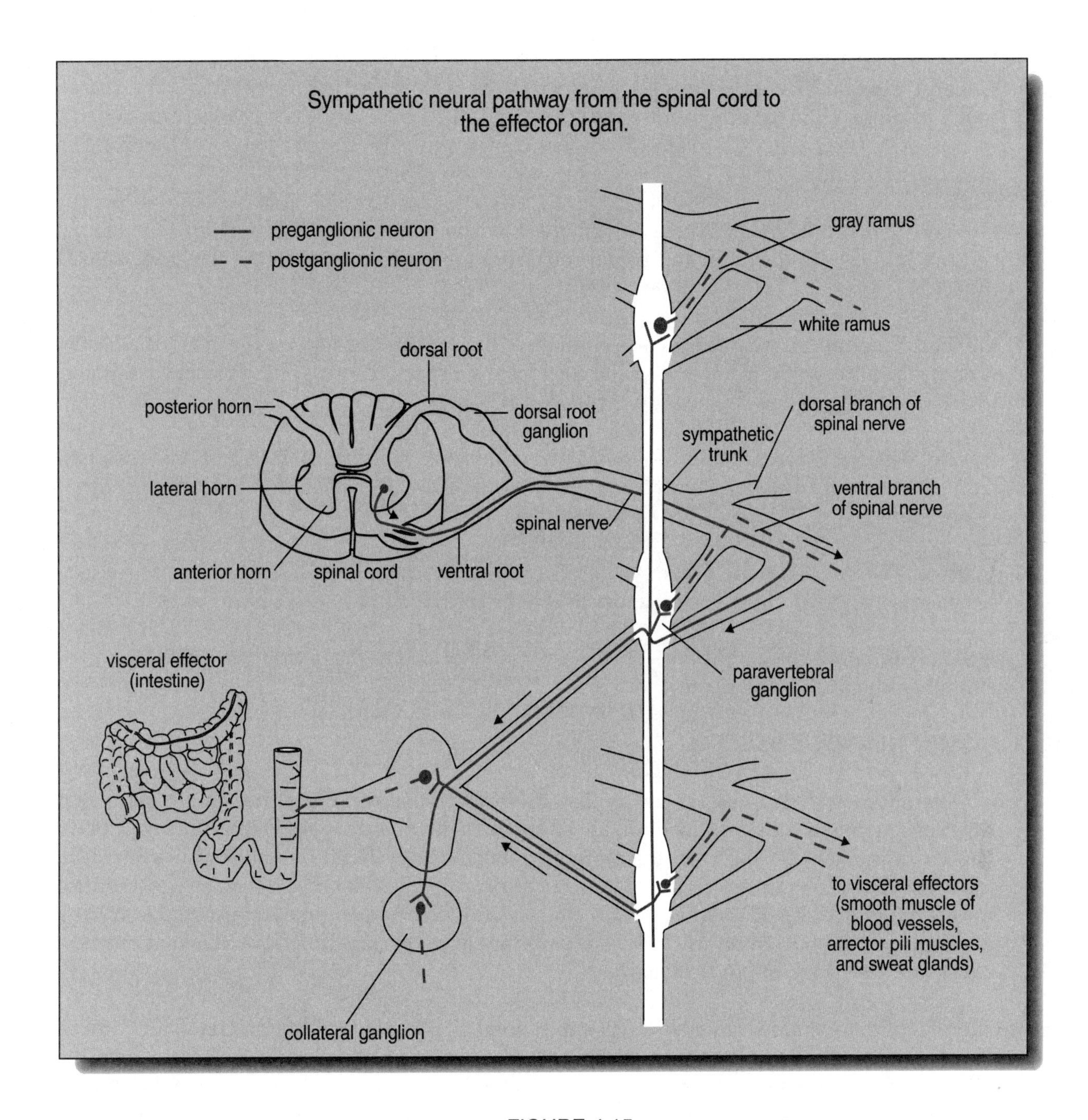

FIGURE 4.15

neurons, the nerve is a *motor nerve,* and if there is a mixture of sensory and motor axons in the bundle, then the nerve is called a *mixed nerve.*

Cranial and spinal nerves can be further subdivided as follows:

1) **General Somatic Efferent Fibers** — (a branch of the Somatic nervous system); they carry motor impulses from the brain and spinal cord to skeletal muscles resulting in contraction.

2) **General Visceral Efferent Fibers** — (a branch of the Autonomic nervous system); they carry motor impulses from the brain and spinal cord to various smooth muscles, cardiac muscles and glands associated with the internal organs, causing muscle contraction or relaxation, or glandular secretion.

3) **General Somatic Afferent Fibers** — (a branch of the Somatic nervous system); they carry sensory impulses to the brain and spinal cord from sensory receptors in the skin and skeletal muscles.

4) **General Visceral Afferent Fibers** — (a branch of the Autonomic nervous system); they carry sensory impulses to the brain and spinal cord from sensory receptors in the blood vessels and internal organs (such as the heart, GI tract, kidneys and so on.)

5) **Special Visceral Efferent Fibers** — (associated with cranial nerves); they carry motor impulses from the brain to muscles involved in chewing, swallowing, speaking and forming facial expressions.

6) **Special Visceral Afferent Fibers** — (associated with cranial nerves); they carry sensory impulses to the brain from the olfactory and taste receptors.

7) **Special Somatic Afferent Fibers** — carry sensory impulses to the brain from the receptors of sight, hearing and equilibrium.

AUTONOMIC NERVOUS SYSTEM

Function. The autonomic nervous system is that division of the peripheral nervous system that functions without conscious volition. It is centrally involved in the maintenance of body homeostasis and it accomplishes this through either direct or indirect control of all systems in the body. It directly influences contraction of cardiac and smooth muscles and also can control the secretory activity of various glands. The autonomic nervous system is activated during periods of emotional stress, and it is also important in preparing the body to combat physical stress and to engage in physical activity.

Organization. The autonomic nervous system is subdivided into the sympathetic and parasympathetic divisions which act together, but whose effects are also antagonistically expressed on the effector organs. Thus, stimulation of the sympathetic fibers on the heart increase heart rate, while stimulation of parasympathetic fibers cause heart rate to decrease. However, since fibers from both sympathetic and parasympathetic divisions innervate most organs (dual innervation), the actual effect produced on the organ is due to the combination of stimuli arriving together. For this reason, cutting the parasympathetic nerve to the heart, while leaving the sympathetic branch intact, results in an increase in heart rate because the remaining sympathetic innervation is no longer balanced by the parasympathetic stimuli. Similarly, cutting the sympathetic branch to the heart while leaving the parasympathetic branch intact will result in a decrease in heart rate.

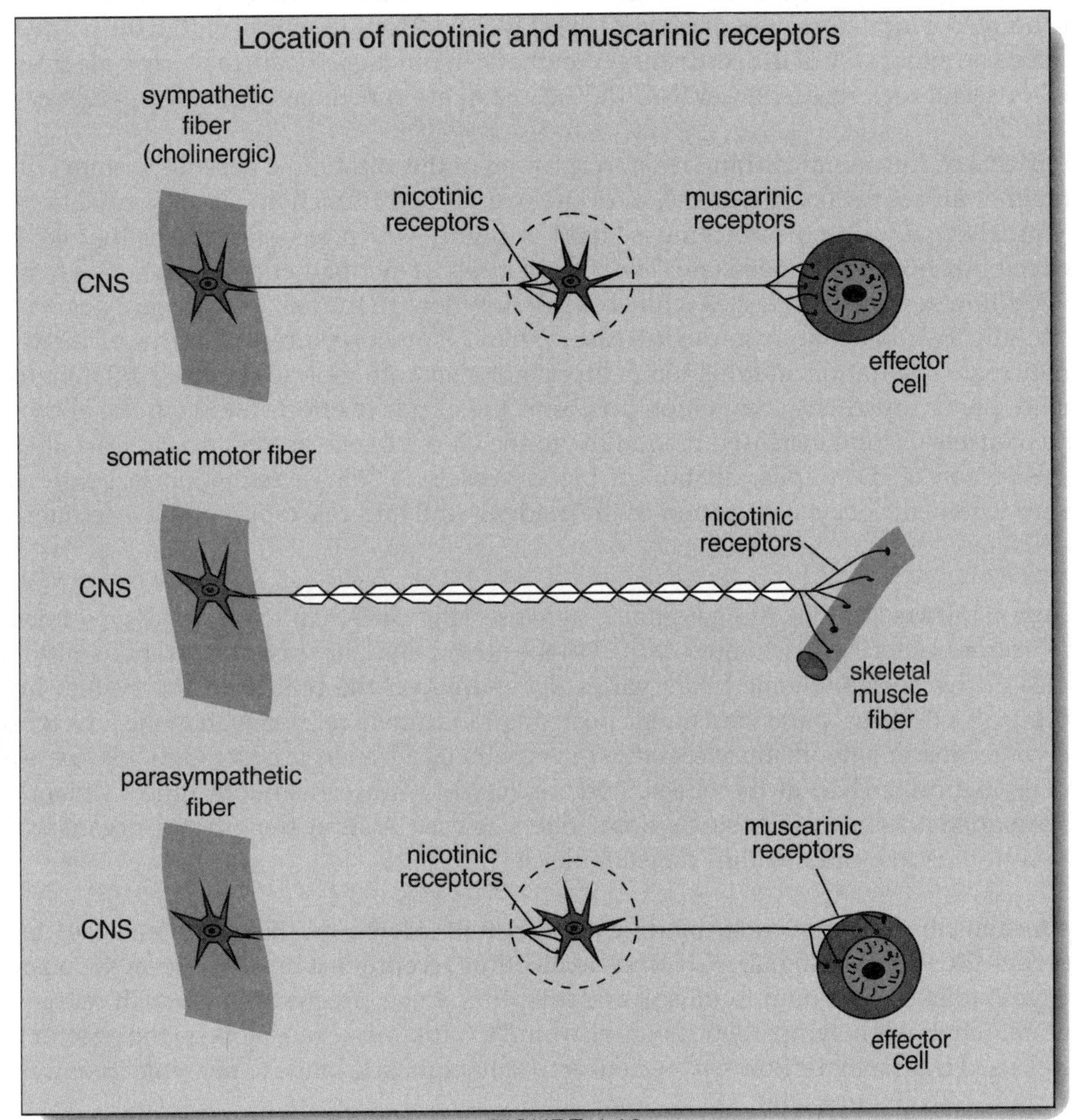

FIGURE 4.16

Fibers of the autonomic nervous system always synapse on autonomic ganglia after leaving the CNS. They then continue to the effector organ where they release transmitter to regulate actions at these structures **(Figure 4.14)**. Autonomic ganglia that are located close to the spinal cord, and are connected together by the sympathetic trunk, are called *paravertebral ganglia* and are part of the sympathetic division. However, autonomic ganglia can also be located farther away from the CNS and these are termed *collateral ganglia* (these may be part of either division). Autonomic ganglia that are part of the parasympathetic division are frequently located directly on the organ that is innervated **(Figure 4.15)**.

Generally speaking, within the sympathetic division there are short preganglionic fibers and long postganglionic fibers, while in the parasympathetic division the preganglionic fibers are usually long

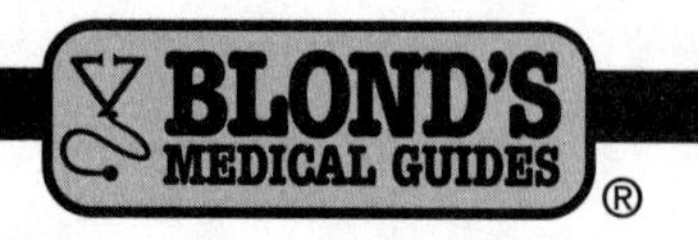

while the postganglionic fibers are short **(Figures 4.14 and 4.15)**. In addition, fibers that originate from either the thoracic or lumbar regions of the cord (T1-T12, L1, L2) (and synapse within the paravertebral ganglia chains on either side of the cord) are sympathetic in function. Fibers that originate from either the cranial or sacral regions (C-III, C-VII, C-IX, C-X, S2-4) are functionally parasympathetic.

Physical Effects of Autonomic Stimulation. Activation of the autonomic nervous system will result in either an activation or depression of the overall body function. Upon stimulation by the sympathetic division the outcome is "fight or flight" which basically means that the body is placed in a hyper-activated state. The overall effect of sympathetic stimulation is to increase blood flow to skeletal muscles while constricting flow to the skin, to increase heart rate and strength of cardiac contraction, to increase systemic blood pressure, to cause pupil dilation, and to increase respiration and dilation of bronchiolar smooth muscle **(Figure 4.17)**. On the other hand, parasympathetic stimulation promotes the opposite effects allowing the body to rest. Such effects include increased blood flow to the GI tract, increased GI tract peristaltic action, constriction of the pupils, dilation of blood vessels to skin, a reduction in heart rate and contraction efficiency, contraction of the bladder wall and relaxation of the internal urethral sphincter.

Neurotransmitters. Within the ganglionic synapses the presynaptic fibers always release the neurotransmitter Acetylcholine (ACh). On the other hand, the type of neurotransmitter that is released by postganglionic fibers varies depending on the branch of the system involved **(Figure 4.14)**. Thus, parasympathetic postganglionic fibers release ACh at the effector organs (cardiac muscle, smooth muscle, glands); sympathetic adrenergic postganglionic fibers release Norepinephrine (NE) at the effector organs (cardiac muscle, smooth muscle, glands); and sympathetic cholinergic postganglionic fibers release ACh at the effector organs (vascular smooth muscle located within skeletal muscles).

These neurotransmitters function by binding to receptors located on the ganglionic cells or in the effector organs. Within the ganglia, ACh binds to nicotinic receptors, while ACh from the postganglionic parasympathetic fibers bind to muscarinic receptors at the effector organs. ACh released from postganglionic cholinergic sympathetic fibers also bind to muscarinic receptors on the effector organs, but ACh released from somatic fibers at the neuromuscular junction binds to nicotinic receptors on the skeletal muscle cells **(Figure 4.16)**.

With respect to adrenergic transmitters, two kinds of receptors mediate the response on the effector organs: Alpha receptors and Beta receptors. While NE only binds to Alpha receptors, Epinephrine (Epi) binds to both Alpha and Beta receptors. Perusal of the effects outlined in the **Figure 4.17** would indicate that for the cardiovascular system, constriction of blood vessels results from stimulation of Alpha receptors, while dilation is mediated by Beta receptors.

Effects of autonomic stimulation on various effectors		
effector location	response to sympathetic stimulation	response to parasympathetic stimulation
integumentary system		
apocrine glands	increased secretion	no action
eccrine glands	increased secretion (cholinergic effect)	no action
special senses		
iris of eye	dilation	constriction
tear gland	slightly increased secretion	greatly increased secretion
endocrine system		
adrenal cortex	increased secretion	no action
adrenal medulla	increased secretion	no action
digestive system		
muscle of gallbladder wall	relaxation	contraction
muscle of intestinal wall	decreased peristaltic action	increased peristaltic action
muscle of internal anal sphincter	contraction	relaxation
pancreatic glands	reduced secretion	greatly increased secretion
salivary glands	reduced secretion	greatly increased secretion
respiratory system		
muscles in walls of bronchioles	dilation	constriction
cardiovascular system		
blood vessels supplying muscles	constriction (alpha adrenergic) dilation (beta adrenergic) dilation (cholinergic)	no action
blood vessels supplying skin	constricted	no action
blood vessels supplying heart (coronary arteries)	dilation (beta adrenergic) constriction (alpha adrenergic)	dilation
muscles in wall of heart	increased contraction rate	decreased contraction rate
urinary system		
muscle of internal urethral sphincter	relaxation	contraction
muscle of bladder wall	contraction	relaxation
reproductive systems		
blood vessels to clitoris and penis	no action	dilation leading to erection
muscles associated with male internal reproductive organs	ejaculation	

FIGURE 4.17

Chapter 5
SPECIAL SENSES

SENSORY RECEPTORS AND TRANSDUCTION

The main function of the afferent division of the peripheral nervous system is to keep the CNS informed of what is happening inside or outside of the body. To accomplish this the afferent sensory nerves have receptors that receive information in the form of energy and convert this information to action potentials. For example, light energy is converted to action potentials by the retinal receptors; sound energy (mechanical vibration) is converted to action potentials by the hair cells in the inner ear. And chemicals activate an energy-releasing process in taste buds that promotes action potentials and so on. This conversion of diverse energy forms (or modalities) to action potentials is called *transduction*.

Each receptor is structurally designed to primarily transduce one form of energy but not others (i.e., heat receptors do not transduce sound, pressure receptors do not transduce light) and this property is known as *the law of specific nerve energies*. However, in some cases there may be a slight overlap as, for example, when pressure applied to the eye promotes action potentials in the retinal cells, even though they are primarily designed to convert light to action potentials.

SENSATIONS REQUIRE THE RECEPTOR, THE TRACK AND THE BRAIN AREA

In order to be consciously aware of a particular sensation, it is necessary that the promoting energy (e.g., light) be transduced by the retinal receptors to action potentials. Additionally, nerve tracks must then carry these action potentials to the brain. Finally, the action potentials must be interpreted by a specific brain area (the occipital lobes) to form meaningful information (the sensation of sight). In the event that any of these components are lacking, the effect is blindness (i.e., damage in the retina, damage in the tracks, damage in the occipital lobes).

SENSORY INTERPRETATION DEPENDS ON SPECIFIC BRAIN REGIONS

Specific brain area are responsible for interpreting the action potentials as particular sensations. Thus, the occipital lobes interpret action potentials as sight, the temporal lobe interprets action potentials as sound, and the parietal lobes interpret action potentials as touch, pressure and so forth. Hypothetically, this implies that if the optic tracks were to be disconnected from the visual centers in the occipital lobes and reconnected to the auditory center in the temporal lobe, then light stimuli would produce sensations of sound and not vision.

TYPES OF RECEPTORS

There are five general groups of receptors:

1) **Chemical Receptors (chemoreceptors)** — are stimulated by the presence of chemicals (or changing chemical concentrations). Examples of such receptors are those of taste (taste buds), of smell (olfactory), and the chemoreceptors that measure levels of blood gases and hydrogen ions.

2) **Pain Receptors (nociceptive receptors)** — are stimulated by tissue damage. They respond to excessive mechanical, electrical, chemical or thermal stimuli and act though specific pain pathways. The effect perceived in the brain is an unpleasant pain sensation.

3) **Temperature Receptors (thermoreceptors)** — are stimulated by changes in temperature in the skin. Since temperature is actually a measure of molecular kinetic energy, these receptors are actually sensitive to molecular motion in the skin regions. However, such changes in molecular motion are actually brought about by infrared (or heat) radiation which is a form of electromagnetic waves. There are two different types of thermoreceptors: those sensitive to heat (actually temperatures above that of the body), and those sensitive to cold (actually temperatures below that of the body).

4) **Mechanical Force Receptors (mechanoreceptors)** — are stimulated by changes in pressure, or movement of fluids. Changes in these mechanical forces result in deformation of the receptor which then leads to the generation of action potentials. Included in the list of such mechanoreceptors are:
   a) *Proprioceptors* — designed to sense changes in muscle tone, tension in tendons, and movement of joints.
   b) *Pressoreceptors* (*Baroreceptors*) — designed to detect changes in arterial blood pressure.
   c) *Stretch receptors* — designed to detect the amount of lung inflation, or muscle contraction.
   d) *Auditory receptors* (*hair cells*) — located in the inner ear, they detect pressure vibrations in a conducting medium.
   e) *Equilibrium receptors* — located in the vestibule and in the semicircular canals in the inner ear.

5) **Light Receptors (photo-receptors)** — are the rods and cones located in the retina of the eye. They are stimulated by visible light which is a form of electromagnetic radiation transmitted by photons of specific frequencies (within the visible range).

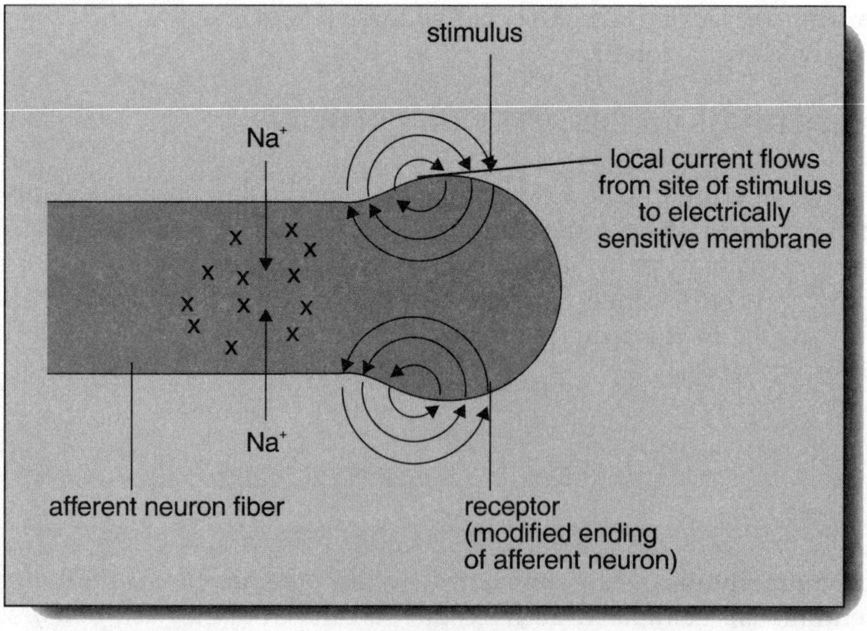

FIGURE 5.2

RECEPTOR AND GENERATOR POTENTIALS

Transduction of the stimulus energy into action potentials results when the stimulus produces a local depolarization in the receptor membrane. This effect can take place by two mechanisms:

1) The stimulus acts on a separate receptor cell. This cell then releases a chemical messenger that opens chemically sensitive Na^+ gates in the afferent nerve cell membrane. The resulting depolarization is known as a *receptor potential* (**Figure 5.1**).
2) The stimulus acts directly on a modified afferent nerve ending. This produces local current flow that opens voltage-gated Na^+ channels. The resulting depolarization is known as a *generator potential* (**Figure 5.2**).

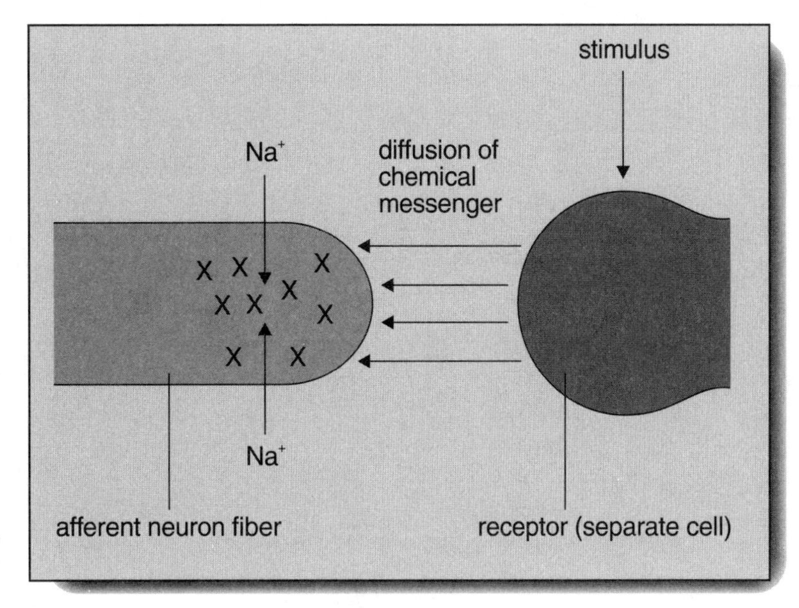

FIGURE 5.1

SENSORY RECEPTOR ADAPTATION

The output of sensory receptors may fluctuate even if the stimuli applied remains constant. Frequently this effect is expressed as a decrease in the receptor potential and is termed *receptor adaptation*. Receptors that undergo sensory adaptation can take two forms (**Figure 5.3**):

1) Tonic receptors where adaptation slowly takes place.

2) Phasic receptors that demonstrate a rapid "on" response, rapidly adapt and then when the stimulus is turned off, produce an additional "off" response.

SOMATIC SENSES

Somatic senses are those that are associated with receptors in the visceral organs, joints, muscles and skin. Specifically they include:

1) **The Exteroceptive Senses** — associated with the body surfaces such as touch, pressure and temperature.

2) **The Proprioceptive Senses** — associated with changes in muscles, tendons and joints, and related to body position.

3) **The Visceroceptive Senses** — associated with changes occurring in the visceral organs. Visceral pain receptors are included in this category.

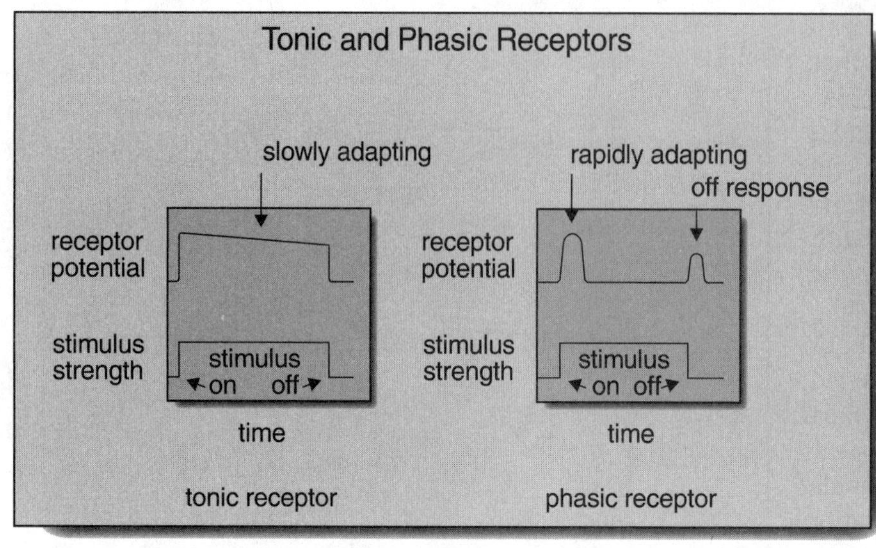

FIGURE 5.3

TOUCH AND PRESSURE SENSES

Receptors that are involved include:

1) **Pacinian Corpuscles**—are located in the deep subcutaneous tissues and in the tendons and ligaments. They are stimulated by heavy pressure and may also detect vibrations in the tissues.

2) **Meissner's Corpuscles** — present in the hairless areas of the skin. They are sensitive to the light motion of objects across the skin surface.

THERMORECEPTORS

1) **Heat Receptors** — are free nerve endings located in the skin. They respond to temperatures above 25°C and become unresponsive above 45°C. At the higher temperatures, pain receptors in the area are also stimulated producing a burning sensation.

2) **Cold Receptors** — are free nerve endings located in the skin. They respond to temperatures in the range of 10°C to 20°C. Below 10°C, pain receptors are stimulated producing a freezing sensation.

Both heat and cold receptors undergo rapid sensory adaptation such that hot or cold sensation which may be unpleasant initially, rapidly fade within a few seconds.

PROPRIOCEPTION

These receptors, designed to monitor the tension or movement in structure associated with muscles, joints, tendons and ligaments, include:

1) **Muscle Spindles** — which detect changes in skeletal muscle length.

2) **Golgi Tendon Organs** — which detect changes in the tension in muscles.

PAIN

Pain receptors provide a protective function because they are stimulated when tissue is damaged. Pain sensations are accompanied by a motivational response such as withdrawal and also frequently by an additional emotional response such as fear. In addition pain perception is partially subjective and learned and thus can be either increased or decreased by previous learning. There are three categories of pain receptors:

1) **Mechanical Nociceptors** — respond to mechanical damage (cutting, crushing, pinching).

2) **Thermal Nociceptors** — respond to extremes of temperature (either heat or cold).

3) **Polymodal Nociceptors** — respond equally to many forms of tissue damage. Stimuli that activate these receptors probably act through irritating chemicals released from damaged cells.

Nociceptors do not undergo sensory adaptation, but their sensitivity can be greatly enhanced by the release of prostaglandins. Because aspirin and similar drugs inhibit prostaglandin synthesis, they provide temporary analgesia.

REGULATION OF PAIN TRANSMISSION BY PAIN GATES

The CNS contains an analgesic system. Preliminary studies suggest that stimulation of the *periaqueductal gray matter* that surrounds the cerebral aqueduct, or stimulation of the *reticular formation* both produce significant analgesia. It is believed that such stimulation blocks the release of substance P (a pain neurotransmitter) by inhibiting its release from afferent pain fibers **(Figure 5.4)**. Inhibition of substance P is mediated by opiate receptors located in the afferent synaptic terminals. While drugs such as morphine can bind to these receptors and thus block pain transmission, endogenous opiates are actually involved in this process. Endogenous opiates are manufactured by the body (they are not drugs administered from the outside) and consist of the substances *endorphins and enkephalins.* It has been suggested that acupuncture and hypnosis may suppress pain by acting to release increased levels

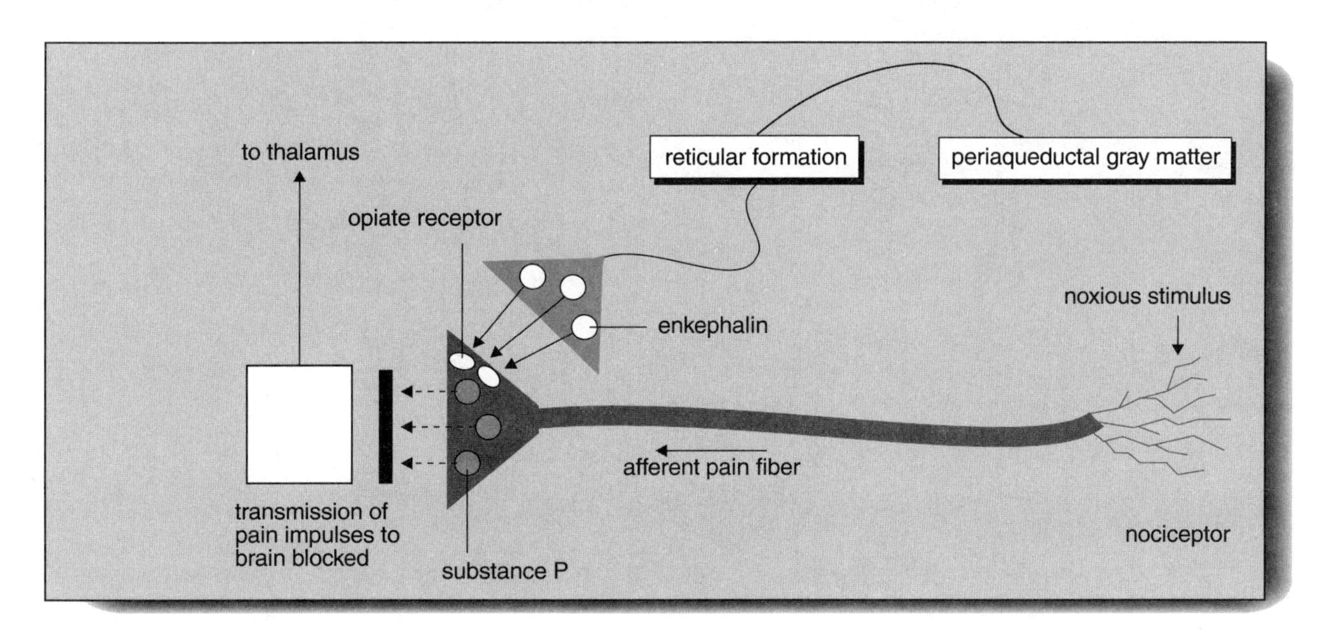

FIGURE 5.4

of the endogenous opiates. Conceivably people with a higher pain tolerance or threshold may release more endorphins while those who are very sensitive to pain may release less of these substances.

PAIN PATHWAYS

Pain stimuli can be classified as follows:

1) **Fast Pain** — These stimuli are transmitted by A-delta (acute pain) fibers and produce a sharp, prickling sensation. Such sensations can be easily localized. Response occurs as a result of the stimulation from mechanical nociceptors and thermal nociceptors. This kind of pain is consciously perceived before the sensations of slow pain.

2) **Slow Pain** — These stimuli are transmitted by C (chronic pain) fibers and produce a dull, aching and burning sensation. Such sensations are poorly localized. Response occurs as a result of the stimulation from polymodal nociceptors. This kind of pain is consciously perceived after the sensations of fast pain and persists for a longer time. It is also more unpleasant than the sensation of fast pain.

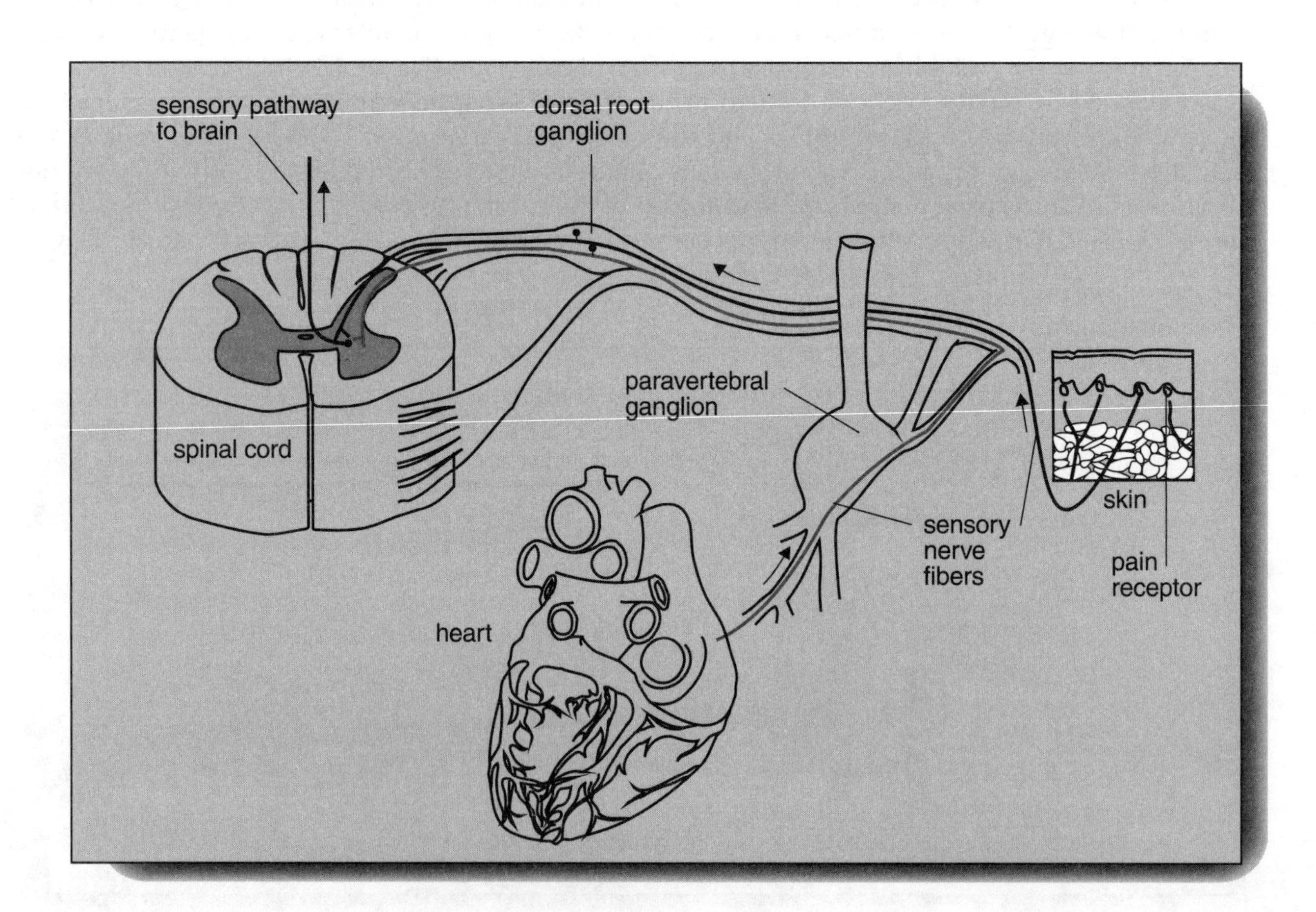

FIGURE 5.5

VISCERAL AND REFERRED PAIN

Pain sensations generated from receptors located in the visceral organs are difficult to localize. Frequently such visceral pain may *appear* to be sensed in one body location in the skin when the actual pain source is generated in another location within the internal organs. Such an effect is termed *referred pain*. For example, pain sensations originating in the heart are perceived to be in the left arm and shoulder, or pain sensations originating in the kidney are perceived to be located in the lower back region. Referred pain is produced because the pain tracks in the CNS that conduct the stimuli receive sensory information from both peripheral sensory nerves originating in the skin, and visceral sensory nerves originating in the internal organs **(Figure 5.5)**. Since the sensory cortex interprets the information as originating in the skin, even if it originates in the visceral organs, it then projects the sensation back to the skin surface. The person thus perceives the pain of a myocardial infarction as originating in the left arm and shoulder.

THE SENSE OF SMELL

The sense of smell (olfaction) is a chemical sense and is probably the least understood of all of our senses. Olfactory receptors are located in patches of olfactory mucosa in the ceiling of the nasal cavity. There are three types of cells that are found in the olfactory mucosa **(Figure 5.6)**:

1) **Supporting Cells** — secrete mucus which coats and protects the nasal mucosa.

2) **Basal Cells** — precursors of new olfactory receptor cells. Replacement of new cells occurs at approximately two month intervals. Since the olfactory cells are actually afferent neurons this is an example of a nerve cell that undergoes cell division. This is quite uncommon since nerve cells almost never divide once differentiation has been completed.

3) **Olfactory Receptor Cells** — are bipolar neurons with specialized chemoreceptors located on the cilia that project into the nasal cavity. Axons of these cells pass through tiny openings in the cribriform plate of the ethmoid bone and synapse in the olfactory bulbs **(Figure 5.7)**. Axons from the nerve cells in the olfactory bulbs then combine to form the olfactory nerve (Cranial Nerve I) and travel by the:

a) *Subcortical route* to the limbic system and the primary olfactory cortical regions located in the lower medial sides of the temporal lobes; and the

b) *Thalamic-Cortical route* into the hypothalamus. Thus, sensations of smell can be coordinated with behavioral reactions associated with such actions as direction orientation, feeding and mating. This route is also of importance for conscious perception of odors and discrimination of odors.

Olfactory neurons are in direct contact with the external environment. This contact eventually results in permanent damage to these nerves, which are then usually not replaced, especially with age. It has been estimated that as people grow older they lose 1% of their olfactory sensitivity each year. Some very old individuals may be quite uninterested in eating as a direct result of their inability to smell the food. Apparently taste and smell are processed by some combined mechanism in the cerebrum and thus the lack of smell, produces a significant reduction in the taste of food as well.

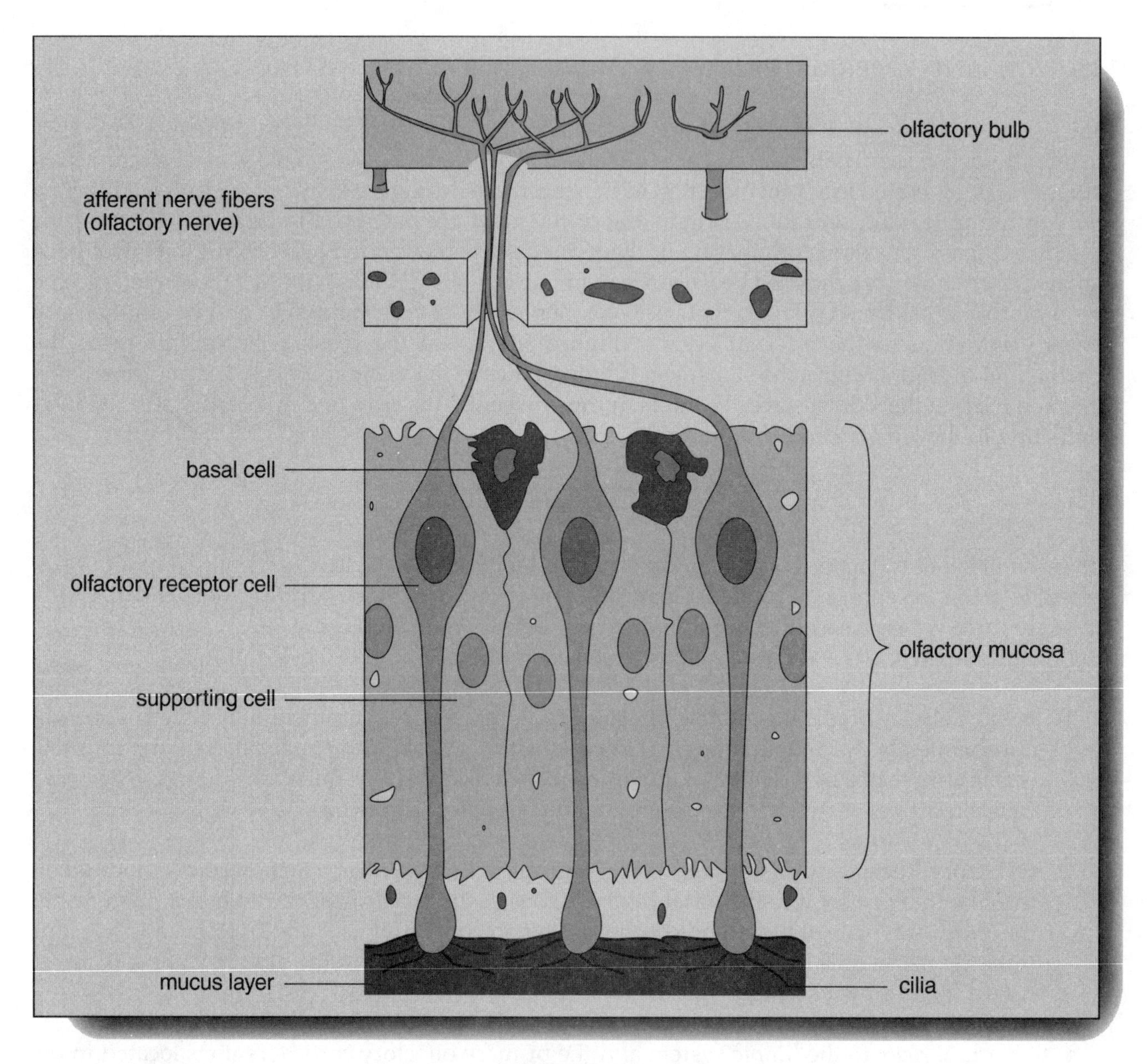

FIGURE 5.6

MECHANISMS INVOLVED IN THE RECEPTION OF ODORS

The ciliary projections from the olfactory cells contain the receptors for odor molecules. During normal respiration there is no air movement over the olfactory patches which are located above the normal air flow path. This limits olfaction sensitivity since odor molecules can only reach the receptors by diffusion. Sniffing increases the movement of odor molecules onto the olfactory receptors because it acts to increase air movement to the top of the nasal cavity.

For a molecule to be sensed as an odor, it must be mixed with the inspired air as a volatile substance. In addition, the molecule must also be water soluble so that it can dissolve in the mucus layer that covers the olfactory mucosa. According to current understanding, the molecule then binds to specialized receptors located on the olfactory cilia. This then opens Na^+ and K^+ channels in the membrane and generates a depolarizing receptor potential. Current flow then acts to generate action potentials in the electrically sensitive nerve membrane. Action potential frequency depends on the concentration of the initiating chemical's molecules.

SENSITIVITY AND SENSORY ADAPTATION OF OLFACTORY CELLS

In comparison to other species, humans have a very poorly developed sense of smell. Nevertheless human olfactory receptors can distinguish very minute concentrations of odor molecules. For example, the chemical methyl mercaptan (which smells like very strong garlic), at a dilution of 1 molecule to 50,000 molecules of air, can be sensed by human olfactory receptors.
Olfactory cells undergo rapid sensory adaptation and thus within one second of perceiving a particular odor, response has decreased by almost 50%, and after 1 minute the receptors become almost insensitive to the odor; however, their sensitivity to other odors remains unchanged.

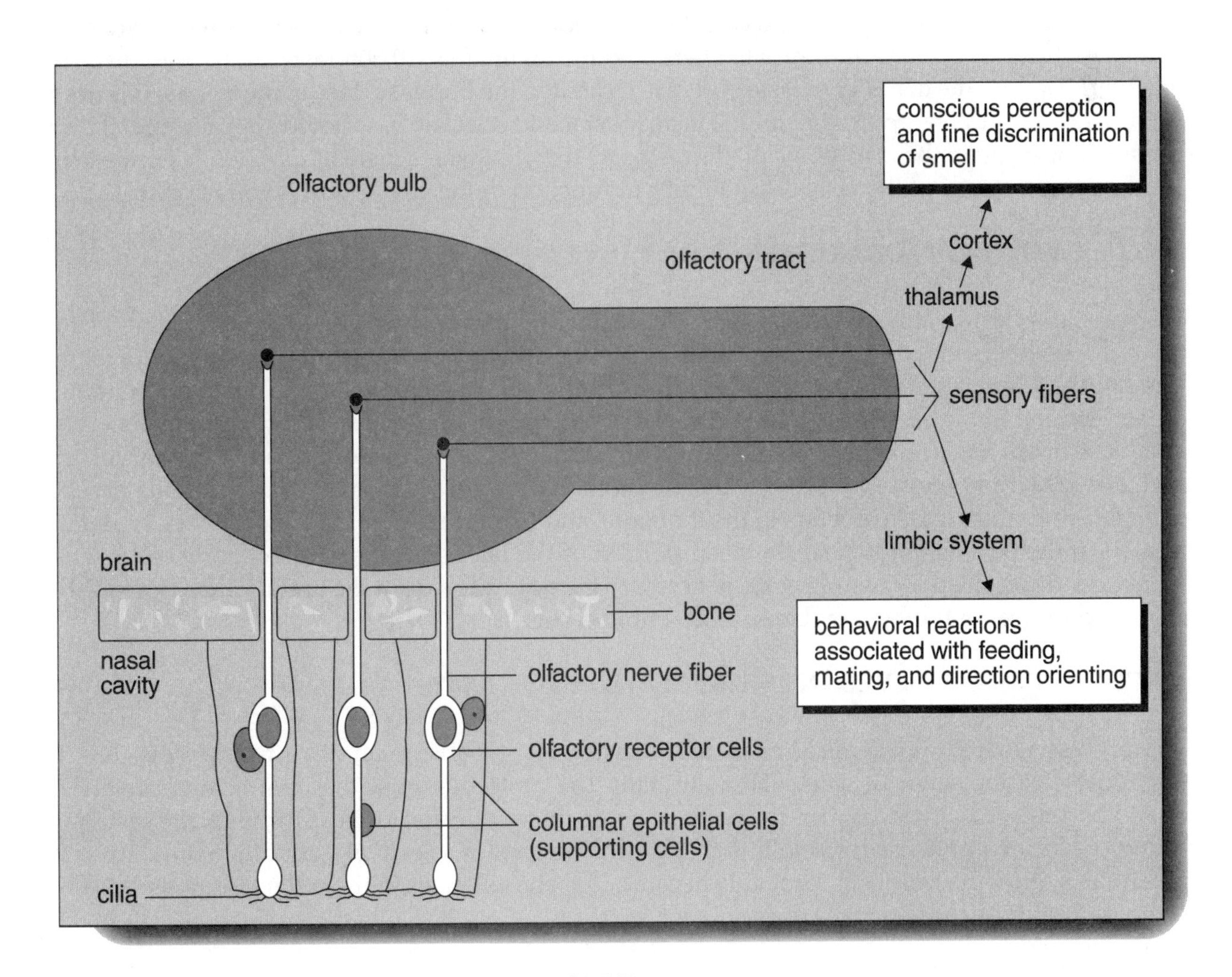

FIGURE 5.7

DISCRIMINATION OF ODORS

The human olfactory system can discriminate between thousands of odors. It has been suggested that the perception depends on combining primary odor sensations. The primary odors are believed to be the following:

Camphoraceous	**Pepperminty**
Musky	**Pungent**
Floral	**Putrid**

It is believed that molecules with similar odors share similar three-dimensional configurations. They are therefore able to attach to the binding sites of a particularly shaped receptor and stimulate the associated olfactory cell.

THE SENSE OF TASTE

The sense of taste (gustation) depends on chemoreceptors located in the taste buds which are present within the oral cavity and the throat. There are approximately 10,000 taste buds present in these regions; however, the majority of taste buds are located in the upper surface of the tongue **(Figure 5.8)**. Each taste bud contains approximately 50 taste (gustatory) cells and associated supporting cells. There is a small opening in the taste bud called a *taste pore* through which taste hairs extend. The taste hairs are the receptor ends of the taste cells (similar in function to the cilia found on olfactory cells).

NERVE PATHWAYS FOR THE SENSE OF TASTE

Each gustatory cell is attached to a sensory nerve fiber that travels in a particular cranial nerve to the brain. Those fibers that originate from receptors in the anterior two-thirds of the tongue travel in the facial nerve (cranial nerve VII); those fibers that originate from receptors in the posterior one-third of the tongue and in the back of the mouth travel through the glossopharyngeal nerve (cranial nerve IX); and those fibers that originate from receptors located at the base of the tongue and the pharynx travel in the vagus (cranial nerve X). Sensory information on taste passes through these cranial nerves and enters the medulla. It then ascends to the thalamus and from there enters the gustatory cortical areas located in the parietal lobes near the deep portion of the lateral sulcus. Unlike many other sensory pathways, these olfactory pathways do not cross to the opposite side of the brain. Fibers from the brain stem also project to the hypothalamus and limbic systems.

MECHANISMS INVOLVED IN THE RECEPTION OF TASTE

The process whereby a chemical molecule is tasted is probably much like the process described previously for the sense of smell. Thus, initially the molecule must dissolve in the saliva. These dissolved molecules then contact the taste hairs or microvilli that project through the taste pores. Receptors located on these microvilli then bind the molecules selectively. Binding stimulates action potentials within the taste cells, probably because it opens ion gates in the cell membranes producing receptor potentials.

Receptors for the sense of taste are frequently damaged by exposure to potent chemicals, but these receptors (unlike the olfactory receptors) are capable of rapid regeneration. Taste receptors appear to be replaced every ten days. Accordingly, epithelial cells differentiate into both supporting cells and gustatory cells to maintain functional taste buds.

DISCRIMINATION OF TASTES

There are at least four primary tastes:

Sweet	**Sour**
Salty	**Bitter**

In addition, there are possibly two other taste sensations:

Alkaline	**Metallic**

Receptors for these tastes are grouped on the tongue as shown in **Figure 5.11**.

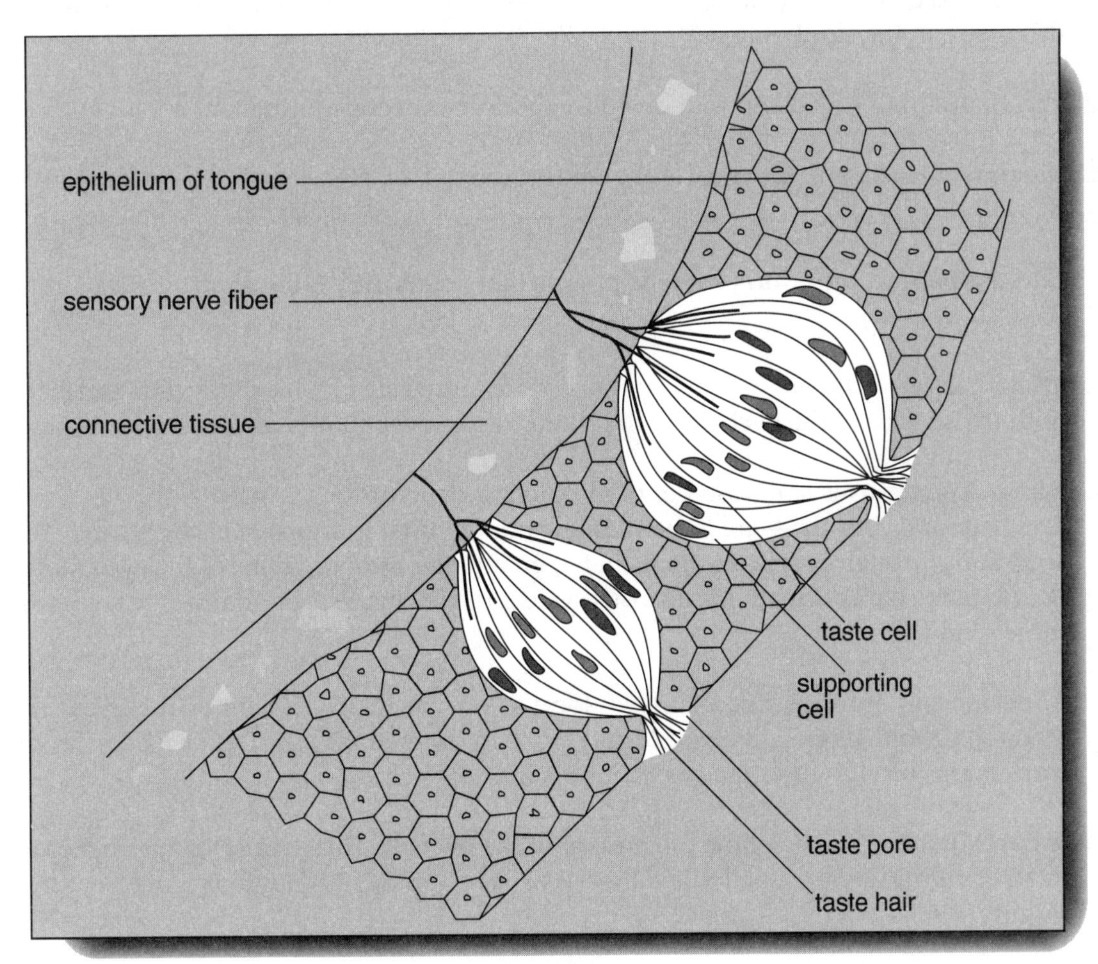

FIGURE 5.8

Sweet receptors are concentrated near the tip and are stimulated by organic substances such as sugars and polysaccharides.

Sour receptors are located along the margins of the tongue and are stimulated by the H^+ ions in acids.

Salt receptors are present near the tip of the tongue and in the upper front portion and are stimulated by ionized inorganic salts. It is believed that the positively charged ions (Na^+) are primarily responsible for stimulating the receptors.

Bitter receptors are located across the rear upper section of the tongue and are stimulated by various chemicals, especially organic compounds. Certain inorganic ions including magnesium and calcium also stimulate these "bitter" receptors. Other substances that are capable of stimulating the "bitter" receptors are the plant alkaloids, such as nicotine or morphine.

VISION

The eye transduces light into action potentials. Associated with the eye are a series of *visual accessory organs* as follows:

The Orbital Cavity — this cavity, formed by the bones of the skull is lined with periosteum from the bones, and contains fat, blood vessels, nerves and connective tissue. The eyeball is located in this space in the skull bones.

The Eyelid (palpebra) — is composed of skin, muscle, connective tissue and conjunctiva.

The Orbicularis Oculi — is the muscle that closes the eyelid; it is controlled by Cranial Nerve VII (Facial).

The Levator Palpebrae Superioris — is the muscle that opens the eyelid; controlled by Cranial Nerve III (Oculomotor).

The Conjunctiva — is a mucous membrane that lines the inner surfaces of the eyelids and is continuous with the surface of the eyeball. In the center of the eyeball it is replaced with the *cornea.*

The Lacrimal Apparatus is composed of the following structures:

The Lacrimal Glands — are located on the upper, outer margins of each eye; they release tear secretions through a series of ducts that open into the upper regions of the eye. Stimulation by nerves of the parasympathetic nervous system results in the production of tear secretions from these glands.

Superior and Inferior Canaliculi — located on the inner corners of the eyes. Tears enter through the openings called *puncta* and are then drained into the *Lacrimal Sacs*. From here the tears drain through the *Nasolacrimal Ducts* into the nose.

Extrinsic Eye Muscles — control the movement of the eyeball. These muscles are anchored into the outer covering of the eyeball called the *Sclera*. Here are six muscles that control eye movements **(Figure 5.9)**:

Superior Rectus — rotates the eye upward and inward. Innervated by Cranial Nerve III (Oculomotor).

Inferior Rectus — rotates the eye downward and inward. Innervated by Cranial Nerve III (Oculomotor).

Medial Rectus — rotates the eye inward. Innervated by Cranial Nerve III (Oculomotor).

Lateral Rectus — rotates the eye outward. Innervated by Cranial Nerve VI (Abducens).

Superior Oblique — rotates the eye downward and outward. Innervated by Cranial Nerve IV (Trochlear).

Inferior Oblique — rotates the eye upward and outward. Innervated by Cranial Nerve III (Oculomotor).

ANATOMICAL FEATURES OF THE EYE

The eyeball is hollow and is composed of three layers or tunics **(Figure 5.10)** as follows:

1) **The Outer Tunic (fibrous tunic)** — is composed of connective tissue plus epithelium. Approximately 1/6 is *cornea* and 5/6 is *sclera*. The cornea is the transparent front (anterior)

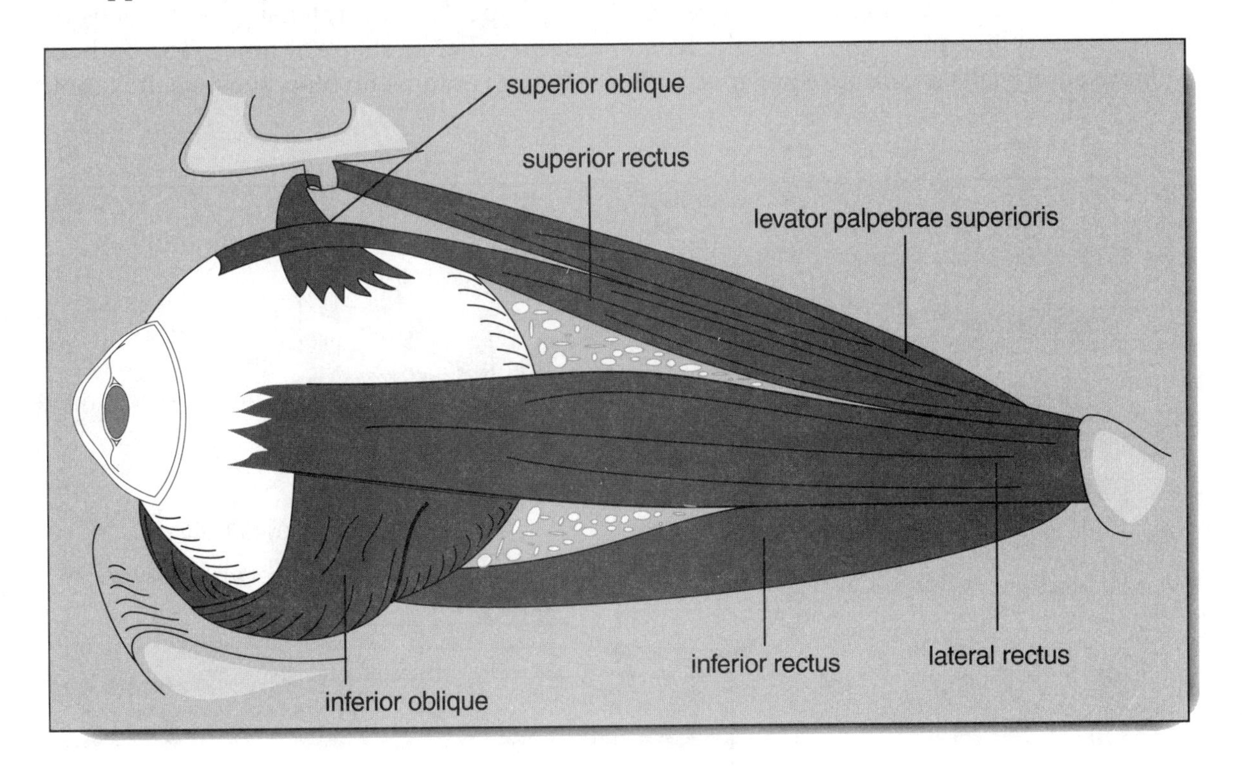

FIGURE 5.9

portion of the eyeball and allows light to pass into the interior. In addition, the cornea also partially focuses the image. The cornea is well supplied with nerves and many pain receptors, but contains few cells and no blood vessels. The sclera is the white portion of the eyeball and is composed of collagenous and elastic fibers. It serves as the attachment for the eye muscles. The *optic nerve* exits from the eye by passing though the sclera on the back (posterior) side of the eyeball. In addition, blood vessels enter and leave the eyeball from the rear.

2) **The Middle Tunic (vascular tunic); uveal layer**—is composed of *the choroid coat*, and *the ciliary body* (including the lens).

a) The choroid coat covers the posterior 5/6 on the inside of the eyeball. It is highly vascularized and contains numerous melanocyte. The production of the melanin pigment by these cells gives the choroid a brown/black color and prevents internal light reflection.

b) *The ciliary body* extends from the choroid and forms a ring-like structure in the front of the eye that surrounds the lens. Radiating folds called *ciliary processes* originate in the ciliary body and extend towards the central lens. Ciliary muscles also originate in the ciliary body. Attached to the ciliary processes are *suspensory ligaments* (or zonular fibers) which hold the lens in place. The suspensory ligaments are attached at their distal ends around the circumference of the lens capsule.

c) *The Lens* is covered by a thin transparent lens capsule. This capsule is elastic and is thus under constant tension forcing the lens into a globular shape. However, tension on the lens by the suspensory ligaments can cause the lens to assume a flatter shape at times. The body of the transparent lens is composed of specialized cells but contains no blood vessels. It is centrally

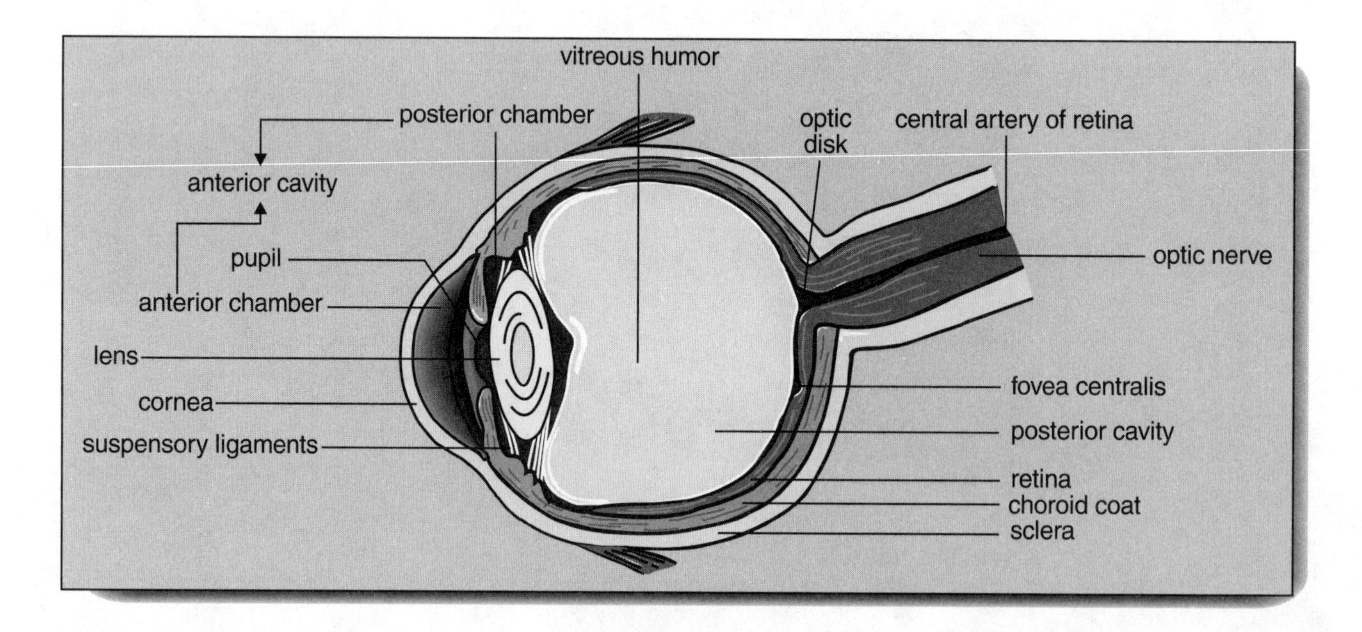

FIGURE 5.10

positioned behind the iris. Cells in the lens undergo differentiation into columnar cells called lens fibers.

d) *Shape changes in the lens* can be accomplished by changing the tension on the lens capsule through the action of the suspensory ligaments. Relaxation of these ligaments will cause the lens to thicken and become more convex, while increasing the tension on these ligaments will flatten the lens making it less convex. The lens becomes more convex during the process of close focusing (termed *accommodation*). When the eye views a close object, the ciliary muscles contract. This results in relaxation of the suspensory ligaments causing the lens to become thicker and more convex.

e) *The Iris Diaphragm* is the colored part of the eye composed of connective tissue. It separates the *anterior chamber* from the *posterior chamber*. (Note that in the anterior chamber and posterior chamber are together classified as the *anterior cavity*). The *posterior cavity* is located at the rear of the eye behind the lens. A watery fluid termed *aqueous humor* fills the anterior cavity, maintains intraocular pressure, and supplies nutrients to the various structures. Within the posterior cavity is found the jelly-like *vitreous humor* that supports and maintains the shape of the eyeball and holds the retina in place. The *pupil* is the circular opening through which light enters the eye to fall on the retina. The pupil can open and close to regulate the amount of light that falls on the retina. Size changes of the pupil are under the control of the iris diaphragm. Smooth muscles of the iris diaphragm are arranged as circular and radial fibers. Stimulation by the parasympathetic nerves that innervate the circular fibers results in constriction of the pupil. This type of pupillary reflex takes place in bright light. On the other hand, stimulation by sympathetic nerves that innervate the radial fibers results in pupil dilation (as in dim light).

f) *The production and removal of the aqueous humor* is a continuous process. Aqueous humor is formed by the epithelium on the inner surface of the ciliary body. It then flows out of the posterior chamber by passing around the lens and though the pupil into the anterior chamber. The pressure of the aqueous humor against the inner side of the cornea maintains the curvature of the cornea and of the anterior chamber. Aqueous humor also nourishes the cells of the cornea and of the lens, since it supplies oxygen and nutrients to these structures and removes waste products. However, since it is clear, it does not impede the transmission of light into the eye (as would blood). The aqueous humor is then removed by passing into veins and also into the *Canal of Schlemm*. If the production of aqueous humor exceeds its removal, this will result in an increase in the intraocular pressure, leading to the disease *glaucoma*. In advanced cases of glaucoma this pressure increase is transmitted to the rear of the eye, damaging the retina and blood supply and causing blindness.

3) **The Inner Tunic (nervous tunic)** — is the retina containing the photoreceptor cells and the lining of the inner side of the eyeball. The retinal layer is actually composed of a pigmented layer of epithelium, a neuronal layer containing nerves and nerve fibers, and a limiting membrane.

There are five major groups of retinal neurons:

The light receptor cells, which are the rods and cones
The bipolar neurons **The ganglion cells**
The horizontal cells **The amacrine cells**

The receptors, bipolar and ganglion cells form a direct pathway for impulses triggered in the receptors to pass into the optic nerve. The horizontal and amacrine cells are involved in further information processing within the retina. Axons derived from the ganglion cells form the optic nerves that exit from the rear of each eye and enters the brain.

a) *The Optic Disk* is an area in the retina which is also called the *blind spot* because there are no rod or cone cells here. It is the region where the nerve fibers from the retinal cells leave the eye to become the optic nerve. The central artery and vein that supply the capillary network of the retina also pass out of the eye at this location.

b) *Macula lutea* is centrally located in the retina and appears as a yellowish spot approximately 1 mm^2. In the center of the macula is a depression called the *fovea centralis*. The fovea centralis contains only cone cells, while the macula lutea contains a mixture of rod and cone cells. The concentration of cones decreases as they move farther away from the macula while the concentration of rods increases. Cone cells only function in bright light, for sharp color day vision, while rods only function in dim light, for gray indistinct night vision. The sharp image of an object seen in bright light is produced mainly by the receptors in the macula and especially the fovea. The remainder of the retina is not utilized to any great extent for day vision. In addition, the image of an object located in the center of the visual field will be projected by the lens directly on the fovea, thus facilitating a sharp color image of object. On the other hand, if you wish to view an object (such as a star) in very dim light, looking directly at this object will cause its image to fall directly on the fovea. Since the fovea is not sensitive to dim light, the object will be difficult, or impossible to see. However, if you view the object with peripheral vision (from the corner of the eye), the image will fall on the part of the retina that contains a high concentration of rods and you will be able to see the object (although not in sharp detail, and not in color).

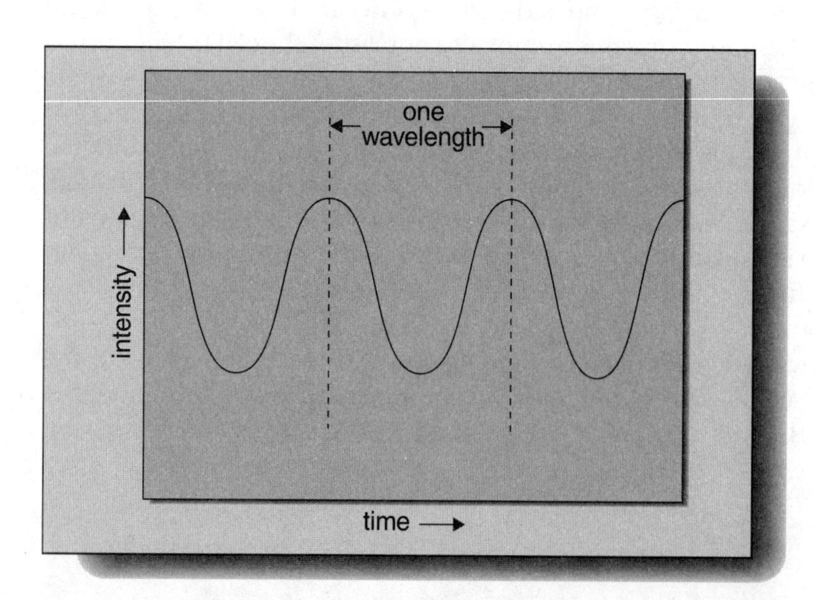

FIGURE 5.11

THE MECHANISMS INVOLVED IN VISION

Light and refraction. Light is a form of electromagnetic radiation composed of particle/packets of energy called *photons*. Photons travel in a wave-like pattern of vibrations, thus the distance between two wave peaks is termed the *wavelength* **(Figure 5.11)**. The wavelength of the photon determines the electromagnetic spectrum which includes a variety of other forms of electromagnetic radiation (radio, TV, microwave, infrared, visible, UV, X-ray, gamma ray, cosmic ray). The visible wavelengths (from 700 nm to 400 nm) are the only wavelengths that stimulate the photoreceptors in the retina. In addition, the color of the light perceived is dependent on the wavelength (red = 700 nm range, yellow = 600 nm range, green = 500 nm range, blue = 400 nm range). Different kinds of cone receptors are sensitive to different wavelengths, thus, the presence of red, green and blue cones (see following material). To rephrase for clarity, short waves are sensed as violet and blue, while longer waves are sensed as orange and red. Very long waves (infrared = 1000 nm range) are sensed as heat by receptors in the skin, but cannot be "seen" as "light" by the retinal cells.

Light intensity. This relates to the size or amplitude of the wave. Thus a low amplitude wave (dim light) would only stimulate the rod cells but not the cone cells, while a larger amplitude wave (bright light) would stimulate certain cone cells (depending on the "color" of the light).

Refractions. Light (in the form of photons) is generated by a light source. The photons radiate from the source and their forward motion in a particular direction is termed a light ray. These light rays may pass directly into the eye, or first bounce off some object (reflection) and then enter the eye. In order to produce a sharp image, divergent rays that reach the eye must first be bent so that they will all converge to a single point of focus on the retina. This bending of the rays is termed refraction. Refraction of light rays takes place when the rays pass from one kind of conducting medium (air for example) into another (glass, water, the lens of the eye and so on). The degree of refraction is determined by the density of the medium and the angle that the ray strikes the second medium. In a lens, the type of curvature determines the direction of the refraction. Thus, a concave lens causes the light rays to diverge and never reach a point of focus, while a convex lens causes the light rays to converge to a focal point.

Light entering the eye must pass through both the cornea and the convex lens and thus the light is refracted and focused on the retina. The resulting image of the object is projected upside-down on the retina, but is reversed during information processing within the occipital lobes. If the focal point falls in front of, or behind the lens, the image will be blurred and will require correction with glasses or contact lenses. On the other hand, if the focal point of the image falls directly on the retina, it will be sharply defined.

1) In a *normal eye (Emmetropia),* the image of a distant object will be focused on the retina without accommodation, while the image of a close object will be focused on the retina with accommodation.

2) If the eyeball is too long or the lens is too strong, this causes *nearsightedness (Myopia)* because the image of a distant object will be focused in front of the retina (where the retina would be in a normal length eye) with no accommodation. However, the image of a close object will be

focused on the retina, but accommodation is not required, as it would be in a normal length eye. To correct this, a concave lens is placed in front of the eye. This allows the image of a distant object to be focused on the retina with no accommodation, while the image of a close object will also be focused on the retina, but now accommodation is required.

3) If the eyeball is too short or the lens is too weak, this causes *farsightedness (Hyperopia)* because the image of a distant object is focused on the retina with accommodation, but the image of a close object is focused behind the retina, even with accommodation. To correct for this, a convex lens is placed in front of the eye. This allows the image of a distant object to be focused on the retina with no accommodation and the image of a close object to be focused on the retina with accommodation.

4) Other ocular disorders include:

   a) *Presbyopia* in people age 45-50 or so, when the lens loses elasticity. As a result it can no longer undergo accommodation and the individual must use corrective lenses for close vision.

   b) *Cataracts* when the transparent fibers of the lens become opaque. The defective lens is removed by surgery and an artificial lens is substituted.

   c) *Astigmatism* where the curvature of the cornea is uneven. This means that light rays are refracted in a disorganized way and cannot come to a common focal point. Special corrective lenses can be used to correct this problem.

VISUAL RECEPTORS

The photoreceptors of the retina are designed to transduce light energy into action potentials. These receptor cells, which are the rods and cones, are actually an extension of the CNS. The neurological arrangement of the retina is such that light entering from the lens must first pass through the various layers and cells (ganglion, amacrine, bipolar, horizontal) before actually impacting on the rods and cones located at the extreme back of the retina. In addition, an individual cone cell synapses with an individual bipolar cell and then an individual ganglion cell from which an individual sensory fiber is projected into the CNS. Each cone cell thus has a "private line" into the CNS through which it can send information. On the other hand a number of rod cells must share the same bipolar cells and a single ganglion cell. Thus, information from all these rod cells passes though the same "party line" into the CNS and brain. This means that if a photon falls on a cone cell the brain knows exactly where this stimulus originates, but if a photon falls on a rod cell the brain has only a vague knowledge of where the stimulus originated on the retina. For this reason, cone vision (color, bright light) produces sharp images in the brain, but rod vision (gray, dim light) produces only a blurred image in the brain.

Both the rod and cone cells possess certain common structural features, but rods have a larger outer segment than do cones. In addition to the outer segment, photoreceptor cells also contain an inner segment where cellular metabolism takes place, and a synaptic terminal where the stimulus is transmitted to the bipolar cell. In orientation, the outer segments of both the rods and cones face the choroid coat (pigmented layer) of the retina. These outer segments are composed of a series of stacked discs where the light sensitive photopigments are concentrated. There are four types of photopigments,

and all are composed of the same *Retinal*, but four different kinds of *Opsins*. In the rods this combination of retinal plus the specific opsin is called *Rhodopsin*, while in the cones the retinal is combined with three different opsins that can discriminate between the different wave lengths (determined as the colors red, green, and blue). In rods the rhodopsin cannot discriminate between different wave lengths in the visual spectrum, but is instead able to distinguish between varying levels of light intensity. Thus, rod vision is perceived only in shades of grays.

The opsin portion of the photopigment is actually an inactive enzyme protein which can be activated when light strikes the pigment. Here are some steps.

1. When a photon in the visual range strikes the photopigment (retinal-opsin) this produces a conformational change in the retinal.
2. The conformational change in the retinal activates the enzymatic activity of the opsin.
3. This results in an enzymatic cascade and eventually results in the activation of the substance cyclic-GMP.
4. The cyclic-GMP closes Na^{+} channels in the membrane of the outer segment.
5. This leads to a hyperpolarization which propagates to the synaptic terminal.
6. Ca^{++} channels in the synaptic terminal close due to the hyperpolarization.
7. Neurotransmitter release decreases because the Ca^{++} levels fall in the synaptic terminal.
8. Neurotransmitter binding is reduced at the postsynaptic receptors of the bipolar cells.
9. The greater the light intensity, the greater is the hyperpolarization and the subsequent inhibition of neurotransmitter release.
10. In the absence of neurotransmitter binding to the receptors of the bipolar cells, there is no longer any inhibitory influence on the bipolar cells.
11. The bipolar cells become excited and stimulate the ganglion cells. Action potentials are now conducted down the optic nerves and into the brain.

RODS AND DARK ADAPTATION

There are 100 million rods and only 3 million cones in each eye. Because of the differential sensitivity of the rods versus the cones, rods are used for vision in dim light while cones are used for color vision in bright light. In addition, cones provide high resolution because individual cone cells are linked to the vision centers of the brain via individual nerve fibers (they have a very small receptive field as compared to rods).

In bright light nearly all of the rhodopsin in the rods is decomposed and thus the sensitivity of the rods is greatly reduced. In dim light, rhodopsin can be regenerated faster than it is broken down. This process of regeneration (or dark adaptation) requires the use of cellular ATP and takes between 15 and 30 minutes. As a result, light sensitivity increases about 100,000 times when dark adaptation is completed.

Since retinal (a component of rhodopsin) is synthesized from vitamin A, a person with a vitamin A deficiency, may have less rhodopsin than normal in their rods and may suffer from a condition known as night blindness. This is usually correctable with vitamin A treatment.

COLOR VISION

There are three types of opsin that compose the photopigments in the three types of cones. Thus the three types of photopigments are sensitive to three different wavelengths of light. When white light strikes an object, certain wavelengths are absorbed and certain ones are reflected. If red light is reflected, the object will appear red, assuming that this light stimulates the cones that respond in the red range. Similarly if blue light is reflected, this will stimulate the blue cones and the object will appear blue. Of the three types of cones (red or *erythrolabe*, green or *chlorolabe*, blue or *cyanolabe*), each cone is maximally stimulated by a particular wavelength. Perception of colors is actually accomplished because combinations of cones are stimulated together (known as the *ratio of stimulation*).

For example, the sensation of red results because only red cones are stimulated (ratio of stimulation: red 100%, green 0%, blue 0%), but the sensation of yellow results because red and green cones are stimulated together (red 83%, green 83%, blue 0%).

COLOR BLINDNESS

This is a genetic disorder and results in the lack of one or more of the types of color cones. Such individuals cannot distinguish colors. In one form the individual cannot distinguish red and green. In complete color blindness the individuals may only see shades of blue or gray. Color blindness is more frequently present in males than in females because certain of the genes involved are present on the X chromosome.

STEREOSCOPIC OR BINOCULAR VISION

This is the process by which an individual can perceive depth of focus, distance of an object from the observer and height and width of the object. The ability to accomplish this results from the fact that the images generated by the two eyes overlap but are not exactly identical. The difference in images results from the fact that the two pupils are about 7 cm apart and thus close objects (less than 20 feet distant) produce slightly different images on the retinas. The primary visual cortex is organized into functional columns that separately process information from different areas of the retina. Information from the two eyes processed in different columns are then merged to enhance depth of perception. If the two views are not successfully merged, this results in double vision or *diplopia*. In addition, the brain can also use other information such as size or position to determine the distance of objects.

VISUAL NERVE PATHWAYS

The retina in each eye is split into inner and outer segments (or left and right halves) . Nerve fibers from the outer segments (left half of the left eye and right half of the right eye) pass into the brain but do not cross, remaining on the same side and synapsing in the *lateral geniculate bodies* on the same side. Thus, the fibers from the left half of the retina in the left eye synapse in the left lateral geniculate body, while fibers from the right half of the retina in the right eye synapse in the right lateral geniculate body. Fibers from the inner sides of each retina (the right half of the left eye and the left half of the right eye) pass into the brain and cross to the opposite sides at the optic chiasma. From the lateral geniculate bodies on each side, the information passes through the *optic radiations* to the visual cortexes, located on each side in the occipital lobes, where it is processed.

Information originating from the inner halves of each retina ends up in the opposite visual cortex, while information from the outer halves of each retina end up in the visual cortex on the same side. The overall outcome is that:

1) An object present in the right visual field is projected into the left visual cortex because it stimulates the left half of the retina in the left eye (which stimulates the left visual cortex), and the left half of the retina in the right eye (which also stimulates the left visual cortex). An object seen in the left visual field will likewise be projected into the right visual cortex.

2) Peripheral vision results from stimulation of the inner halves of the retinas in both eyes. Thus, if the optic chiasmas are damaged the individual will have no peripheral vision but only tunnel vision. This is because the outer halves of each retina will still be functional, but the inner halves will be effectively blind since the tracks are cut.

HEARING AND BALANCE

HEARING

Sound is created by the mechanical vibration of some object. These vibrations then generate pressure waves in a conducting medium (as, for example, air). Such waves are then conducted to the ear where they are interpreted as sound. For most sounds the common conducting medium is air, however solids and liquids can also act as conducting media. In the absence of such conducting media, pressure waves would not be transmitted from the source to the ear (as, for example, in the complete vacuum of space).

PITCH (OR TONE) OF THE SOUND

The pitch of the sound is determined by the frequency of the vibrations as expressed in cycles/second (or Hertz). The greater the frequency of the vibrations, the higher is the pitch of the sound. The human ear can be stimulated by frequencies of sound from 20 Hertz to 20,000 Hertz. It is most sensitive to frequencies of sound in the range of 1000 Hertz to 4000 Hertz (where normal speech sounds are generated).

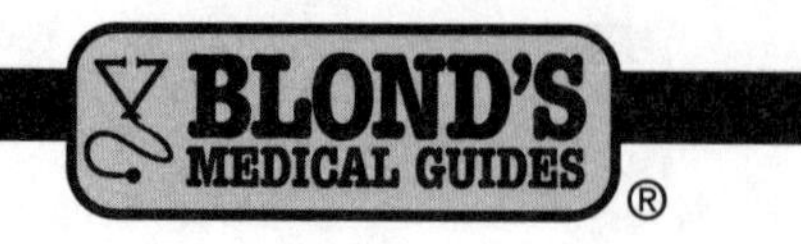

INTENSITY OF THE SOUND

The intensity or loudness of the sound is determined by the amplitude of the vibration. The greater the amplitude, the louder is the sound sensed by the receptors in the inner ears. Amplitude is measured in decibels (dB) which are measured on a logarithmic scale. Thus, a sound that is 20 dB is 100 times louder than a sound that is 10 dB.

The ear is divided into external, middle and inner parts.

EXTERNAL EAR

The outer funnel-like structure is the *auricle* which traps and concentrates the vibrations from the air. Attached to this is the *external auditory meatus* which is approximately 2.5 cm long, passes through the temporal bone and ends at the *tympanic membrane*. Modified sweat glands (*ceruminous glands*) secrete wax that protects the walls of the external auditory meatus. Pressure waves that enter the external auditory meatus are transmitted to the tympanic membrane through the air that fills this tube. These pressure waves then generate vibrations of the tympanic membrane (eardrum) which are then transmitted into the middle ear.

MIDDLE EAR

The middle ear is composed of the tympanic membrane, the tympanic cavity filled with air, which lies directly behind the eardrum, and the three auditory ossicles. In addition, the middle ear cavity is connected to the back of the pharynx by the eustachian (or auditory) tube. The tympanic membrane stretches over the end of the external auditory meatus and on its inner side is attached to one of the first auditory ossicles named the *malleus* (or hammer). The tympanic membrane is a semitransparent layer of skin and mucous membrane, and when sound waves strike it, the vibrations are transferred to the malleus. Connected to the malleus is the second auditory ossicle called the *incus* (or anvil), and in contact with this bone is the third ossicle called the *stapes* (or stirrup). Thus, when the malleus vibrates it causes the incus to vibrate and this causes the stapes to vibrate. Since the stapes is in contact with the *oval window* of the inner ear, these vibrations then enter the inner ear where they are converted to action potentials.

SOUND AMPLIFICATION DURING TRANSMISSION

In addition to transmitting vibrations from the tympanic membrane to the inner ear, these bones are arranged to form a lever type system resulting in amplification of the vibrations. Amplification also takes place because the surface area of the tympanic membrane is much larger than that of the oval window. Thus, a smaller pressure at the tympanic membrane is converted to a larger pressure at the oval window because:

$$\text{pressure} = \text{force}/\text{unit area}.$$

THE TYMPANIC REFLEX

This is a protective reflex and is necessary to prevent the delicate hair cells of the inner ear from damage due to loud sounds (above 70 dB). Tiny skeletal muscles attached to two of the auditory ossicles mediate this reflex. One of these is the *tensor tympani* muscle which is attached to the malleus bone, and the other is the *stapedius* muscle attached to the stapes bone. These muscles contract when a loud sound is sensed by the hair cells in the inner ear. Upon contraction, the auditory ossicles rigidly contact each other, and this reduces vibrations conducted through these bones to the inner ear. In addition, the tensor tympani muscle maintains constant tension on the eardrum to allow it to vibrate effectively.

THE EUSTACHIAN OR AUDITORY TUBE

This tube connects the middle ear cavity to the back of the pharynx and is closed by a valve-like flap at the pharyngeal end. It allows pressure equalization to take place between the middle ear cavity and the outside atmosphere. If there is unequal pressure on the two sides of the eardrum, the membrane will bulge either inward or outward. This will limit the ability of the eardrum to vibrate effectively and will thereby reduce hearing sensitivity. Swallowing usually equalizes the pressures (causing the ears to "pop"). In addition, chewing and yawning also opens the valve and allows the pressure to equalize.

Because the mucous membranes are contiguous throughout the entire system, an infection in the pharynx may be passed up into the middle ear cavity. Blowing the nose very strongly may also force infected material into the eustachian tubes and thus promote ear infections.

THE INNER EAR

Within the temporal bone of the skull is a series of complex interconnected chambers known as the *bony or osseous labyrinth*. Centered within the bony labyrinth is the membranous labyrinth **(Figure 5.12)**. The

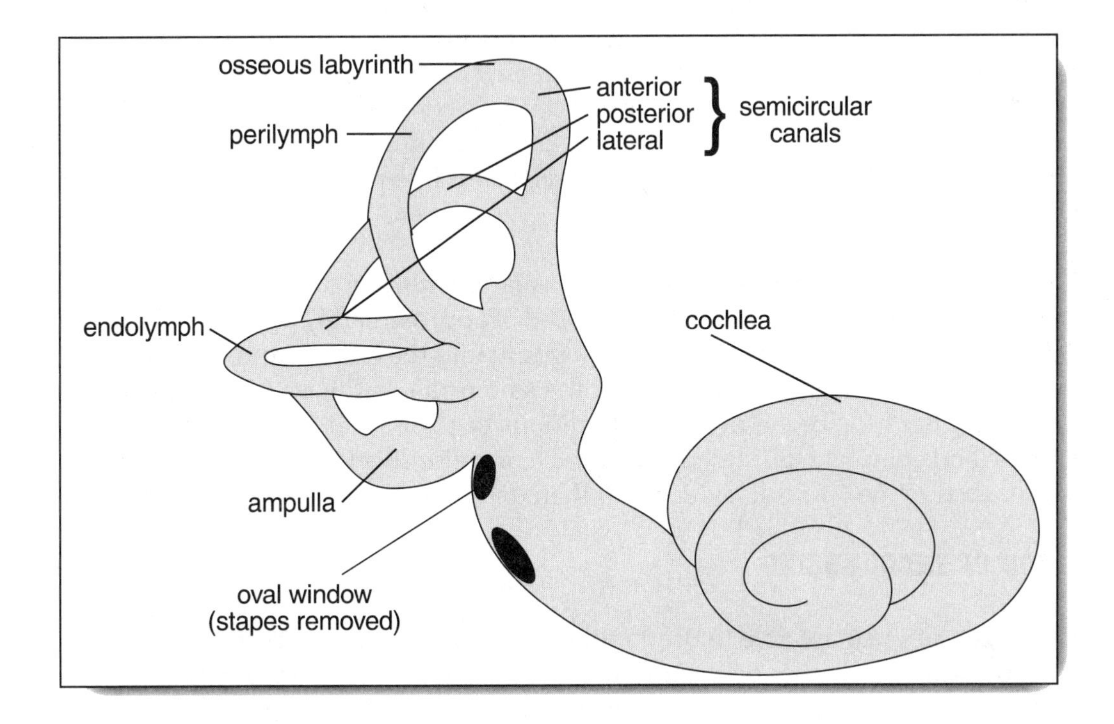

FIGURE 5.12

space between the osseous and membranous labyrinth is filled with a watery fluid called *perilymph*, while the membranous labyrinth itself contains the fluid *endolymph*. The labyrinths can be divided into the part that is involved in hearing (*the cochlea and cochlear duct*) and the part that is involved in equilibrium (*the semicircular canals and the vestibule*).

THE COCHLEA AND COCHLEAR DUCT

The cochlea and cochlear duct are involved in the process of transduction of the sound vibrations into action potentials. Specifically, the cochlea is the bony labyrinth filled with perilymph and the cochlear duct, located inside the cochlea, is the membranous labyrinth filled with endolymph. The cochlea and cochlear duct is coiled up like a snail shell. The top and bottom are bony labyrinths filled with perilymph. The top bony labyrinth is the *Scala vestibuli* and the bottom bony labyrinth is the *Scala tympani*. In the center is the membranous labyrinth filled with endolymph. The upper membrane that separates the scala vestibuli from the cochlear duct is the *vestibular membrane* and the bottom membrane that separates the scala tympani from the cochlear duct is the *basilar membrane*.

The basilar membrane extends from the base (where the oval and round windows are located), to the apex (where the helicotrema is located) of the cochlear duct. The basilar membrane contains thousands of stiff, elastic fibers, with shorter fibers towards the base of the basilar membrane and longer fibers towards the apex. This arrangement means that the base of the basilar membrane is narrow and stiff, becoming wider and more flexible towards the apex.

The bony labyrinth and the membranous labyrinth extend from the base, to a point at the apex where the top and bottom bony labyrinth connect to each other. This area is called the *Helicotrema* and here the perilymph of the scala vestibuli and the scala tympani are in contact. At the bases of the labyrinths are found the openings to the middle ear cavities. The opening into the scala vestibuli, which is covered by a membrane, is called the *oval window*. The stapes bone is in contact with the membrane of the oval window, and it is here that the vibrations enter into the scala vestibuli. There is also an opening, covered by a membrane, in the scala tympani called the *round window*. Here pressure waves leaving the scala tympani can pass back into the middle ear cavity.

Extending all the way along the upper surface of the basilar membrane from the narrow, stiff base to the wider, flexible apex is the *organ of Corti*. It is composed of approximately 16,000 hair cells, arranged in four parallel rows that have hair-like *stereocilia* extending into the endolymph of the cochlear duct. Located above these hair cells, and in contact with the stereocilia, is the *tectorial membrane* that also extends the entire length of the organ of Corti. Vibrations of the basilar membrane will vibrate the tectorial membrane and stimulate the stereocilia of the hair cells. This will generate action potentials in nerve fibers attached to the hair cells which will then enter the brain to be interpreted as sound.

DISCRIMINATION OF FREQUENCIES

Hair cells in the organ of Corti are able to differentiate between different frequencies (or pitches) of sound. As vibrations from the stapes bone enter the scala vestibuli by the oval window, they set up vibrations in the perilymph. These pressure waves can migrate all the way around through the helicotrema, enter the scala tympani and exit through the round window. In addition they can also cross from the scala vestibuli into the cochlear duct by vibrating the vestibular membrane (which will generate additional vibrations in the endolymph). These pressure waves in the endolymph then vibrate the basilar membrane, cross in the scala tympani and exit by the round window. The different parts of the basilar membrane vibrate best at different frequencies due to the differences in the width and stiffness of the basilar membrane at the various locations. The narrow end at the base vibrates best at high frequencies (20,000 CPS) while the wider end at the apex vibrates best at low frequencies (20 CPS). The intermediate frequencies are sorted out along the length of the basilar membrane, with a particular part being most responsive to a particular frequency. Individual nerve tracks from all the various hair cells carry this information into the brain, and the brain then interprets different pitches of sound according to the location of the hair cells that are being stimulated.

DISCRIMINATION OF INTENSITY

The sound intensity (or loudness) depends on the amplitude of the sound waves. With increased intensity there will be greater oscillations of the basilar membrane in the regions of the peak response in the organ of Corti. Greater oscillations stimulate hair cells more intensely and increase the frequency of action potentials (number/time) that are conducted to the auditory centers of the brain.

NERVE PATHWAYS THAT CONDUCT AUDITORY STIMULI

1. The cochlear branch of the vestibulocochlear nerve (VIII Cranial Nerve) enters the medulla where it synapses on nuclei. Some fibers remain on the same side, however, the majority of fibers cross to the opposite side and synapse again.

2. Fibers originating from the nuclei in the medulla on both sides then pass up to the midbrain where they synapse on midbrain nuclei.

3. Fibers originating from these midbrain nuclei on both sides then pass to the thalamus where they synapse on the medial geniculate bodies located in the thalamus on the left and right sides.

4. Tracks originating from these thalamic nuclei on both sides then radiate to the auditory cortex of the left and right temporal lobes where they synapse on cortical nuclei.

5. Within the auditory cortex, the information from these tracks is interpreted as sound of a particular pitch and loudness.

DEAFNESS

Deafness can be partial or complete and can be classified as either *conductive deafness* or *nerve deafness*. Conduction deafness occurs because of a problem in the ability to conduct the sound waves through the external, middle or inner ear structures. Nerve deafness occurs because the receptor cells, tracks or CNS structures are damaged.

BALANCE OR EQUILIBRIUM

While the cochlea and cochlear duct of the inner ear are involved in the sense of hearing, the *vestibular apparatus* of the inner ear is involved in the sense of equilibrium. The vestibular apparatus consists of the *semicircular canals* and the *otolith organs*.

THE OTOLITH ORGANS

These structures help maintain static equilibrium because they provide information about the position of the head relative to gravity and also detect changes in motion. The otolith organs are located inside the *utricle* and *saccule* (part of the membranous labyrinth filled with endolymph). The utricle and saccule are found in the portion of the bony labyrinth called the *vestibule* which is itself filled with perilymph. The otolith organs (also known as the *maculae*) consist of hair cells, a gelatinous matrix and otoliths. The hairs of the receptor cells protrude into the overlying gelatinous matrix. Imbedded in the top regions of the gelatinous matrix is a layer of tiny calcium carbonate crystals called *otoliths* (ear stones) that make the gelatinous matrix top heavy. Tilting the head causes the gelatinous matrix to shift, which mechanically deforms the hairs (the *stereocilia and kinocilia*) of the receptor cells and generates action potentials in the attached nerves. In addition, changes in horizontal linear (starting and stopping) motion also displaces the gelatinous matrix and are sensed by the utricle, while changes in vertical (up and down) motion are sensed by the saccule.

THE SEMICIRCULAR CANALS

The semicircular canals are mainly involved in detecting changes in dynamic equilibrium, such as rotational or angular acceleration or deceleration of the head. There are three semicircular canals in the vestibular apparatus of each inner ear arranged three-dimensionally in planes that lie at right angles to each other. Each semicircular canal is composed of an outer bony labyrinth filled with perilymph, and an inner membranous labyrinth filled with endolymph. At the base of each semicircular canal is an enlarged region called an *ampulla* where the ampulla receptors (or *crista ampullaris*) are found. These receptors consist of hair cells with stereocilia and kinocilia that project in the gelatinous cupula. A swaying motion of the cupula results from changes in the direction of flow of the surrounding endolymph. As the head is rotated the endolymph, in at least one of the semicircular canals, undergoes acceleration or deceleration. This fluid movement mechanically deforms the cupula and stimulates the hair cells. Action potentials generated by these hair cells then travel into the brain.

NERVE PATHWAYS FROM THE VESTIBULAR APPARATUS

Action potentials from the various receptors of the vestibular apparatus are transmitted through the vestibular division of the vestibulocochlear nerve to the vestibular nuclei in the brain stem. This vestibular information is processed and integrated in conjunction with information from the other senses (sight, proprioception), is used to maintain balance and posture. It also controls the eye muscles which allow the eyes to remain fixed on an object as the head is moved.

Chapter 6
MUSCLE PHYSIOLOGY

The primary function of muscle tissue is to provide movement for body structures. Muscle tissue is classified into three subgroups: smooth, skeletal and cardiac. This chapter will consider all three varieties, but will concentrate primarily on skeletal muscle because this is the most completely understood of all the muscle types.

Skeletal muscles function in association with the bones of the skeletal system to produce movement of the limbs and the body proper in three dimensions, and to allow the individual to manipulate the surrounding environment. In addition, skeletal muscles also stabilize the body, maintain body posture and produce heat to maintain the body temperature above that of the surroundings.

STRUCTURE OF SKELETAL MUSCLE

Skeletal muscles are connected to bones through tendons that are continuous at one end with the *fascia* of the skeletal muscle and with the *periosteum* of the bone at the other **(Figure 6.1)** An outer connective tissue covering on the skeletal muscle is called the *epimysium*. Skeletal muscles are divided into bundles of muscle cells (*fascicles*) which are each enclosed in a connective tissue covering (the *perimysium*). Finally within a fascicle are found groups of individual skeletal muscle cells (also called *skeletal muscle fibers*) and each fiber is separated from other fibers by a thin connective tissue layer called *endomysium*.

Individual skeletal muscle cells (or fibers) are long and multinucleated and are covered by a plasma membrane (the *sarcolemma*). Present within the cytoplasm (*sarcoplasm*) of the muscle cell are tubule-like *myofibrils*. These myofibrils are composed of bundles of thick and thin filaments arranged to create a pattern of darker and lighter bands **(Figure 6.2)**. This same banding pattern can also be seen in an intact muscle cell and in skeletal muscle tissue under magnification. It is for this reason that skeletal muscles are also called *striated muscles*.

THE BANDING PATTERN OF SKELETAL MUSCLE

If a section of skeletal muscle tissue is viewed under a light microscope, the classic banding pattern can be seen **(Figure 6.2)**. The banding pattern is actually based on the organization of the myofibrilla proteins within individual skeletal muscle fibers (cells). This pattern of lines repeats over and over down the entire length of the skeletal muscle cell (and internal myofibrils). Muscle physiologists named the various bands using letters as follows:

1. **Z lines** which appear as a clearly defined zig-zag pattern. The region between two Z lines is the *sarcomere* (about 2.5 um in length) and is the smallest functional unit of muscle contraction.

2. **I band** is a region surrounding the Z line that appears lighter in color.

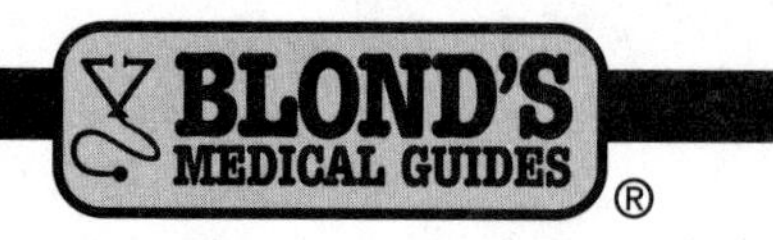

3. **A band** begins where the I band ends and extends to the next I band. In the center of the A band is a lighter zone called the *H zone* in the center of which is the *M line.*

Note that the banding pattern just described is only found in a relaxed sarcomere. Remember that a relaxed sarcomere contains one whole A band and two half I bands (each located next to the Z lines at each end of the sarcomere). If the skeletal muscle is contracted, the sarcomere will appear as follows **(Figure 6.6)**:

1. The Z lines will be closer together, thus the sarcomere has shortened (by approximately 1 um).

2. The I band shrinks significantly.

3. The A band has not changed in length.

In a myofibril composed of many thousands of individual sarcomeres, upon contraction each individual sarcomere will shorten a minute amount. But because they are all attached together, the overall shortening of the entire skeletal muscle cell will be significant. (If there are 30,000 sarcomeres in the myofibril and each shortens by 1 um, the complete myofibril will shorten 30,000 x 1 um = 30,000 um = 30 mm = 3 cm.)

THE THICK AND THIN FILAMENTS

Purification of the filaments from skeletal muscle yields two major proteins: *myosin and actin.* The individual myosin molecules (which possess a long tail and a globular head) can combine into myosin thick filaments **(Figure 6.4)**, while the actin sub-units (which appear as small globular structures) can polymerize into actin thin filaments **(Figure 6.3**). In addition, associated with the thin actin filaments are two other proteins: *troponin and tropomyosin.*

The arrangement of the thick and thin filaments as seen in **Figure 6.2** is responsible for the banding pattern of the relaxed myofibrils. Accordingly, the actin thin filaments extend from the Z lines inwards towards the middle of the sarcomere, but do not reach completely across. The space between the thin filaments is the H zone. The length of the myosin thick filaments is equivalent to the A band, and the end of one thick filament to the beginning of the next is the I band. In the center of the I band is the Z lines. Thus, there are only thick filaments and no thin filaments in the H zone of the A band, and there are only thin filaments and no thick filaments in the I band.

Myosin and the Thick Filament. As shown in **Figure 6.4** a single myosin molecule has a long tail and a globular head (also called a myosin *cross bridge*). Present within the structure of the head is the enzyme ATPase that catalyzes the breakdown of *ATP to ADP + Pi + Energy*. The release of energy from ATP acts on the globular head and results in a swiveling or bending motion of the head protein (see following material). In addition, this globular head is also designed to bind to a specific site on the actin sub-unit. When single myosin molecules are bound together to form the thick filament the tails lie parallel and the heads protrude out at either end from the bundle.

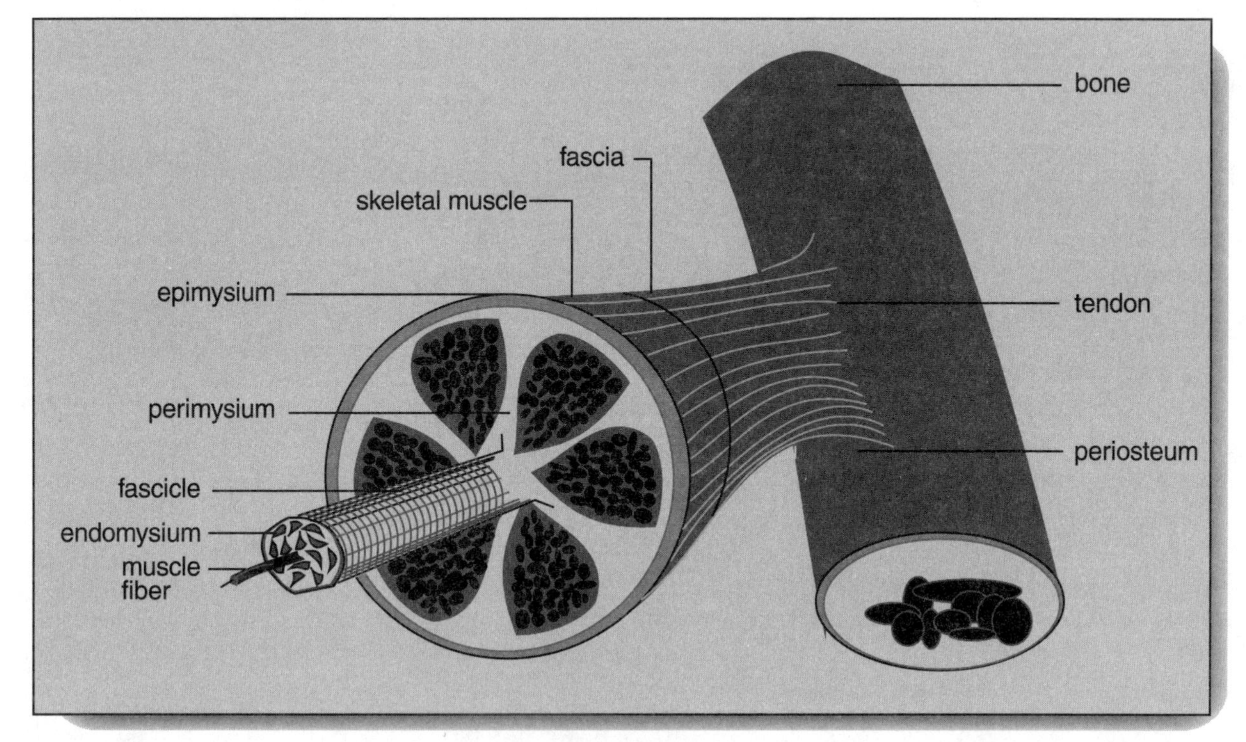

FIGURE 6.1

Actin, Troponin, Tropomyosin and the Thin Filament. Single actin sub-units as shown in **Figure 6.3** are spherical, and at one location on the molecule possess a binding site for the attachment of the globular myosin head or cross bridge. Polymerization of individual actin sub-units produces a twisted double chain called the *actin helix*. In addition, associated with the actin helix are two other proteins, troponin and tropomyosin. The tropomyosin forms an additional twisted strand on the surface of the actin helix. Located at intervals on the tropomyosin strand are molecules of troponin, a calcium binding protein.

Because actin and myosin actively generate movement (see following material) they are described as the *contractile proteins*. Troponin and tropomyosin are the regulatory proteins since they control the movement of the actin and myosin.

THE SLIDING FILAMENT MODEL OF CONTRACTION

As has been mentioned, the banding pattern of skeletal muscle changes upon contraction. The appearance of the banding pattern upon contraction results from the arrangement of the thin and thick filaments, and the fact that the thin filaments slide inward over the thick filaments. This sliding results in the ends of the thin filaments moving into the H zone thereby shortening the sarcomere. However, because the thin filaments slide inward over the thick filaments, the thick filaments do not shorten and thus the A band remains the same length **(Figure 6.6)**.

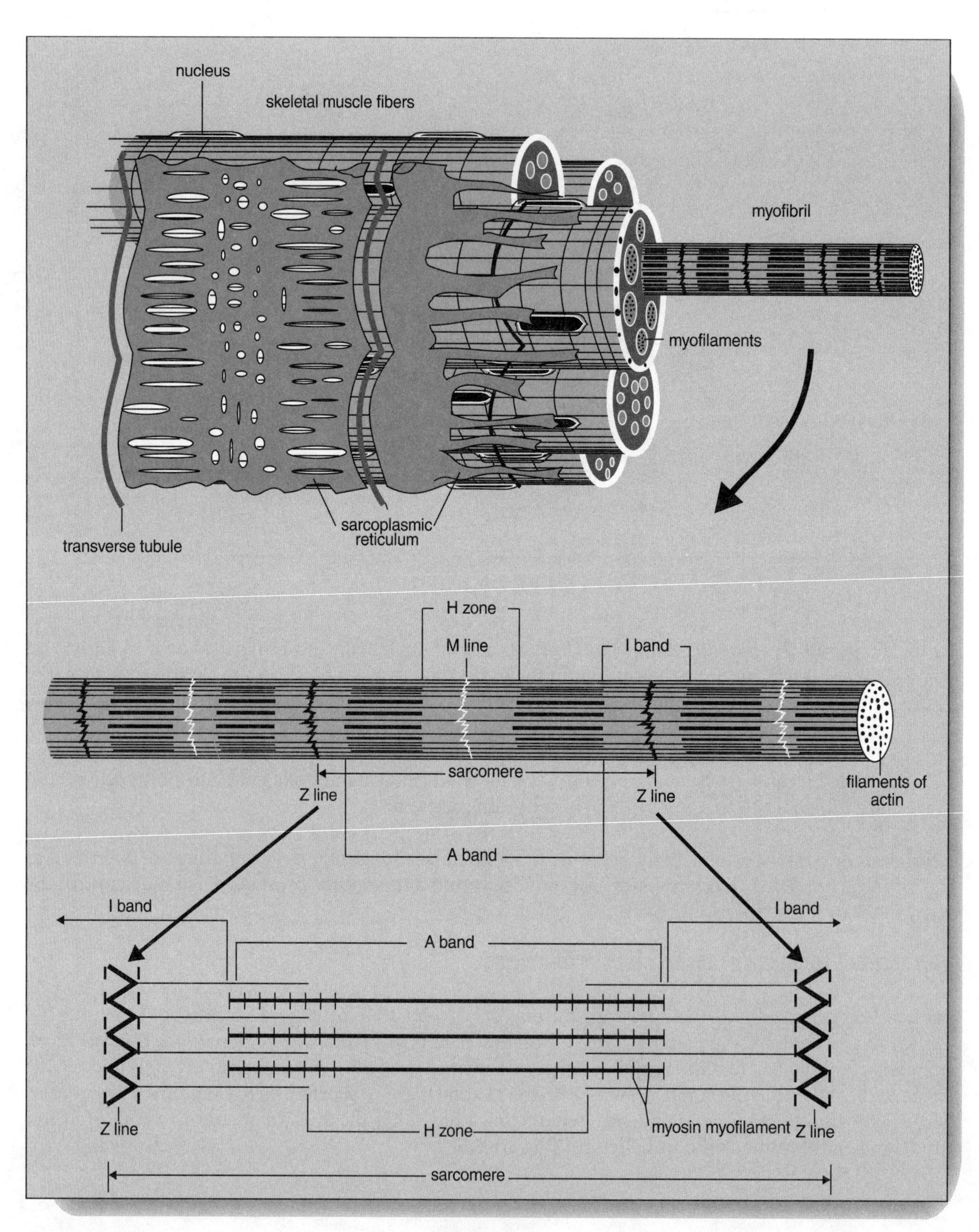
nucleus
skeletal muscle fibers
myofibril
myofilaments
sarcoplasmic reticulum
transverse tubule
H zone
M line
I band
sarcomere
Z line
Z line
filaments of actin
A band
I band
I band
A band
Z line
H zone
myosin myofilament
Z line
sarcomere

FIGURE 6.2

MOLECULAR EVENTS AND SHORTENING OF THE SARCOMERE

The sliding of the thin filaments over the thick filaments results from the swiveling of the myosin cross bridge after it binds to the myosin binding site on the actin sub-unit. This process is catalyzed by ATPase that hydrolyses ATP to release energy. The energy causes a conformational change in the protein of the myosin head. The sequence of steps that lead to the swiveling of the cross bridges and shortening of the sarcomere are outlined here and in **Figure 6.7**.

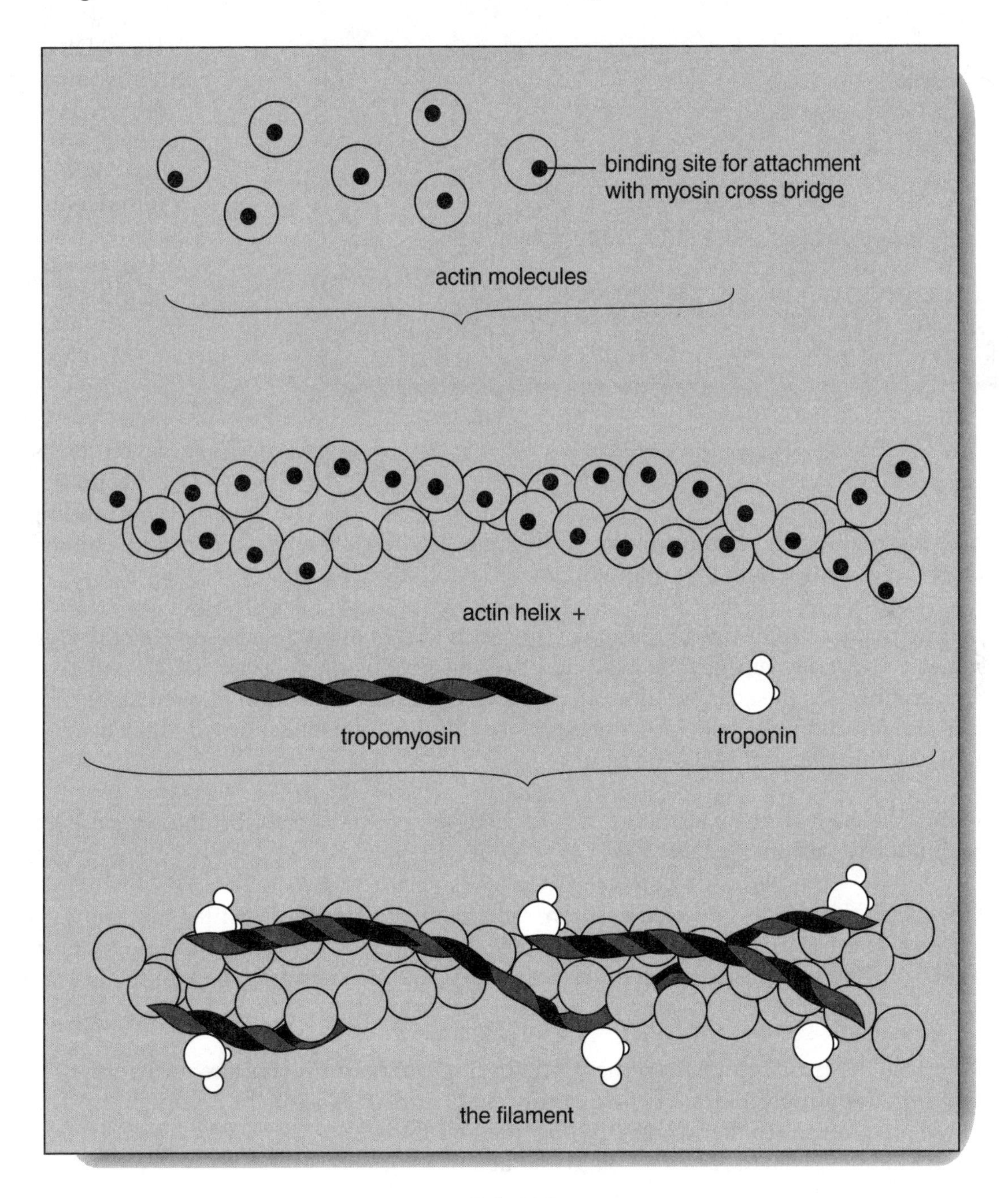

FIGURE 6.3

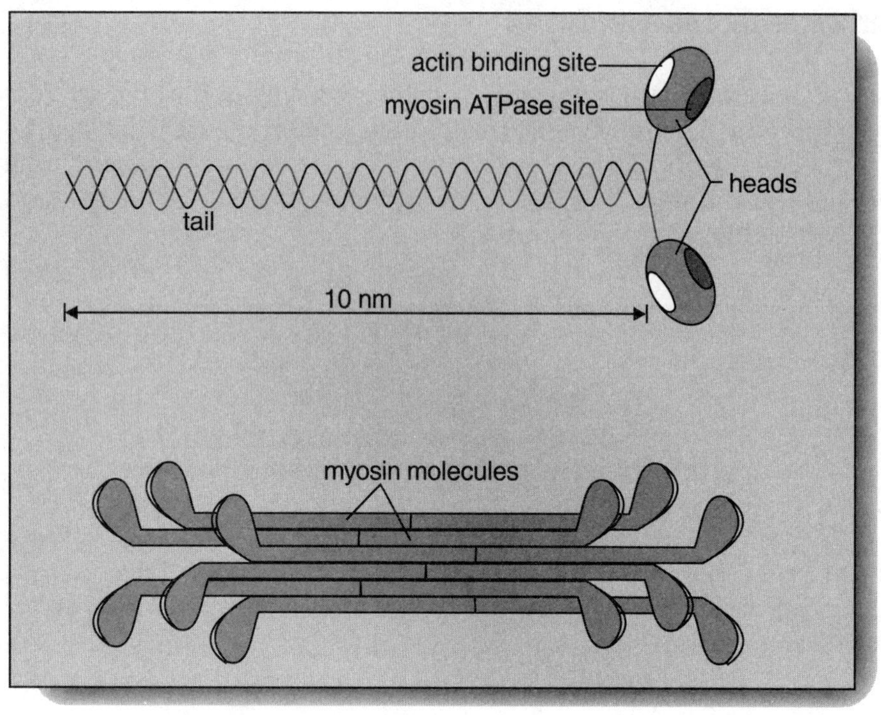

FIGURE 6.4

1. $ATP(Mg^{++})$ binds to the ATPase on the myosin cross bridge.

2. ATP is hydrolyzed to ADP+Pi+Energy. The ADP and Pi remain attached to the cross bridge and a portion of the energy is stored in the cross bridge (some is also released as heat).

3. The myosin cross bridge binds to the myosin binding site on the actin sub-unit. This binding only takes place if tropomyosin uncovers the binding site. Tropomyosin uncovers the binding site if Ca^{++} binds to troponin (see following materialand **Figure 6.5**).

4. Binding of the cross bridge to the site on the actin thin filament releases the energy stored up in the myosin cross bridge. The head now swivels producing the *power stroke*, and because it remains linked to the actin binding site, the swiveling causes the thin filament to be moved in the same direction. (If the head swivels left, the thin filament will slide left; if the head swivels right, the thin filament will slide right.)

5. As the head swivels the ADP and Pi which were bound to the cross bridge, are now released into the sarcoplasm.

6. If the cross bridge, in its swiveled position, remains attached to the thin filament, this will result in a *Rigor Complex*. This results in the sarcomeres being locked together unable to contract or to relax. In a living muscle this will result in major muscle cramping. In a dead muscle this is termed *Rigor Mortis*.

7. To release the cross bridge from the actin binding site (and prevent the formation of a rigor complex), a new molecule of $ATP(Mg^{++})$ now binds to the ATPase in the myosin head. The cross bridge now returns to the straight up position and the ATP is then hydrolyzed. This process (go to #1 above) is then repeated again and again.

8. The only difference in this new cycle (from the previous cycle) is that the myosin head now binds to a different actin sub-unit. Thus with each cycle the myosin cross bridge is located at a different position on the actin thin filament. This means that the thin filament slides inward in short bursts with a ratchet-like motion.

9. This cycle will continue as long as:
 a. There are new molecules of ATP to supply the driving energy and release the myosin from the actin, and
 b. Ca^{++} remains bound to the troponin (see following material).

10. If Ca^{++} is removed from the troponin, the myosin head can no longer bind to the actin and the contraction process stops (see following material).

In order for the sarcomere to shorten, the thin filaments on the left half of the sarcomere must slide towards the right and the thin filaments on the right half of the sarcomere must slide towards the left. Thus, the cross bridges on the left side of the sarcomere must swivel right and the cross bridges on the right side must swivel to the left.

Regulation of Cross Bridge Movement. Muscle contraction can be inhibited if the cross bridge binding to the actin site is blocked. Under these conditions the potential energy bound up in the myosin head cannot be released and the cross bridge does not swivel. To block the binding of the myosin cross bridge to the actin, the protein tropomyosin covers the binding site on the actin sub-unit. This prevents the muscle from contracting **(Figure 6.7)**.

To activate a contraction the tropomyosin must unblock the site by drawing out of the way. This is accomplished through the action of the other regulatory protein, troponin. Troponin is a Ca^{++} binding protein. If Ca^{++} is available, it will bind with high affinity to the troponin altering the shape of the troponin. This causes the tropomyosin to move out of the way exposing the binding site on the actin. The myosin cross bridge now binds to the actin and the cycle of swiveling begins (see prior material).

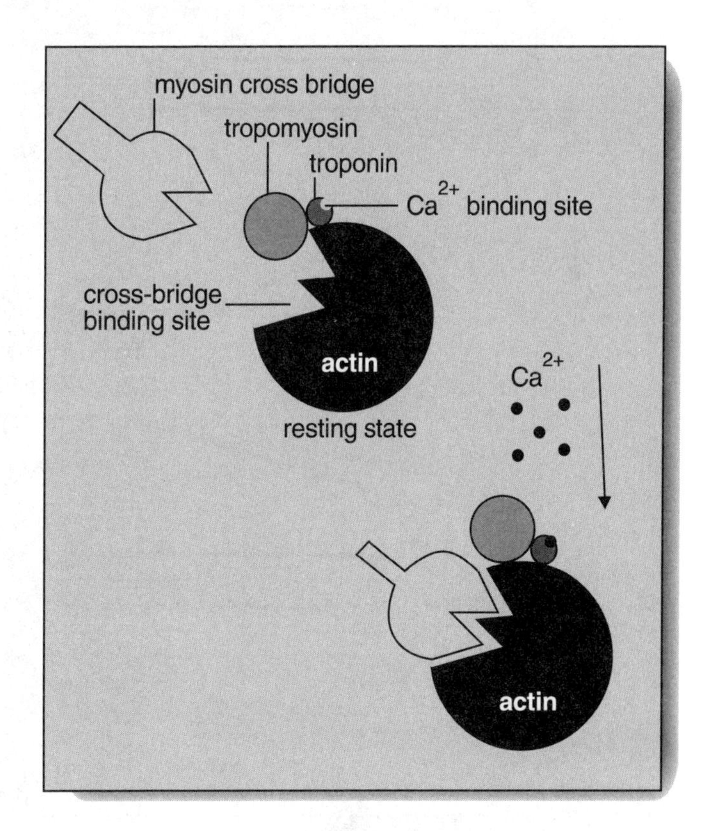

FIGURE 6.5

CA++ AND EXCITATION CONTRACTION COUPLING

Regulation of Ca^{++} is the key to controlling muscle contraction. Ca^{++} is stored in the *Sarcoplasmic Reticulum (SR)*, a series of

membranous sacs that cover the myofibrils **(Figure 6.8)**, and is released when Ca^{++} channels in the SR membrane open. Because Ca^{++} is at a higher concentration inside the SR than outside, it will leave the SR by passive diffusion and enter the sarcoplasm. In the sarcoplasm the Ca^{++} binds to the troponin causing the tropomyosin to unblock the actin binding sites and the contraction commences. Thus, muscle contraction depends on opening Ca^{++} channels in the SR membrane.

These Ca^{++} channels are under the control of an electrical stimulus (or depolarization) initiated in the muscle membrane (*sarcolemma*). This stimulus is conducted from the outer sarcolemma to the SR membrane through the *transverse tubules* (*T-tubules*) **(Figure 6.9)**. These T-tubules extend from the outer sarcolemma into the sarcoplasm of the muscle cell and are closely associated with the SR membranes. Therefore, an action potential that is initiated in the outer sarcolemma can penetrate to the SR

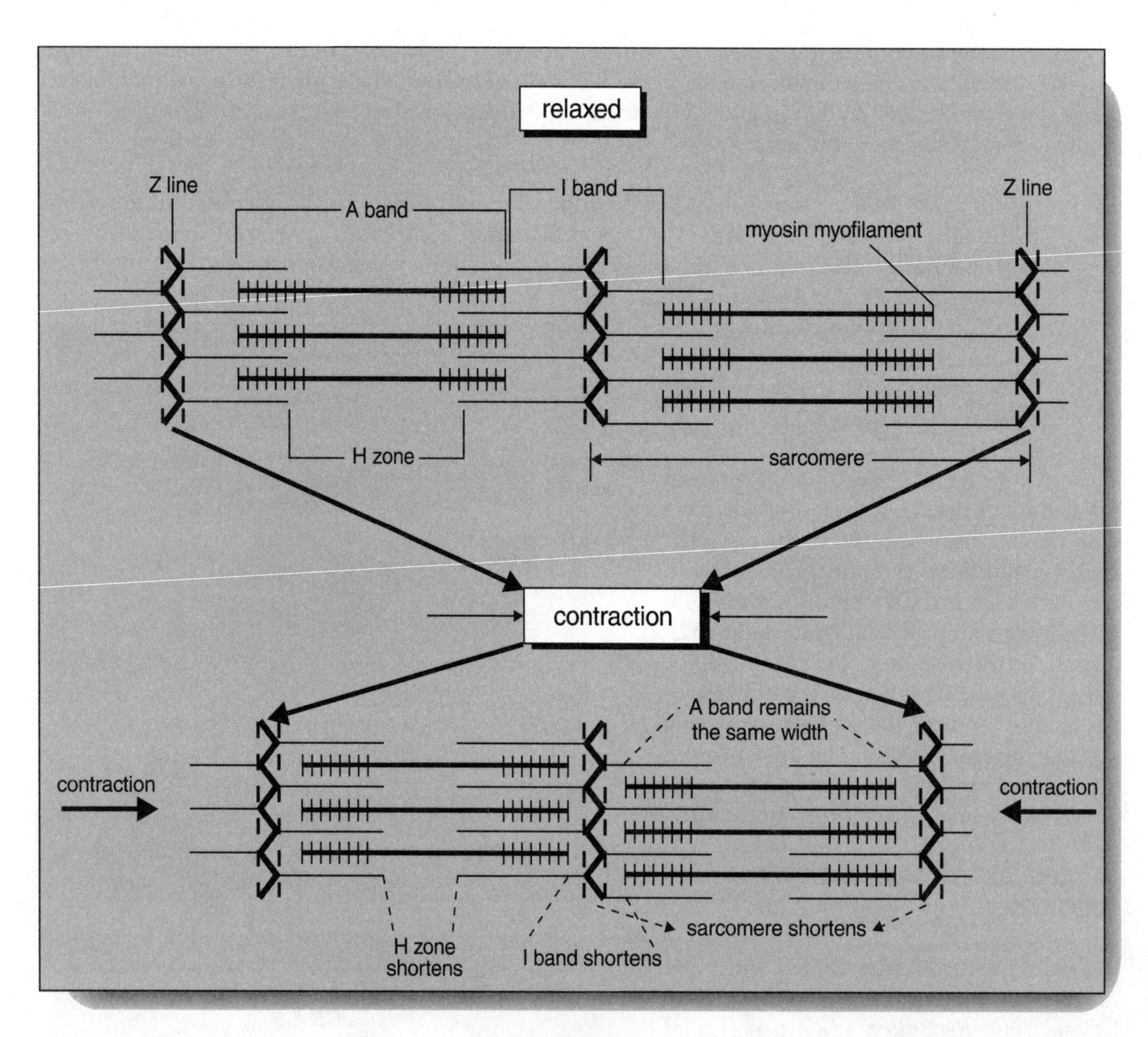

FIGURE 6.6

membrane where the resulting depolarization opens the Ca^{++} channels. The term *Excitation Contraction Coupling* thus refers to the coupling between the stimulus and contraction through the release of Ca^{++}.

THE SR AND Ca^{++} UPTAKE

To reduce the levels of Ca^{++} in the sarcoplasm and to stop muscle contraction, a Ca^{++} pump—which is operating all the time in the SR membrane—actively transports the Ca^{++} back into the SR. Like other active transport pumps, the energy source is from the hydrolysis of ATP, and the direction of transport is against a concentration gradient. In order to sustain a muscle contraction the amount of Ca^{++} released from the SR must be greater than its reuptake by active transport.

NEUROMUSCULAR JUNCTION

The neuromuscular junction is constructed much like a nerve to nerve synapse **(Figure 6.9)**. Accordingly, the presynaptic fiber is the axon of the motor neuron that terminates in the synaptic knob. Present within this region are the vesicles containing the transmitter, *Acetylcholine* (*ACh*), along with many mitochondria to supply energy. A synaptic cleft (or neuromuscular cleft) separates the presynaptic and postsynaptic structures. On the post synaptic side (the muscle cell side) the membrane is folded. This folded sarcolemma of the neuromuscular junction is termed the *motor end plate*. Located

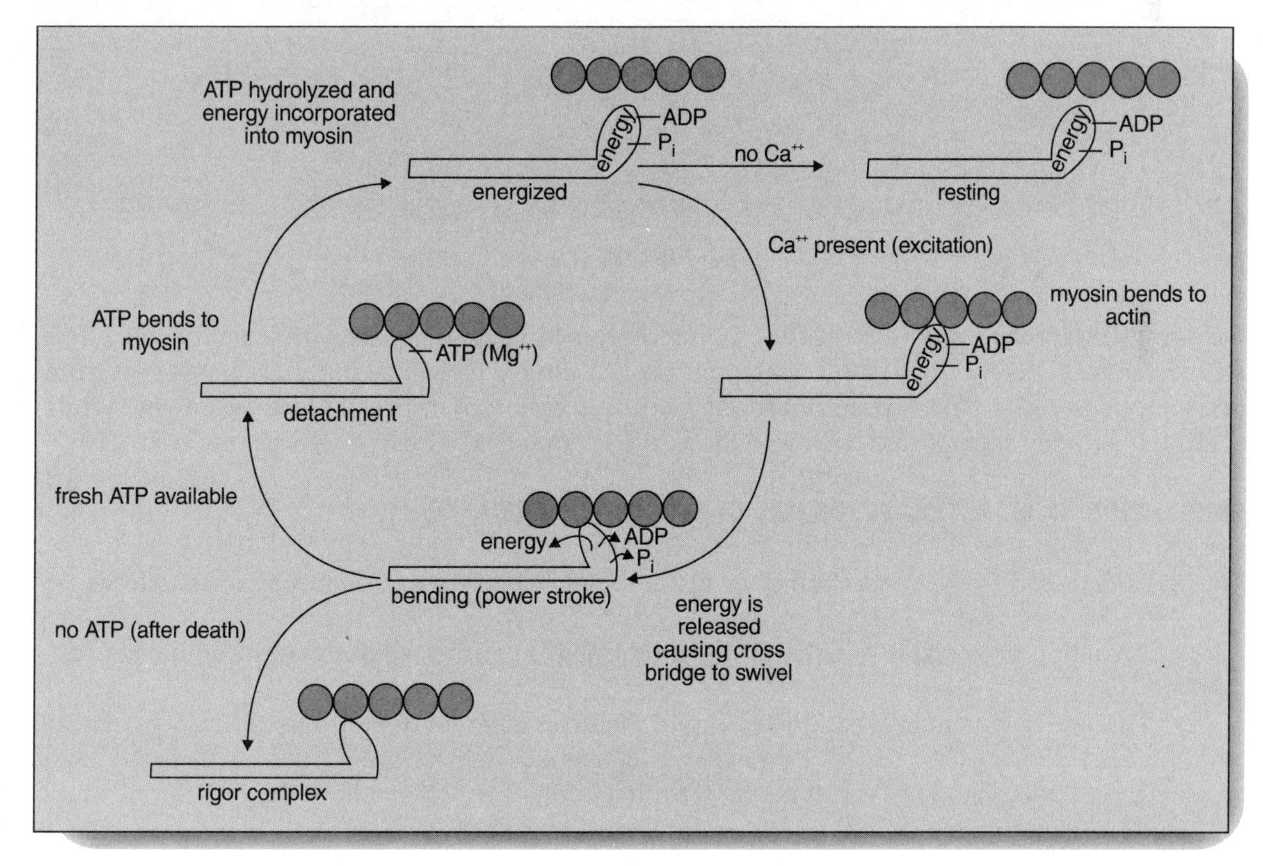

FIGURE 6.7

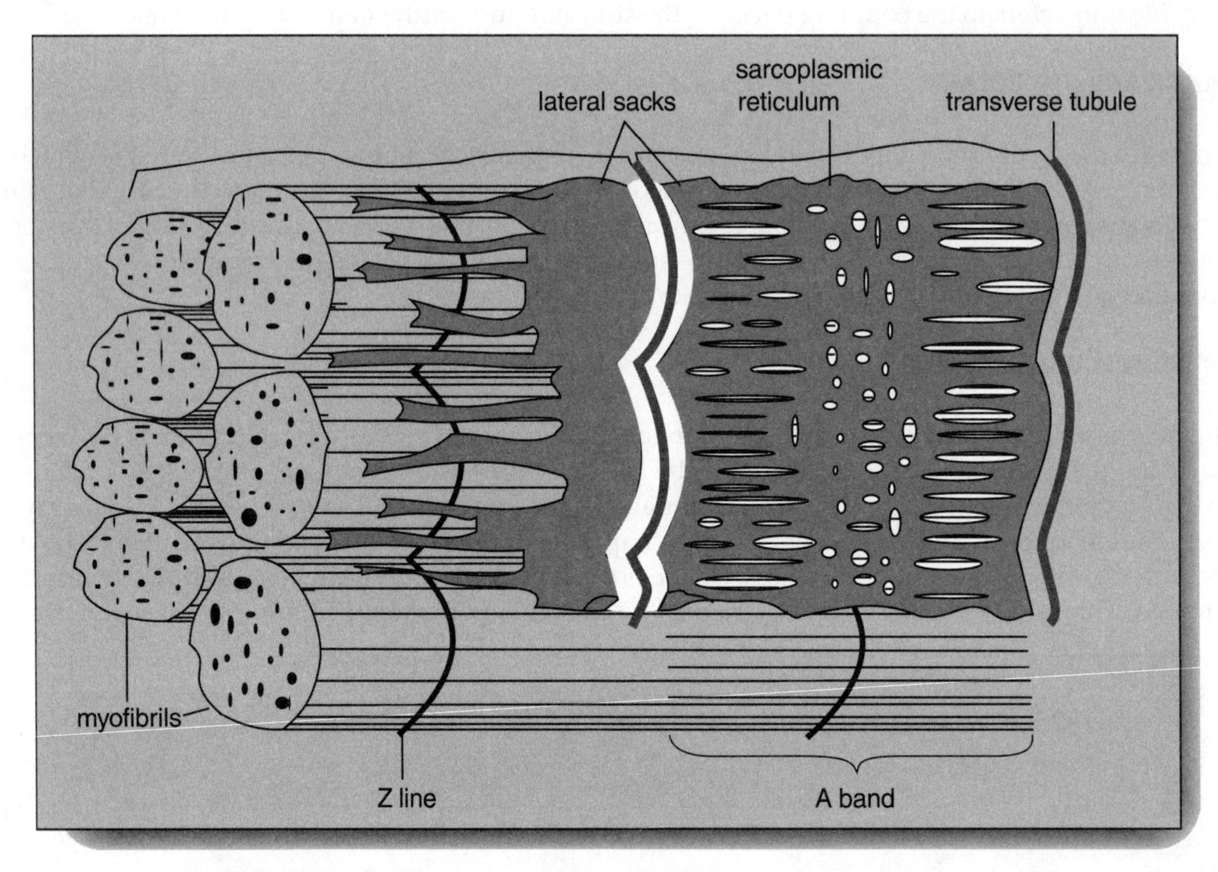

FIGURE 6.8

here are *ACh receptors*, and *ACh esterase*. ACh receptors bind ACh released presynaptically, while the ACh esterase splits the ACh into acetate and choline almost as fast as the ACh is released from the presynaptic vesicles. The acetate produced from this breakdown process diffuses away, while the choline is actively transported back into the nerve terminal to be resynthesized into new ACh.

GENERATION OF AN ACTION POTENTIAL IN THE SARCOLEMMA

The steps that lead to the generation of an action potential on the sarcolemma are as follows:

1. An action potential is initiated in the motor neuron at the cell body in the spinal chord.

2. This action potential is conducted down the axon where it depolerizes the synaptic terminal.

3. The depolarization at the synaptic terminal opens Ca^{++} channels in the membrane and Ca^{++} diffuses into the terminal.

4. ACh containing vesicles merge with the outer membrane of the nerve and release ACh into the synaptic cleft.

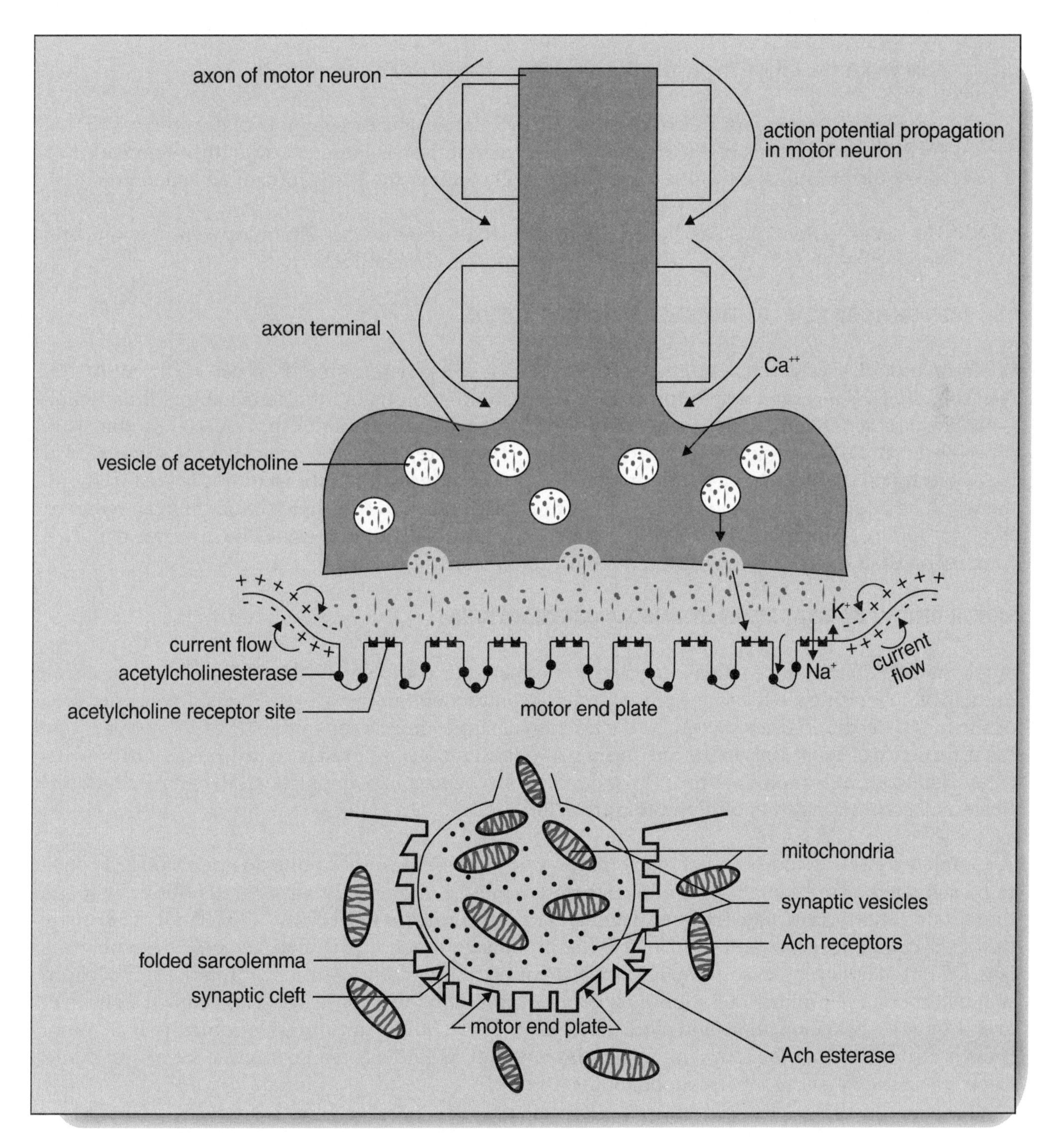
axon of motor neuron
action potential propagation in motor neuron
axon terminal
Ca++
vesicle of acetylcholine
K+
current flow
acetylcholinesterase
Na+
current flow
acetylcholine receptor site
motor end plate
mitochondria
synaptic vesicles
Ach receptors
folded sarcolemma
synaptic cleft
motor end plate
Ach esterase

FIGURE 6.9

5. ACh diffuses across the synaptic cleft and binds to the ACh receptors in the motor end plate.

6. Na^+ and K^+ channels open in the motor end plate and Na^+ diffuses into the motor end place while K^+ diffuses out.

7. This produces a local depolarization called an *End Plate Potential* (EPP).

8. The EPP generates a local current flow that depolarizes the sarcolemma of the muscle cell that surrounds the motor end plate. If the local current flow is large enough, this will result in a threshold stimulus at the sarcolemma and will result in the generation of an action potential.

9. The action potential is conducted down the sarcolemma to the T-tubules were it enters and depolarizes the SR. Ca^{++} is released and the muscle cell contracts.

THE "ALL-OR-NONE LAW" OF MUSCLE CELL CONTRACTION

When ACh is released at the neuromuscular junction, it will generate an EPP. When sufficient ACh is released, the EPP is large enough to produce a threshold stimulus on the sarcolemma; thus, action potentials are generated. Under these conditions there will be sufficient Ca^{++} released by the SR to promote a complete contraction of the muscle fiber. Therefore, if a single muscle cell is stimulated at threshold it will undergo a complete contraction. On the other hand, if the stimulus is subthreshold, there will be no contraction at all. The result of this "all or none law" is that muscle cells can either be "On" or "Off"; or completely contracted or not contracted at all; but they *cannot* be in a state of partial contraction (half contracted, slightly contracted, and so on).

MOTOR UNITS AND GRADATION OF MUSCLE CONTRACTIONS

Single muscle fibers can be either completely contracted or completely relaxed, but full muscles are capable of generating a wide range of force (slight contractions all the way up to maximal generation of tension). This difference between the contraction of single muscle cells and of the full muscle is due to the presence of *motor units* in the full muscle. A motor unit is composed of many muscle fibers (cells) connected to a single motor neuron. Thus, if this motor neuron is stimulated, all the attached muscle fibers will contract together as a unit **(Figure 6.10)**.

A complete muscle is composed of many motor units, but all the motor units do not necessarily have to be activated at the same time. For example, if a muscle contains 50 motor units but only 10 are undergoing contraction, then the tension generated in this muscle will be only 1/5 (10 out of 50) of the maximum possible. For example if 25 were undergoing contraction, the muscle would be generating 50% (25 out of 50) of the maximum tension. Activation of the motor neurons within the spinal chord will determine the number of motor units that are stimulated within the muscle. To elicit a stronger and stronger contraction, more and more motor units are *recruited* (stimulated to contract). If all motor units are stimulated at the same time, then the muscle will undergo a maximum contraction that is known as a *tetanic contraction* (see following material).

The number of muscle fibers that constitute a motor unit can vary greatly. For example, muscles designed for fine or delicate contractions (such as those used for eye movement) contain many motor units with only a few muscle fibers in each. On the other hand, muscles designed for powerful movements (such as those of the legs) contain few motor units but each has 1,500 or more muscle cells. Contraction by such muscles generates much force but is not finely regulated, and recruitment of motor units produces large increases in tension.

In those muscles that are responsible for sustained contractions (such as muscles that maintain body posture), *asynchronous recruitment* of motor units takes place. Here some motor units are switched "on" while others are turned "off" to rest. This allows the muscles to continue to remain under partial contraction for extended periods because some motor units are resting while others are actively maintaining the contraction.

FREQUENCY OF STIMULUS V. TENSION GENERATED

The amount of tension generated in a muscle fiber (cell) is proportional to the frequency of stimulation applied to that fiber (even though the fiber contracts to the greatest extent possible). Accordingly, when a single action potential is applied to the fiber, a single twitch can be recorded as an increase in tension generated **(Figure 6.11)**. If the fiber is allowed to return to 0 tension and then restimulated, an identical twitch, at the same level of tension, will again be produced. However, if after the first twitch, a second stimulus is applied before the fiber returns to 0 tension, a larger amount of tension will be recorded for the second twitch. This effect, known as *wave summation*, can only be elicited because the duration of the action potential (1-2 msec) is much shorter than the duration of the tension generated (100 msec). An increased level of tension can be produced as the frequency of the applied action potentials is increased. The level of tension will eventually reach a maximum known as *tetanus*.

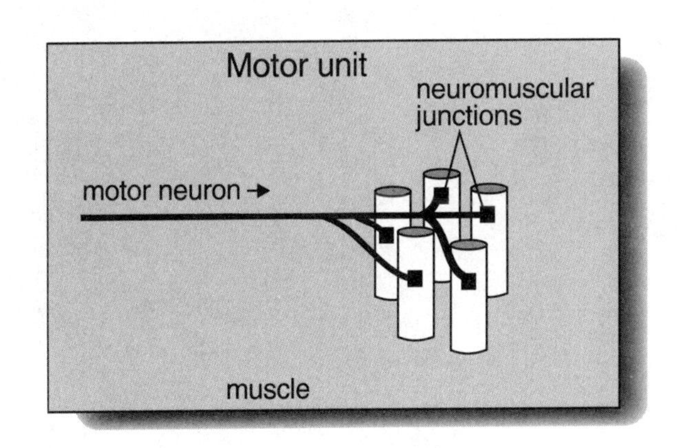

Figure 6.10

The mechanism responsible for this wave summation is not due to an increase in Ca^{++} released. Indeed, with a single action potential, enough Ca^{++} is released to saturate all the available troponin binding sites in the muscle cell and activate all cross bridges. Instead the mechanism relates to the presence of a hypothetical *series elastic element* in the myofibril. This series elastic element is actually the elasticity of the myofibrilla proteins. Thus, as the cross bridges begin to swivel and initially produce *internal tension*, a portion of the force is absorbed in the spring-like quality of the myofibrils. This means that the force transmitted to the ends of the muscle fiber is reduced resulting in a smaller *external tension* which can be measured. However, with repeated stimulation, this "internal spring" is stretched until it no longer can absorb any more force. This means that now all the internal tension generated by the moving cross bridges will be transmitted to the ends of the muscle fiber. Therefore, with repeated simulation the amount of tension generated increases.

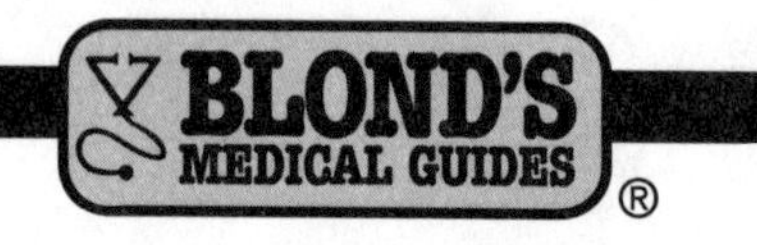

As mentioned previously, in a complete muscle composed of many motor units, recruitment can also account for increased tension generated. Under these conditions with increased stimulation, motor units that respond at higher thresholds will be activated. This will result in an increase in tension generated in the muscle, as the frequency or intensity of the applied stimuli increases.

THE LENGTH/TENSION RELATIONSHIP IN MUSCLE

Intact muscles are designed to operate *in vivo* (in the body) at a particular length. This length is optimum for the particular location and function that each muscle must fulfill. At its *optimum length* (l_o), maximum force can be generated upon contraction. If the length is physically altered by either

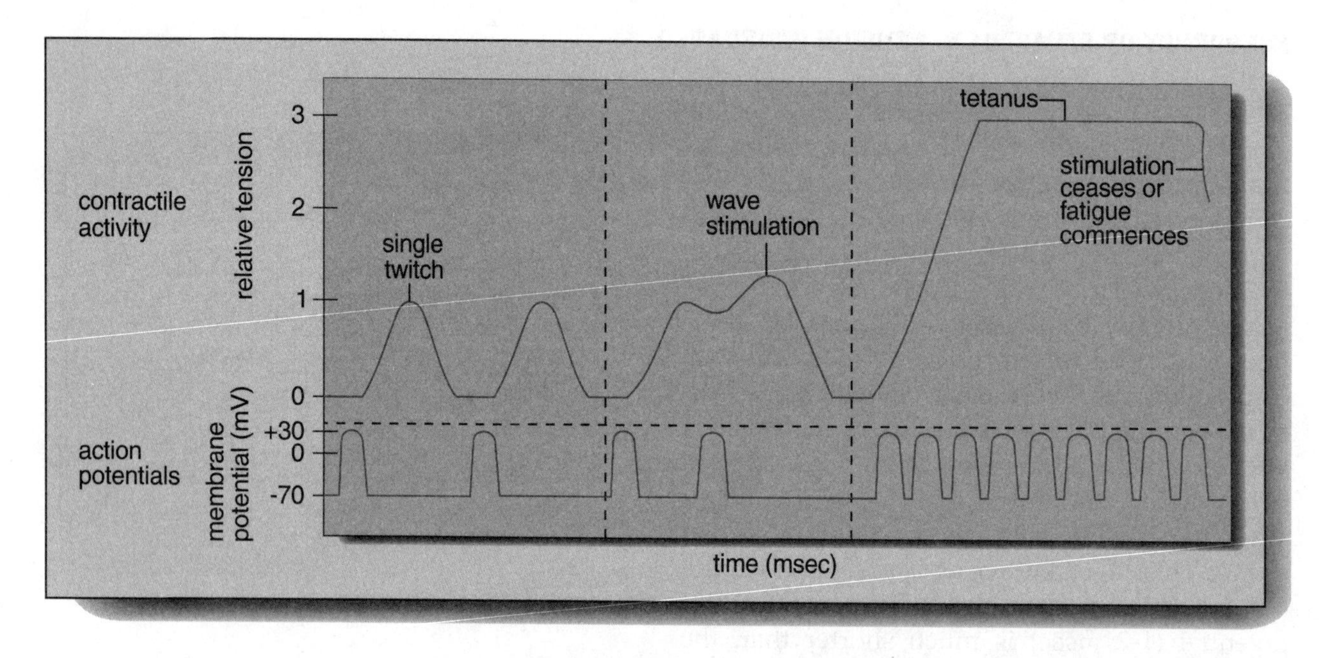

FIGURE 6.11

stretching or compressing the muscle prior to contraction, the force generated will be reduced **(Figure 6.12)**. The reason for this length-tension relationship relates to the organization of the thick and thin filaments. For example:

1) If the muscle is stretched to 130% of its optimal length (where l_o = 100% in **Figure 6.12**), fewer cross bridges can link with the thin filaments and thus the tension that the muscle can generate during a contraction is reduced to 70% of maximum.

2) If the muscle is compressed to 70% of its optimal length the organization of the cross bridges is disrupted because the thin filaments block each other and thus the tension generated is 60% of maximum.

MUSCLE ENERGETICS

The initial synthesis of ATP from glucose can be divided into the *anaerobic reactions* and the *aerobic reactions*. Anaerobic reactions do not require the input of O_2 and take place within the cytoplasmic compartment, while the aerobic reactions require the input of O_2 and take place within the mitochondria.

1. During anaerobic respiration, glucose is converted to pyruvic acid with a net output of 2 ATP molecules. The glucose used in the reaction is derived from:

   a. Blood glucose which originates in the liver where it is stored as liver glycogen and is converted to glucose upon demand.
   b. Blood glucose which originates in the liver where it is synthesized from other noncarbohydrate sources (fats or less commonly, proteins).

   c. Glycogen stores within the muscle that is converted to glucose upon demand.

2. The output of these anaerobic respiratory reactions is pyruvic acid which can be converted to:

   a. *Acetyl coenzyme A* and then it takes parts in aerobic respiration in the mitochondria when O_2 is present.

   b. *Lactic acid* which is produced in the absence of O_2.

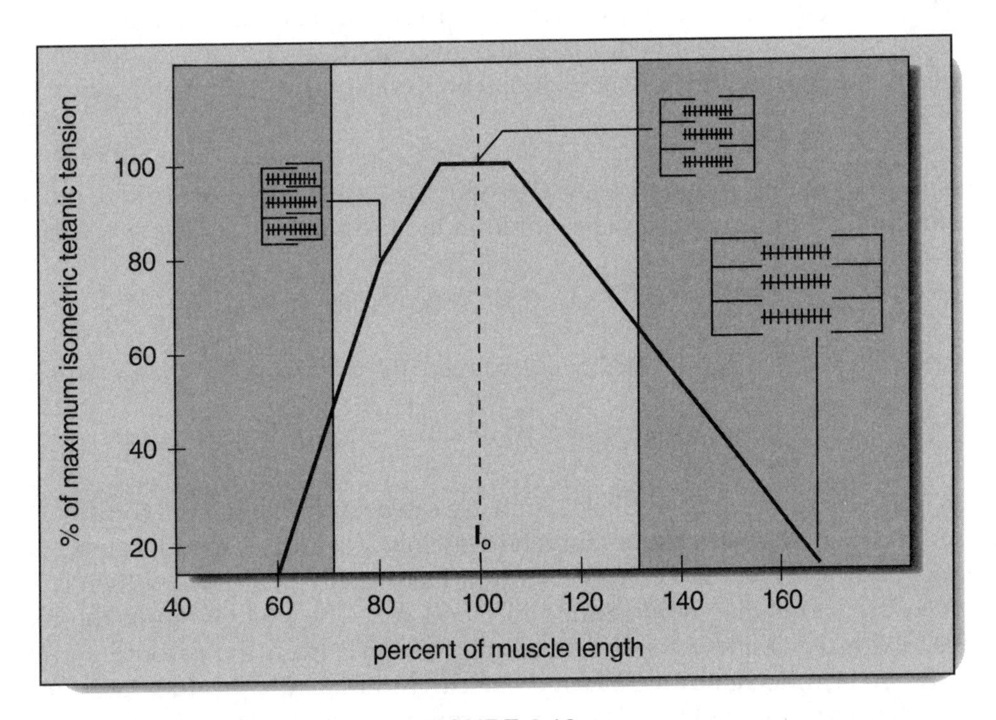

FIGURE 6.12

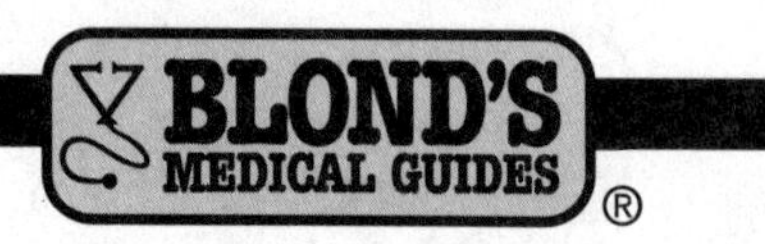

3. In the presence of O_2, Acetyl coenzyme A is converted to ATP in the mitochondria. The pathways involved are:

   a. The citric acid cycle (or Krebs cycle), and
   b. The electron transport chain.

Here an additional 36 molecules of ATP are synthesized. In addition to the requirement for pyruvate (converted to Acteyl coenzyme A), this process also requires the input of O_2 which acts as the final electron acceptor in the electron transport chain, and results in the production of H_20 as a byproduct of this process. CO_2 is also generated from the reactions in the citric acid cycle.

Thus, from one molecule of glucose the maximum amount of ATP that can be synthesized is 2 molecules from anaerobic respiration and 36 from aerobic respiration, or a grand total of 38 molecules of ATP.

In the event that O_2 is not available in sufficient quantities, the pyruvic acid is converted to lactic acid. Under these conditions only 2 molecules of ATP are then synthesized from one molecule of glucose. Such a scenario might be envisioned during extreme exercise when O_2 levels are limiting. After the exercise has ended, the lactic acid is then reconverted to glucose in the liver.

OTHER SOURCES OF ENERGY FOR MUSCLE METABOLISM

Although glucose acts as the primary energy source in muscle cells, it is possible to utilize fats and even proteins to supply energy in place of glucose. Fats (broken down to fatty acids and glycerol) and proteins (broken down to amino acids) can enter the citric acid cycle and generate ATP. However, the amount of ATP produced is less then if pyruvic acid (from glucose) is utilized. In addition, during fat metabolism ketone bodies are generated, which at high concentrations can lead to metabolic acidosis. In diabetes mellitus the production of such ketone bodies can lead to diabetic comma and possibly death.

Creatine Phosphate. Energy derived from ATP production is used for cross bridge movement and to drive the SR Ca^{++} pump, or it can be stored in the form of *Creatine Phosphate* for later use:

Creatine + ATP ——> Creatine Phosphate + ADP

and ATP can then be resynthesized from creatine phosphate when required:

Creatine Phosphate + ADP ——> Creatine + ATP

Fast and Slow Muscle and Myoglobin. Skeletal muscle can be divided into red (or slow) and white (or fast) muscle. Red muscle cells contain relatively low levels of stored glycogen, a large number of mitochondria, and a high concentration of the protein *myoglobin*, which is an O_2 binding protein related to hemoglobin in the red blood cell. The myoglobin in the red muscle cell supplies extra O_2 to allow the cell to synthesize ATP from the oxidative phosphorylation pathways in the mitochondria. Since red muscle can efficiently produce large amounts of ATP this way, it is very resistant to fatigue and can maintain contractions for long periods. It is thus the muscle that maintains body posture. However, it also is slow to contract after a stimulus is applied.

White muscle cells contain large amounts of stored glycogen, low levels of myoglobin and few mitochondria, and synthesize the majority of their ATP from anaerobic glycolysis. White muscle cells can contract quickly after stimulation, but also rapidly fatigue since anaerobic glycolysis is very inefficient and ATP is quickly used up when the glycogen stores run out.

MUSCLE FATIGUE

After sustained contractile activity, the tension in the muscle begins to decline. This is termed *muscle fatigue* and results from:

1. Accumulation of lactic acid which may inhibit enzyme action within the muscle.
2. Depletion of energy reserves such as glucose or glycogen.
3. Depletion of ACh at the neuromuscular junction which prevents transmission of stimuli from the nerve to the muscle.
4. Psychological factors (psychological fatigue) when the CNS no longer activates the motor neuron supplying the working muscles.

MUSCLE MECHANICS

There are two primary types of contractions that can be generated in skeletal muscles:

1) An *isometric contraction* takes place if the muscle is not allowed to shorten and instead simply experiences an increase in the tension. Such a contraction can be produced if both ends of the muscle are anchored in place.

2) An *isotonic contraction* takes place if the muscle is allowed to shorten during a contraction while the tension remains constant. In this case only one end of the muscle is anchored to a support while the other is freely moving or may be attached to a small weight which it is capable of being lifted. There are two types of isotonic contractions:

   a. A *concentric isotonic* contraction takes place if the muscle shortens during contraction.

   b. An *eccentric isotonic* contraction takes place if the muscle lengthens or is stretched while contracting. Such an effect might take place if the muscle resists stretching. As a weight is being lowered muscle stretch takes place and the length increases. To resist this stretch, the muscle contracts and the tension supports the weight of the object.

VELOCITY OF SHORTENING V. LOAD

The velocity of shortening (V) is the speed (change in length/time) with which the muscle shortens during an isotonic contraction. The velocity of shortening depends on the weight the muscle is lifting. Thus, if the muscle is lifting no weight V will be maximum, and as more and more weight is added V is reduced **(Figure 6.13)**. Eventually so much weight is added that V becomes 0. It can be seen that at

0 weight, when V is maximum, this represents a *pure isotonic contraction*. On the other hand, when the weight is maximum such that V is 0, tension in the muscle is increasing but the length is not changing, and this represents a *pure isometric contraction*.

MUSCLES OPERATE IN PAIRS

Because the sarcomere in a muscle fiber can only shorten, muscles can only undergo active contraction and shortening but never active relaxation and lengthening. Thus, for any movement a pair of muscles must be involved; one to make the movement and the other to reverse the movement made by the first muscle.

1) *Prime movers or agonists* are those muscles that actually make the movement. For example, consider the act of flexing the arm. Here the biceps contract and shorten and in so doing produce flexion of the arm. At the same time, the agonist will stretch the antagonist muscle (the triceps).

2) *Antagonists* are those muscles that reverse the movement made by the prime movers. Thus, the triceps cause extension of the arm and in addition stretch the agonist muscle (the biceps).

Note that if the movement of interest was not flexion, but instead was extension of the arm, then the triceps would be considered as the prime mover and the biceps would be the antagonist. For many movements more than two muscles are involved. These additional muscles that assist the agonist or antagonist are called the *synergists*.

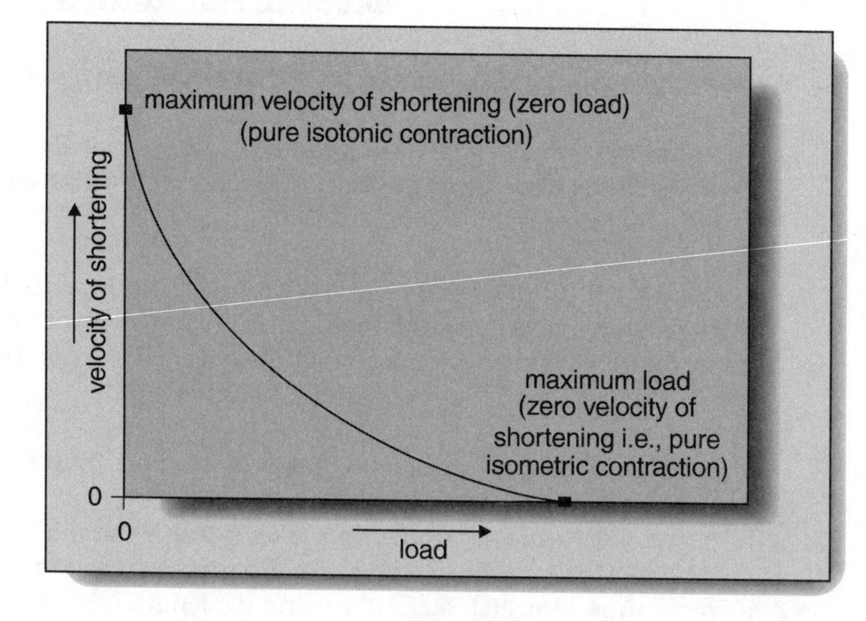

FIGURE 6.13

SMOOTH MUSCLE V. CARDIAC MUSCLE

SMOOTH MUSCLE

This type of muscle is quite different in many ways from skeletal muscle. The name *smooth muscle* indeed defines its appearance, since it has no banding pattern as does skeletal muscle. Smooth muscle is involved in maintenance of body homeostasis since it is found in the walls of hollow organs. In this location, it can control the movement of substances within these regions by constriction or dilation of the organ. Thus, it controls air flow in the respiratory system, food movement in the digestive system, blood flow in the circulatory system, and so on. Major differences between smooth and skeletal muscle are listed below:

1. Smooth muscle is innervated by the autonomic nervous system (the involuntary portion of the nervous system), while skeletal muscle is innervated by the somatic nervous system (the voluntary portion).

2. Smooth muscle cells are small and spindle shaped with a single nucleus per cells. Skeletal muscle cell are long and are multinucleated.

3. Smooth muscle cells have actin thin filaments, myosin thick filaments and filaments of intermediate size that may not be directly involved in the process of contraction. However, smooth muscle cells do not contain any troponin and tropomyosin. Instead smooth muscle cells have the substance called *calmodulin* that regulates the contraction process (see following material).

4. Smooth muscle cells have a poorly organized sarcoplasmic reticulum.

5. Smooth muscle cells have no specific location for a single neuromuscular junction, as do skeletal muscle cells. Instead receptors for neurotransmitter are positioned over the entire membrane.

6. A single smooth muscle cell can receive stimuli from more than one autonomic fiber at the same time. A single skeletal muscle cell, on the other hand, can be controlled by only a single branch of a motor neuron, and never by more than one motor neuron.

7. Stretching the membrane of a smooth muscle cell can cause it to contract.

8. Smooth muscle cells can be either stimulated to contract, or inhibited and prevented from contracting by the action of different kinds of neuro-transmitters or by hormones acting directly through the receptors on the cell membrane (see following material).

9. The rate of contraction of smooth muscle cells can be controlled by pacemaker cells (see following material).

NEUROTRANSMITTERS AND HORMONES CONTROL SMOOTH MUSCLE

Smooth muscle cells are under the control of neurotransmitters from the autonomic nervous system. Commonly the two transmitters involved are norepinephrine released from the post-ganglionic sympathetic adrenergic fibers, and Ach released from the postganglionic parasympathetic fibers. However, hormones from the adrenal gland, such as epinephrine (adrenalin) and norepinephrine can also affect smooth muscles. In comparison, skeletal muscle cells can only be simulated to contract by the release of Ach from somatic fibers. Skeletal muscle cells cannot be inhibited directly through transmitter binding to receptors on the cell surface. Any inhibition of contraction takes place in the spinal chord through the action on the motor neuron cell bodies.

MECHANISM OF CONTRACTION OF SMOOTH MUSCLE CELLS

Activation of the smooth muscle cell can be accomplished when:

1. Neurotransmitter binds to the membrane receptors.
2. Hormones or other substances bind to membrane receptors.
3. Physical stretch of the membrane takes place.
4. Electrical stimuli are conducted from one cell to the next through gap junctions.

Upon activation Ca^{++} levels rise in the cytoplasm of the smooth muscle cell (an increase in *cytosolic* Ca^{++} takes place). This Ca^{++} is released from the vesicular sarcoplasm reticulum, and it also enters the cytosol from the outside by passing through Ca^{++} channels that open in the outer membrane. The Ca^{++} binds with *calmodulin* that is structurally similar to troponin **(Figure 6.14)**. The *Ca^{++}-calmodulin complex* now activates an *inactive myosin kinase.* The activated myosin kinase *phosphorylates myosin* binds with the actin causing the cross bridge cycle to begin.

To terminate the contraction, the Ca^{++} is removed by the action of Ca^{++} active transport pumps which transport the Ca^{++} back into the vesicular SR and also out of the cell. In the absence of cytosolic Ca^{++}, the myosin is dephosphorylated and can no longer interact with the actin, and thus the muscle cell relaxes.

It should be noted that in smooth muscle the Ca^{++} turns on the cross bridge cycle by producing a change in the myosin thick filaments, while in skeletal muscle it turns on the contraction by inducing a change in the thin actin filaments.

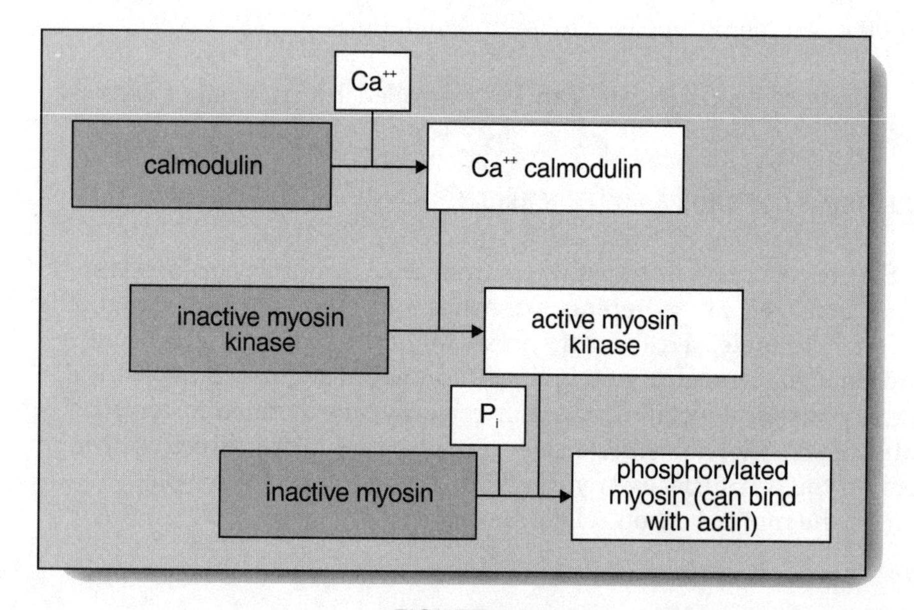

FIGURE 6.14

SINGLE-UNIT AND MULTI-UNIT SMOOTH MUSCLE

Smooth muscle cells can be found in two different arrangements termed *single-unit* or *multi-unit.* The cells in both arrangements function as described above but there are some differences that are outlined here.

Multi-unit. This refers to the fact that each smooth muscle cell functions independently, but all of the cells in the group are *neurogenic,* meaning that they are all controlled by neurotransmitter released by nerve fibers of the autonomic nervous system as well as hormones circulating in the blood **(Figure 6.16)**.

For example, the smooth muscle in the walls of the bronchioles will undergo constriction if Ach is released by parasympathetic nerve fibers, and will undergo relaxation (and dilation) if norepinephrine is released by sympathetic fibers. In addition it will also relax if circulating epinephrine is present.

This type of smooth muscle is found in the walls of large airways in the lungs, in the walls of large blood vessels, in the ciliary body of the eye that causes accommodation of the lens, in the iris of the eye that controls the size of the pupil, and at the base of hair follicles in the *Arrector pili muscle*.

Single-unit. Also known as *visceral smooth muscle*, is found in the walls of many hollow organs. Most smooth muscle tissue in the body is single-unit smooth muscle. Because the cells in this tissue are electrically connected to each other through *gap junctions*, electrical excitation of one cell in the "unit" causes all the other cells in the unit to also become electrically excited and to contract as a "single unit" **(Figure 6.16)**. Such a unit of interconnected cells is a *functional syncytium*.

In addition, single-unit smooth muscle is *self-excitable* meaning that there are *pacemaker cells* that control the contraction rate for all the other cells in the unit, since the pacemaker is connected to the other cells through a functional syncytium. However, the depolarization rate of the pacemaker is under the control of neurotransmitters released from autonomic fibers as well as other factors such as hormones circulating in the blood.

Thus, in the GI tract, peristaltic activity of smooth muscle in the walls is controlled by pacemakers. Peristalsis can be increased when ACh is released by parasympathetic fibers and decreased when norepinephrine is released by sympathetic fibers.

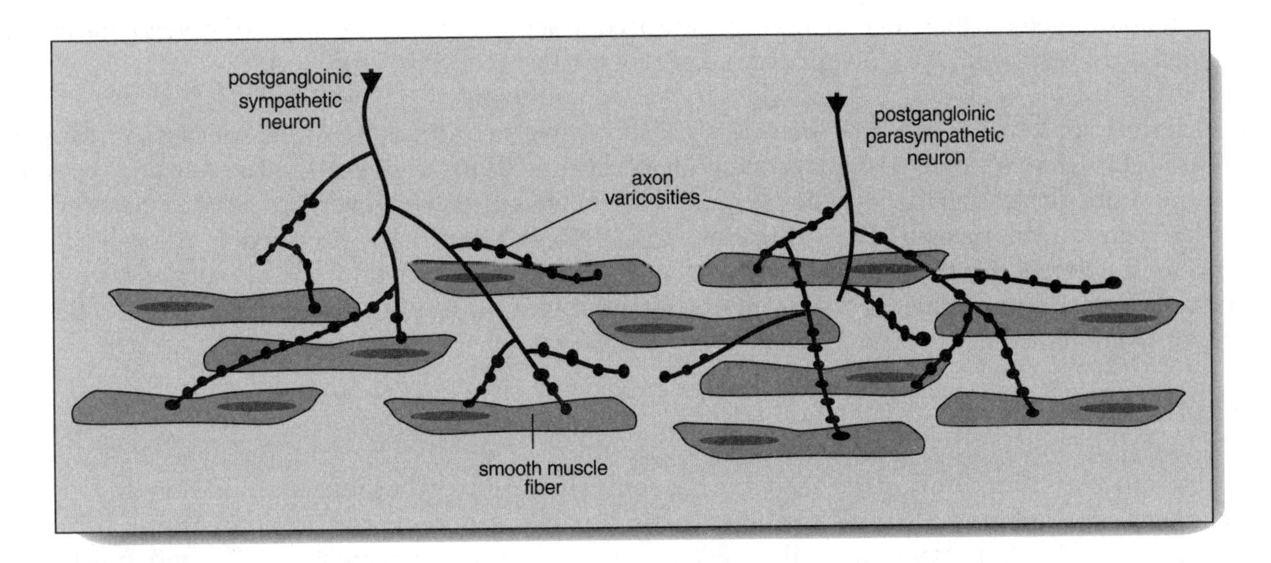

FIGURE 6.15

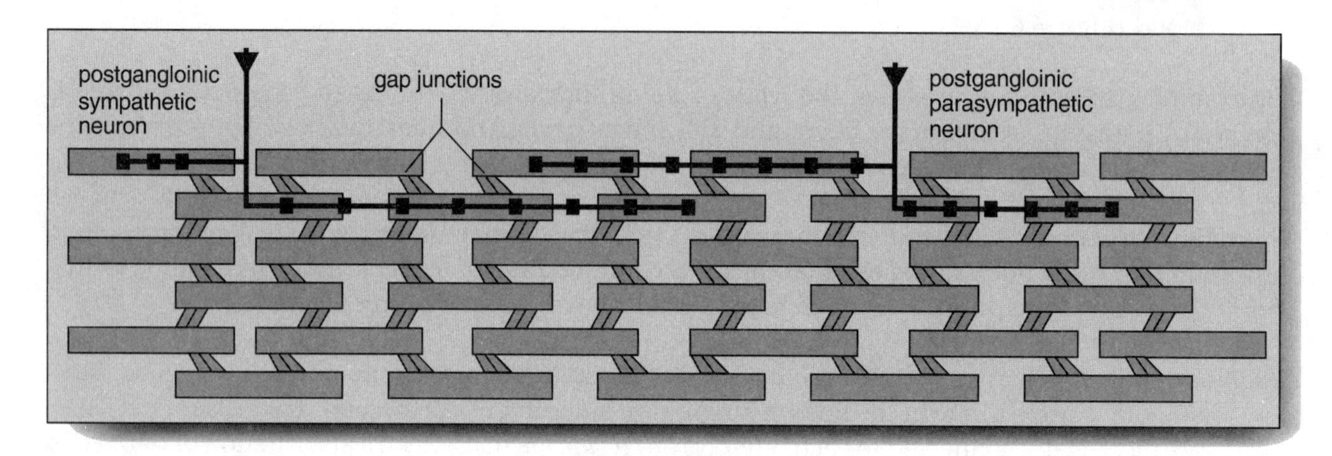

FIGURE 6.16

PACEMAKERS

Pacemaker cells have an unstable resting membrane potential (*pacemaker potential* or activity) that continuously depolarizes to threshold, and then repolerizes **(Figure 6.17)**. Passive ionic leakage across the membrane of the pacemaker cell results in a rise of the resting membrane potential until it reaches threshold. At the threshold level the active opening of ionic channels in the membrane then leads to the generation of an action potential and contraction. The cyclic rate of depolarization and repolerization is different in different pacemaker cells due to the different rates of ionic leakage (permeability) into the cell. Thus some pacemakers undergo rapid contractions while others undergo slower contraction cycles.

If in a functional syncytium two pacemakers are linked together along with other non-pacemaker cells, the faster pacemaker in the unit will drive the rate of contraction of the entire unit. However, if the faster pacemaker is destroyed, the slower pacemaker will take over and drive the unit.

In addition to pacemaker potential, *slow-wave potential* can also be found in single-unit smooth muscle **(Figure 6.18)**. Here a slowly rising and falling electrical potential is present in the cells, but even at maximum depolarization the level never quite reaches threshold. However, the peak of the depolarization cycle can be pushed above threshold if additional stimuli are supplied to the cell. If this happens, it will result in the generation of action potentials and contraction. Factors such as hormones and neurotransmitters influence the rise of the potential above threshold to promote the generation of action potentials.

CARDIAC MUSCLE

Cardiac muscle is found only in the heart and has some similarities to both skeletal and smooth muscle. It contains an organized SR with T-tubules, actin, myosin, troponin and tropomyosin, a banding pattern and a single nucleus per cell. However, it also contains pacemaker cells, and functional

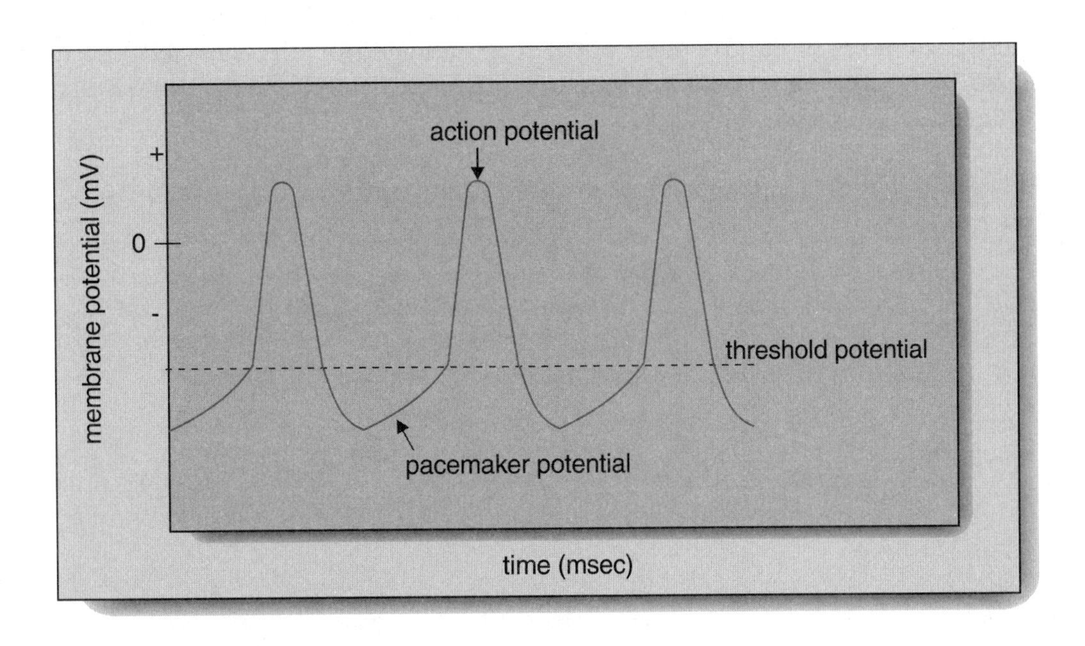

FIGURE 6.17

syncytium in both the atria and ventricles. Cardiac cells are connected to each other through *intercalated discs* which contain gap junctions and thus can be controlled by pacemaker activity. In turn the rate of depolarization of the pacemaker is regulated by fibers of the autonomic nervous system and by circulating hormones.

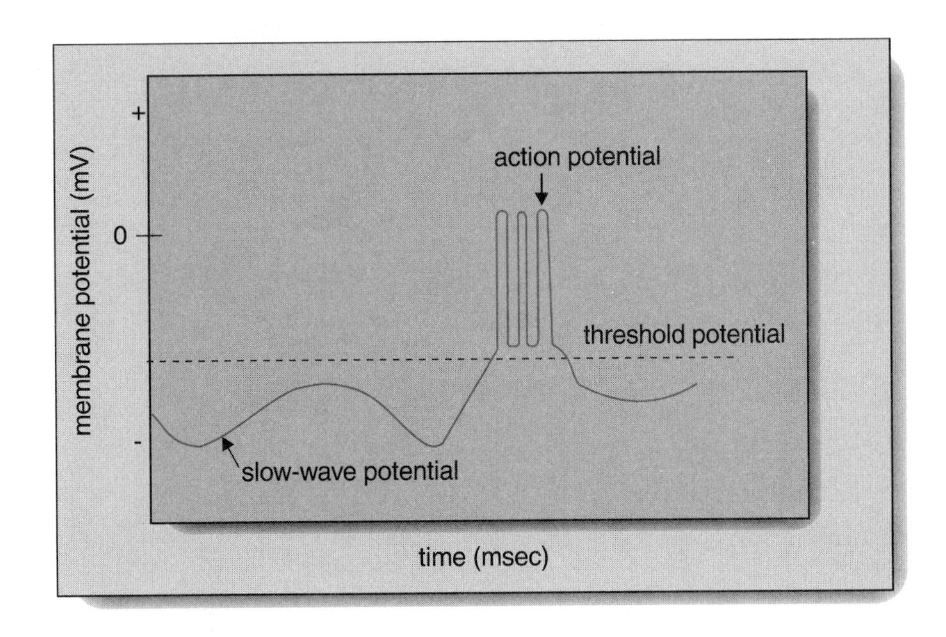

FIGURE 6.18

Like pacemakers in single-unit smooth muscle, in the heart the fastest pacemaker drives the system. Here the fastest pacemaker is the SA node but if it is damaged or disconnected, the next fastest pacemaker is the AV node which can take over to drive the ventricles. If the AV node is damaged, the *Purkinje* pacemaker drives the ventricles.

This will all be discussed fully in the chapters devoted to the cardiovascular system.

Chapter 7
CARDIAC PHYSIOLOGY

THE CARDIOVASCULAR SYSTEM: AN OVERVIEW

Ever since the first creature crawled out of the ancient oceans to live on dry land, land dwellers have had to carry a little piece of the primordial ocean around with them in order to maintain fluid, electrolyte and gas homeostasis. The cardiovascular system is central to the maintenance of these homeostatic conditions. The main components of this system are:

1. The circulating liquid blood.
2. The various blood vessels that carry the blood to all locations in the body, and also allow for exchange of gases, nutrients and waste products with the cells.
3. The heart that supplies the pumping action to maintain hydrostatic pressure within the cardiovascular system.

The following chapters will consider all aspects of the cardiovascular system.

THE HEART

WALL STRUCTURE

The heart is a muscular pump, approximately 14 cm long by 9 cm wide, located in the mediastinum of the thorax cavity. The heart is covered by an outer *fibrous pericardium* surrounding two layers of inner pericardium (the *parietal pericardium* and the *visceral pericardium*). The parietal and visceral pericardium are composed of serous membrane and the inner visceral pericardium (also known as the *epicardium) is* closely associated with the muscle tissue (*myocardium*) of the heart wall. The outer parietal pericardium forms the inner layer of the fibrous pericardium and is separated from the inner visceral pericardium by a *pericardial cavity.* A small volume of serous fluid, secreted by the pericardial membranes is present in this cavity. This fluid acts to reduce friction between the parietal and visceral pericardium when the heart is moving during a contraction.

ARRANGEMENT OF THE CHAMBERS, VESSELS AND VALVES

The heart is a four-chambered structure and contains two thin walled upper chambers called the *atria* and two thick walled lower chambers called the *ventricles* **(Figure 7.1)**. The right and left ventricles are separated by the *interventricular septum.* Lying along the inner margins of the interventricular septum are the right and left bundle branches which are part of the conduction system of the heart (see following material).

Both the *superior and inferior vena cava* are attached to the right atrium, and it is through these great veins that blood from the systemic circuit returns to the heart. Blood exits the right ventricle and enters the attached right and left pulmonary arteries and then after passing through the pulmonary circuit returns to the left atrium through the right and left pulmonary veins. Finally the blood exits the left ventricle by entering the massive aortic arch and from there passes into the systemic circulation.

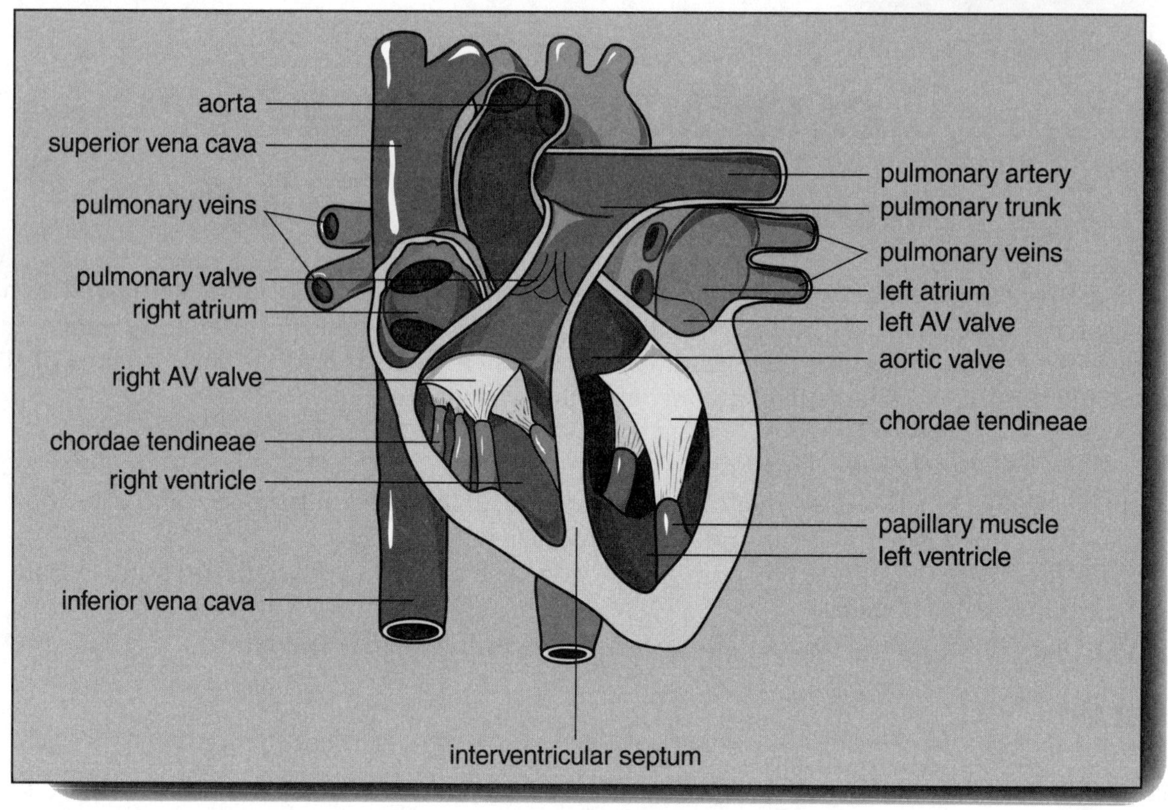

FIGURE 7.1

The *atrioventricular (AV) valve* that separates the right atrium from the right ventricle is the *tricuspid valve,* so named because it is composed of three leaf-like flaps. The left AV valve separates the left atrium from the left ventricle and is named either the *bicuspid valve* (it has two leaf-like flaps) or the *mitral valve*. The two *semilunar valves,* so named because of the half moon appearance, allow blood to exit from the ventricles and enter the attached arteries. The right valve is called the *pulmonic semilunar,* and the left valve is named the *aortic semilunar* valve.

Attached between the margins of the two AV valves are thin string-like chords called the *Chordae tendineae*. These Chordae tendineae extend to the *Papillary muscles* located in the inner walls of the ventricles. During a contraction of the ventricles (*ventricular systole*) the papillary muscles also contract and apply tension to these chordae tendineae. This increased tension applied to the margins of the valves supports the valves and prevents them from collapsing backwards into the atria. Therefore even in the left ventricles where the pressure can rise to at least 120 mm Hg, the AV valve does not buckle and allow blood to leak into the left atria.

Surrounding all four valves are rings of dense fibrous connective tissue. This connective tissue is located between the atria on the top of the heart and the ventricles at the bottom and is known as the *skeleton of the heart*. The function of this skeleton is to act as an anchor point for the valves, preventing then from tearing away from the heart during a contraction. In addition, because this connective tissue is electrically non-conductive this arrangement effectively prevents electrical stimuli generated in the atria from passing directly into the ventricles during a contraction.

BLOOD FLOW THROUGH THE HEART

The direction of the blood flow through the heart is outlined in **Figure 7.2**. Since this is a one-way system, the valves located between the atria and ventricles, and between the ventricles and arteries allow blood to pass only in one direction and inhibit back flow. The valves open or close depending on the pressure on the two sides as follows:

1. The right AV valve opens when the pressure in the right atrium is greater than the pressure in the right ventricle.

2. The right AV valve closes when the pressure in the right ventricle is greater than in the right atrium.

3. The pulmonic semilunar closes when the pressure in the pulmonary artery is greater than in the right ventricle.

4. The pulmonic semilunar opens when the pressure in the right ventricle is greater than in the pulmonary artery.

5. The left AV valve opens when the pressure in the left atrium is greater than in the left ventricle.

6. The left AV valve closes when the pressure in the left ventricle is greater than in the left atrium.

7. The aortic semilunar closes when the pressure in the aorta is greater than in the left ventricle.

8. The aortic semilunar opens when the pressure in the left ventricle is greater than in the aorta.

Although there are no valves between the atria and the attached veins, back-flow of blood into the veins (especially the vena cava) is limited because:

1. The opening between the veins and the atria are partially compressed during an atrial contraction, and

2. The pressure in the atria normally is not much greater than the pressure in the attached veins.

However, should one of the AV valves begin to leak during ventricular systole (known as *valvular insufficiency*), this may cause blood to also enter the attached veins in the wrong direction resulting in major clinical problems including an increase in venous pressure and a drop in cardiac output.

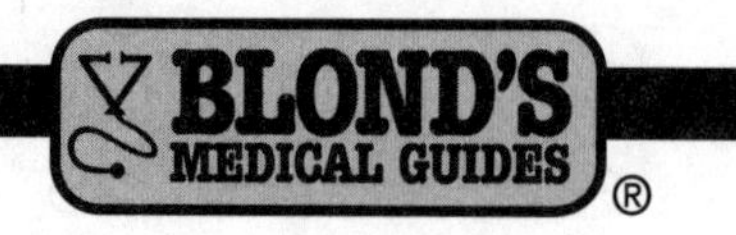

CORONARY CIRCULATION

The *right and left coronary arteries* supply blood to the heart tissues. These vessels attach to the base of the aortic arch immediately after it leaves the left ventricle and lie within the pericardial cavity on the surface of the epicardium.

1. The left coronary artery branches into the:

   a. *Circumflex artery* that supplies the walls of the left atrium and left ventricle.

   b. *Anterior interventricular artery* (or *left anterior descending artery*) that supplies the walls of both ventricles.

2. The right coronary artery branches into the:

   a. *Posterior interventricular artery* that supplies the walls of both ventricles.

   b. *Marginal artery* and supplies the walls of the right atrium and right ventricle.

After passing through the capillary beds that supply the cardiac tissues, the blood drains into a network of coronary veins and then into the coronary sinus, located on the posterior side of the heart, which then empties into the right atrium.

CORONARY BLOOD FLOW IS MATCHED TO OXYGEN NEEDS

The blood vessels in the heart are capable of undergoing a form of autoregulation regulated by the need for oxygen. Thus:

1. As metabolic activity of the cardiac muscle increases, levels of O_2 fall.
2. Adenosine levels increase.
3. Coronary vessels undergo dilation.
4. Blood flow in the vessels increases.
5. Increased levels of O_2 are delivered to the cardiac cells.
6. Cells function more efficiently.

THE MYOCARDIUM AND INTERCALATED DISCS

Cell (fibers) of the myocardium are arranged in planes separated by layers of connective tissue. The interlacing bundles of muscle fibers are organized is such a way that during a contraction the twisting and shortening of the chambers forces the blood upward towards the valves located at the base of ventricles. Individual cardiac muscle cells are interconnected by *intercalated discs* that consist of both *gap junctions and desmosomes*. The gap junctions provide an electrical pathway through which action potentials can travel from one cell to the next creating a functional syncytium.

A *functional syncytium* is a group of muscle cells that function electrically and mechanically as a unit. This term has been used previously in relation to single unit smooth muscle.

PACEMAKER ACTIVITY IN THE HEART

Much like single unit smooth muscle, cardiac muscle cells contain pacemakers that initiate and regulate the process of contraction. The majority of the cardiac cells (99%) are not autostimulatory; that

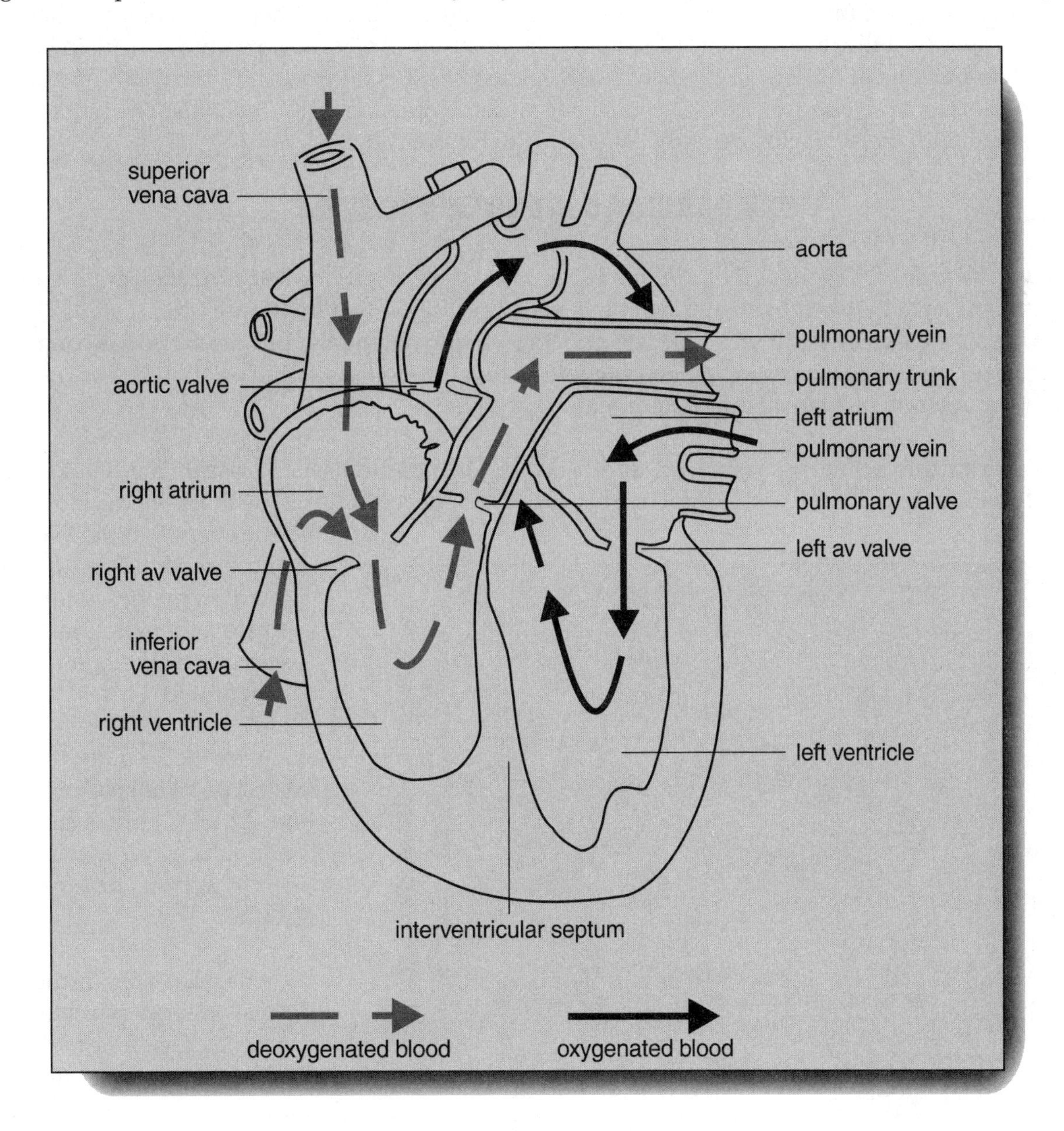

FIGURE 7.2

is, they are not pacemakers and must be stimulated by the pacemakers to contract. These pacemaker cells are cardiac muscle cells with unstable membrane potentials that continuously depolarize to threshold, generate an action potential and then repolarize again. The action potentials thus generated then spread throughout the heart to cause rhythmic contractions without the need for nerve stimulation.

Although it has not been absolutely established, it is believed that in pacemaker cells the depolarizing drift of the membrane potential towards threshold results from a slow and cyclic decrease in the outflow of K^+. In this scenario, the membrane permeability to the K^+ ion decreases due to closure of K^+ channels, but a continued slow leak of positively charged Ca^{++} or Na^+ causes the inside to gradually become less negative. Eventually threshold is reached and at this point activation of Ca^{++} channels causes Ca^{++} permeability to increase. The resulting large inflow of Ca^{++} now produces the rising phase of the action potential (unlike a nerve cell where the rising phase is due to an inflow of Na^+). The falling phase results from the closure of the Ca^{++} channels and the opening of K^+ channels. The whole cycle begins again with the slow closure of the K^+ channels **(Figure 7.3)**.

THE CARDIAC CONDUCTION SYSTEM

The *Sinoatrial node or SA node* is the primary pacemaker of the heart, located near the opening of the superior vena cava in the right atrium wall. It is from this region that heart rate is regulated. The SA node is the fastest pacemaker with a base rate of 70 to 80 action potentials/minute. Since it is connected to the remaining structures through a conduction system (see following material), it thus drives the entire heart at a rate of 70 to 80 beats/minute.

In addition to this primary pacemaker, additional autorhythmic tissues exist in the heart. These alternate pacemakers are not normally active because their effect is eclipsed by the SA node. However, in the event that the SA node is nonfunctional or disconnected from the conduction pathway, then these alternate pacemakers take over to drive the ventricles.

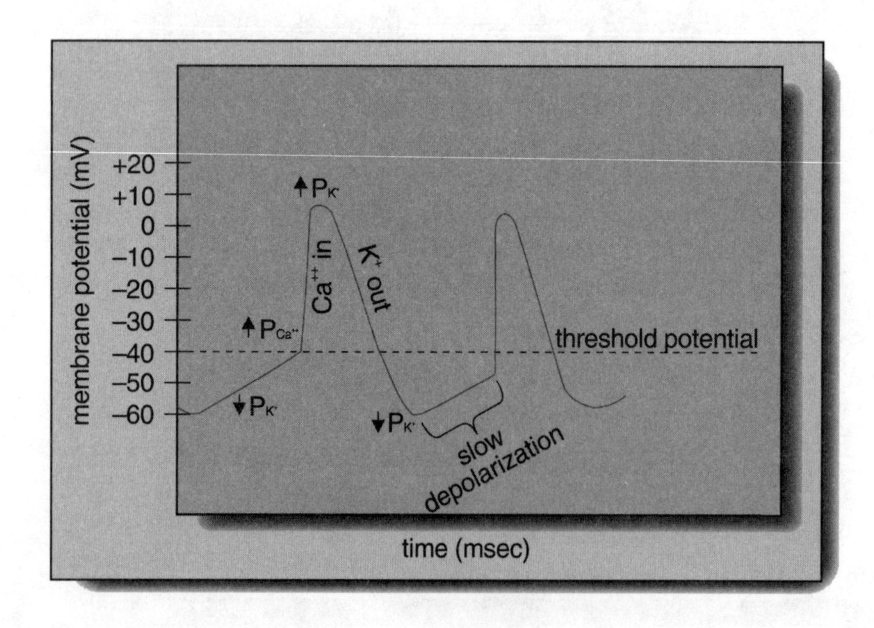

FIGURE 7.3

1) After the SA node, the next fastest potential pacemaker is the Atrial Ventricular (AV) node which can generate 40 to 60 action potentials/minute.

2) After the AV node the next fastest potential pacemaker is found in the Bundle of His and Purkinje fibers which can generate 20 to 40 action potentials/minute.

Under normal conditions, the SA node generates action potentials 70 to 80 times/minute which then passes via the atrial syncytium to all atrial cells resulting in contraction of both atria almost simultaneously.

ATRIOVENTRICULAR (AV) NODE

In order to bridge the non-conductive fibrous skeleton of the heart that divides the atrium from the ventricles, the atrial depolarization wave then passes into the *junctional fibers* that connect to the AV node located on the floor of the right atrium close to the atrial septum. These *junctional fibers* are of small diameter resulting in a delay in impulse conduction into the AV node. Further delay in conduction also takes place within the AV node itself. The presence of this delay allows the atrium to complete contraction before the ventricles begin to contract (see following material).

AV BUNDLE OF HIS, BUNDLE BRANCHES AND PURKINJE FIBERS

Having passed through the AV node, the impulses then enter the large diameter fibers of the *AV bundle of His* where they are then rapidly conducted. The AV Bundle then branches at the top of the septum into the *Left and Right Bundle Branches* lying just beneath the endocardium on the septum walls and then passes into the *Purkinje fibers*. The Purkinje fibers spread out from the interventricular septum into the papillary muscles of the ventricle walls, and also pass into the *apex* (or tip) of the heart, and then into the ventricle walls **(Figure 7.4)**. The impulses can then enter the ventricular syncytium in the walls of the heart resulting in a wave of contractions passing from the apex towards the base. Since the myocardial fibers in the ventricles are arranged in a cross circular pattern, the resultant contraction appears as a twisting, or "wringing out" motion.

ECTOPIC FOCUS

Under abnormal conditions (such as after damage from a myocardial infraction or in response to stress, nicotine, caffeine or other drugs) one of the alternate potential pacemakers may begin to fire faster than the normal SA nodal pacemaker. Such would be the case if the alternate Purkinje pacemaker began to fire at 140 action potentials/minute. This is then termed an *ectopic focus* and the result is a *premature beat* (*extrasystole*), or in some cases a number of such beats, of the ventricles.

HEART BLOCKS

In the event that damage is present in the SA node, or in the AV node or conduction pathways to the ventricles, the atrium and ventricles may not beat at the same rate. This may be due to an increase in the impulse delay through the AV node resulting in a normal atrial rhythm but a reduced ventricular rhythm. For example, a *2:1 block* (*or partial block*) means that there are two beats of the atrium for one beat of the ventricles. Here the atrium would beat at the normal 70 beats/minute, but due to the abnormally long delay in conduction through the AV node, the ventricles would operate at 40 beats/minute. The two rhythms are clearly associated since both the atrium and ventricles are still driven by the SA node. On the other hand in a *complete heart block* the atrium, driven by the normal SA nodal pacemaker, beats at the usual 70 times/minute; but the ventricles, driven by some alternate pace-

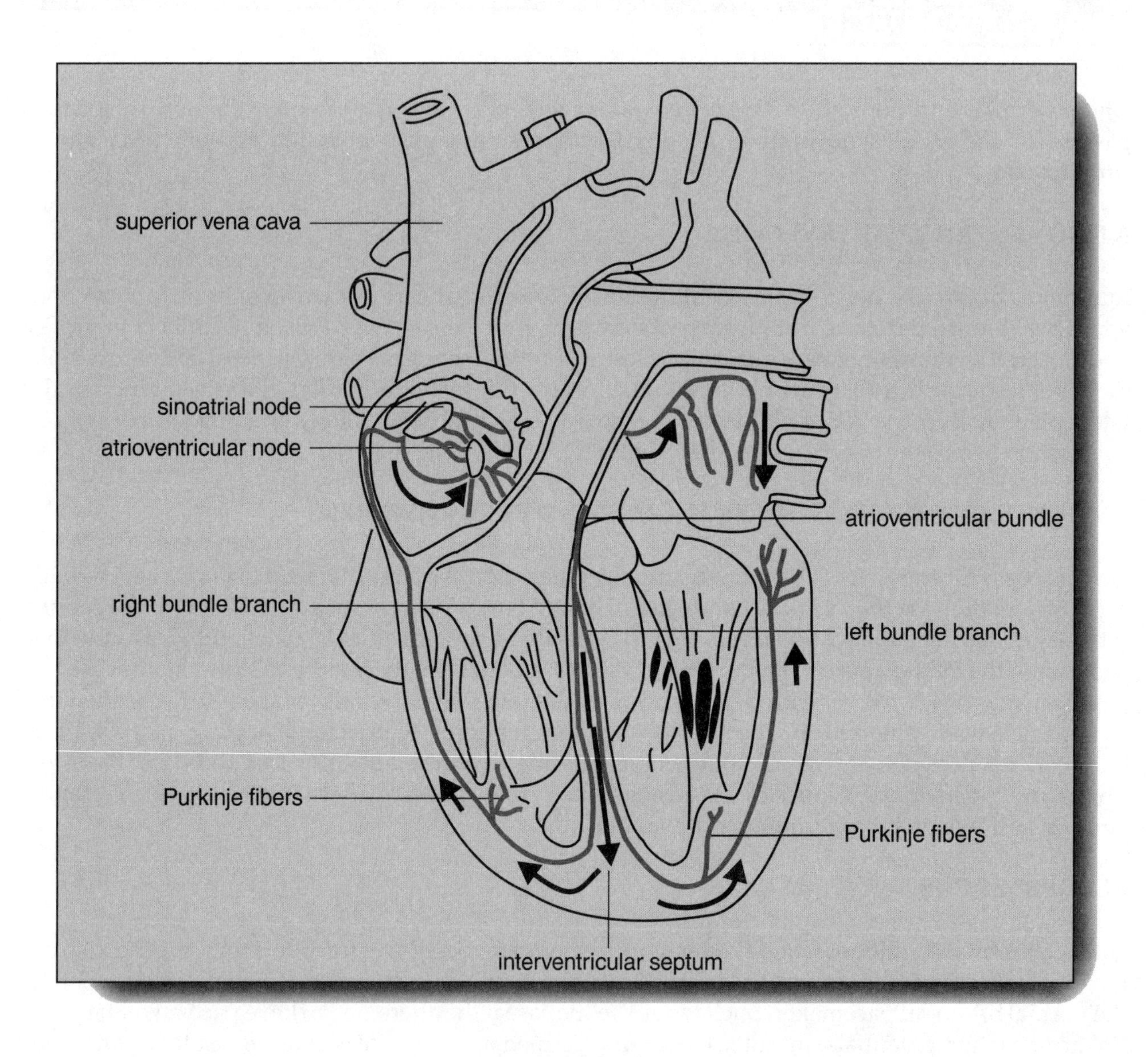

FIGURE 7.4

maker, will beat slower. In this case the two rhythms are not associated at all and an electrocardiogram tracing appears as a mixture of conflicting wave patterns (see following material).

ACTION POTENTIALS OF PACEMAKER AND NON-PACEMAKER CELLS

As just mentioned, the action potential in the pacemaker cells is unstable driven by changes in the permeability of the ions through the membrane. Comparison between the action potential generated in a pacemaker **(Figure 7.3)** and non-pacemaker cell **(Figure 7.5)** clearly demonstrates the presence of the unstable membrane potential in the pacemaker cell resulting in cyclic threshold stimuli. In the non-pacemaker cell, the stable resting membrane potential of -90 mV is maintained unless the cell receives a threshold stimulus from another non-pacemaker through a gap junction or from the pacemaker

acting through the conduction system. The events leading to the action potential in a non-pacemaker cell are as follows:

1) The rapidly rising phase of the action potential results from an increase in Na^+ permeability. The resultant large inflow of Na^+ into the cell causes the inside to depolarize to +30 mV.

2) Voltage dependent opening of "slow" Ca^{++} channels results in a slow inward diffusion of Ca^{++}. Voltage dependent closing of K^+ channels results in a decrease in the outflow of K^+. The outcome is to prolong the internal positive charge creating the plateau that lasts 250 mSec. This is unlike the

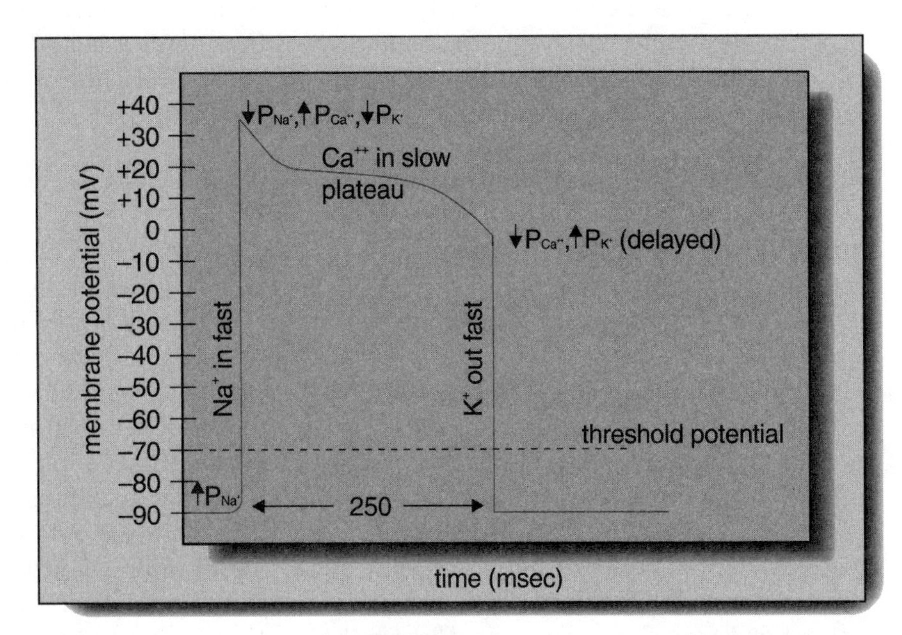

FIGURE 7.5

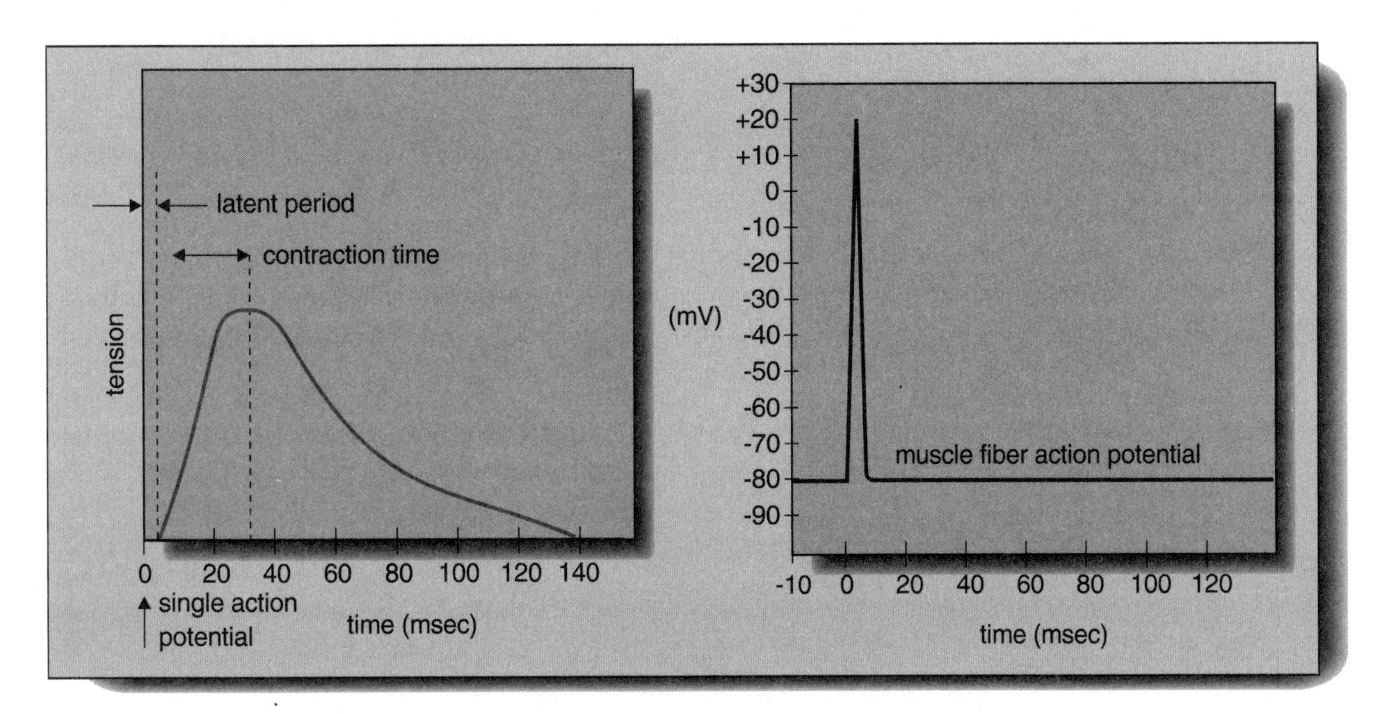

FIGURE 7.6

action potential present in a nerve cell where the repolarization phase begins immediately after completion of the inflow of Na^+. As a direct result of this prolonged plateau the cardiac cell remains in absolute refractory period for almost the entire period of the action potential (see following material).

3) The repolarization phase of the action potential results from the closure of the Ca^{++} channels in the membrane and the decrease in the inflow of Ca^{++}. In addition, the opening of K^+ channels now promotes outflow of K^+ from the cell. This results in the rapid repolarization of the cell back to -90 mV resting membrane potential.

THE REFRACTORY PERIOD IN CARDIAC MUSCLE CELLS

During the refractory period, stimuli applied to a muscle cannot produce an additional contraction. In skeletal muscle the duration of the action potential (and thus the length of the refractory period) is about 2 msec, but the generation of tension in the cell lasts for more than 100 msec. Thus, in skeletal muscle if an additional stimulus is applied at the end of the refractory period (after 2 msec) but before the muscle relaxes (during the next 98 msec), an increase in the internal tension results **(Figure 7.6)**. It is for this reason that additional stimuli (or increased frequency of stimulation) to a skeletal muscle can produce greater and greater tension eventually resulting in a tetanic contraction. On the other hand, within cardiac muscle the duration of the action potential is 250 msec, and therefore the refractory period also lasts for 250 msec. **(Figure 7.7)**. Since the generation of tension in cardiac muscle lasts 300 msec, during most of this period of contraction, application of additional stimuli cannot cause an increase in tension. In other words, cardiac muscle cannot undergo summation of stimuli and tetanic contraction as can skeletal muscle. Thus, in the heart even if pathological stimuli are present, blood will continue to be pumped, and homeostasis and life will be maintained.

ELECTROCARDIOGRAM

Like other muscles the heart generates electrical stimuli during its contraction process. Since the tissues of the body can act as an electrical conductor, with appropriate equipment this electrical activity of the heart can be monitored through surface electrodes attached to amplifying equipment. The resultant wave-form graphed on paper or projected onto a cathode ray tube is termed an *electrocardiogram* (ECG).

1. The ECG measures only the electrical activity present at the surface of the body and not directly on (or in) the heart muscle itself.

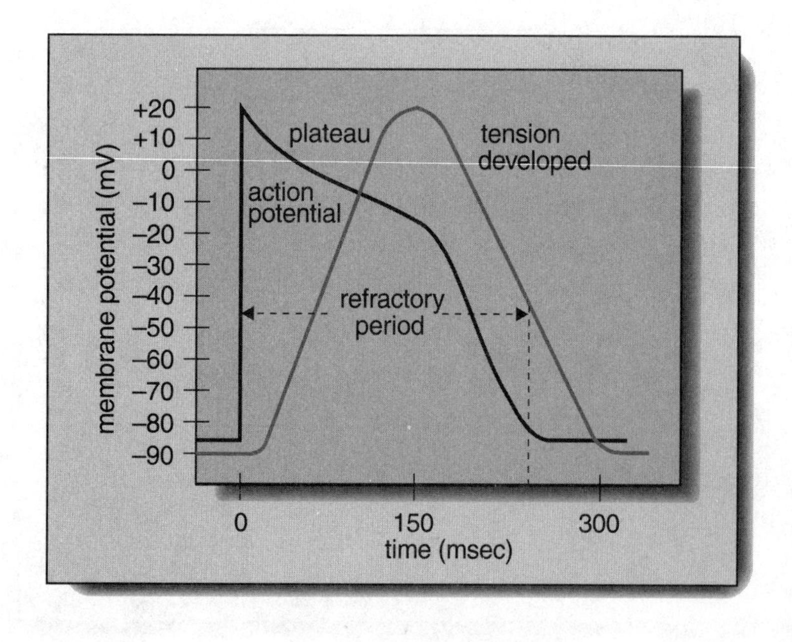

Figure 7.7

2. The ECG is not a record of a single action potential, but is generated from the combined activity of all active cells in the heart, including the pacemaker cells.

There are six limb leads (I, II, III, aVR, aVL, aVF), and six chest leads (V_1 - V_6) utilized for an ECG recording.

1. Leads I, II and III are bipolar leads because the signal that is recorded from them requires two electrodes to be used together. Actually what is measured here is the potential difference between pairs of these leads (or electrodes).

   Lead I records the potential difference between the right arm and left arm, *lead II* records the potential difference between the right arm and left leg, and *lead III* records the potential difference between the left arm and left leg.

2. The aVR, aVL and aVF leads are unipolar electrodes because they measure the signal only under the lead itself (with respect to the rest of the body).

3. The six chest leads (V_1 - V_6) are also unipolar leads but actually represent six positions on the chest around the heart. The electrode position actually records the electrical activity of the heart directly under the lead.

The tracing of an ECG wave pattern **(Figure 7.8)** is for a single heart beat (atrial and ventricular systole) and consists of the following components:

1. The *P wave* which represents the depolarization of the atrial muscle.

2. The *PR segment* which represents the AV nodal delay.

3. The *QRS complex* which represents the ventricular depolarization accompanied simultaneously by atrial repolarization.

4. The *ST segment* which represents the time during which the ventricles are contracting and emptying.

5. The *T wave* which represents ventricular repolarization.

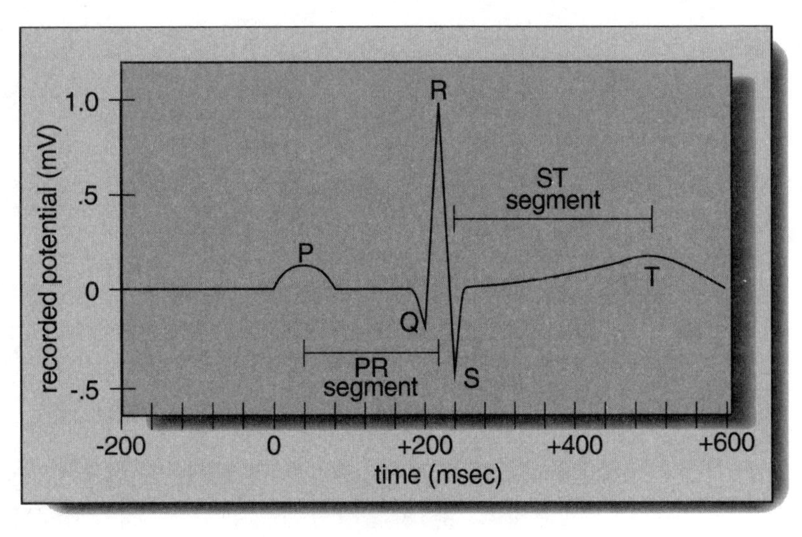

FIGURE 7.8

6. The *TP interval* which represents the time during which the ventricles are relaxing and filling.

The appearance of various ECG patterns are presented in **Figure 7.9**.

Tachycardia — abnormally fast heart beat (over 100 beats/min).

Bradycardia — abnormally slow heart beat (less than 60 beats/min).

Extrasystole(or premature heartbeat) — takes place before the normal beat of the cardiac cycle and originates from an ectopic focus that is not the normal SA node.

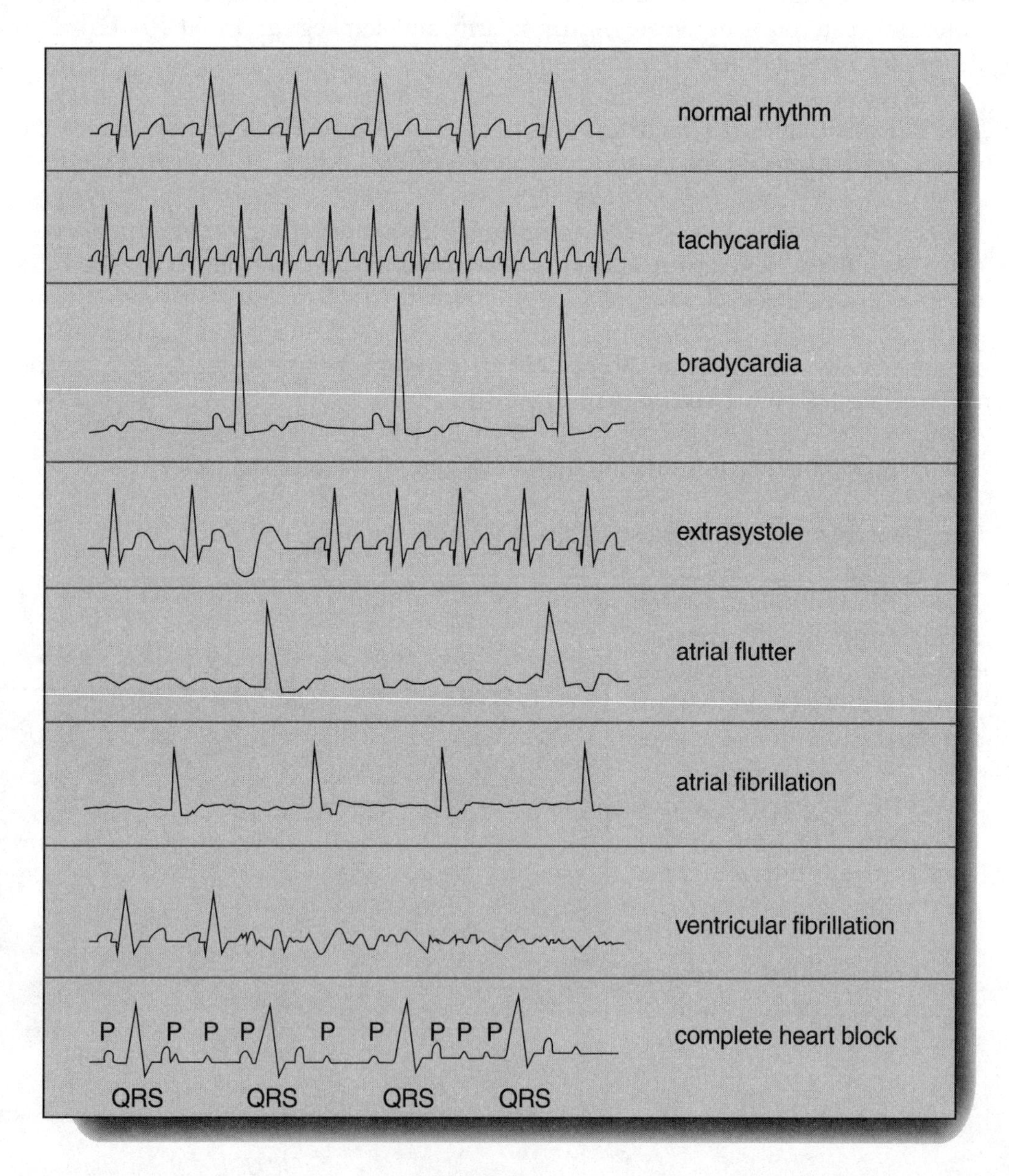

FIGURE 7.9

Flutter(atrial or ventricular flutter) — takes place when a heart chamber contracts regularly but at a very rapid rate (250 - 350 contractions/min). This is usually due to damage to the myocardium.

Fibrillation — characterized by rapid, but disorganized, heart action where small regions of the heart are contraction independently. A chamber that is undergoing fibrillation will not be able to pump any blood. Atrial fibrillation is not life threatening but ventricular fibrillation is life threatening.

Heart blocks — The various types just described.

THE CARDIAC CYCLE

The cardiac cycle describes the events that take place during one complete contraction (*systole*) and relaxation (*diastole*) in the heart. A diagrammatic representation of the process for the *left atrium and ventricle* can be found in **Figure 7.10**. A graph of these same events for the right atrium and ventricle would appear quite similar, with only a few differences as described below. The elements in this figure outline the changes in volume in the left ventricle, the changes in pressure in the left ventricle, the changes in the pressure in the aorta, the opening and closing of the various valves, the ECG, and some major heart sounds. Since this is a cycle, we will begin to study it at the time when the heart has completed about half of the diastolic phase and the ventricles are filling with blood (the numbers refer to the various parts depicted in **Figure 7.10**):

1. Pressure in the left atrium is slightly greater than in the left ventricle. Thus the left AV valve (the Bicuspid) is open and blood is flowing out of the atrium into the ventricle.

2. The volume of blood in the ventricle has already reached approximately 100 ml and continues to rise as blood flows from the atrium to the ventricle but the rate of filling has slowed down (*reduced filling phase*). In addition, blood continues to enter the atrium due to the driving force of blood from the venous system (*venous return*).

3. Blood pressure in the aorta is approximately 90 mm Hg and is greater then the pressure in the ventricle (which is only a few mm Hg). Thus the aortic semilunar valve is closed.

4. The SA node fires and the atrium depolarize. This generates the P wave of the ECG and results in contraction of the atria.

5. Left atrial pressure rises to about 10 mm Hg and this is accompanied by a similar incremental rise in left ventricular pressure.

6 Volume in the ventricle rises to 135 ml. Since the end of diastole has now been reached, this volume is termed the *End Diastolic Volume* (*EDV*). The EDV is closely associated with cardiac function and this will be considered in detail later.

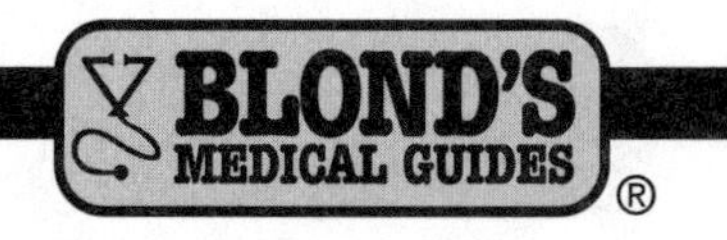

7. The impulse is now conducted through the AV nodal bundle and experiences a delay. It then passes into the bundle branches and Purkinje fibers, and then to the ventricles which now contract, leading to the QRS complex.

8. Contraction of the left ventricle results in an increase in the pressure in the ventricle which begins to rise above the pressure in the atrium. As soon as the pressure is greater in the ventricle than

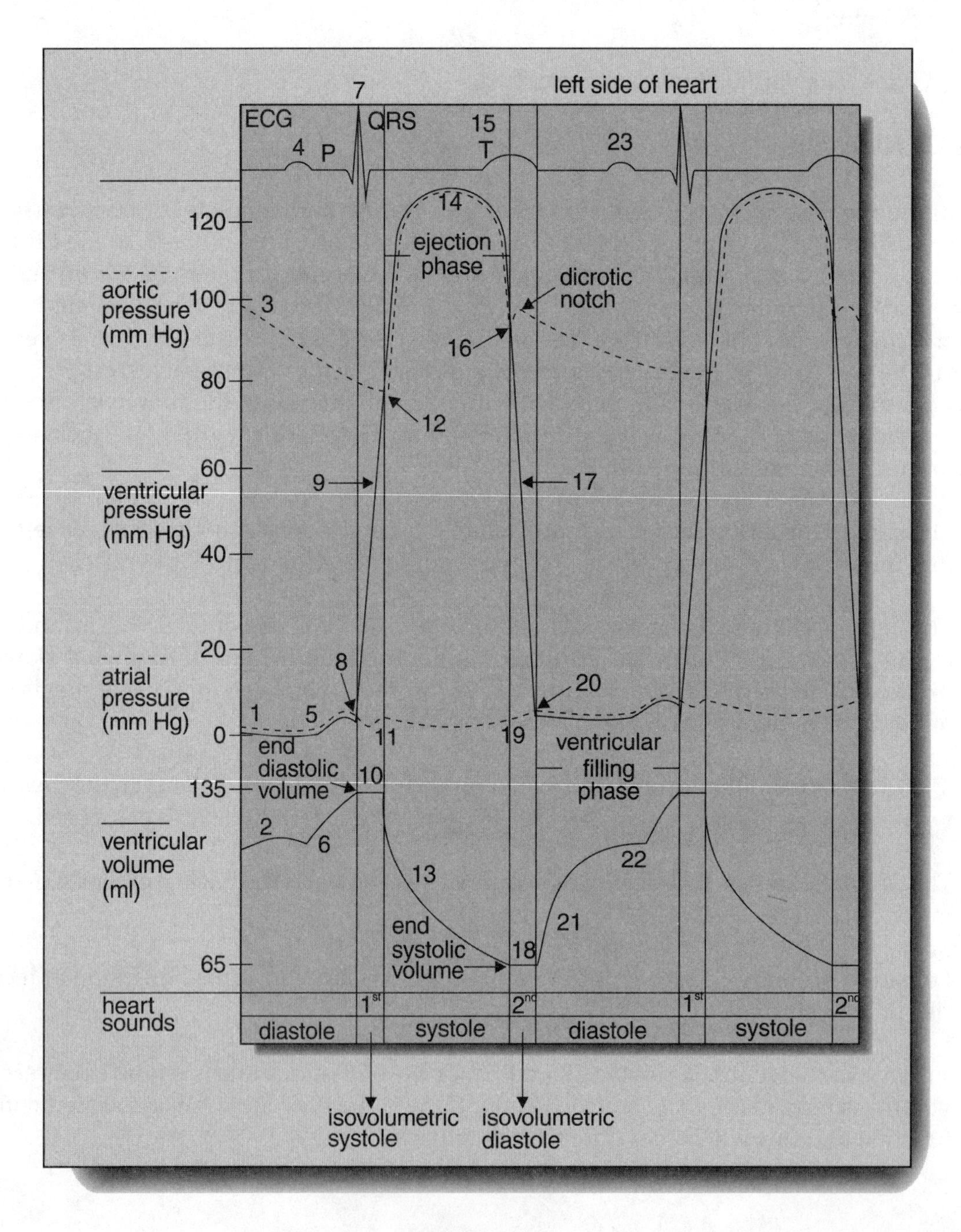

FIGURE 7.10

in the atrium, the left AV valve now closes. This represents the onset of *ventricular systole.* Note that the pressure in the aorta is still greater than in the ventricle.

9. Although ventricular pressure is rising, it has not yet reached the aortic pressure level. Thus the aortic semilunar valve remains closed. Since the AV valve is now also closed, the ventricle is contracting (systole) but against two closed valves. The pressure in the ventricle is rising as the blood is becoming more compressed, but the volume remains the same. This phase is therefore termed *isovolumetric ventricular systole* and can be compared to an isometric contraction of skeletal muscle where the length remains the same but the tension increases. Here the length of the cardiac muscle fibers remains the same, but the tension (in the form of ventricular blood pressure) increases.

10. Volume of blood in the ventricle remains constant because both the AV valve and the semilunar valve are closed.

11. As the ventricular contraction begins and the AV valve closes, the valve bulges into the left atrium. This causes the left atrial pressure to rise slightly.

12. As the pressure in the ventricle continues to rise, it will eventually reach a point where it becomes slightly greater than the pressure in the aorta. The aortic semilunar valve now opens and the *ejection phase of ventricular systole* begins.

13. As the blood is rapidly ejected from the ventricle through the open semilunar valve, ventricular volume decreases from the initial 135 ml to approximately 65 ml which is called the *end systolic volume* (ESV). The*stroke volume* (SV) is thus the volume of blood ejected during a single beat. The SV is approximately 70 ml/beat (135 ml - 65 ml = 70 ml) under conditions where the ventricle is contacting during rest or only very mild exercise.

14. The ventricle continues to contract causing the blood pressure to increase. This pressure increase is mirrored in the aortic pressure which also rises as the blood is ejected into the aorta.

15. The T wave of the ECG is temporally associated with the end of ventricular systole. Remember that the T wave represents ventricular repolarization.

16. The pressure in the ventricle falls as the contraction force slowly dissipates. When the pressure in the ventricle falls below the pressure in the aorta, the aortic semilunar valve closes. This represents the onset of *ventricular diastole.* Closure of the aortic semilunar valve results in a slight surge in the pressure in the aorta called the *dicrotic notch.*

17. The aortic valve has now closed and the AV valve remains closed so that the ventricle is relaxing; both valves are closed and blood pressure is decreasing, but blood volume in the ventricle remains constant. This period is termed *isovolumetric ventricular diastole* (and is similar to isometric relaxation in skeletal muscle where there is a decrease in tension but no change in muscle length).

18. Since both valves are closed the volume of blood in the ventricle remains constant.

19. Venous return continues to cause blood to enter the atrium. Thus atrial pressure continues to rise.

20. The pressure in the ventricle continues to fall until it is less than the pressure in the left atrium. Therefore the AV valve now opens. This ends the *isovolumetric* phase of diastole.

21. Blood rapidly enters the ventricle through the open AV valve from the atrium. This is termed the period of *rapid filling*.

22. Blood continues to fill the ventricle but the rate of filling is reduced with the decreasing pressure gradient between the atrium and ventricle. This is termed the period of *reduced filling*.

23. The SA nodal pacemaker generates a new action potential and the process of atrial and ventricular systole now begins again.

HEART SOUNDS

During the cardiac cycle **(Figure 7.10)**, two major heart sounds can be heard with a stethoscope.

1. Heart sound 1 occurs at the time that the two AV valves close and is low pitched.

2. Heart sound 2 occurs at the time when the semilunar valves close and is higher pitched.

These sounds are sometimes described as 'lub' (1st) and 'dub' (2nd) and result from the vibration in the ventricles and attached vessels as the blood decelerates against the closed valve orifice. In addition, the second heart sound can be split into two heart sounds, if the pulmonic and aortic semilunar valves do not close simultaneously. Such a splitting effect takes place upon inspiration because the amount of blood entering the right ventricle is greater than in the left ventricle, due to an increase in venous return (VR) to the right atrium during inspiration. This splitting of the second sound does not take place during expiration because VR decreases (see following material).

PRESSURE ON THE RIGHT SIDE

The cardiac cycle on the right side is much like that on the left side. Naturally the names of the valves are different and the blood flows out of the right ventricle into the pulmonary artery, however, the SV generated by the right ventricle is exactly the same as for the left ventricle. On the other hand, the maximum pressure produced by the right ventricle during systole is much less than in the left ventricle. Right ventricular pressure may reach 30 mm Hg during the ejection phase of systole, and thus the pressure in the pulmonary artery will also rise to about 30 mm Hg.

Why is the pressure generated on the right side significantly less than on the left side? The answer lies in the difference in resistance in the systemic circuit v. the pulmonary circuit. The systemic circuit has a very high resistance because of the "miles" of blood vessels here. In the pulmonary circuit, the resistance is much lower and therefore less pressure is required to promote blood flow.

Flow = pressure gradient/resistance

Thus, with a lower resistance the pressure required to drive the blood flow only needs to be about 30 mm Hg in the pulmonary circuit. In the systemic circuit a greater pressure (120 mm Hg) is required to overcome the greater resistance to maintain a similar flow.

EFFECTS OF HEART RATE ON VENTRICULAR FILLING TIME

During diastole the ventricles are filling with blood. Thus, if diastole is shortened, the ventricular filling time will also be reduced. Stimulation of the SA node by nerves or hormones can increase the rate of pacemaker depolarization and lead to an increase in heart rate (HR). As HR increases the number of ventricular systoles become more frequent and the time spent by the heart in diastole is shortened. Thus, as HR increases, filling time decreases.

During early diastole the ventricle is filling very rapidly and therefore, when HR increases during exercise, ventricular filling is not significantly affected. Thus, during exercise the amount of blood pumped (*cardiac output or CO*) will remain unchanged or may even increase (see following material). However, if HR exceeds 200 beats/min, filling time is greatly shortened and CO is reduced.

FACTORS THAT EFFECT THE CARDIAC OUTPUT

Cardiac output is the amount of blood that is pumped out of each ventricle of the heart and is measured as a volume/time—usually in Lt/min. The CO from the right ventricle and from the left ventricle will usually be almost exactly the same, although transient differences may be present for 1 or 2 beats. Intrinsic mechanisms built into the heart muscle itself are responsible for maintaining the equality of CO for both ventricles (see following material).

CO can be calculated from the relatively simple formula:

$$CO = HR \times SV$$

where: CO is *cardiac output* in Lt/min
HR is *heart rate* in beats/min
SV is *stroke volume* in ml/beat

Since both HR and SV are independent variables that effect CO, each will be considered separately.

HEART RATE (HR)

HR is controlled by the SA node. The rate of depolarization of the SA nodal cells is regulated by sympathetic and parasympathetic innervation and by circulating hormones.

1. Stimulation of the SA node by the postganglionic sympathetic fibers releases norepinephrine. The rate of rise of the membrane potential in the SA nodal cells increases. The cells reach threshold more frequently and more action potentials/minute are generated. Therefore HR increases **(Figure 7.11)**.

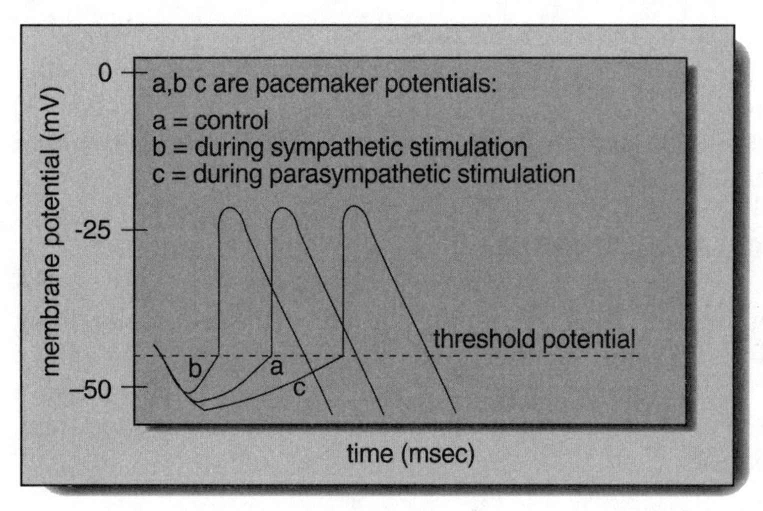

FIGURE 7.11

2. Stimulation of the SA node by the postganglionic parasympathetic fibers releases acetylcholine. The rate of rise of the membrane potential in the SA nodal cells decreases. The cells reach threshold less frequently and fewer action potentials/minute are generated. Therefore HR slows down.

3. Stimulation of the adrenal medulla by preganglionic sympathetic fibers causes the cells in the adrenal medulla to release a large quantity of epinephrine and a smaller quantity of norepinephrine into the blood. These adrenergic hormones now circulate to the heart and by acting on the SA node cause HR to increase. **(Figure 7.12)**.

STROKE VOLUME (SV)

Stroke volume is the volume of blood that is ejected from the ventricle during each contraction.

$$SV = EDV - ESV$$

where: SV is the *stroke volume* in ml/beat

EDV (or *end diastolic volume*) is the amount of blood in ml present in the ventricle at the end of diastole (just prior to the beginning of the systolic contraction)

ESV (or *end systolic volume*) is the amount of blood in ml remaining in the ventricle at the end of the systolic contraction

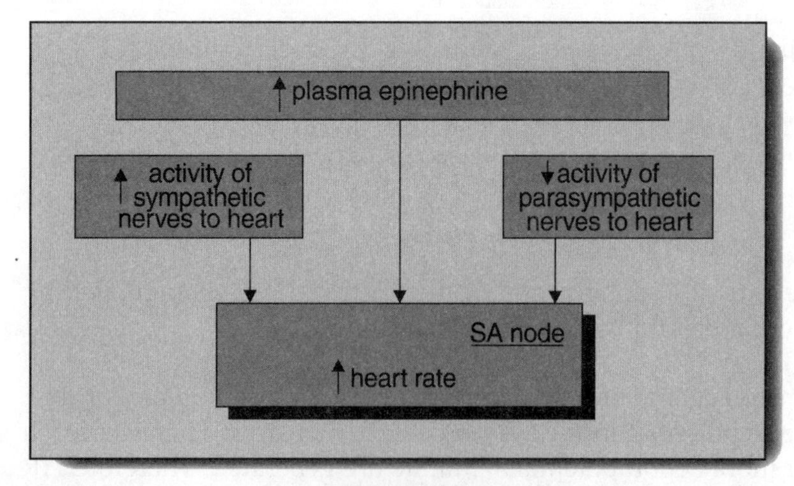

FIGURE 7.12

EDV. EDV (or preload) is directly effected by the amount of blood that enters the ventricle while the heart is in diastole. Thus, EDV is due to Venous Return (VR).

ESV. ESV is directly affected by the strength of ventricular contraction. The greater the strength of the contraction (or contractility), the more blood will be ejected during systole which will reduce ESV for that contraction. In addition, ESV also depends on the pressure against which

the ventricle is ejecting the blood. Thus, on the left side this would be the average pressure already present in the aorta, while on the right side this would be the average pressure present in the Pulmonary artery (such pressures are termed the afterload).

RESISTANCE

Resistance of the vessel affects these arterial pressures because it influences the rate of blood run-off. In addition, the resistance of the blood passing through the semilunar valves may also be a factor in certain diseases where the valve orifice itself is narrowed.

THE FRANK-STARLING LAW OF THE HEART

SV is effected by preload, afterload and contractility. While afterload relates mainly to the blood pressures in the arteries; preload, hormones and nerves stimulation determines the contractility and thus SV. The relationship between preload (EDV), hormone and nerve stimulation, contractility and SV is described by the *Frank-Starling Law of the Heart* which can be further subdivided into the *intrinsic* and *extrinsic control mechanisms.*

1. The intrinsic control mechanism of the Starling Law states that as VR increases, EDV increases leading to an increasing stretch on the ventricular cardiac muscle fibers at the end of diastole **(Figure 7.13)**. When these fibers are stimulated to contract at the onset of systole, they will therefore contract harder and more blood will be pumped out of the heart: thus SV will increase. (Keep in mind that the intrinsic control mechanism will operate in a heart that is disconnected from any outside influences except EDV, since it is affected by only the stretch of the muscle fibers.) Since SV is related to CO, the relationship is as follows:

VR
EDV
Contractility
SV
CO

This intrinsic control of SV is directly responsible for keeping the CO on the right side equal to the CO on the left side, although it may take one or two beats to equilibrate. Thus, as VR increases and the CO on the right side increases, CO on the left side also increases to match that on the right. In the event that CO is not equal on the two sides (as would take place after a pulmonary embolism), the out-

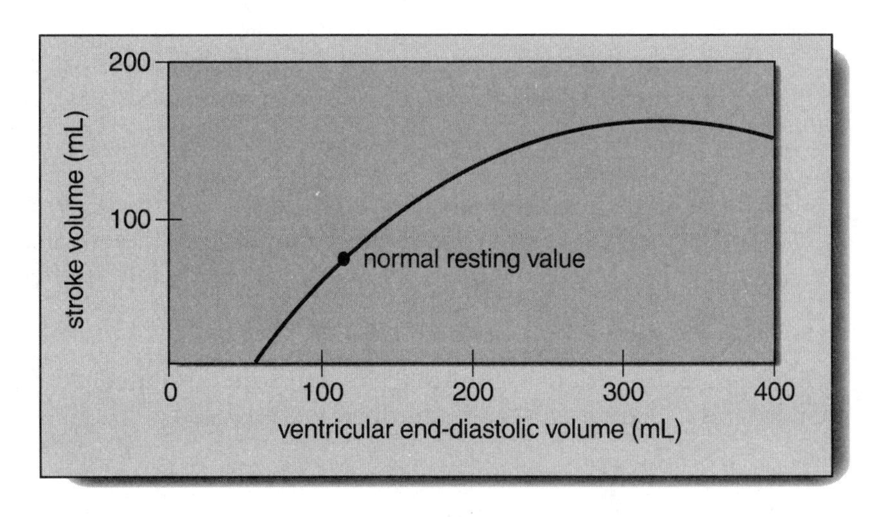

FIGURE 7.13

come would certainly be life threatening.

2. The extrinsic control mechanism of the Starling Law relates the SV to external factors. These factors are sympathetic stimulation provided by postganglionic innervation of the ventricular muscle, and the circulating adrenal hormones epinephrine and norepinephrine. Accordingly, in the presence of any or all of these external factors the efficiency of the contraction is increased **(Figure 7.14)**, but the intrinsic control mechanisms still function. In other words, the heart is now functioning on a more efficient Starling Curve.

Observation of **Figure 7.14** indicates that without sympathetic stimulation (control Starling curve) at an EDV of 200 ml the SV is about 90 ml, however with sympathetic stimulation an identical EDV of 200 ml generates a SV of about 150 ml. Thus, for the same amount of ventricular stretch the contractile efficiency is increased.

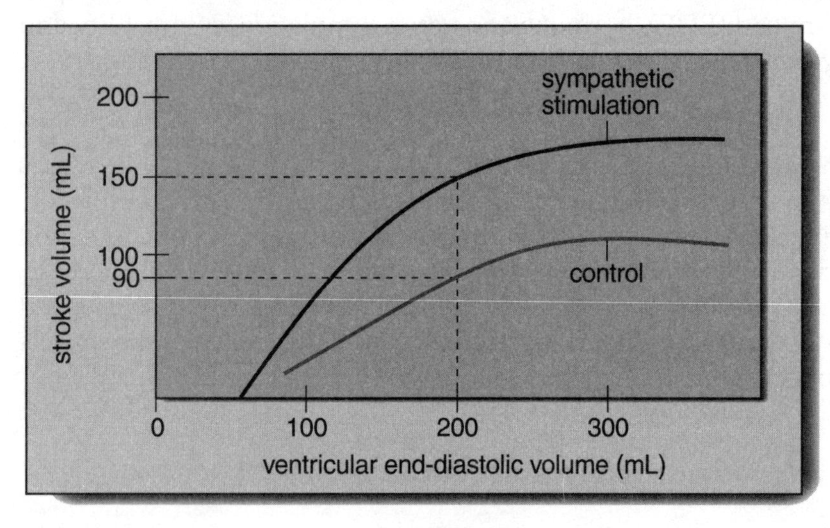

FIGURE 7.14

OTHER FACTORS THAT EFFECT CO

In addition to the various factors already mentioned that effect SV and thus CO, here are some things that also play a role.

1. Sympathetic influence on the AV node decreases the delay time and increases the rate of impulse conduction through the AV node. Thus, the delay between atrial and ventricular contraction is reduced.

2. Parasympathetic influence on the AV node increases the delay time and slows down the rate of conduction through the AV node. Thus the delay between atrial and ventricular contraction is increased.

3. Parasympathetic influence on ventricular contractility acts in reverse of the sympathetic influence. Thus, parasympathetic stimulation on the ventricular muscle reduces the contractility and efficiency of the heart. However, this is a minor effect as compared with the effects of sympathetic stimulation.

In **Figure 7.15** the various factors that effect CO are presented. Along with items discussed above, note that SV and CO are increased if afterload is decreased.

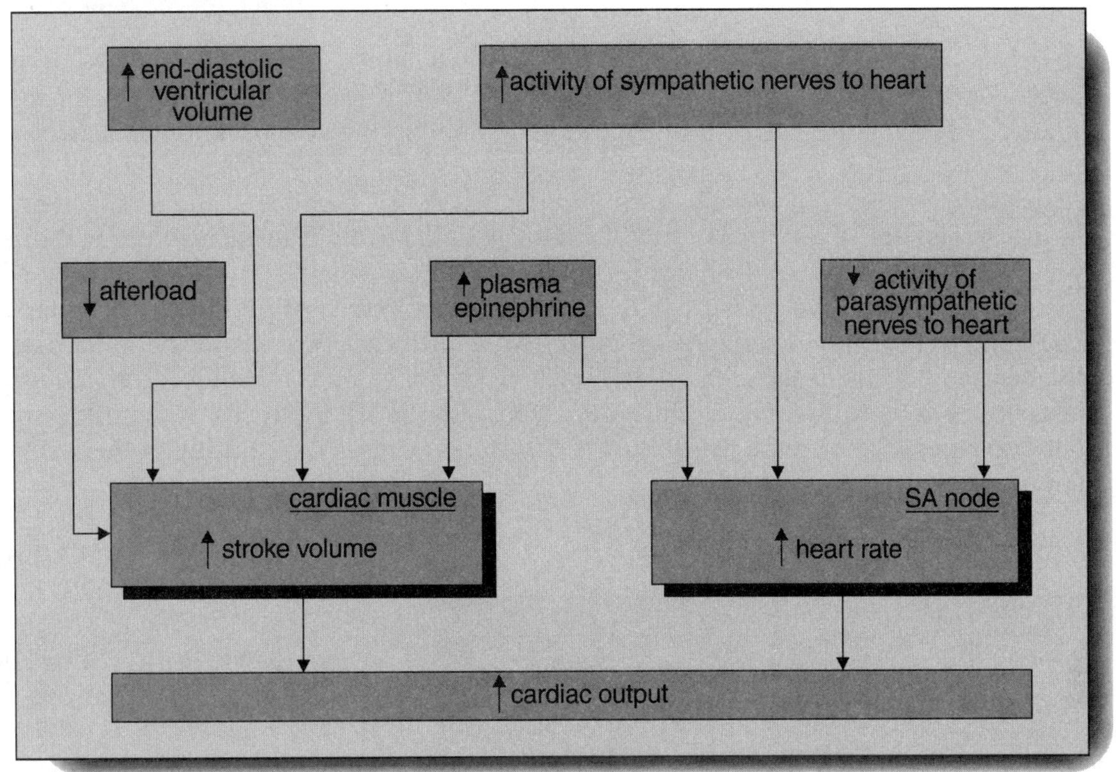

FIGURE 7.15

FACTORS INVOLVED IN VENOUS RETURN

VR effects CO as just described. VR itself is the amount of blood that returns to the right atrium from the body. VR is affected by peripheral venous pressure and peripheral venous pressure is related to the following components **(see Figure 7.16)**:

1. *Blood volume (BV)* — Total BV present in the circulation and especially in the venous side of the circulation can effect VR. At any given moment there is much more blood present in the venous side than in the arterial side of the system. Increasing BV on the venous side will force more blood to return to the heart. Thus, increasing BV in the veins will increase VR.

2. *Sympathetic nerve stimulation* — Pooled blood on the venous side can act as a reservoir for additional VR if needed. Stimulation of the sympathetic nerves to the multiunit smooth muscle in the walls of the veins will cause contraction and force the blood to return to the heart and increase VR. (Remember that blood cannot move away from the heart in the veins due to the presence of one-way valves.)

3. *Skeletal muscle pumps* — Veins have relatively thin walls and frequently are surrounded by skeletal muscles. As these muscles undergo contraction, they squeeze the veins and force blood towards the heart (valves prevent back flow). Thus during exercise, skeletal muscle pumping of the veins increases VR.

4. *Arterio-venous coupling* — Large arteries frequently lie in the same connective tissue sheath as large veins. As the arteries pulse with blood the compression-relaxation cycle of the artery squeezes the vein that lies next to it and forces blood in the vein back to the heart.

5. *Respiratory Pumps* — Increasing respiration increases VR because of the arrangement of the large vessels (Vena Cava) that return blood to the right atrium. During inspiration the negative pressure in the thoracic cavity becomes more negative (and thereby the lungs are inflated). This increase in the negative pressure in the thoracic cavity decreases pressure in the interthoracic veins and right atrium. This increases the pressure difference between the peripheral veins and the heart and tends to act as a suction force to increase blood flow in the Vena Cava from the abdominal to thoracic cavities. Thus, during inspiration VR increases, and during expiration VR decreases. The increase in VR during inspiration causes the splitting of the second heart sound.

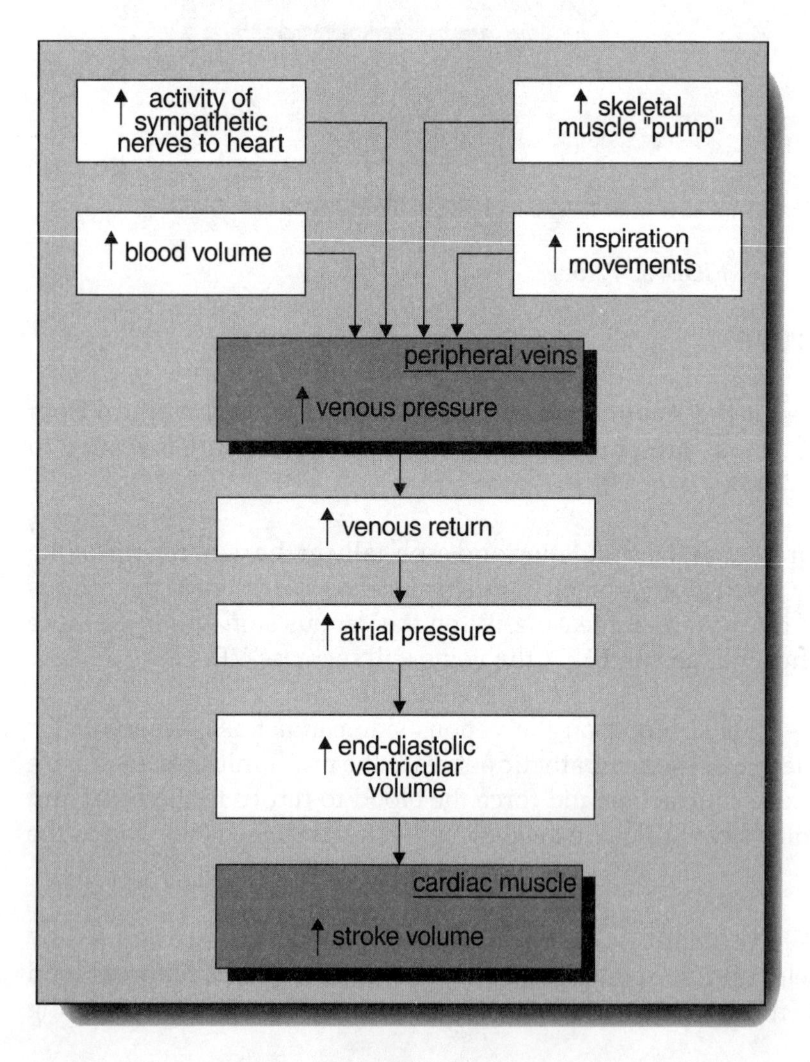

FIGURE 7.16

EXAMPLE OF A CALCULATION FOR CARDIAC OUTPUT

Given the fact that: CO = HR x SV
and if: HR = 72 beats/min,
and SV = 70 ml/beat

then: CO = 72 beats/min x 0.07 Lt/beat = 5.0 Lt/min

Chapter 8
THE VASCULAR SYSTEM AND CONTROL OF BLOOD PRESSURE

PATHWAYS OF CIRCULATION

The circulatory system consists of the pulmonary circuit and the systemic circuit **(Figure 8.1)**. The cardiac output at rest within the systemic circuit is distributed as follows:

	Rest
Abdominal Organs	25%
Skeletal Muscle	21%
Kidneys	19%
Brain	13%
Skin	9%
Heart	4%
Other Tissues	10%

However, this distribution will change during moderate exercise, with an increase to the skeletal muscles, heart and skin, and a decrease to the viscera. There will be no change to the brain.

BLOOD VESSELS

ARTERIES AND VEINS

The walls of the larger blood vessels (arteries and veins) contain the following layers:

1. On the outside, there is the *Serosa*.

2. An outer layer is called the *tunica adventitia*. This is a relatively thin layer of connective tissue containing some elastic and collagenous fibers.

3. A middle layer is called the *tunica media*. This is a very thick layer, especially in arteries, and is comprised of circular layers of smooth muscle along with a thick layer of elastic and connective tissue.

4. An inner layer is called the *tunica intima*. This is comprised of a layer of simple squamous epithelium called *endothelium*, supported on a connective tissue membrane, also containing many elastic and collagenous fibers.

Veins also possess flap-like structures along the inner wall called valves. These valves allow blood to flow towards the heart but prevent back-flow away from the heart. In addition, the walls of large arteries contain very small blood vessels called the *vasa vasorum* and capillary beds that nourish the more external cells in the vessel wall.

Arteries — function to rapidly transport blood *away* from the heart with low resistance due to the large radius of these vessels. The elastic wall structure is designed to withstand a very high pressure. The elastic recoil of the wall acts as a pressure reservoir (see following material).

Arterioles — function as the *major resistance element* of the circulatory system. The walls of the arterioles are comprised of a single layer of endothelium surrounded by smooth muscle cells that act as sphincters to regulate blood flow through the arterioles into the attached capillary beds (see following material).

Veins — function to transport blood from the body back *towards* the heart. Valves in the veins promote blood flow only towards the heart. The wall structure of veins are more compliant than the walls of arteries.

Compliance — this term refers to how easily the wall structure can be stretched. Because the walls of the veins are more compliant than the walls of arteries, they stretch farther for a given increase in volume. Thus, if 100 ml of blood entered an artery, there would be a greater increase in pressure (than if 100 ml of blood entered a vein) because the vein wall would be able to stretch more easily than would the artery wall. Because the venous side of the circulation is more compliant than is the arterial side, the veins act as a reservoir for blood storage. Sympathetic nervous system stimulation of the smooth muscle in the walls of the veins, however, will cause the walls to contract and push additional blood back to the heart (VR will increase).

Capillaries — blood flows very slowly through the capillary beds due to the large cross-sectional area covered by these vessels. This very slow rate of flow promotes exchange of materials between the circulatory system and the cells. Capillary walls consist of a single layer of endothelium, and filtration and reabsorption of materials across the walls are generally through slit-like gaps formed where these cells join each other.

FLOW AND PRESSURE

Regardless of the type of vessel, blood flow is dependent on the pressure gradient within the vessel and on the resistance, as follows:

$F = \Delta P/R$

where: F = Flow rate in Volume/Time
ΔP = Pressure Gradient (Usually in mm Hg)
R = Resistance

Observation of this relationship indicates that F is directly proportional to the pressure gradient (ΔP), and is inversely proportional to the resistance (R). In other words, as the pressure gradient increases, F increases; while as the resistance increases, F decreases.

The Pressure Gradient (ΔP). ΔP results from the difference in pressure at the two ends of a closed tube. Thus, if the pressure P_1 at one end of a tube was 70 mmHg and the pressure P_2 at the other end was 30 mmHg, then ÄP would be 40 mm Hg ($\Delta P = P_1 - P_2 = 70 - 30 = 40$ mmHg). In the circulatory system the driving force, or hydrostatic Pressure results from the action of the heart.

Resistance. The term *resistance* defines various factors that act to hinder the movement of blood through the vessel. As resistance increases it becomes more difficult for blood to move through the vessel. A number of physical factors act to effect resistance (R), as follows:

1. *The length (L) of the vessel*: As L increases, R also increases. However since the length of the vessels in the circulatory system do not change, L is not important for minute to minute regulation of blood flow.

2. *The viscosity (η) of the blood*: Viscosity is a measure of the thickness, or stickiness of the blood. Viscosity is affected by the amount of protein in the plasma, as well as the number of RBCs in

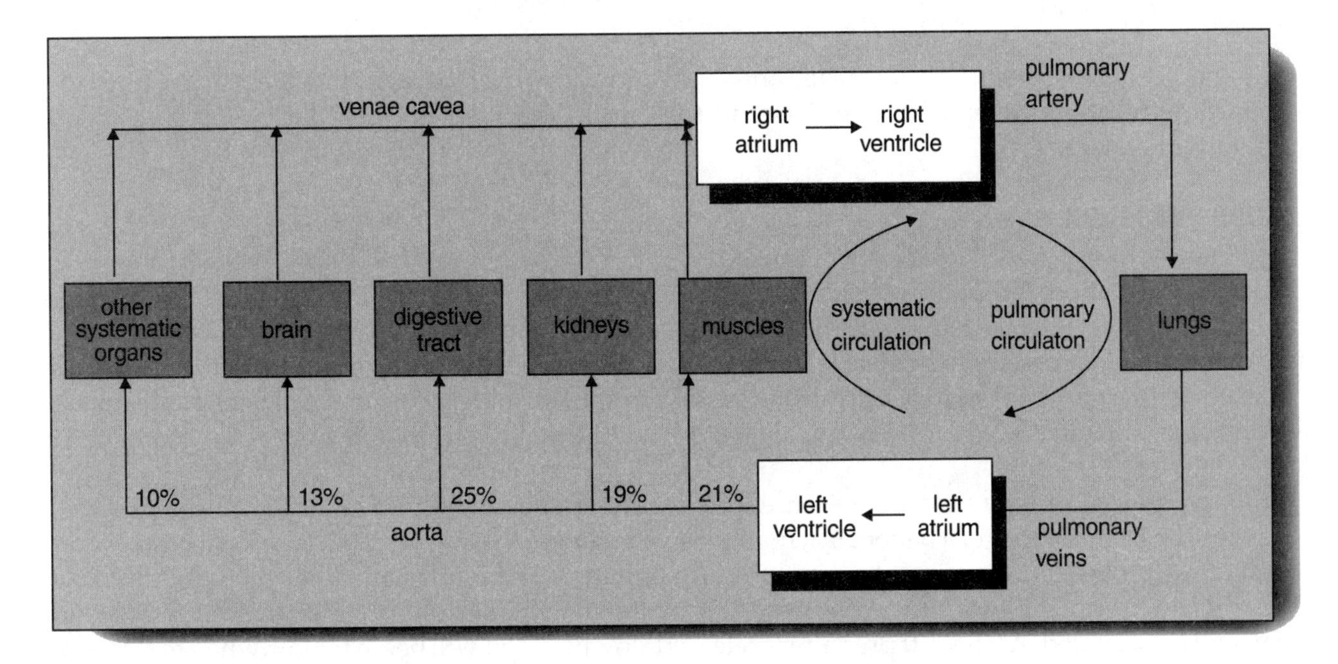

FIGURE 8.1

the blood. However, since these factors do not change (except in certain diseases), viscosity cannot be used to regulate minute to minute flow of blood in the circulatory system.

3. *The radius (r) of the vessel*: Resistance is proportional to $1/r^4$, or: $R=1/r^4$

What this means is that if the radius doubles (for example, by going from 1 cm to 2 cm), the resistance decreases sixteen fold:

when: r = 1 cm, $R = 1/1^4$, or $R = 1/1$, or $R = 1$

when: r = 2 cm, $R = 1/2^4$, or $R = 1/16$

Thus, a small change in the radius (r) produces a very large change in the resistance (R). Clearly, if the resistance decreases sixteen fold, the flow must increase sixteen fold given the formula: $F = \Delta P/R$

Assuming, of course, that ΔP remains constant.

The various factors affecting flow can be integrated into a single equation (*Poiseuille's law*):

$$F = \pi \Delta P r^4 / 8 \eta L$$

However, when calculating flow through a particular organ (for example, the liver, kidney or heart), then this equation can be slightly altered as follows:

$$F_{organ} = MAP/R_{organ}$$

Here the MAP is the *mean arterial pressure* and is used in place of ÄP because it is the driving force for blood flow through the organ. To understand the factors involved in MAP, let us consider blood pressure in general.

BLOOD PRESSURE

Pressure fluctuation within the arteries is due to the action of the heart. Pressure in the large arteries fluctuates between the bottom of the diastolic pressure (when the left ventricle is in diastole) and the top of the systolic pressure (when the left ventricle is in systole). In a normal individual, these changes fall in the range of approximately 120 mm Hg systolic to approximately 80 mm Hg diastolic and appear as depicted in **Figure 8.2**.

Although arterial pressure can be measured by insertion of a canula into an artery, these pressure changes can be measured in a patient in a noninvasive manner by using an external *sphygmomanometer* and cuff. After the cuff is wrapped around the arm, inflation of the cuff with air applies pressure to a superficial artery in the arm. If the cuff pressure is greater than the systolic pressure in the artery, blood flow will be occluded. As the cuff pressure is released slowly, blood will begin to flow when the arterial pressure is slightly greater than the cuff pressure. The sound of the blood flow at this pressure can be heard using a stethoscope applied to the arm, and the pressure in the cuff, which is equal to the systolic,pressure, can now be read off of a manometer attached to the cuff. As the air pressure continues to be released, more blood flows in the artery and finally when no more sound is heard, this signifies the diastolic pressure level (also read off of the manometer).

CALCULATION OF PULSE PRESSURE AND MEAN ARTERIAL PRESSURE

Pulse pressure (PP) can be calculated by subtracting the systolic pressure (SP) from the diastolic pressure (DP):

$$PP = SP - DP$$

Thus: if $SP = 120$ mmHg
$DP = 80$ mmHg
$PP = 120 \text{ mmHg} - 80 \text{ mmHg} = 40 \text{ mmHg}$

PP can be used clinically to ascertain if certain problems exist in the circulatory system. For example, if the arterial walls are coated with deposits that reduce their elastic recoil (see following material), the DP can be lower than usual (for example 40 mmHg) while the SP can be elevated (for example, 145 mmHg)—thus there is a widening of the PP to 105.

While information on PP may be useful, knowledge of MAP is frequently more helpful since it is responsible for driving the blood through the tissues. It should be noted that the MAP is not calculated as a simple average (i.e., (120 + 80)/2 = 100 mmHg), because the heart spends more time in diastole than it does in systole. (At a resting heart rate of 70 beats/minute the heart spends 2/3 of the time in diastole and 1/3 in systole). Thus, to calculate MAP the following formula is used:

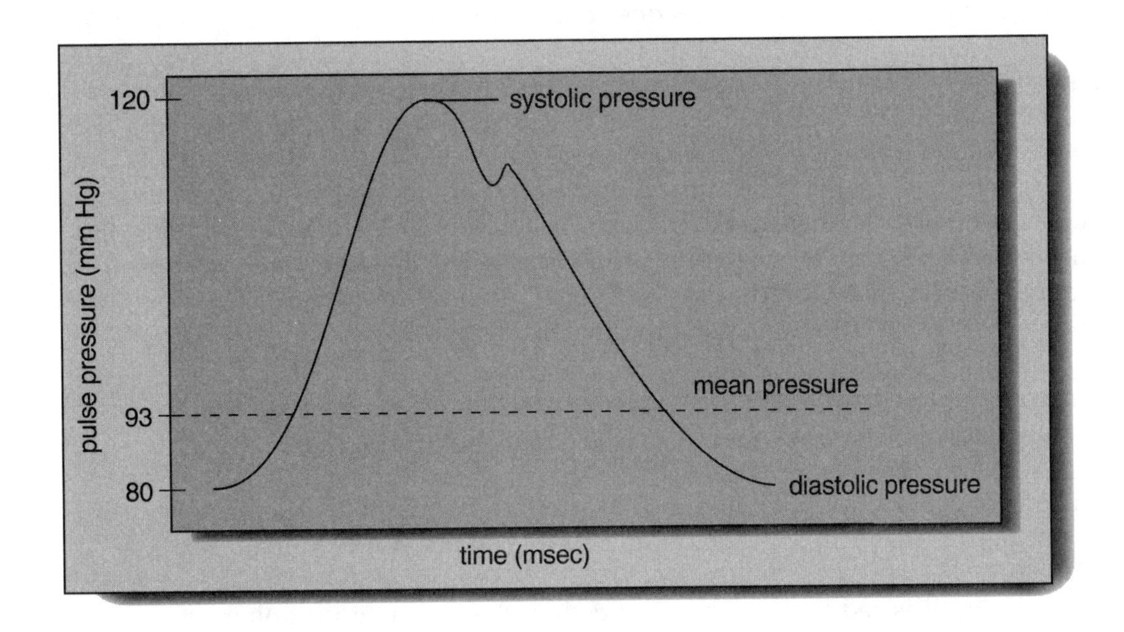

FIGURE 8.2

MAP = DP + 1/3(PP)

thus, when: SP = 120 mmHg
DP = 80 mmHg
PP = 40 mmHg (see prior material)

then: MAP = 80 mmHg + 1/3 (40 mmHg) = 93 mmHg

As can be seen, the MAP of 93 mmHg is less than the simple average pressure of 100 mmHg, since the diastolic component is weighted more heavily in this formula than is the systolic component.

PRESSURES IN DIFFERENT PARTS OF THE CIRCULATORY SYSTEM

Blood pressures change as the blood flows through the various elements of the circulatory system **(Figure 8.3)**. Obvious differences between systolic and diastolic pressure are present in the larger arteries, but these changes become less pronounced as the blood passes into the smaller arteries and significantly less in the arterioles. By the time the blood enters the capillaries these fluctuations disappear almost completely. Within the venous system no pressure fluctuations are present and the pressure is only a few mmHg. A slight negative pressure may, in fact, be present at times in the section of the vena cava that is close to the right ventricle. Within the pulmonary circuit similar, although less pronounced, pressure fluctuations are also present.

ARTERIAL ELASTICITY AND ALTERNATE PUMPING

During the cardiac cycle, pressure in the left ventricle rises to about 120 mmHg during systole and then falls to 0 mmHg during diastole. However, while systolic pressure in the large arteries rises to 120 mmHg, diastolic pressure in the arteries does not fall to 0 mmHg but only to 80 mm Hg.

DP does not fall to 0 mm Hg in the arteries because the arterial walls act as alternate pumps. As previously mentioned, the walls of the arteries contain a thick layer of elastic tissue. This elastic tissue is stretched as blood from the heart enters the aortic arch during systole **(Figure 8.4)**. Thus, a portion of the force, generated by the heart muscle in the form of blood pressure, is absorbed in the elastic stretch of the arterial wall. When the heart enters diastole, the wall now recoils, forcing the blood within to be pushed farther down the system. A continuous wave of stretch and recoil proceeds down the entire arterial system, and this acts as an alternate pump when the heart is resting during diastole, maintaining the diastolic pressure at approximately 80 mmHg.

To return to PP again, it may be remembered that if the walls of the arteries lose their ability to elastically recoil, the PP widens. This is because in the absence of wall elasticity:

1. The force of blood pressure entering the arterial system from the heart is not cushioned by wall stretch; thus the systolic pressure can be elevated above 120 mmHg, and

2. Elastic recoil is reduced or absent and thus the alternate pumping effect is negated; thus diastolic pressure can be reduced below 80 mmHg.

CAPILLARY BLOOD FLOW

Blood flow in the capillary beds depends on the MAP and the resistance within the arterioles that supply the capillaries, as well as the resistance in the capillaries themselves. In order for blood to move into the capillaries, it must flow from the larger arteries into the arterioles **(Figure 8.5)**. Constriction or dilation of the arteriolar precapillary sphincters acts to regulate this flow into the attached capillaries. The constriction and dilation of the smooth muscle cells that comprise these precapillary sphincters is under two levels of control: *systemic and local*. This will be described below.

METARTERIOLES

Because blood flow into the capillary beds is dependent on the minute to minute requirements of the particular tissue, these sphincters open and close as needed. However, when they close, and flow into the capillaries is reduced, flow must still be maintained from the arterial to the venous side of the system.

This is necessary, otherwise VR to the heart would be disrupted and arterial blood pressure would fall. It is the arterial-venous shunt vessels called *metarterioles* that allow blood to cross from the arteriolar to the venous side of the circulation when the precapillary sphincters are constricted and blood cannot enter the capillary beds.

CROSS-SECTIONAL AREA

As blood flows through the various elements of the circulatory system, the total cross-sectional area (measured in cm^2), increases **(Figure 8.6)**. The cross-sectional area in the arteries is quite small, but as the arteries decrease in size and branch into the arterioles, the cross-sectional area increases from a few cm^2 up to nearly 3000 cm^2. As these arterioles branch into the capillary beds the cross-sectional area increases farther to 3000 cm^2. In addition, as the cross-sectional area increases, the velocity of flow (in cm/sec) decreases dramatically from 30 cm/sec in the arteries to 1 or 2 cm/sec in the capillary beds.

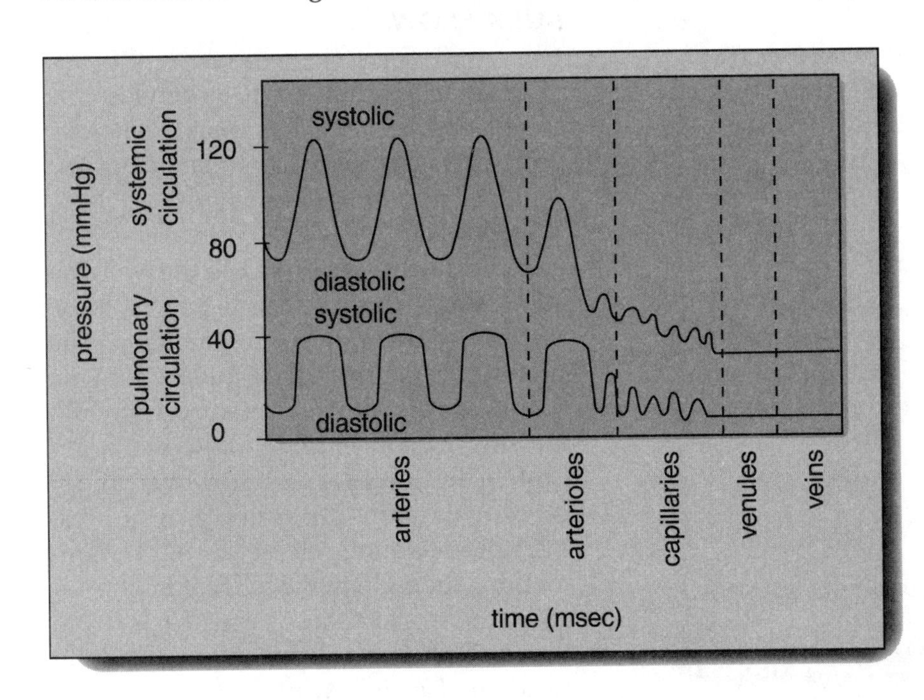

FIGURE 8.3

CAPILLARY FILTRATION AND REABSORPTION

The primary function of the capillary is to support exchange of nutrients, gases and waste products with the tissues. Capillaries accomplish this function because:

1. The large cross sectional area of the capillaries reduces flow to only a few cm/sec.

2. The wall of the capillary consists of a single layer of endothelium containing slit-like gaps between the adjacent cells **(Figure 8.7)**.

3. Water soluble substances (e.g., salts, nutrients, waste products of metabolism, certain hormones) cross through the slit-like gaps, while the gases O_2 and CO_2 traverse the wall by moving directly through the endothelial cell cytoplasm **(Figure 8.8)**.

There are no carrier mediated transport systems in the capillary walls. Thus, any exchange that takes place must be dependent on either *passive diffusion* or *bulk flow*.

PASSIVE DIFFUSION

In the case of passive diffusion, substances such as O_2, CO_2 and glucose move down their individual concentration gradients until there is no longer a difference in the concentration between the blood and surrounding cells. The extent of this exchange process is independently determined by the magnitude of the concentration gradient present for each individual substance.

BULK FLOW

The presence of slit-like gaps between the endothelial cells provides a means whereby protein free plasma can filter out of the capillary into the tissues.

In addition, these openings in the wall also allow fluid from the tissue spaces to be reabsorbed back into the capillary. The filtrate passing from the capillary into the tissues is protein-free because normally the slits are too small to allow the larger protein molecules access. However, under certain conditions (for example, in the presence of histamine release during inflammation), the endothelial cells contract, widening the slits to allow protein to pass from the capillary into the tissue fluid resulting in edema (see following material).

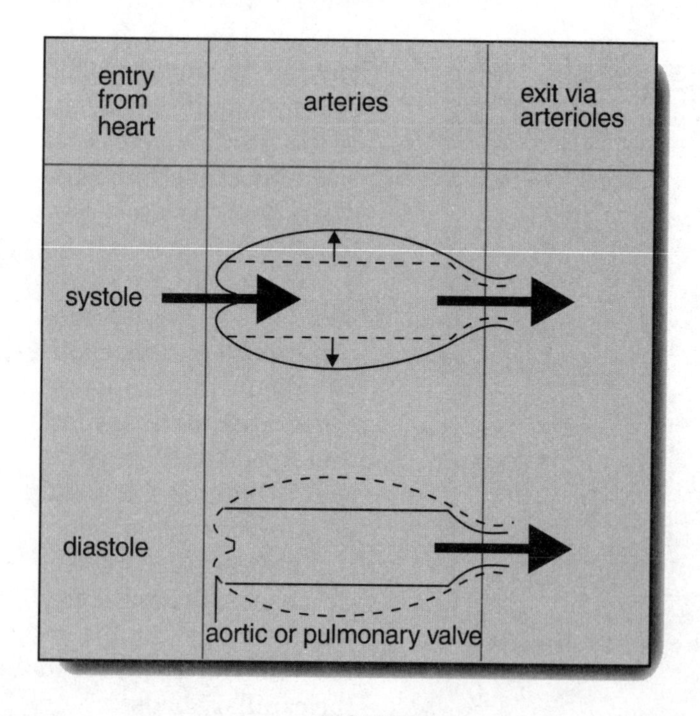

FIGURE 8.4

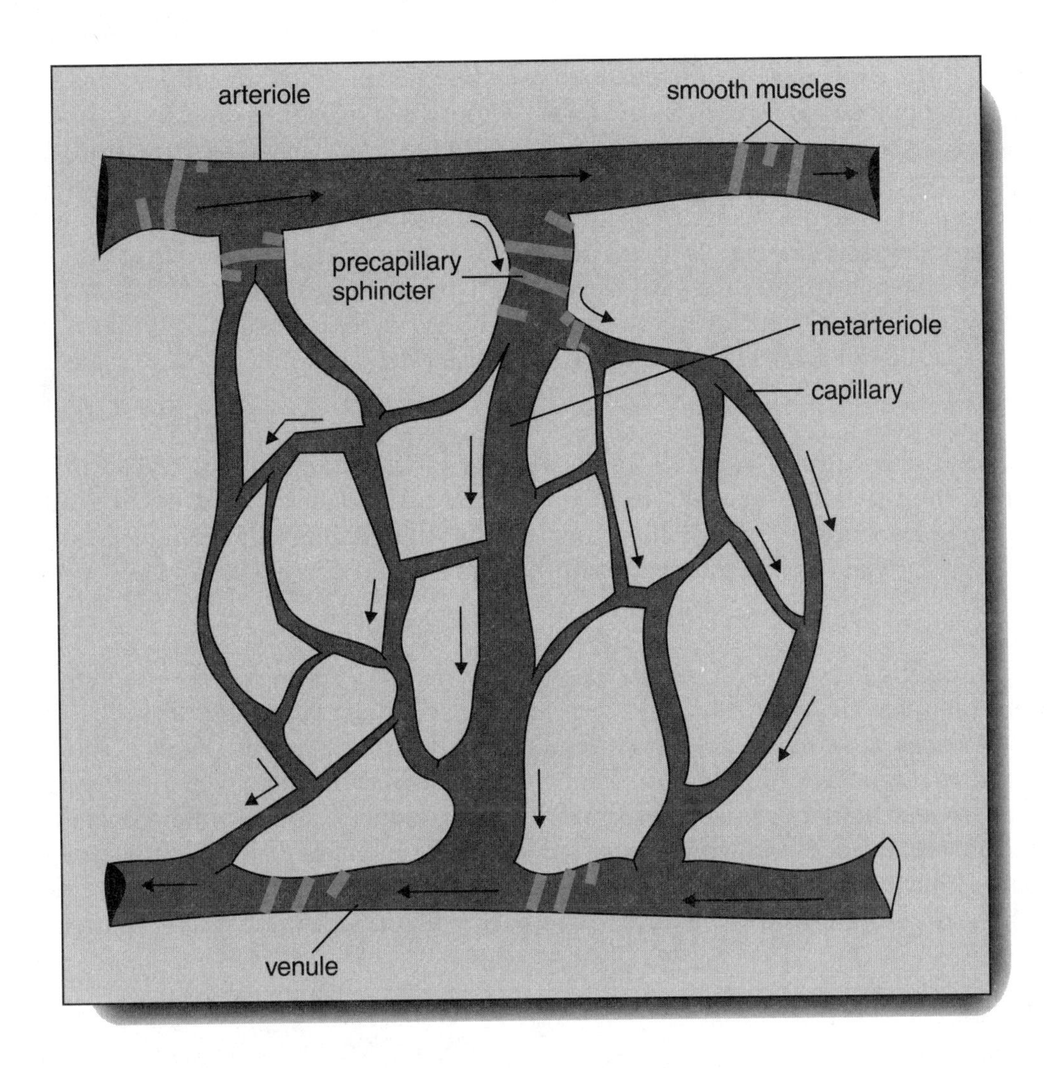

FIGURE 8.5

This process of movement of fluid from the capillary into the tissues and visa versa is termed *bulk flow* because all of the substances presence in the fluid are moving together, and not individually as is the case in diffusion. The importance of bulk flow is to maintain a balance in the distribution of the extracellular fluid between the vascular and interstitial compartments. Bulk flow is not important in regulating the movement of individual solutes from the capillary to the cells and visa versa. Such regulation is mainly dependent on diffusion.

THE STARLING MECHANISM OF THE CAPILLARY

This theory, first proposed by the physiologist Starling, explains how it is possible for bulk flow in the capillary to take place in both directions at the same time (i.e., filtration out of the capillary, and reabsorption into the capillary). To understand the mechanism **(Figure 8.9)**, various forces involved must be considered individually:

Capillary Hydrostatic Pressure (P_c) (or capillary blood pressure) — is the major force responsible for filtration out of the capillary. Arterial blood pressure is the driving force for P_c, and P_c in turn then acts to forces fluid through the slit-like gaps in the wall. The *ultrafiltrate* produced by this process is composed of all water soluble substances that are capable of moving through the slits, but protein, and red and white cells cannot pass across. On the average, P_c is about 37 mmHg at the arteriolar end of the capillary and declines to approximately 15 mmHg at the venous end. Clearly as systemic arterial pressure increases, P_c also increases as does filtration. In addition, when the precapillary arteriolar sphincters dilate, P_c also increases, as does filtration. Constriction of the precapillary sphincters will have the reverse effect, as will a drop in systemic arterial blood pressure.

Interstitial Fluid Hydrostatic Pressure (P_{if}) — is a minor force responsible for reabsorption of fluid back into the capillary. It results from fluid pressure that may build up in the interstitial compartment and push against the outer wall of the capillary. Although it is difficult to measure, it is probably slightly above atmospheric pressure, falling in the range of approximately 1 mmHg.

Blood-Colloid Osmotic Pressure (Π_p) (sometimes called the oncotic pressure) — is the major force responsible for reabsorption of fluid back into the capillary. It results from the presence of plasma protein (mostly albumin) within the blood. To understand how it functions, it is necessary to reconsider the process of osmosis. Recall that in osmosis, water moves by diffusion from an area of higher to lower concentration. To produce a lower water concentration, the water must be diluted with some other substance. In the case of π_p, the presence of plasma protein within the vascular compartment reduces the concentration of water in the blood. Thus, the water concentration in the interstitial compartment (outside the capillary) is higher than the

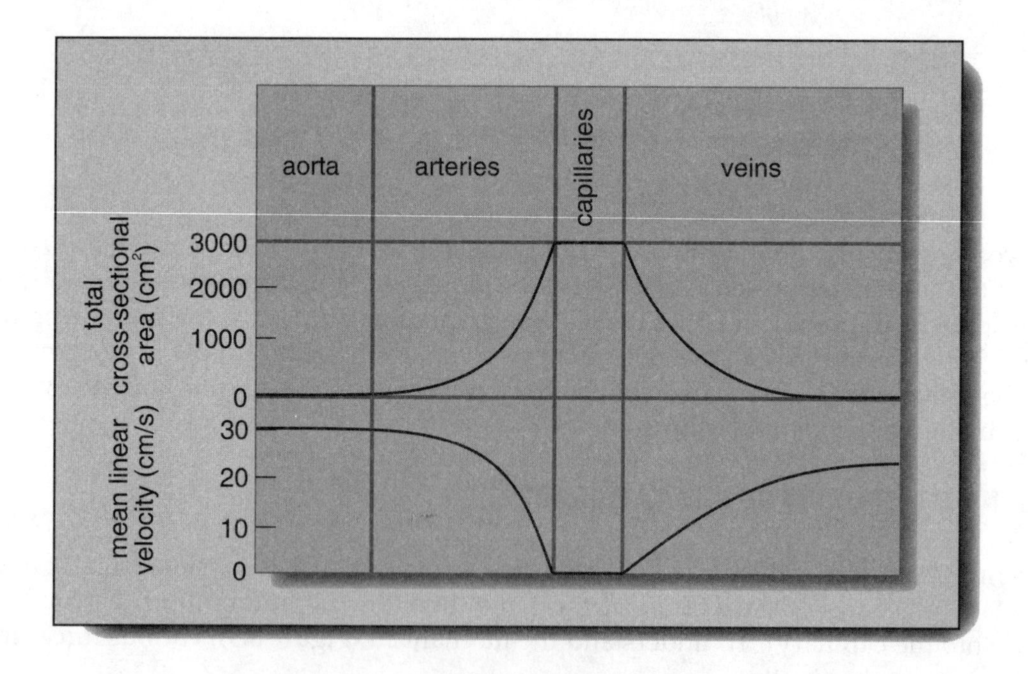

FIGURE 8.6

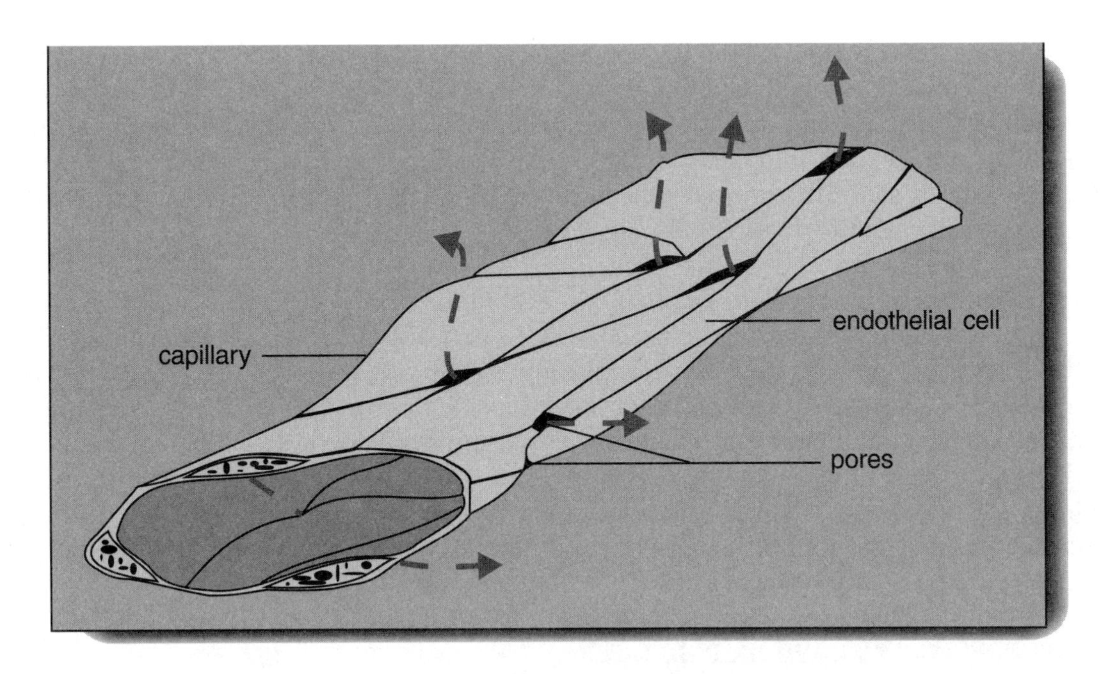

FIGURE 8.7

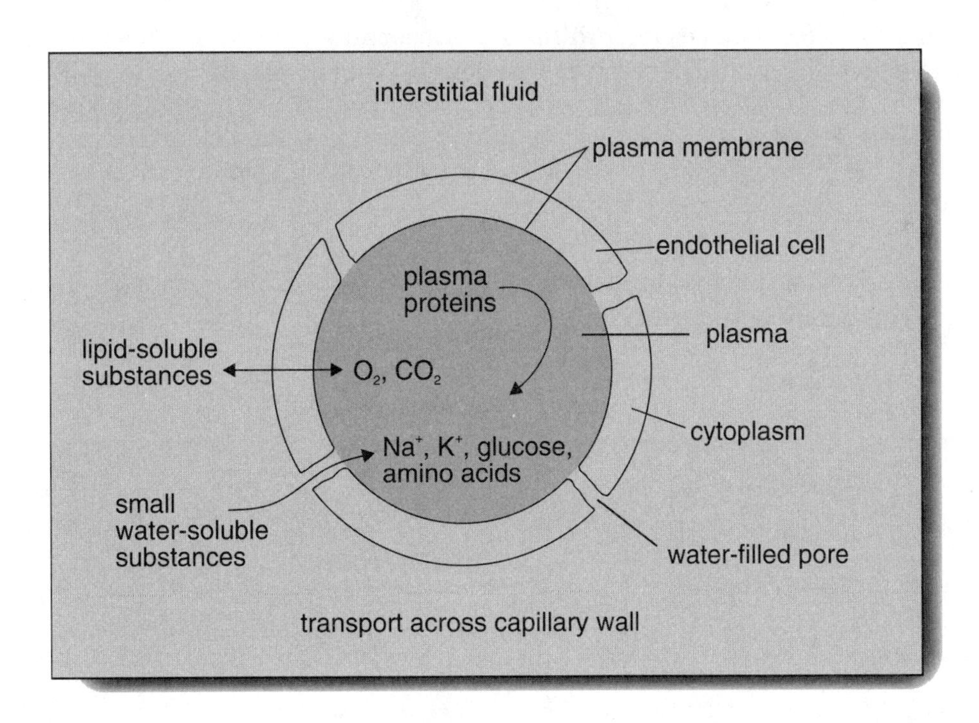

FIGURE 8.8

water concentration inside the capillary, causing water to move into the capillary from the interstitial compartment. Since the driving force is the effective protein concentration in the blood, elevating or decreasing the concentration of plasma protein will affect the process of reabsorption. However, the concentration of plasma protein normally remains constant except in certain diseases. At the normal concentration of plasma protein, the reabsorptive ∂_p force exerted is equivalent to about 28 mmHg inward.

Interstitial-Fluid Colloid Osmotic Pressure (Π_{if}) — is a minor force responsible for filtration of fluid out of the capillary, but under normal conditions probably has little or no affect on filtration. If this force is present, it is because some plasma protein has leaked out of the capillary into the interstitial compartment. Under normal conditions, the very small quantity of plasma protein that may enter the interstitium is removed by the lymphatic system. However, during an inflammatory response (when the slits between the endothelial cells widen due to histamine release), the increased leakage of plasma protein into the interstitial compartment causes ∂_{if} to increase, thereby promoting excessive fluid retention and edema within the interstitium. Under normal conditions, ∂_{if} may be approximately 2 mmHg.

BALANCE OF FORCES AND CAPILLARY LENGTH

The combination of forces responsible for bulk flow leading to filtration and reabsorption vary depending on the location studied along the capillary length. Depending on the balance between these various forces, the net effect will be either filtration, reabsorption, or no movement of fluid across the capillary wall in either direction. The formula that defines Net Filtration Pressure (NFP) is as follows:

$$NFP = (P_c - P_{if}) - (\pi_p - \pi_{if})$$

FILTRATION

Filtration at the arteriolar end of the capillary takes place when the combination of outward forces are greater than the combination of inward forces.

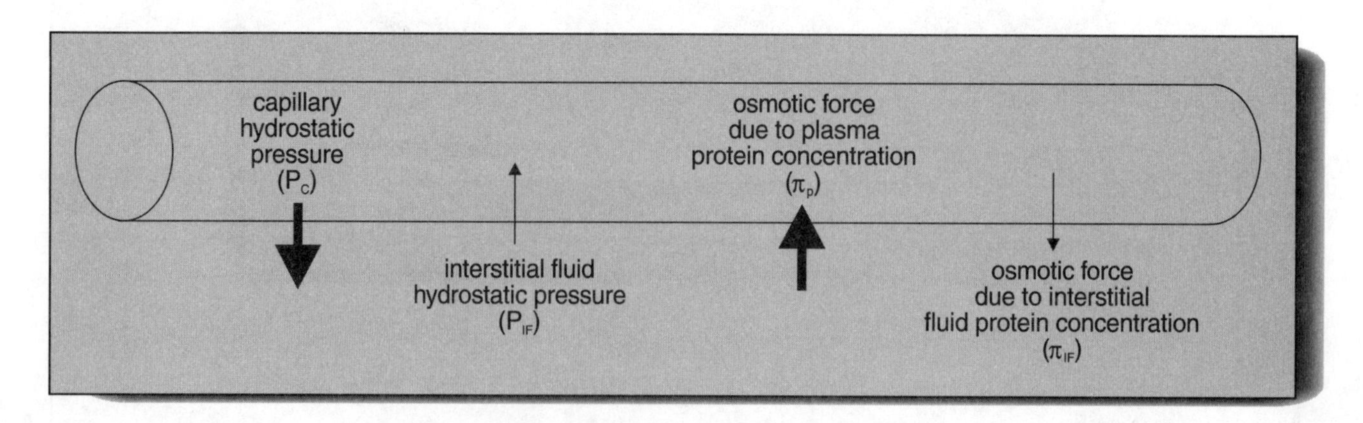

FIGURE 8.9

Forces at the arteriolar end of the capillary:

Outward Forces	Inward Forces
$P_c = 37$ mmHg	$\Pi_p = 28$ mmHg
$\Pi_{if} = 2$ mmHg	$P_{if} = 1$ mmHg

NFP = (37-1)-(28-2) = 10 mmHg

Thus, at the arteriolar end of the capillary there is a net force of 10 mmHg resulting in filtration of fluid out of the capillary and into the interstitial space.

But *reabsorption* takes place at the venular end of the capillary when the combination of inward forces is greater than the combination of outward forces.

Forces at the venular end of the capillary:

Outward Forces	Inward Forces
$P_c = 15$ mmHg	$\Pi_p = 28$ mmHG
$\Pi_{if} = 2$ mmHg	$P_{if} = 1$ mmHG

NFP = (15-1)-(28-2) = -12 mmHg

Thus, at the venular end of the capillary there is a net force of -12 mmHg resulting in reabsorption of fluid back into the capillary from the interstitial space.

CHANGES IN P_C ACROSS THE CAPILLARY LENGTH

As demonstrated, P_c at the arterial end of the capillary is approximately 37 mmHg; however, it decreases to about 15 mmHg at the venular end. This change in P_c is responsible for filtration at the arterial and reabsorption at the venular end of the capillary, since ∂_p remains at a constant 28 mmHg. This drop in P_c is due to resistance in the capillary to blood flow and is linear across the capillary length **(Figure 8.10)**. Since P_c and ∂_p are the major forces involved, we can make the following 'rule of thumb' statements:

when: $P_c > \pi_p$ *filtration* takes place

when: $P_c < \pi_p$ *reabsorption* takes place

when: $P_c = \pi_p$ this is the *transition point*

At the transition point neither filtration nor reabsorption takes place. The transition point is located at the approximate midpoint of the capillary length.

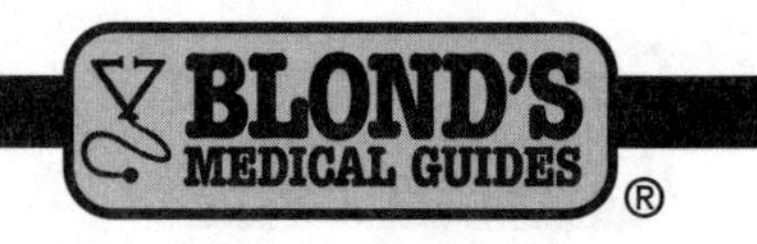

EFFECTS OF ARTERIOLAR RADIUS ON FILTRATION/REABSORPTION

Dilation of the arteriole that feeds the capillary will increase P_c, while constriction of arteriole will reduce Pc. These changes in P_c will then change the extent of filtration or reabsorption that takes place down the length of the capillary.

1. For example, since the drop in P_c due to resistance is linear down the length of the capillary, if dilation of the arteriole increases Pc to 42 mmHg at the arteriole end, P_c will be 22 mmHg at the venular end **(Figure 8.11)**. Clearly, under these conditions, filtration will take place throughout the majority of the length of the capillary.

2. On the other hand, if the arteriolar end undergoes constriction, P_c will drop at the arteriolar end to, for instance, 30 mmHg, while at the venular end, P_c will fall to 8 mmHg **(Figure 8.12)**. Under these conditions throughout most of the length of the capillary, reabsorption will take place.

CHANGES IN THE CONCENTRATION OF PLASMA PROTEIN

Since the reabsorptive force is due to the presence of plasma protein, it is obvious that changing the concentration of this substance will alter filtration/reabsorption. Thus, in an individual with glomerulonephritis, where protein is spilled in the urine (proteinuria), plasma protein will be reduced (hypoproteinemia). This will lead to a reduced reabsorption and an increased filtration of fluid in all capillaries (i.e., $P_c > \pi_p$). Such an individual will, therefore, suffer from overall body edema.

THE INVOLVEMENT OF THE LYMPHATIC SYSTEM

Although both filtration and reabsorption take place in capillaries, under normal conditions more fluid is filtered than is reabsorbed. Thus, there is a buildup of fluid in the interstitial compartment, which if unchecked will lead to edema. This excess fluid is removed by drainage into the lymphatic system by entering lymphatic capillaries present throughout the tissues. It then passes into larger lymphatic vessels (that contain valves like those found in veins) and after being processed in lymph nodes, returns to the circulatory system. A full description of the lymphatic system will be presented in the chapter on immunology.

BLOOD PRESSURE

Blood pressure is the driving force that moves blood through the circulatory system. When the term blood pressure is used it almost always implies *systemic arterial blood pressure (BP)*. The most important factors that affect BP are:

Blood Volume
Cardiac Factors
Total Peripheral Resistance

The relationship between these factors and blood pressure is outlined in **Figure 8.13**.

BLOOD VOLUME

Blood volume (BV) is a measure of the amount of blood (consisting of cells plus plasma) that is contained in the total circulatory system. Observation of **Figure 8.14** demonstrates that the majority of blood volume (64%) is contained on the venous side of the system because veins are more compliant than are other vessels. (Note also that only 5% of the blood is present in capillaries yet this is the blood that accounts for the exchange of materials.)

Increasing blood volume results in an increase in BP because:

CO = HR x SV

BP = CO x TPR

where: CO = Cardiac Output
HR = Heart Rate
SV = Stroke Volume
TPR = Total Peripheral Resistance
BP = Systemic Arterial Blood Pressure

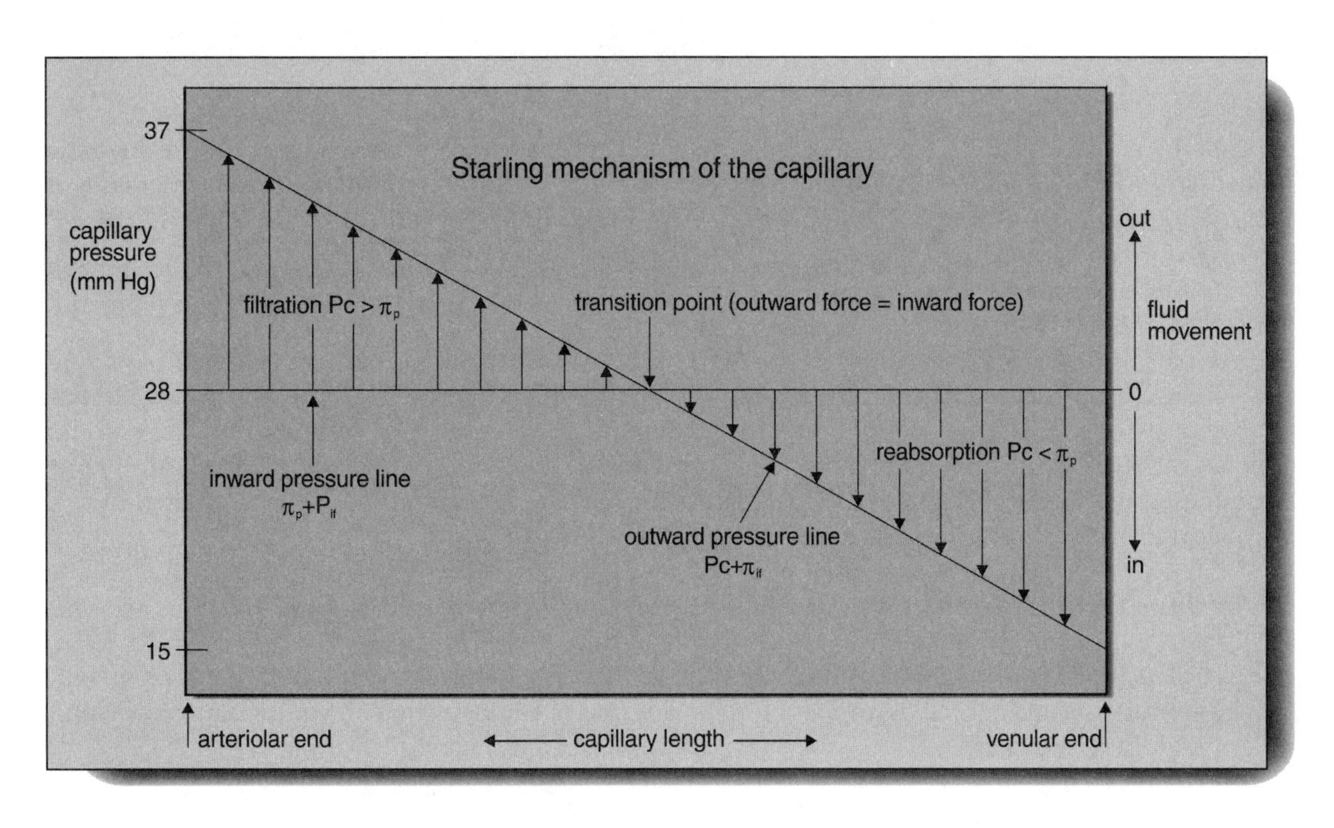

FIGURE 8.10

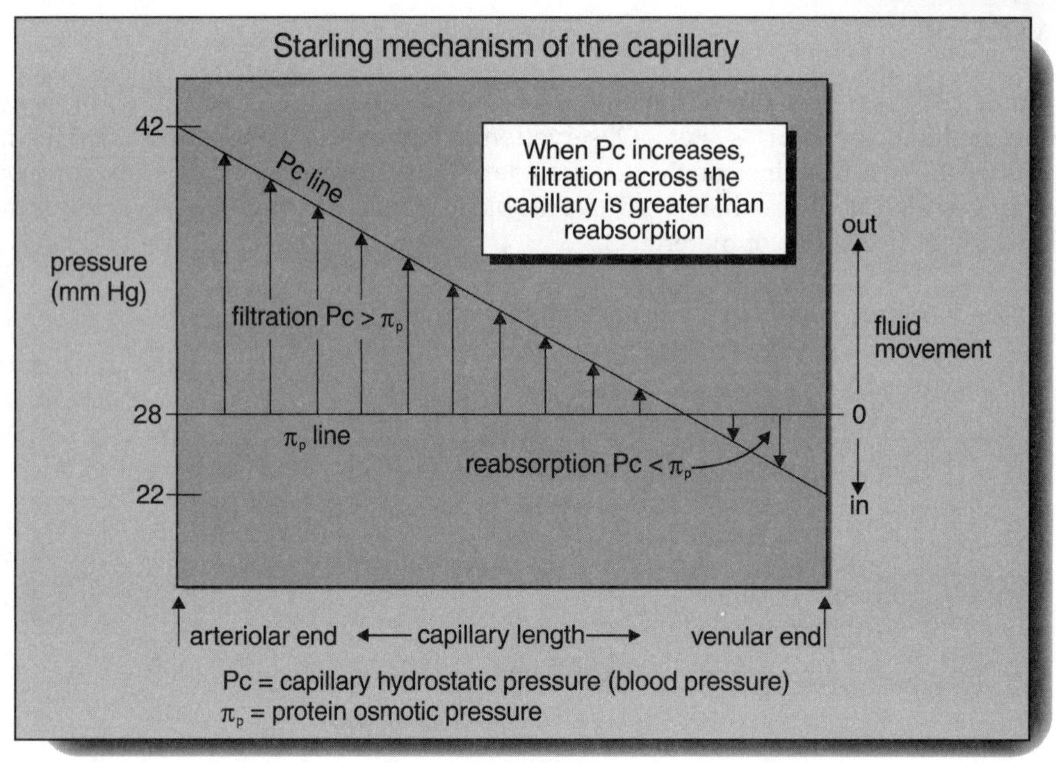

FIGURE 8.11

Formula 1 has already been described in the previous chapter and expresses what could be termed the *cardiac factor*. From this formula, it is evident increasing SV will increase CO. Since SV increases if Venous Return (VR) increases (through the Intrinsic Starling Mechanism), it follows that if BV increases for any reason, then VR increases, SV increases, CO increases and BP increases.

CARDIAC FACTORS

As described , the cardiac factors are heart rate (HR) and Stroke Volume (SV). These factors influence BP because they affect CO. Remember that HR is regulated by the SA node, and SA nodal depolarization is controlled by sympathetic stimulation, parasympathetic stimulation and by the hormones epinephrine and norepinephrine released from the adrenal gland. SV is regulated by the Intrinsic and Extrinsic Starling Mechanisms.

The Intrinsic Starling Mechanism — describes the relationship between VR, EDV and SV.

The Extrinsic Starling Mechanism — describes the relationship between sympathetic nervous system stimulation and/or the hormones epinephrine and norepinephrine on ventricular contraction and therefore on SV.

TOTAL PERIPHERAL RESISTANCE

Factors that affect resistance are outlined above in this chapter. Of all the factors that play a role, the most important is the *radius* of the peripheral precapillary arteriolar sphincters. *Total Peripheral Resistance (TPR)* is the resistance that the blood encounters as it passes through the arteriolar precapillary sphincters. *Local controls* and *extrinsic controls* can alter the radius of these precapillary sphincters through contraction and relaxation, and can therefore effect TPR.

Local Controls depend on mechanisms independent of nerves and hormones that allow individual tissues to regulate their own blood supply. This self-regulation (or *autoregulation*) takes place through *active hyperemia, reactive hyperemia* and *pressure autoregulation.*

1) ***Active hyperemia* (Figure 8.15)** — leading to an increased blood flow, takes place in an organ when there is an increase in the metabolic activity in the organ. This increased blood flow is in response to local changes in the concentration of various metabolites in the tissue during increased activity. For example, the arterioles that supply the skeletal muscle dilate in response to an increased concentration of CO_2 and a decreased concentration of O_2 and this increases blood flow. In addition, changes in other metabolites including adenosine, eicosanoids and bradykinin have been implicated in local control of arteriolar radius.

2) ***Reactive hyperemia*** — is similar to active hyperemia in that it also results in an increased flow in the tissue due to changes in metabolites. However, the immediate cause is occlusion of flow through the organ due to a blockage of the vessels feeding the organ or tissue. As a result of

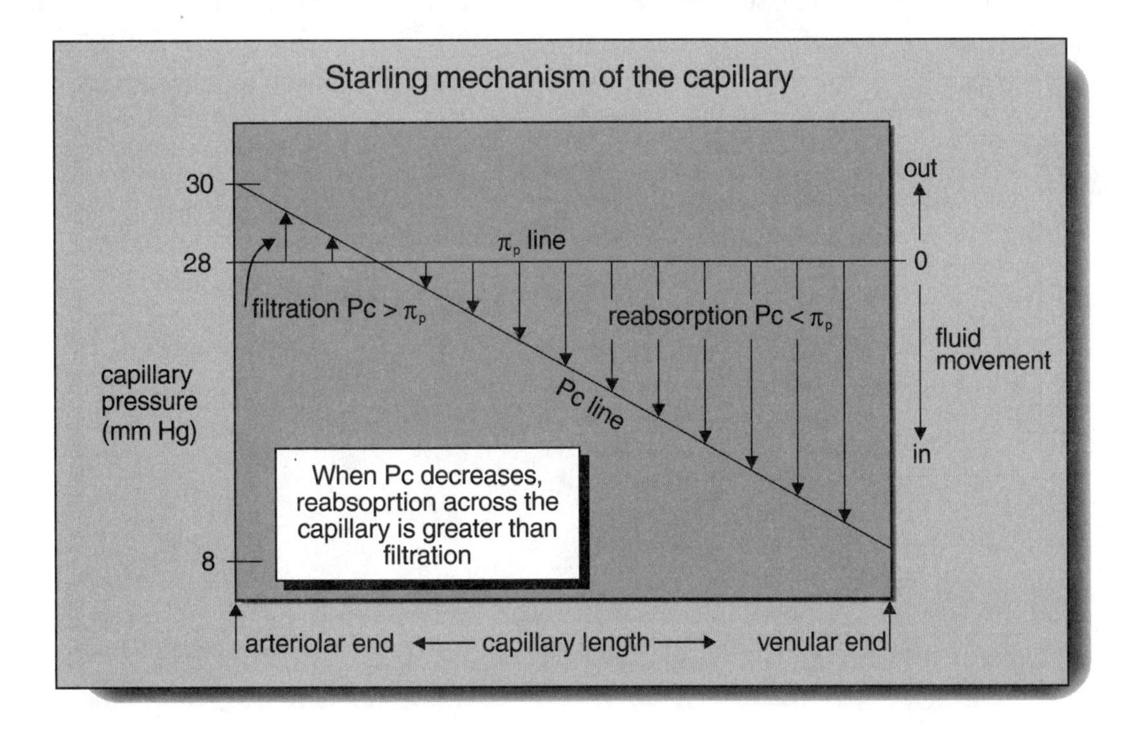

FIGURE 8.12

the blockage, the tissue starves for O_2; CO_2 builds up as does other metabolites and the smooth muscle of the arterioles feeding the tissue dilates. Upon removal of the blockage, there is now a massive increase in flow into the tissues which continues until the levels of the various metabolites normalizes. At this point the arteriolar radius returns to normal and flow decreases.

3) ***Pressure autoregulation* (Figure 8.16)** — operates to maintain a constant flow through an organ in the face of fluctuations in systemic blood pressure. Flow thorough an organ will decrease if MAP drops, or if the blood vessel supplying the organ is partially occluded. To correct this problem, the arterioles supplying the organ dilate causing flow to increase to its previous level. Thus the change in resistance in the arteriole is inversely proportional to the change in MAP.

Pressure autoregulation acts through changing metabolite levels in the tissues. For example, if MAP decreases, flow will initially fall in the kidney. This will lead to changes in the various metabolites described previously for active hyperemia. As a result, the arterioles feeding the kidney tissue will dilate and flow will return to the normal range. Pressure autoregulation also operates when MAP rises, causing flow to initially increase, and under these conditions, the effect would be to decrease flow back to the normal range.

In addition to the role of local chemical factors in autoregulation, another mechanism may also be involved in autoregulation, namely the *myogenic* effect (or the myogenic response). Here, increasing MAP stretches the arteriolar smooth muscle which then responds by constricting. Thus increasing MAP will initially increase flow, but upon arteriolar constriction, flow will decrease. On the other hand, if MAP were to suddenly decrease causing flow to decrease, the reduced stretch on the arteriolar

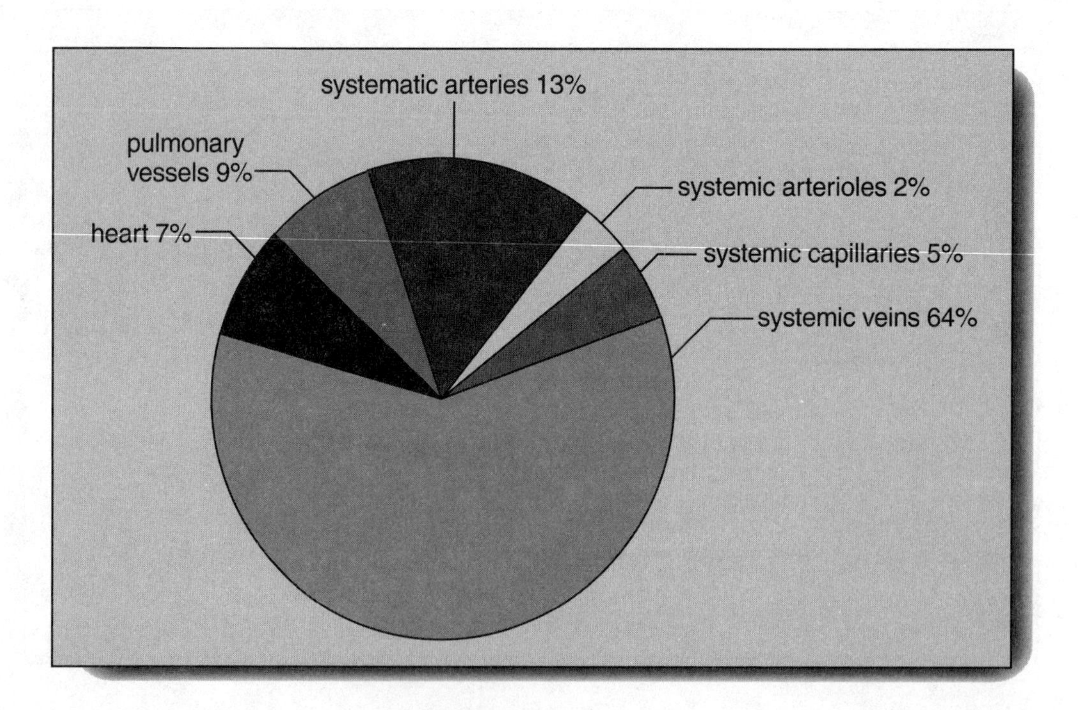

FIGURE 18.14

smooth muscle would then cause sphincter relaxation, increasing the radius of the vessels, reducing resistance and increasing flow.

Extrinsic Controls below depend on regulation of the arteriolar smooth muscle by branches of the post-ganglionic sympathetic nerves, or through circulating hormones.

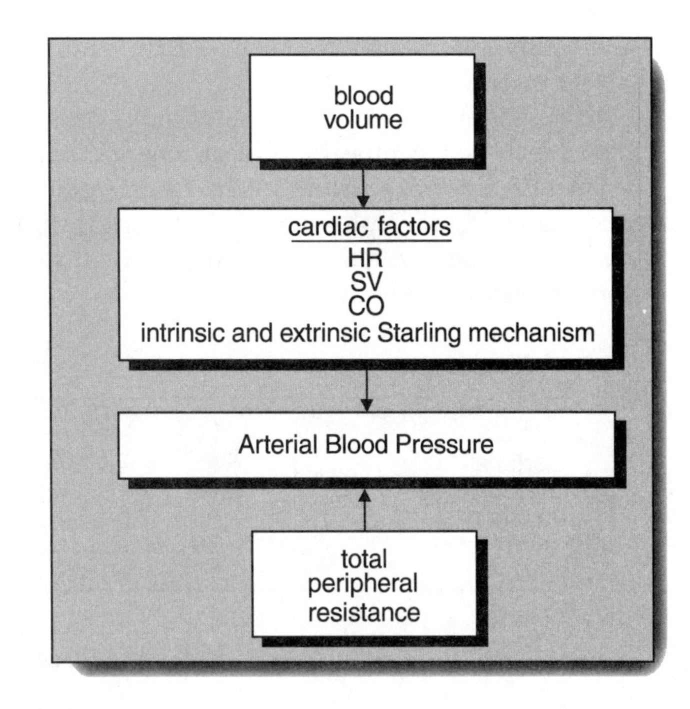

FIGURE 8.13

Innervation by Autonomic Nerves. Since most arteriolar smooth muscle is innervated by adrenergic sympathetic nerve fibers, stimulation of these nerves releases norepinephrine causing arteriolar constriction, leading to increased resistance and decreased flow. Such nerves normally supply tonic stimulation, thus by increasing the stimuli above the tonic level constriction is achieved and by decreasing this stimuli dilation is achieved in the arterioles. Parasympathetic innervation is not present in the majority of arterioles. The only arterioles innervated by parasympathetic nerve fibers are those that supply blood to the external genitals, and here such stimulation results in vasodilation.

Hormonal Regulation. The hormones, epinephrine and norepinephrine, released by the adrenal medulla upon sympathetic stimulation are highly vasoactive on arteriolar smooth muscle. However, the response elicited by these hormones is dependent on the presence of á and â adrenergic receptors in the tissues **(Figure 8.17)**.

α receptors bind both norepinephrine and epinephrine and the response to this binding to produce vasoconstriction of the arteriolar smooth muscle. The majority of this norepinephrine is from adrenergic postganglionic sympathetic fibers, although a small amount is also released from the adrenal medulla upon sympathetic stimulation. Epinephrine, on the other hand, is released only from the adrenal medulla upon sympathetic stimulation.

β receptors bind only epinephrine released from the adrenal medulla upon sympathetic

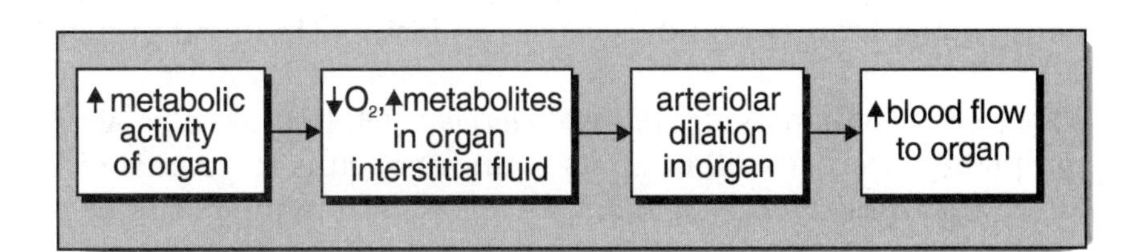

FIGURE 8.15

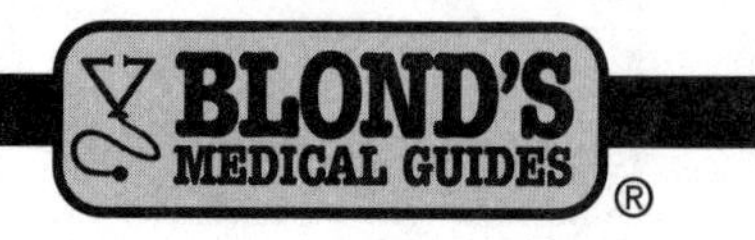

stimulation, and the response of this binding is to produce vasodilation of the arteriolar smooth muscle.

Thus, depending on the type of adrenergic hormone (epinephrine or norepinephrine), and the presence of α or β receptors in the target tissues, the outcome can be either vasoconstriction or vasodilation. Obviously the majority of arterioles found in the systemic circuit contain α receptors since upon stimulation by sympathetic nerves or through release of adrenal hormones, the effect is to promote vasoconstriction.

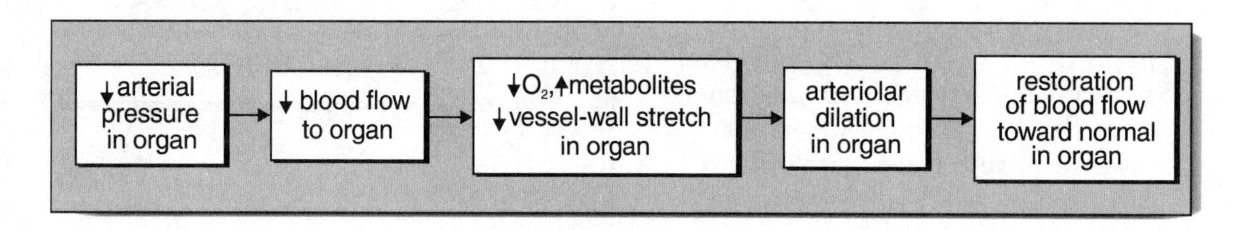

FIGURE 8.16

ANGIOTENSIN AND VASOPRESSIN

In addition to the hormones from the adrenal gland that can regulate arteriolar radius, the hormones *angiotensin II* and *vasopressin (ADH)* can also be involved in arteriolar constriction. They will be discussed in detail elsewhere (with kidney functions), however, a short discussion on their effects is of importance at this point.

Angiotensin II is generated when renin is released by the juxtaglomerular (JG) apparatus of the kidney. Renin itself is released when:

1. There is a decrease in blood flow in the kidney.
2. There is a decrease in blood pressure in the kidney, or
3. When there is a decrease in the levels of Na^+ and Cl^+ in the fluid filtered by the glomerulus.

The released renin then enters the blood and here it enzymatically acts on the precursor plasma protein, angiotensinogen (produced by the liver) to convert it to angiotensin I. Angiotensin I is then converted to angiotensin II by a converting enzyme in the lungs. Angiotensin II is actively vasoconstrictive on peripheral arterioles, and thus causes an increase in blood pressure.

Vasopressin, known as *Antidiuretic Hormone (ADH)*, can also act as a vasoconstrictive substance on peripheral arterioles although its main function is to promote increased water reabsorption in the kidney. Increased levels of ADH are released from the posterior pituitary (neurohypophysis) when:

1. There is an increase in the osmolarity of the body fluids,and
2. During the baroreceptor reflex elicited by a drop in arterial blood pressure.

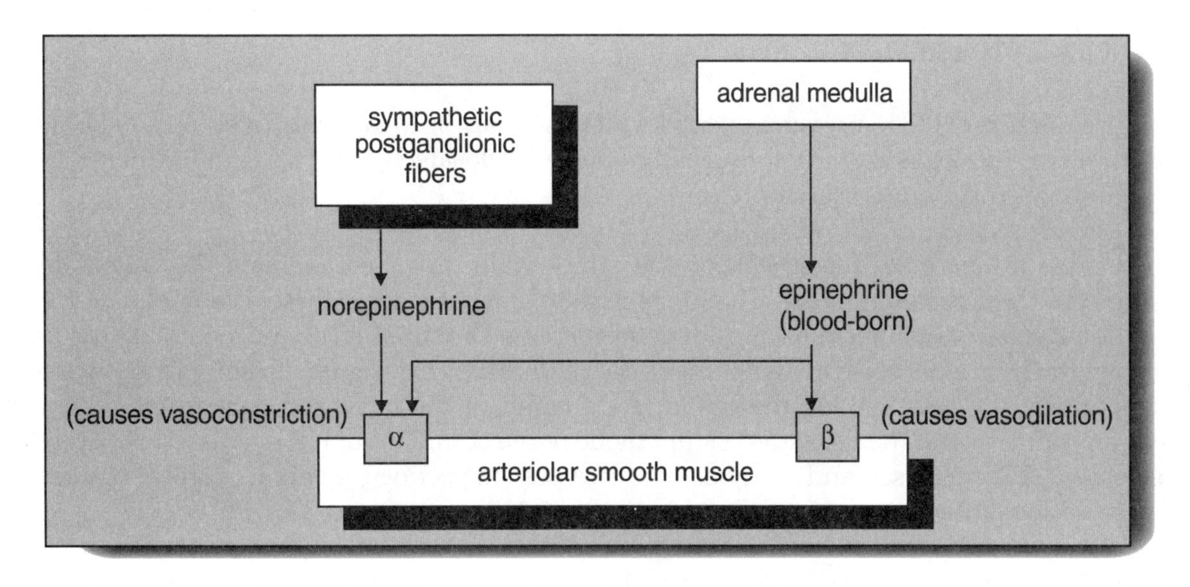

FIGURE 8.17

The various factors that act to alter arteriolar radius are outlined in **Figure 8.18** and the overall relationship between all of the factors that play a role in regulation of systemic arterial blood pressure is outlined in **Figure 8.19**.

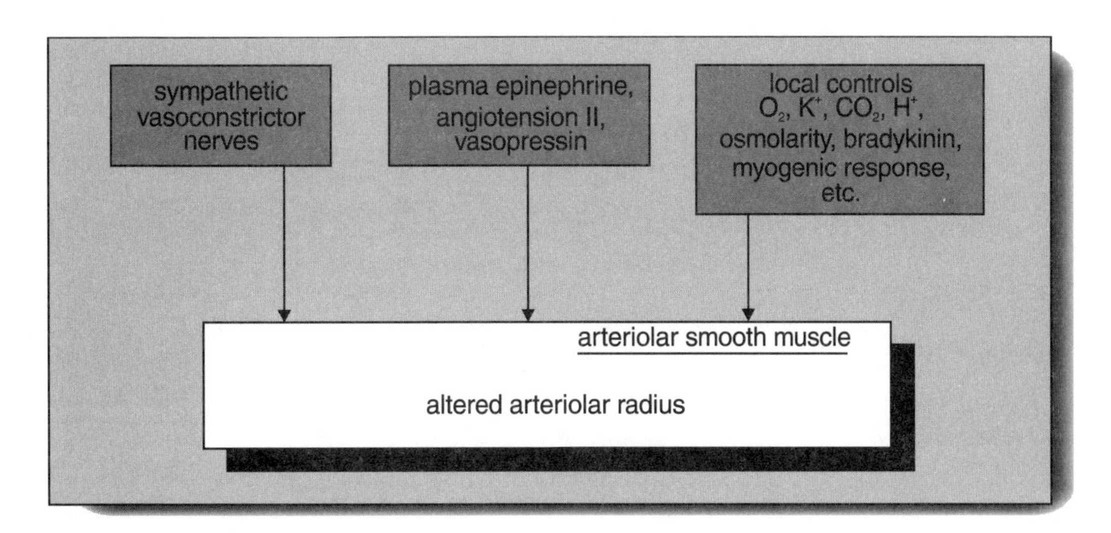

FIGURE 8.18

THE BARORECEPTOR REFLEX

Systemic Regulation of blood pressure overrides all local controls to maintain overall body homeostasis and preserves life. Pressure receptors called *baroreceptors* located in the two carotid sinuses and the aortic arch monitor systemic arterial pressure.

As blood pressure increases the stretch on the artery walls also increases and this stimulates the baroreceptors. An increased number of action potentials generated from these receptors now travel to the *medullary cardiovascular center*. As a result of the increased stimulation on the medullary cardiovascular center, there is a decrease in sympathetic nerve impulses exiting the center and an increase in parasympathetic impulses exiting the center. As a result of the increased parasympathetic nerve stimulation coupled with the decreased sympathetic nerve stimulation, HR decreases, SV decreases, CO decreases, TPR decreases, and systemic BP decreases (correcting the initial problem of increased BP). These various interactions are outlined in **Figure 8.20**.

In **Figure 8.21** the complex interactions between the various elements is presented. Note that the initial cause for the reduced BP is a reduction in blood volume possibly due to a hemorrhage. Note also the interaction that take place in the capillary, as well as in the adrenal medulla. The relationship of vasopressin (ADH) and Angiotensin II is also in this scheme.

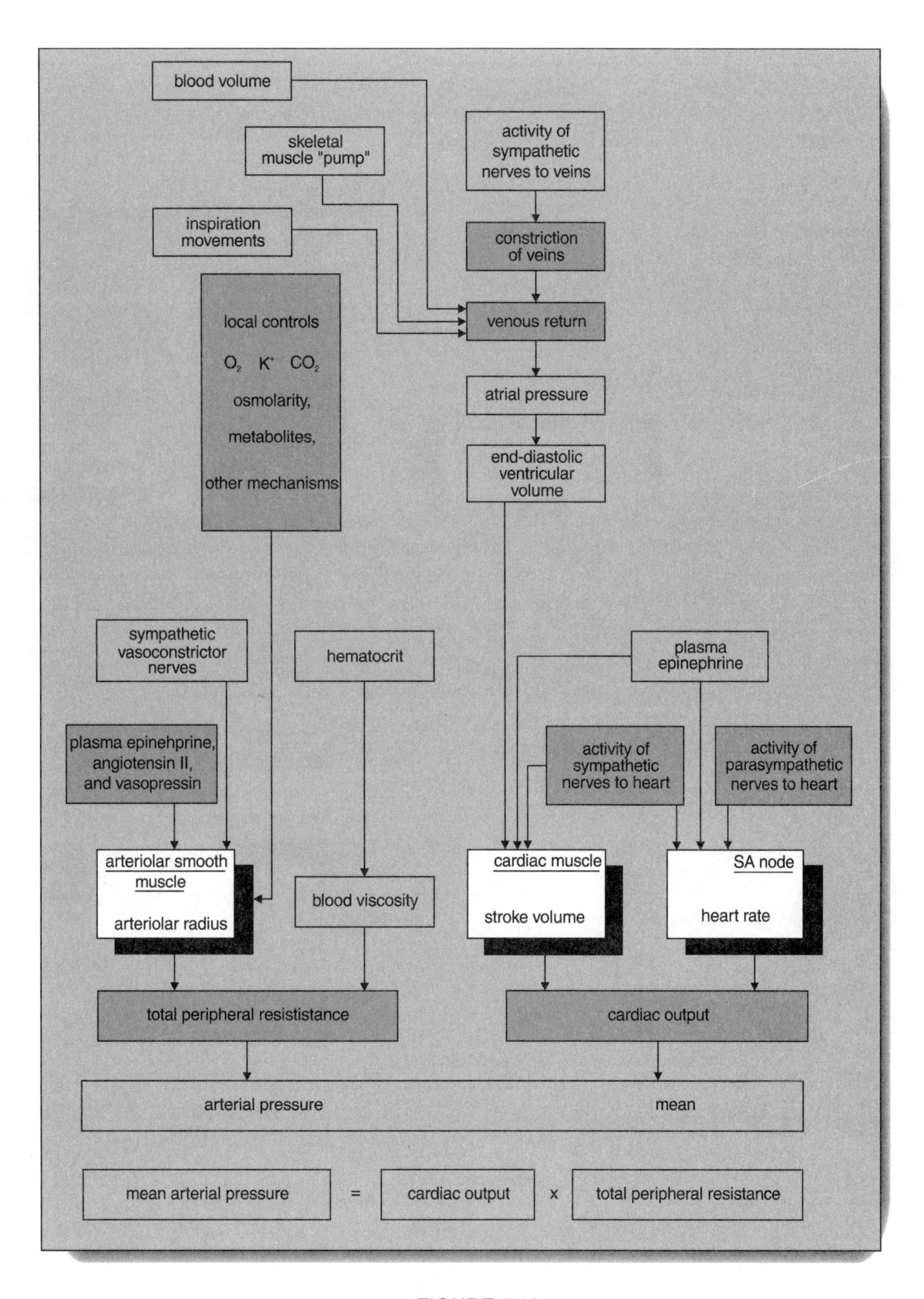
blood volume
skeletal muscle "pump"
activity of sympathetic nerves to veins
inspiration movements
constriction of veins
local controls
O_2 K^+ CO_2
osmolarity,
metabolites,
other mechanisms
venous return
atrial pressure
end-diastolic ventricular volume
sympathetic vasoconstrictor nerves
hematocrit
plasma epinephrine
plasma epinehprine, angiotensin II, and vasopressin
activity of sympathetic nerves to heart
activity of parasympathetic nerves to heart
arteriolar smooth muscle
arteriolar radius
blood viscosity
cardiac muscle
stroke volume
SA node
heart rate
total peripheral resististance
cardiac output
arterial pressure
mean
mean arterial pressure
=
cardiac output
x
total peripheral resistance

FIGURE 8.19

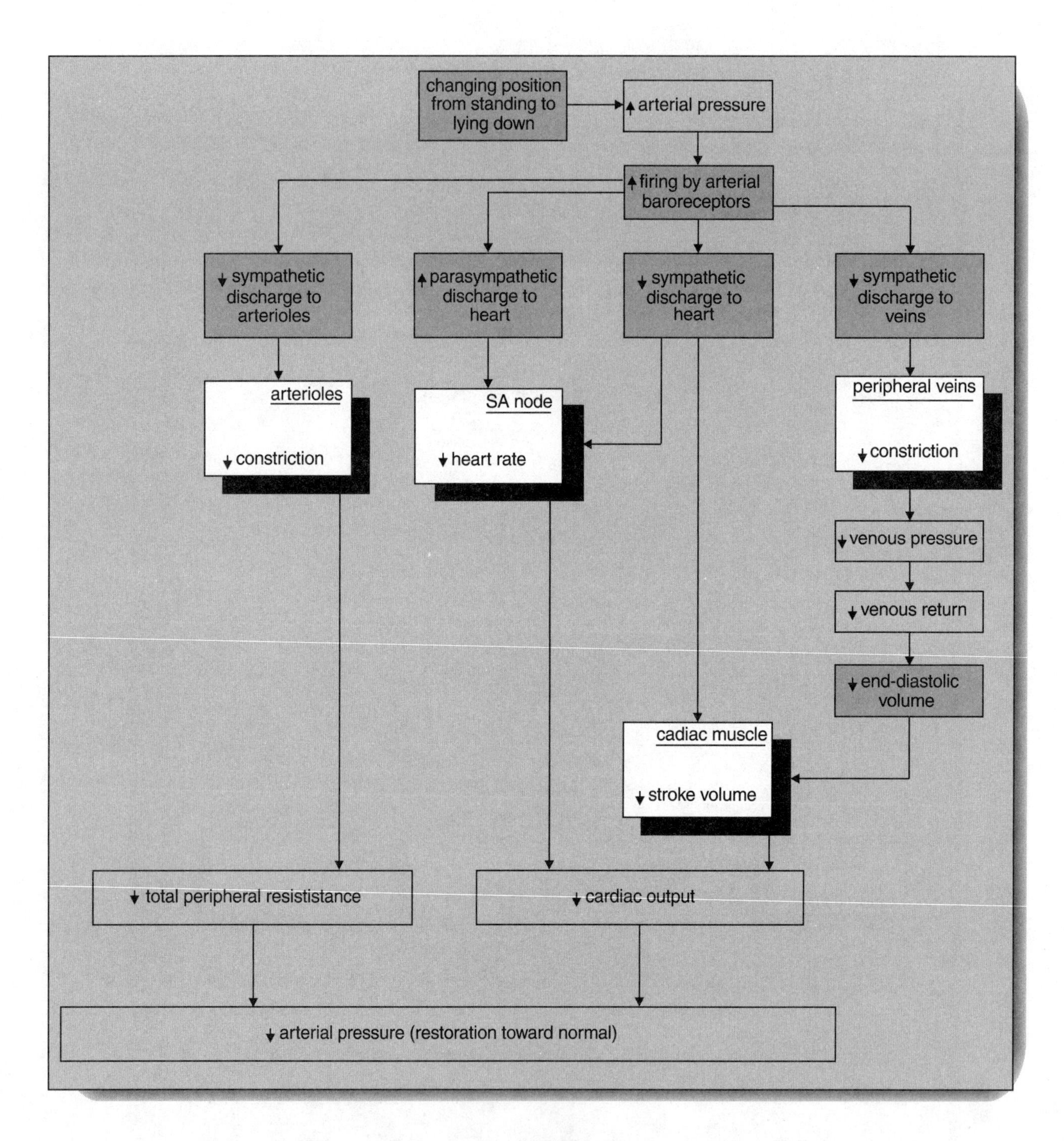

changing position from standing to lying down
↑ arterial pressure
↑ firing by arterial baroreceptors
↓ sympathetic discharge to arterioles
↑ parasympathetic discharge to heart
↓ sympathetic discharge to heart
↓ sympathetic discharge to veins
arterioles
↓ constriction
SA node
↓ heart rate
peripheral veins
↓ constriction
↓ venous pressure
↓ venous return
↓ end-diastolic volume
cadiac muscle
↓ stroke volume
↓ total peripheral resististance
↓ cardiac output
↓ arterial pressure (restoration toward normal)

FIGURE 8.20

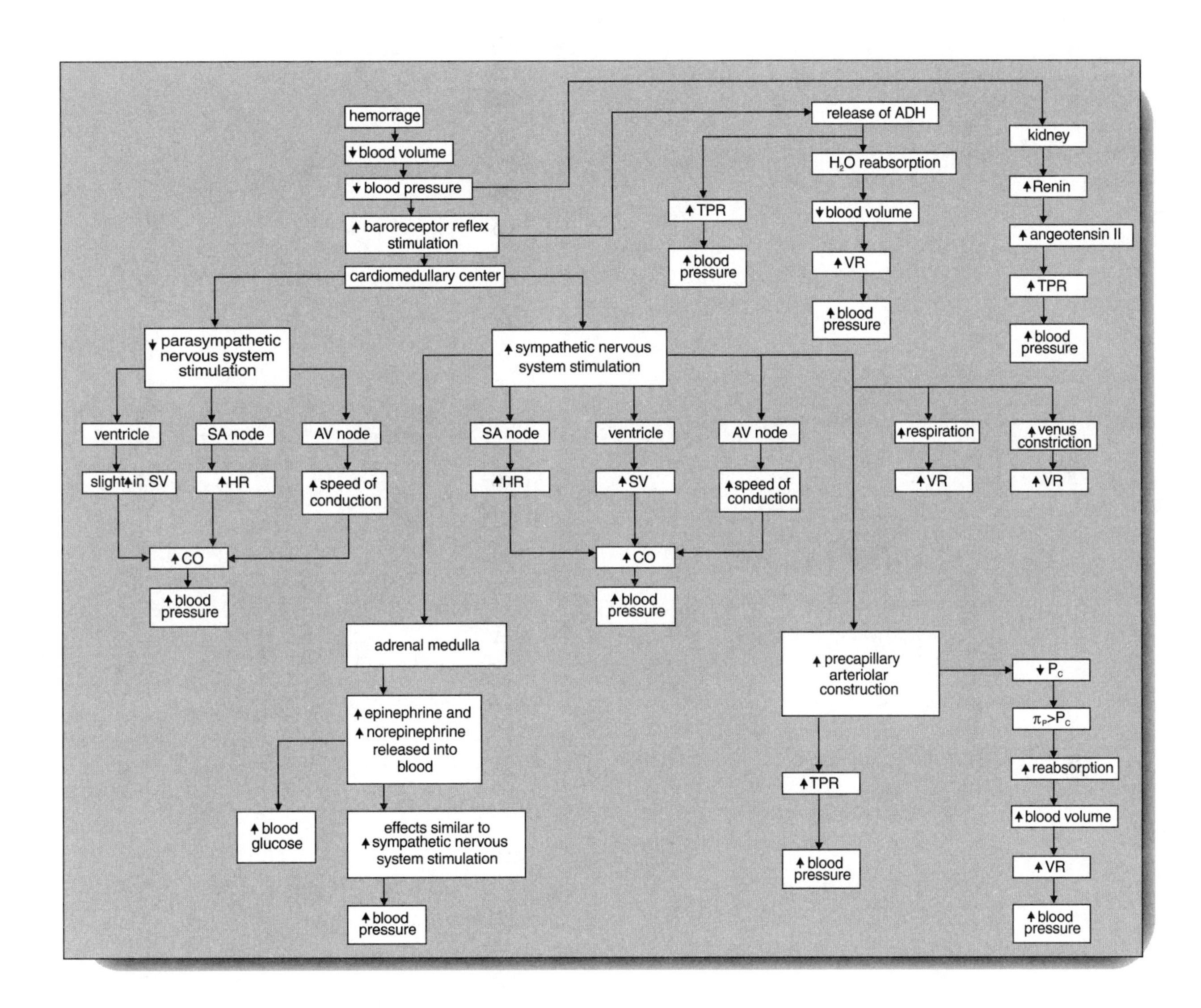

hemorrage
↓blood volume
↓blood pressure
↑ baroreceptor reflex stimulation
cardiomedullary center
release of ADH
H_2O reabsorption
↓blood volume
↑VR
↑blood pressure
↑TPR
↑blood pressure
kidney
↑Renin
↑angeotensin II
↑TPR
↑blood pressure
↓ parasympathetic nervous system stimulation
ventricle
SA node
AV node
slight↑in SV
↑HR
↑speed of conduction
↑CO
↑blood pressure
↑sympathetic nervous system stimulation
SA node
ventricle
AV node
↑HR
↑SV
↑speed of conduction
↑CO
↑blood pressure
↑respiration
↑VR
↑venus constriction
↑VR
adrenal medulla
↑epinephrine and ↑norepinephrine released into blood
↑blood glucose
effects similar to ↑sympathetic nervous system stimulation
↑blood pressure
↑ precapillary arteriolar construction
↑TPR
↑blood pressure
↓P_C
$\pi_P > P_C$
↑reabsorption
↑blood volume
↑VR
↑blood pressure

FIGURE 8.21

Chapter 9
THE BLOOD

INTRODUCTION

Blood is classified as a liquid connective tissue. The liquid "matrix" is comprised of *plasma* while the "formed elements" consist of various cell components—*red cells* (*RBC*), also known as *erythrocytes*; *white cells* (*WBC*), also called *leukocytes*; and *platelets*, also called *thrombocytes*. To determine the relative proportions of the plasma versus cells, a hematocrit is used.

HEMATOCRIT

Put simply, blood is collected from a subject and inhibited from clotting with an anticoagulating agent, such as heparin. It is then centrifuged in a glass tube (or specialized hematocrit capillary tube) to cause the formed elements (mostly red blood cells) to pellet on the bottom with the plasma supernatant fraction on top. Sandwiched between the erythrocytes and the plasma supernatant is the *buffy coat* consisting of a very thin layer of leukocytes **(Figure 9.1)**.

Using a graduated tube, or specialized hematocrit ruler, the percent of the erythrocyte fraction and plasma can now be determined. Thus, as depicted in **Figure 9.1**, the plasma is 55% of the total, and the RBC fraction is 45% of the total, (55% + 45% = 100% as expected). If the RBC volume is 45%, the *hematocrit* (also sometimes called the *packed cell volume*, or *PCV*) is equal to 45. On the average the hematocrit for women is about 42 and for men is 45, and although individuals may vary. Significantly lower hematocrits are indicative of anemias or other blood disorders.

TOTAL BLOOD VOLUME

To calculate a theoretical total blood volume, apply the following rule of thumb:

> **The blood volume (BV) is 8% of body weight in Kg. In a 70 Kg (154 lbs) male, this works out as: BV = 8% x 70 Kg = 5.6 Kg, and since 1Kg = 1L, BV = 5.6 L.**
>
> **In addition, if the patient's hematocrit is 45, it is now possible to calculate the volume of cells and plasma given a total BV of 5.6L, thus:**
>
> RBC volume = 5.6L x 45% = 2.5L
> Plasma volume = 5.6L - 2.5L = 3.1L

Another way to calculate plasma volume is as follows:

> **given the hematocrit = 45 (or 45% of cells),**
> **% plasma volume = 100% - 45% = 55%**
> ***plasma volume* = 5.6L x 55% = 3.1L**

BLOOD CELLS

ERYTHROCYTES

The main function of the RBC is to transport O_2 and CO_2 between the lungs and the tissues, and this process is facilitated by the presence of hemoglobin and the enzyme carbonic anhydrase within the RBC cytoplasm. Erythrocytes or red blood cells (RBC) are biconcave in shape. This flattened shape allows them to slip through narrow capillaries sideways and also increases the efficiency of exchange of gases across the RBC membrane between the RBC cytoplasm and the blood plasma.

Clinically, RBC counts are employed in conjunction with the hematocrit to assess disease states. Although the number of RBCs vary to some degree, the normal range in the population is as follows:

4.6×10^6 - 6.2×10^6 cells/mm^3 in adult males
4.2×10^6 - 5.4×10^6 cells/mm^3 in adult females
4.5×10^6 - 5.1×10^6 cells/mm^3 in children

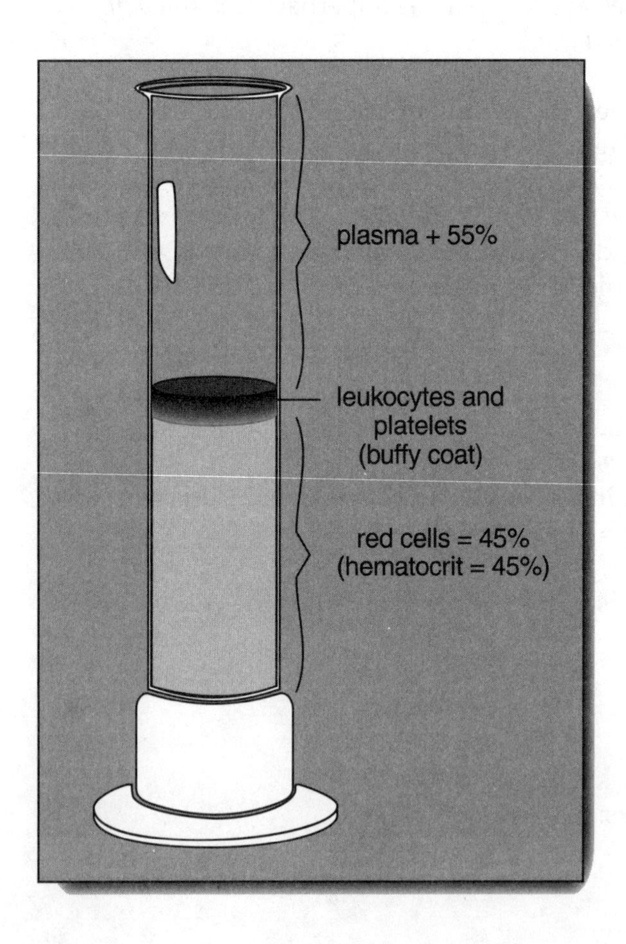

FIGURE 9.1

LIFE CYCLE OF THE RBC

Erythrocytes are produced from hematocytoblast cells in the bone marrow and progressively develop through a series of stages as follows:

Hematocytoblast
Proerythroblast
Erythroblast
Normoblast
Reticulocyte
Erythrocyte

It is interesting to note that as the cell progresses from the normoblast stage into the reticulocyte stage the nucleus disappears. Thus, the adult erythrocyte does not possess a nucleus. This means that RBCs cannot undergo mitosis nor can they produce new forms of mRNA. However, they possess pools of mRNA in their cytoplasm that allows them to generate certain proteins to maintain cell function. In the absence of a functional nucleus, the cell's life is limited to approximately 120 days.

ERYTHROPOIESIS

The process of RBC formation is termed *erythropoiesis* and is under the control of a hormone generated by the kidney called *erythropoietin*. The rate of RBC formation is regulated by a negative feedback mechanism involving the levels of transported O_2 **(Figure 9.2)**. Since RBCs are central in the transport of O_2 to the tissues, a reduction in their number would reduce the availability of this gas. When O_2 delivery to the kidney falls below a certain set-point, the cells in this organ release the hormone erythropoietin. This hormone circulates to the bone marrow where it promotes RBC formation (erythropoiesis). Over a period of a few weeks, the levels of circulating RBCs increase as does the ability of the blood to transport O_2. As the levels of O_2 delivered to the kidney rise, the release of erythropoietin decreases and erythropoiesis in the bone marrow is shuts down.

NUTRIENTS, IRON AND THE PRODUCTION OF RBC

Before erythrocytes can be formed in the bone marrow certain essential nutrients obtained from the digestion of food must be absorbed in the small intestine. In this regard, vitamin B_{12} and folic acid are necessary for the synthesis of DNA in the RBCs while iron is an integral component of hemoglobin production. After absorption in the GI track, these substances are transported by the blood to the bone marrow where they are utilized in the actual production of RBCs.

Within the body, iron balance is maintained by negative feedback mechanisms. Small amounts of iron are lost from the body in the urine, feces, sweat, and in women in the menstrual blood. To maintain constant (and essential) levels of iron, an amount equal to that lost must be replaced by ingestion of iron through the GI track. Regulation of iron absorption is primarily controlled within the intestinal epithelium itself. These epithelial cells absorb a small fraction of the available iron present in the chyme of the small intestinal lumen. As iron levels in the body decrease, the amount in the intestinal epithelium also decreases which promotes increased absorp-

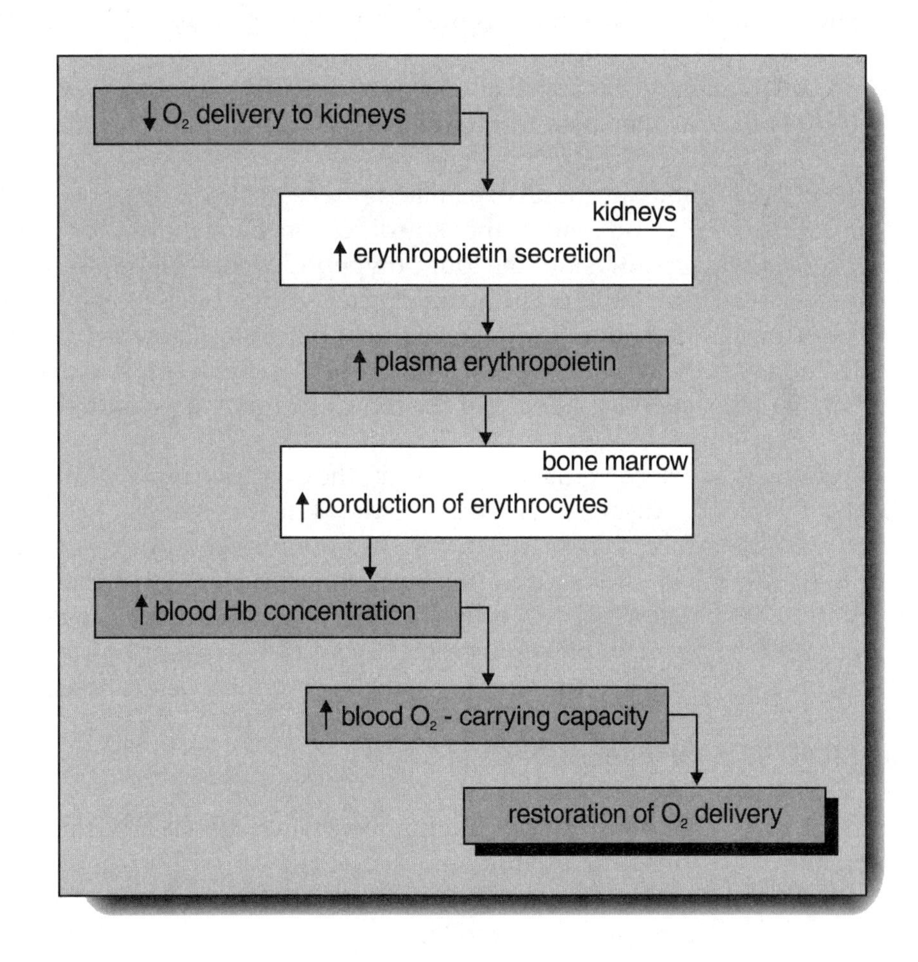

FIGURE 9.2

tion from the intestine. As iron levels in the body increase the amount present in the epithelium increases and iron uptake decreases.

Iron is stored in the liver bound to the protein *ferritin*. Approximately 25% of the total body iron content is present in ferritin, 50% in hemoglobin and 25% in heme containing proteins such as myoglobin in red skeletal muscle tissue. During the destruction of the RBCs in the liver, the released iron is complexed with the carrier protein *transferrin* which transports it back to the bone marrow where it is incorporated into new RBCs.

HEMOGLOBIN

Hemoglobin is composed of four polypeptide subunits that constitute the *globin* portion of the molecule. In the center of each polypeptide is a *heme* group consisting of an iron-nitrogen moiety. Thus, in a complete molecule of hemoglobin (with four heme groups), the four iron molecules can bind four molecules of O_2. Hemoglobin bound to O_2 (oxyhemoglobin) has a reddish color (as in the arterial blood), while hemoglobin that is not bound to O_2 has a bluish color (as in venous blood). In addition to binding O_2, hemoglobin also binds CO_2 (carbamino-hemoglobin), and carbon monoxide. The distinct binding parameters of hemoglobin will be discussed in the chapter on respiration.

Since RBCs only live for approximately 120 days, the old, damaged and dying cells are removed from the circulation as they pass through the sinusoids of the liver (and spleen).

Within these organs are found fixed macrophages (called *Kupffer cells* in the liver), which phagocytosize the damaged cells and remove them by digestion. Each molecule of hemoglobin extracted during this process is broken down into the four polypeptide globin subunits, and the four heme groups. Iron is disassembled from the heme to be recycled and the globin is further catabolized into the green pigment biliverdin, and then into the orange pigment bilirubin. These two pigments are excreted into the liver bile and enter the small intestine to be excreted in the feces. It is these pigments that give the bile its characteristic yellowish color, and the feces its brown appearance.

In certain liver disorders, such as hepatitis, the excretion of the pigments is disrupted. As a result, they enter the blood causing such tissues as the skin and whites of the eyes to take on a yellowish cast (known as jaundice). About 33% of newborn infants also appear jaundiced because the immature liver is less efficient in clearing the pigments into the bile, and because of the increased load of fetal erythrocytes being destroyed at this time. This condition, known as *physiologic jaundice*, is treated by feedings that promote bowel movements, and by exposure of the infant to fluorescent lights that activate mechanisms in the skin to reduce pigment concentrations.

DISORDERS OF ERYTHROCYTES

Anemia is a disorder in which the number of functional RBCs is reduced, thus reducing the total amount of O_2 transported to the tissues. Anemia can result from the following causes:

1. *Hemorrhagic Anemia* due to loss of blood.

2. *Aplastic Anemia* due to a malfunction in the bone marrow, synthesis of RBCs is reduced. The bone marrow disorder may result from exposure to radiation, chemicals, or drugs, or can arise from cancer. In some cases the cause of this disorder remains unknown.

3. *Hemolytic Anemia* results from an abnormally high rates of RBC rupture (hemolysis). This type of anemia may result from genetic defects, or from parasitic infections (such as malaria), or adverse drug reactions. In some cases the cause remains unknown.

Polycythemia is a condition in which the individual has an abnormal increase in the number of RBCs in the circulation. This condition is due to the increased production of RBCs in the marrow. In some cases the marrow functions abnormally, while in others the production results after the individual ascends to high altitudes and remains there for an extended period.

As a result of the increased number of RBCs in the circulation, there is an increase in the viscosity of the blood, with a concomitant increase in resistance to flow. This causes increased stress on the heart, which has to work harder to compensate for the increased systemic resistance. If blood flow through the smaller vessels is reduced by the increased resistance, the deoxygenated blood may cause the skin to become *cyanotic* (appear bluish).

LEUKOCYTES

Leukocytes, or white blood cells (WBCs), come in various types **(Figure 9.3)**. Like erythrocytes, all classes of leukocytes develop in the bone marrow initially from the hematocytoblast. WBCs can be subclassified into those with granules that stain in their cytoplasm (the *granulocyte*), and those with no granules (the *agranulocyte*). The total number of leukocytes fall between $5 \times 10_6$ - $10 \times 10_6$ /ml of blood with an average of 7×10^6 WBCs/ml. (This is expressed as an average WBC count of 7,000/mm^3.)

Under abnormal conditions, leukocyte numbers may be above or below the average range and this will cause significant problems. A serious reduction in WBCs is termed leukopenia and can result from exposure of the bone marrow to drugs, chemicals or radiation. The outcome is an inability of the immune system to fight off invader organisms. On the other hand, over-production of one or more types of WBCs, as in the disease leukemia, will also lead to an altered immune response because the cells produced are immature and thus almost nonfunctional. (In leukemia the WBC count can rise as high as 500,000 cells/mm^3 as compared with the normal 7,000 cells/mm^3.) In addition, the abnormal proliferation of one or more subclasses of WBCs can displace other hematopoietic cell lines from the bone marrow resulting in anemia due to the reduction in RBC production, and in hemorrhage due to the reduction in platelet production.

THE GRANULOCYTE PATHWAY

Here the hematocytoblast differentiates into the Myeloblast, which then differentiates into the Progranulocyte. From here the pathway diverges into three possible routes producing the neutrophil, the eosinophil and the basophil.

Neutrophils — have a large nucleus with two to five lobes and the granules in the cytoplasm stain pink in a neutral pH stain. The granules are lysosomes containing catabolic enzymes used to digest

material that has been phagocytized by the cell. The major function of the neutrophil is phagocytosis of bacteria and other foreign particles. Neutrophils are chemotactically attracted to areas of infection and exit from the vascular compartment to the tissue spaces by crossing the capillary wall by pushing through the spaces between adjacent endothelial cells. The process is called *diapedesis*. As measured by a differential WBC count, neutrophils comprise 54% to 62% of the total WBC population in a typical blood smear.

Eosinophils — have a bilobate nucleus and cytoplasmic granules that stain red in an acid pH stain. The eosinophil functions to kill various parasites, such as worms, in the intestine. It may also be involved in certain aspects of inflammation and allergic reactions. Eosinophils comprise 1% to 3% of the total WBC population in a typical blood smear.

Basophils — have a lobed nucleus and the cytoplasm contains granules of histamine (a vasoactive substance) and heparin (an anticoagulant) that stain blue with a basic pH stain. The major

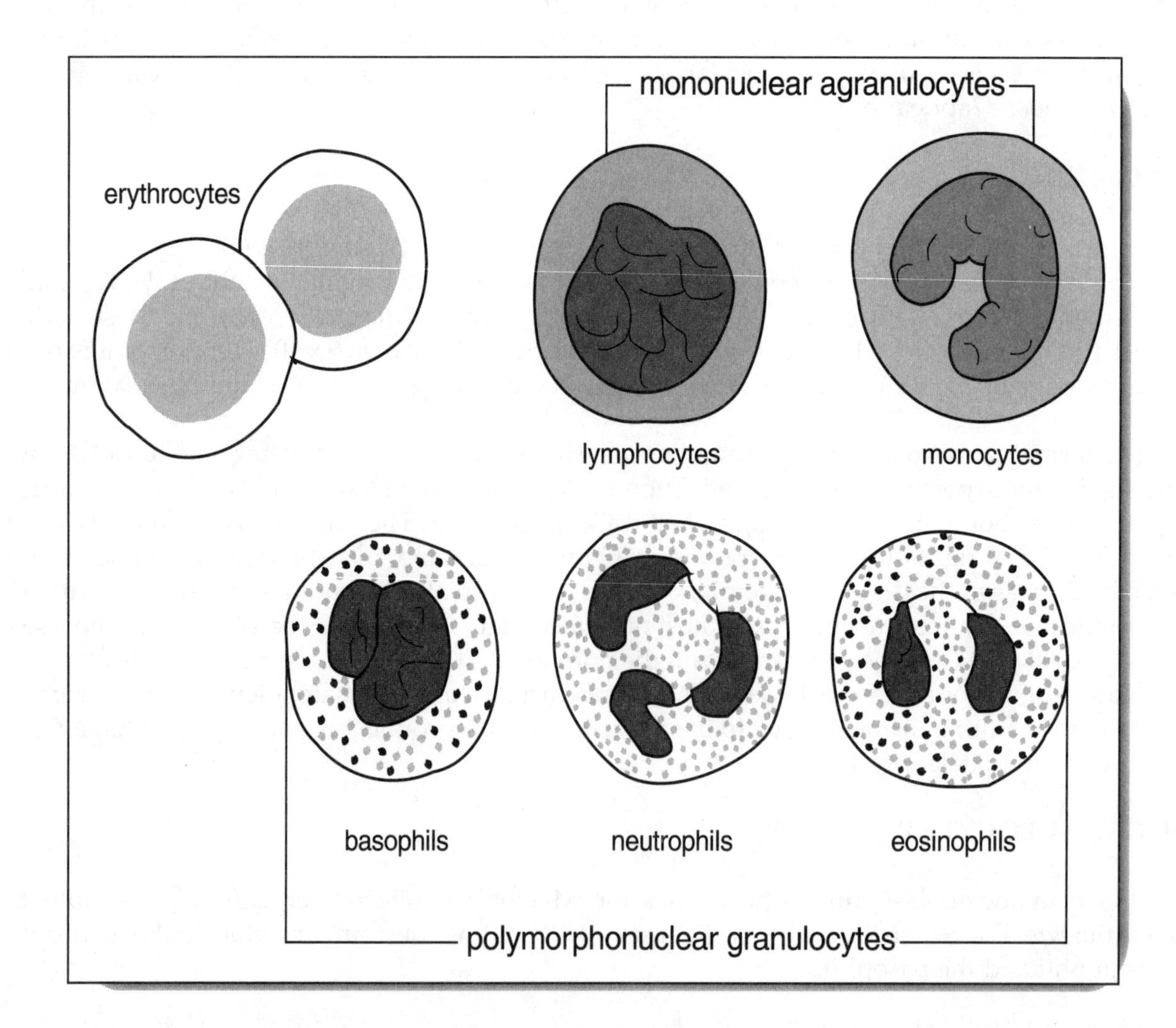

FIGURE 9.3

function of the basophil is to release histamine and promote an inflammatory response (increased blood flow, increased filtration, edema). The mast cell found in the connective tissues is closely related to the basophil and both probably originate from a common precursor cell. Basophils account for less than 1% of the total WBCs present in a typical blood smear.

THE AGRANULOCYTE PATHWAY

Here the hematocytoblast differentiates into the Monoblast, Lymphoblast and Megakaryoblast and these cells will continue to differentiate into the adult forms as follows:

Monocytes — develop from monoblasts and are two to three times larger than erythrocytes. The nucleus in this cell is variable in shape and the major function of the cell is to complete development and become a macrophage. Initially the monocytes leave from the bone marrow and circulate for a few days after which they exit the vascular space and enter the tissues by diapedesis. Here they complete their differentiation into *macrophages*. As macrophages, they are avidly phagocytic and along with neutrophils are involved in removal of bacteria and other foreign particles from the body. Macrophages also play an important role in antigen presentation to lymphocytes, but this aspect of their function will be discussed in detail in the chapter on immunity. Monocytes make up from 3% to 9% of the WBCs present in a typical blood smear from a normal person.

Lymphocytes — develop from lymphoblasts and are only slightly larger than a RBC with a nucleus that fills almost the entire cellular volume. Functionally and by origin, lymphocytes can be subclassified as thymic derived (T-cells) and bone marrow (or bursa) derived (B-cells). Both T-cells and B-cells are involved in the various aspects of specific immunity. T-cells are responsible for mediating cell mediated immunity and their functions encompass both cellular cytotoxicity and regulation. B-cells are responsible for mediating humoral immunity and are the source of immunoglobin. Lymphocytes account for 25% to 33% of the WBCs present in a typical blood smear from a normal person. A complete discussion of the functions of these cells types can be found in the chapter on immunity.

Thrombocytes(or Platelets) — develop from the megakaryocyte (or giant cell) in the bone marrow. Initially, the megakaryocyte itself develops from the megakaryoblast. The megakaryocyte forms the platelets by pinching off small cytoplasmic fragments enclosed by a plasma membrane and containing vasoactive granules. Since platelets do not contain a nucleus, they cannot promote new protein synthesis or cell division. However, they are capable of undergoing limited metabolic activity, are able to demonstrate pseudopod-mediated movement and can release their vasoactive granules during the events leading to hemostasis. Normally in the blood there are 130,000 to 360,000 platelets/mm^3.

PLASMA

Plasma is the liquid component of the blood and as such is comprised primarily of water. However, it also contains a variety of proteins, salts and metabolites. A partial list of plasma constituents is presented here:

PLASMA CONSTITUENTS

Water:		92%
Electrolytes:		
Na+	145 mM	
K+	4.0 mM	
Ca++	2.5 mM	
Mg++	1.5 mM	
Cl^-	103 mM	
$HCO3^-$	24 mM	
$HPO4^-$	1.0 mM	
$SO4^-$	0.5 mM	
		1.5%
Proteins:		7.3%
Albumin		4.5%
Globins		3.0%
Fibrinogen		0.3%
Gases:		
CO_2	2.0 mL/100 mL	
O_2	0.2 mL/100 mL	
N^2	0.9 mL/100 mL	
Nutrients:		
Carbohydrates (mainly glucose)	100 mg/100 mL	
Amino Acids	40 mg/100 mL	
Lipids	500 mg/100 mL	
Cholesterol	150 - 250 mg/100 mL	
Vitamins	up to 2.5 mg/ 100 mL	
Waste Products:		
Urea	34 mg/100 mL	
Creatinine	1 mg/100 mL	
Uric Acid	5 mg/100 mL	
Bilirubin	up to 1.2 mg/100 mL	
Hormones:	up to 0.05 mg/100 mL	

PLASMA GLOBINS

The globin fraction of the plasma protein consists of: α globin 0.7% - 1.3%; β globin 0.6% - 1.1%; and δ globin 0.7% - 1.7%. *α globin and β globin* transport lipids and fat soluble vitamins in the plasma. Antibodies make up the *δ globin* fraction of the plasma protein. These antibodies are manufactured by activated B-cells called *plasma cells* and are involved in various aspects of the specific humoral immune response.

PLASMA LIPOPROTEINS

Depending on their composition, the plasma lipoproteins are classified on the basis of their density as follows:

Chylomicrons — consist of a high concentration of triglycerides which are being transported through the blood to the adipose tissues and to muscle cells. These chylomicrons originate from the small intestine where they were involved in the process of fat absorption. They are initially synthesized in the intestinal epithelium and then enter branches of the lymphatic system (lacteals) located in the villi of the small intestine. They are then transported through the lymphatic system and enter the blood along with the lymphatic drainage. After the chylomicrons have delivered their triglycerides to the adipose and muscle cells, the remnants of the chylomicrons are transferred to the liver by High Density Lipoproteins (HDL).

Very Low Density Lipoproteins (VLDL) — comprised of a high concentration of triglycerides which were synthesized in the liver from carbohydrates. These VLDL will then be transported in the blood to adipose cells.

Low Density Lipoproteins (LDL) — synthesized from VLDL molecules that have released their triglycerides at adipose cells. The LDL molecule contains a high concentration of cholesterol which is transported to target cells possessing LDL receptors. Cholesterol is slowly released at these cells through the mechanism of receptor mediated endocytosis. The cholesterol thus supplied is used by the cells to synthesize various substances.

High Density Lipoproteins (HDL) — synthesized in the liver and small intestine. HDLs transport the chylomicron remnants to the liver where they enter the liver cells by receptor mediated endocytosis. HDL molecules contain a high concentration of protein and a low concentration of lipid.

BLOOD TYPING

To effectively make a tissue (for example, a kidney) transplant between a donor and a recipient, certain immunological parameters between the donor organ and recipient must be known. In this way it is possible to determine if the transplant will remain viable or will be rejected. A parallel situation exists when making a blood transfusion in that the blood types between the donor and recipient must be appropriate or the transfusion will be "rejected." If the transfused blood from the donor and recipient are not compatible, this will lead to a transfusion reaction, causing the RBCs to clump in the capillaries, possibly killing the recipient.

THE ABO SYSTEM

Blood types in the ABO system are based on antigens (or agglutinogens) present on the surface of the RBC membranes. Structurally there are four possible types of RBC surface antigens, and depending

	ABO System(Table 9.1)		
	Blood Type Plasma	RBC Surface Antigen	Plasma Antibody
#1	A	A antigen	Anti-B antibody
#2	B	B antigen	Anti-A antibody
#3	AB	A antigen B antigen	None
#4	O	None	Anti-A antibody Anti-B antibody

on the type of antigen, this will determine the blood type. In addition, at about six months of age, circulating antibody (or agglutinin) specifically directed against these antigens appears in the plasma.

Note that the terms *agglutinogen* and *agglutinin* are used only in relation to the ABO blood group system. Thus, an agglutinogen is actually the antigen (Ag) on the surface of the RBC and an agglutinin is the circulating plasma antibody (Ab) directed against the agglutinogen.

Regarding the presence of the plasma Ab directed against the RBC Ag, as will be evident in the immunology chapter, antibody Ab is generated in response to an initial stimulus by a foreign substance called an *antigen* (Ag). However, in the ABO system, anti-RBC Ab appears in the circulation by the age of about six months, without the individual being previously stimulated by the "foreign" RBC antigen. (In other words, in a type-A person, Anti-B Ab appears in the circulation without the person ever being exposed to type B blood, and so on.) The mechanism to account for this is still under investigation, however, it has been suggested that the presence of bacteria in the intestine may be responsible because these organisms may have surface Ag similar in structure to the Ag of the ABO system **(Table 9.1)**.

In order to make a successful transfusion, only the following rule should apply **(Table 9.2)**:

The Donor RBC Ag must **NOT** *be bound by the Recipient plasma Ab.*

In this rule *DO NOT* consider the donor plasma Ab and the recipient RBC Ag; they will not affect the outcome of the transfusion. The reason is that any plasma Ab that is transfused from the donor to the recipient will not circulate long enough to cause any difficulties because it will be bound up by the tissues in the recipient almost immediately. Only the donor RBCs will circulate long enough to react with any recipient plasma Ab and cause problems.

The numbers listed for donor and recipient here refer to the **Table 9.1**.

Table 9.2		
Donor	**Recipient**	**Outcome**
#1	#2	A Ag will bind with anti-A Ab, not compatible.
#2	#1	B Ag will bind with anti-B Ab, not compatible.
#1	#1	Compatible.
#2	#2	Compatible.
#3	#3	Compatible.
#4	#4	Compatible.
#1,2,3	#4	RBC Antigens will react with plasma Ab. None of these are compatible.
#1,2,4	#3	All compatible. No plasma Ab present in recipient. RBC Ag on donor cells does not react. Thus, type AB is the universal recipient. Type AB can receive from any other group, but can only donate to another type AB.
#4	#1,2,3	All compatible. No Ag is present on donor RBCs. These RBCs will not react with any plasma Ab that might be present in the recipient. Thus, type O is the universal donor because it can give to any other blood type. However, type O can only receive from another type O.

RH FACTOR

In addition to the ABO system just described, there are another set of RBC surface antigens that must **ALSO** be taken into account when making a transfusion. These are the antigens of the Rh system. (These Rh antigens were first identified in the Rhesus monkey which is where the term *Rh* originated.) In the Rh system, there are two possible types of RBCs **(Table 9.3)**; those expressing the Rh antigen (Rh+) and those that do not express the Rh antigen (Rh-). In addition, unlike the ABO system where plasma Ab is present, in an Rh- person the appearance of anti-Rh+ Ab in the plasma takes place only if the Rh- person was previously exposed to the Rh+ Ag on RBCs. Such exposure of a Rh- individual to Rh+ blood can happen in only two ways **(Table 9.4)**.

1. When a Rh- individual is accidently transfused with Rh+ blood instead of Rh- blood. Such a mistransfusion will stimulate an immune response in the Rh- individual, who will then become presensitized to the Rh+ antigen and generate anti-Rh+ Ab. If this person is again given a second mistransfusion of Rh+ blood at a later date, the secondary immune response will generate very high levels of anti-Rh+ Ab which will rapidly react with the Rh+ Ag on the transfused RBCs. The result will be a transfusion reaction that can be life threatening.

2. When a Rh- mother has a Rh+ baby some of the baby's Rh+ RBCs may enter the mothers circulation and stimulate her immune system. She will then be presensitized to Rh+ Ag. If in a subsequent pregnancy she again has a Rh+ baby and more of the baby's RBCs enter the mother's circulation, her secondary immune response will generate high levels of anti-Rh+ Ab that will cross the placenta (IgG class Ab cross the placenta) and will attack the baby's RBCs. The baby's RBCs will be ruptured and the baby will become extremely anemic (a *blue baby*). This disorder is termed *erythroblastosis fetalis*, or *hemolytic disease of the newborn*.

To treat this problem, the Rh- mother receives an injection of *Rogan* which is actually a preparation of *anti-Rh+ Ab*. This is believed to react with any Rh+ RBCs present in the mother and thus inhibit the generation of an active immune response against Rh+ Ag in the mother. Thus the Rh- mother is able to have additional Rh+ babies without becoming presensitized to the Rh+ Ag so long as she receives Rogan treatment immediately after each delivery.

Note: Rh+ people can never have anti-Rh+ Ab (just as a type-A person can never have anti-A Ab, or a type-B person have anti-B Ab) because if they did, the Ab would react with their own RBCs and cause death. In immunological terms, they are tolerant of their own Ag and cannot react immunologically against it. If their immune system does react against their own cell or tissue Ag, they have an *autoimmune disease*.

Since the Rh antigen and the ABO antigens can both be expressed simultaneously by a RBC, the blood type of each person will consist of one of the ABO blood groups and either the Rh+ or Rh- factor. Thus, A Rh+, A Rh-, B Rh+, B Rh-, AB Rh+, AB Rh-, O Rh+, and O Rh- are all the possible blood types that can be found in the population. Therefore, when making a transfusion it is necessary not only to be certain of the ABO compatibility between the donor and recipient, but also to insist that the Rh compatibility is correct. Rh compatibility is determined as follows:

Table 9.3	
Donor	**Recipient**
Rh+	Only another Rh+
Rh-	Either Rh- or Rh+ (Since there is no Rh Ag on the Rh- RBC it will not stimulate an immune response in any recipient regardless of type)

THE UNIVERSALIST OF DONORS AND OF RECIPIENTS

As mentioned, a type O person can donate to all other types (A, B, AB, O) since the type O RBC has no Ag on the membrane. However, the Rh factor MUST ALSO BE CONSIDERED WHEN MAKING A TRANSFUSION. Thus, only Rh+ donors can give blood to Rh+ recipients, but Rh- donors can donate to both Rh+ and Rh- recipients:

As can be seen, *O Rh- is the universalist of donors* since it can donate to all ABO blood types and to both Rh+ and Rh-. This is because the O Rh- RBC has no antigen on the surface. In addition, *AB Rh+ is the universalist of recipients* since it can receive from any other ABO and Rh blood type. This is because there is no circulating anti-RBC antibodies in the plasma to react to any RBC Ag that is transfused.

Cross Matching. Even given the information above on compatible transfusions, it is still safest to first test the blood of donor and recipient to be absolutely certain that there will not be a transfusion reaction. This is because in addition to the major ABO and Rh antigens there are an additional 12 minor RBC antigen systems that may also effect the compatibility of the transfusion. Thus, the mixing of some of the donor RBCs with recipient plasma in the presence of the proper reagents is termed a *cross matching* and is used to determine if the transfusion is compatible.

HEMOSTASIS

Hemostasis is the control of blood loss. After an injury three important mechanisms act to limit bleeding from the damaged vessels. The first is blood vessel spasm, the second is formation of a platelet plug, and the third is clot formation.

BLOOD VESSEL SPASM

When the smooth muscle cells in the walls of arterioles or venules is damaged, cut or broken they react immediately by completely constricting. Smooth muscle constriction may also be elicited by local or systemic reflex activity activated by pain receptors in the damaged tissues. Reduction in the vessel radius through such constriction limits the loss of blood because it increases the resistance to flow in the damaged vessel. Blood vessel spasm is further maintained through the release of chemicals from platelets (see following material).

Table 9.4

Donor	Recipient	Commen
A Rh+	A,AB only Rh+	Never A or AB Rh-
B Rh+	B,AB only Rh+	Never B or AB Rh-
AB	Rh+ AB Rh+	Never AB Rh-
O Rh+	A,B,AB,O only Rh+	Never A,B,AB,O Rh-
A Rh-	A,AB both Rh+ & Rh-	
B Rh-	B,AB both Rh+ & Rh-	
AB Rh-	AB both Rh+ & Rh-	
O Rh-	All blood types, both Rh+ and Rh-	

FORMATION OF A PLATELET PLUG

When a blood vessel is damaged, the interior endothelial surface is disrupted exposing the underlying collagen matrix. Platelets are attracted to the exposed collagen and proceed to aggregate at this location **(Figure 9.4)**.

Adherence of platelets to the vessel wall is facilitated by the presence of the plasma protein called *von Willebrand factor* (*vWF*) which is synthesized by endothelial cells and acts as a bridge between the collagen and the first layer of platelets. During this aggregation process, the platelets release various chemicals from their storage granules. One such substance released is Adenosine Diphosphate (ADP) which causes the surface of additional platelets to become sticky and promotes the build up of a platelet mass at the site of vessel wall damage. Additional ADP is also released during this process and this acts as a positive feedback mechanism stimulating more platelets to adhere and form a platelet plug.

During this aggregation process, *thromboxane* A_2 is synthesized from membrane fatty acids and released by the platelets. Thromboxane A_2 is structurally related to the prostaglandins and as such, is a fatty acid derivative of arachidonic acid which acts locally to promote additional platelet aggregation, discharged from platelet granules, acts on the smooth muscle of the vessel wall to promote and maintain constriction.

Since the formation of a platelet plug is mediated through a positive feedback system, an inhibitor must also be available to prevent inappropriate platelet aggregation on normal vessel epithelium once the process has been initiated at the damage site. Such inhibition is in fact provided by the substance *Prostacyclin* (also termed *Prostaglandin* I_2; PGI_2). Prostacyclin is synthesized by the adjacent normal epithelium which contains an enzyme that converts arachidonic acid to prostacyclin instead of thromboxane A_2. Since prostacyclin inhibits platelet aggregation, this prevents the spread of the platelet plug into the normal regions of the vessel wall.

The platelet plug performs the following important functions:

1. It initially seals the break in the vessel wall limiting blood loss.

2. Formation of actin-myosin protein complexes present within the platelets after aggregation contracts the platelet plug squeezing the loose platelet complex into a compact and leakproof mass.

3. Through the release of serotonin, epinephrine, and thromboxane A_2 by the platelet plug, the vascular spasm is maintained and reinforced.

4. The platelet plug releases other chemicals that enhance the process of blood coagulation (clot formation).

THE BLOOD CLOTTING SYSTEMS

While the formation of a platelet plug is effective in sealing small tears in capillaries, damage in larger vessels requires the formation of a blood clot (or *thrombus*) to effectively limit blood loss.

The clotting mechanism is initiated at the same time that the platelet plug is forming and results in the activation of a series of soluble plasma proteins (originally produced by the liver). The outcome is the formation of an insoluble gel-like matrix called a *clot* that traps circulating red blood cells. The strands of the protein fibrin that form the clot collect around the platelet plug and complete the sealing process begun when the platelets aggregated.

There are two separate, but interrelated pathways that can result in clot formation, namely the *intrinsic clotting mechanism* and the *extrinsic clotting mechanism* **(Figure 9.5)**.

However, regardless of how the process was initiated, the final outcome is the enzymatic conversion of the soluble plasma protein *prothrombin* (*factor II*) to *thrombin*. The thrombin (which now possesses enzyme activity) acts on the soluble plasma protein *fibrinogen* (*factor I*) and converts it into fibrin that polymerizes into a loose meshwork of insoluble and sticky strands. The loose fibrin is stabilized and covalently linked through the action of the plasma protein factor XIII, which was itself activated by thrombin.

THE INTRINSIC CLOTTING MECHANISM

The intrinsic clotting mechanism **(Figure 9.5)** is one of the two pathways that can initiate clot formation.

To activate the intrinsic pathways, the blood must contact a foreign or abnormal surface such as the underlying collagen exposed in the wall of a damaged vessel. However, the wall of a glass test tube will act equally well to activate the intrinsic pathway and it is for this reason that blood drawn into a glass test tube also clots. In any event, the presence of such an abnormal surface activates the plasma

protein factor XII (previously named Hageman factor) and this in turn then activates a series of other plasma proteins in a cascade effect as follows:

Factor XII activates factor XI
Factor XI + Ca^{++} activates factor IX
Factor IX + Ca^{++} + factor VIII + PF_3 activates factor X
Factor X + Ca^{++} + factor V + PF_3 activates Prothrombin (factor II to form Thrombin
Thrombin then converts fibrinogen (factor I) to fibrin and the clot forms.

In the above reactions, factor VIII and V are soluble plasma proteins that do not form linkages in the cascade, but are required as co-factors in the two reactions. The activation of both factor VIII and V are stimulated by thrombin. PF_3 is phospholipid exposed in the plasma membrane of platelets when they aggregate and it must also be present for the two steps in the cascade to be functional.

THE EXTRINSIC CLOTTING MECHANISM

This pathway is simpler than that of the intrinsic clotting mechanism and is initiated by tissue damage. The damaged tissue releases a protein substance called *factor III* also known as *tissue factor*, or *tissue thromboplastin*, and this substance + the plasma protein factor VII + Ca_{++} activates factor X. Active factor X then continues the cascade and the clot is eventually formed.

CLOT RETRACTION AND PLASMA V. SERUM

After the fibrin clot has been formed and stabilized through the action of factor XIII, platelets trapped in the clot undergo contraction. This process causes the fibrin strands to tighten down and pulls the edges of the hole in the vessel walls closer together. The effect is to create a leakproof seal that prevents all loss of fluid from the damaged vessel. If clot retraction takes place in the blood in a glass test tube, the fluid squeezed from the clot is free of fibrinogen and other clotting factors and is now called *serum*. While serum is free of the clotting factors, it contains essentially all of the other soluble substances previously present in the plasma.

THE ANTICLOTTING SYSTEMS

Pathologic formation of excessive clotting is opposed by three mechanisms, namely: *Prostacyclin* (PGI_2) that inhibits platelet aggregation (and thus clot formation) on normal endothelium (see previous material), *Protein C* and *Antithrombin III.*

PROTEIN C

Protein C is initially activated by thrombin itself and functions to *deactivate* or inhibit factors VIII and V which are required as co-factors for two steps in the cascade **(Figure 9.6)**. Since thrombin also activates these co-factors, the process requires an exquisite balance between the activation and the deactivation processes to maintain a functional clotting system, but one that does not produce excessive clot formation.

ANTITHROMBIN III/HEPARIN

Antithrombin III inactivates thrombin and functions in conjunction with heparin. Heparin itself is generated by mast cells and basophiles and is also present on the surface of endothelial cells. Heparin acts to bind antithrombin III which then blocks the formation of thrombin and down-regulates the clotting mechanism.

THE FIBRINOLYTIC SYSTEM

While new clot formation is inhibited by the various substances that comprise the anticlotting system, the enzyme *plasmin* acts to promote the catabolism and removal of clots that have already formed (i.e., *fibrinolysis*, or *thrombolysis*). Plasmin is derived from the proenzyme *plasminogen* which is converted to active plasmin through the action of *plasminogen activators*. One plasminogen activator involved in this process is generated by endothelial cells and is the *tissue plasminogen activator*. In addition, factor XII (Hageman factor), which activates the intrinsic clotting cascade leading to clot formation, also activates a cascade leading to the formation of plasminogen activator that forms plasmin and digests fibrin into soluble fragments.

Normally, small quantities of fibrin are continuously formed, but through the activation of plasmin are also being continuously removed maintaining the system in a delicate balance.

If this equilibrium is disrupted, the outcome could be the generation of inappropriate clots in the tissues or on the vessel walls. Such intravascular clots are called *thrombi* and can result in the blockage of blood flow through a vessel. In addition, if the thrombi break loose and float through the vessels, they are called *emboli*. Emboli can cause great problems, especially if they get stuck and occlude flow in vessels that supply important organs such as the lungs, brain or kidneys. Intra-vascular thrombus formation can result from:

1. A disequilibrium between the clotting and anticlotting systems due to disease, chemicals or drug reactions.

2. Roughened endothelial surfaces present in atherosclerosis.

3. Inappropriate release of tissue thromboplastin after tissue trauma.

4. A reduction in the velocity of blood flow through a partially occluded vessel leading to the accumulation of increased levels of fibrin.

ANTICLOTTING DRUGS

Various drugs can be used to inhibit clot formation such as:

1. *Aspirin* which acts by blocking the formation of *thromboxane* in platelets. This prevents platelet aggregation and activation of the clotting system.

2. *Coumarin* which blocks the synthesis of the liver-derived clotting factors by interfering with the

action of Vitamin K.

3. *Heparin* which acts as the co-factor for antithrombin III.

4. *Streptokinase* which acts as a plasminogen activator and can thus stimulate plasmin formation. Upon administration of streptokinase at the site of a blocked vessel, such as a coronary artery, the formation of plasmin at this local site may dissolve the clot and reestablish blood flow.

BLEEDING DISORDERS

The bleeding disorder *Hemophilia* can be caused by a deficiency in any one of the clotting factors synthesized by the liver. However, the majority of the hemophiliacs (80%) suffer from an inability to synthesize soluble *factor VIII*. In the absence of one or more of the clotting factors, the individual suffers from profuse, and in some cases, life threatening blood loss. Platelet deficiency also exists, and here the loss of large quantities of blood is uncommon. Instead such individuals repeatedly develop multiple small hemorrhagic areas over their whole body. When these bluish bruise-like regions are present in the skin, the individual is said to suffer from *thrombocytopenia purpura*, which may be autoimmune in origin. *Vitamin K* deficiency can also cause excessive bleeding since this vitamin is a necessary co-factor in the synthesis of the clotting factors by the liver.

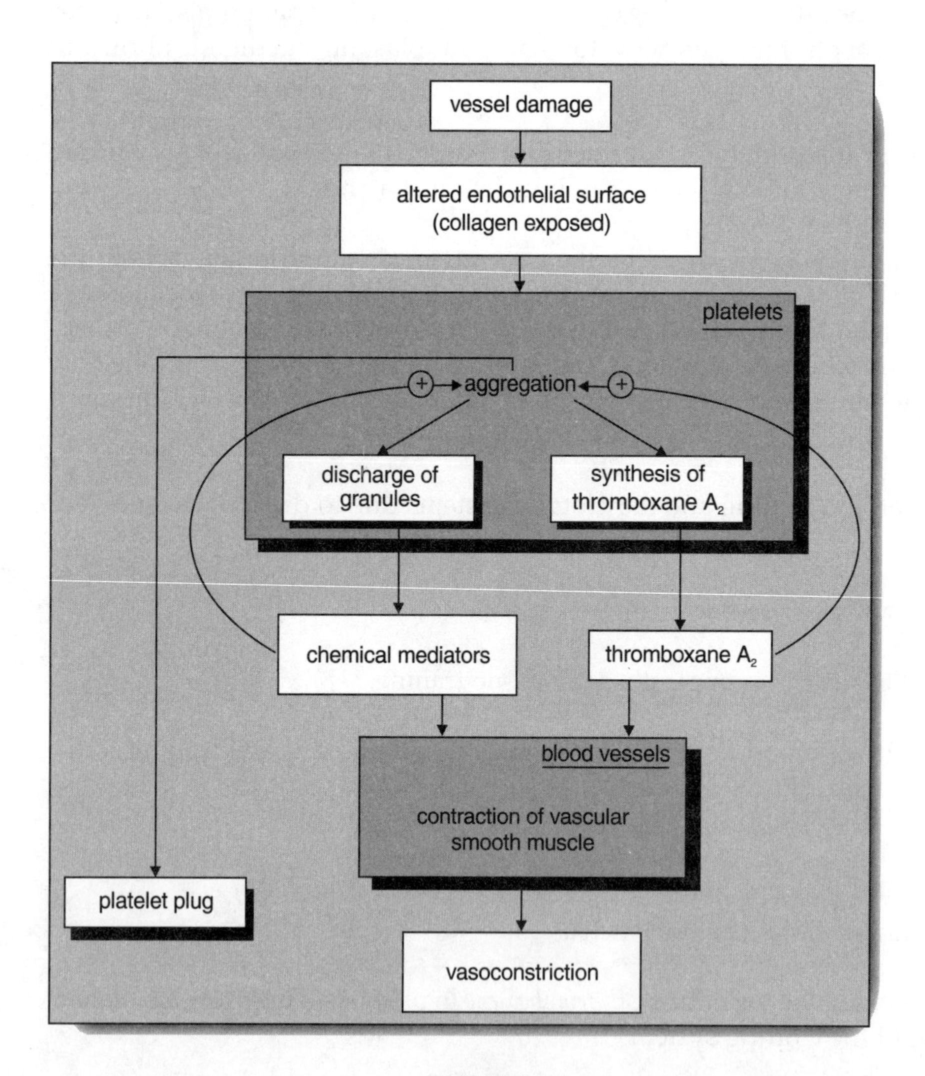

FIGURE 9.4

Chapter 10
DEFENSE MECHANISMS

LYMPHATIC SYSTEM

Drainage of excess fluid from the interstitial compartment is accomplished by the lymphatic capillaries. These structures are located in the interstitium and here fluid containing tissue protein enters by means of an osmotic gradient. Movement of this fluid from the lymphatic capillaries into larger lymphatic vessels takes place when these vessels are compressed during skeletal muscle contraction. This process is similar to venous return during exercise; and like veins, the lymphatic vessels also contain one-way valves so that the system is unidirectional. The lymph flow passes into attached lymph nodes where it is processed (see following material), after which it continues towards the heart.

Finally the fluid is collected in two large lymphatic vessels: on the left side in the thoracic duct, and on the right side in the right lymphatic duct. The thoracic duct empties into the left subclavian vein near the junction of the left jugular vein, while the right lymphatic duct empties into the right subclavian vein near the junction of the right jugular vein **(Figure 10.1)**.

Overall, the thoracic duct drains the majority of the lymphatic system since fluid originates from lymphatic vessels of the entire left side and the middle and lower right side. The right lymphatic duct thus only drain from the upper right region of the chest, back right arm and right half of the head.

LYMPHATIC FLUID

Tissue, or lymphatic fluid, originates mostly from capillary filtrate that is not reabsorped back into the capillaries (more filtration than reabsorption).

The composition is slightly different from capillary filtrate and contains various nutrients, gases and smaller proteins, but lacks larger proteins. The increased concentration of these smaller proteins in the tissue fluids is responsible for the reabsorption of tissue fluid into the lymphatic capillaries. Movement of tissue fluid into the lymphatic capillaries is facilitated by flap-like valves in the lymphatic capillary wall that allow fluid to enter, but not to leave.

PRIMARY V. SECONDARY LYMPHATIC TISSUES

Organs of the lymphatic system are classified as either primary or secondary.

Primary lymphatic organs are those involved in the development of effector lymphocytes. Thus, the bone marrow and thymus are considered to be primary lymphatic tissues, and in birds the bursa is also classified as a primary lymphatic organ. Secondary lymphatic organs or tissues are those areas of the lymphatic system where mature lymphocytes are found. These cells having been "educated" within the primary lymphatic tissues, migrate into, and then colonize the secondary lymphatic tissues, such

as the lymph nodes, spleen, tonsils, and *gut associated lymphatic tissues* (GALT), also called *MALT* (*mucosa associated lymphatic tissues*). GALT/MALT is present in the walls of the stomach and intestine **(Figure 10.2)**.

LYMPH NODES

Lymph nodes are masses of cells, primarily of immune system origin, enclosed in a fibrous collagenous capsule. The node is organized with an outer subcapsular (marginal) sinus region, a cortex, a deeper paracortex, and a medullary region in the center **(Figure 10.3)**. Afferent lymphatic vessels enter the node through the fibrous capsule carrying lymph collected from the tissues. The lymphatic fluid filters through the node, passing through a series of internal sinuses and eventually exits at a central point called the *hilus* through an efferent vessel. Additionally, an artery and vein connect to the node also at the hilus.

Located within the cortical region, and extending partly into the paracortex, are primary follicles comprised primarily of B-lymphocytes (see following material). During an active immune response, these areas develop into secondary follicles containing germinal centers. Within these germinal centers the process of clonal formation is taking place with the resultant production of antibodies from the activated B-cells also known as *plasma cells* (see following material).

SPLEEN

The spleen filters blood, just as the lymph node filters lymph. Antigen present in blood is trapped by splenic macrophages, or is transported into the spleen in antigen-processing cells, and here clonal activation of both B-cells and T-cells takes place. The spleen is surrounded by a colloganous capsule; internally it is supported by a reticular framework.

Within the spleen are found two main types of tissue—the red pulp and the white pulp. Red pulp is composed mainly of RBCs and within the red pulp region destruction of old and damaged RBCs is taking place. White pulp contains the lymphoid tissue, and the cells here are arranged around central arterioles known as the periarteriolar lymphoid sheath (PALS). In the PALS, T-cells are located close to the central arteriole, while B-cells surround the T-cell zone further away from the arterioles. B-cell germinal centers can also be found in the PALS—they can be either primary (unstimulated) germinal centers, or secondary (antigen-stimulated) follicles **(Figure 10.4)**.

THYMUS

The thymus is a primary lymphatic organ located within the mediastinum.

It is quite large in children and is positioned above the heart, but regresses in size after puberty and is usually very small in the adult. Changes in thymic mass and function are regulated by various endocrine hormones elaborated from the gonads, adrenals and pituitary, but many other factors also impact on thymic structure and function. In mammals, the thymus is bilobed with each lobe organized into lobules, and here are found lymphoid cells called *thymocytes*. The outer margins of the lobules are called the *thymic cortex*, while the inner regions are the *thymic medulla*. Immature thymocytes in various stages of development and programming are found within the cortex, while the more mature cells are

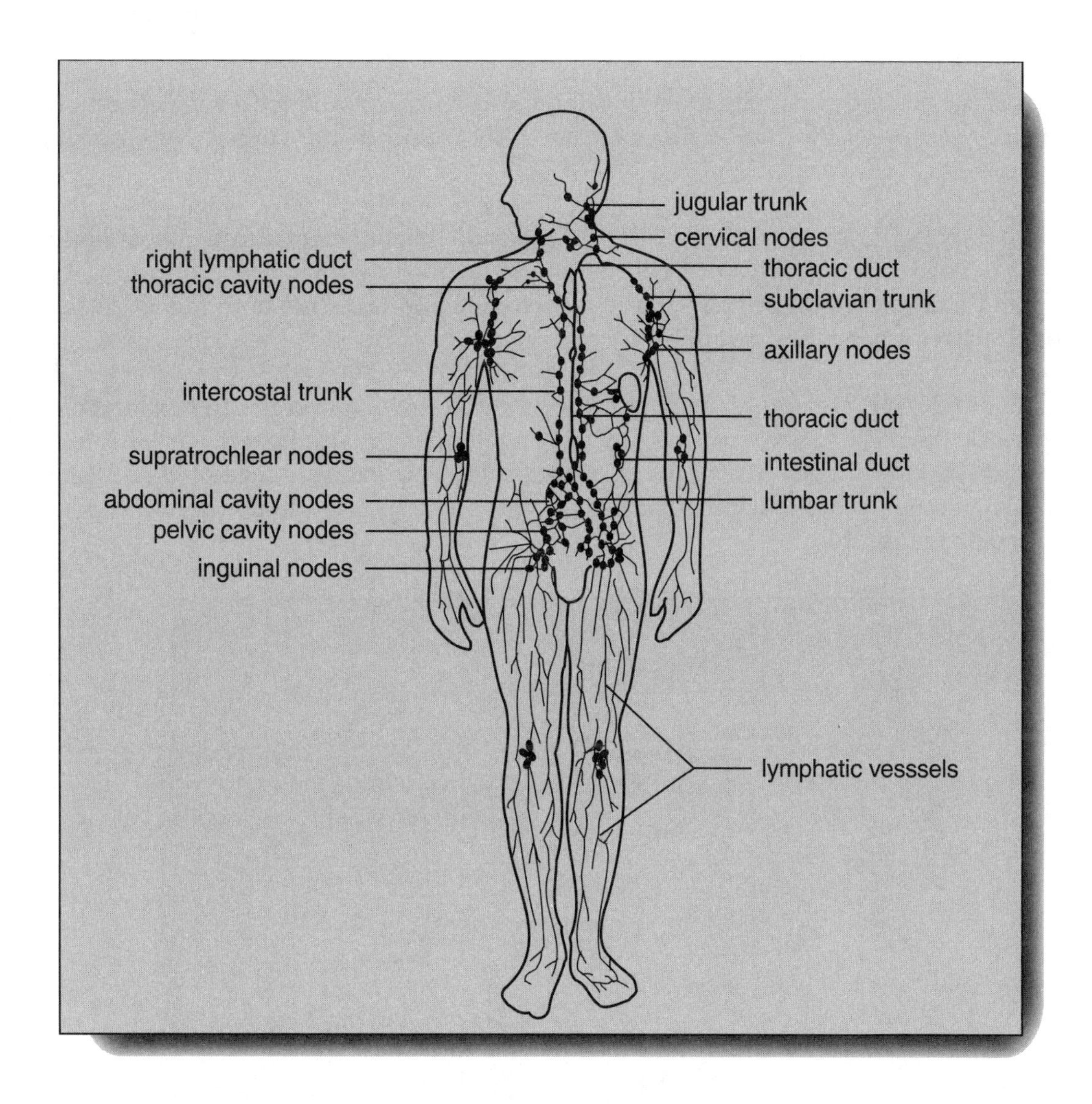

FIGURE 10.1

located within the medulla. Mature thymic derived lymphocytes (or T-cells) leave the medulla to colonize the thymic dependent regions of the secondary lymphatic tissues.

DEVELOPMENT OF IMMUNE EFFECTOR CELLS

Currently, it is believed that all immune effector cells initially originate from some common hematopoietic stem cell present early in development. As can be seen in **Figure 10.5**, three major developmental pathways diverge from the common hemopoietic stem cells. One through the common *myeloid progenitor* produces the *graulocytic* series (i.e., *basophil, neutrophil, eosinophil*) as well as *mast cells* and *monocytes/macrophages*. A second pathway via the common lymphoid progenitor results in the lymphocytes which can be further subdivided into *T-lymphocytes* and *B-lymphocytes*, and an additional pathway thought to lead to *antigen presenting cells* and *third population/NK cells*.

Functionally, basophiles and mast cells contain histamine and heparin and are intimately involved in the inflammatory response. But mast cells are commonly found in the connective tissues, while basophile are present in the circulation.

Eosinophils are responsible for providing protection against parasitic worm infestation, although this process is also dependent on a variety of additional interactions involving macrophages, B and T lymphocytes and mast cells. Neutrophils and monocyte/macrophages are both highly phagocytic, and additionally macrophages are required for activation of lymphocytes via antigen presentation.

Third population/NK cells are involved in cytotoxic responses which lead to direct killing of target cells. Such direct killing is also accomplished by subpopulations of lymphocytes called *cytotoxic T-lymphocytes*. Furthermore, antigen activation of lymphocytes is specifically mediated and generates memory cells. This accounts for the acquired or specific immune response which is quite different from that of non-specific immunity.

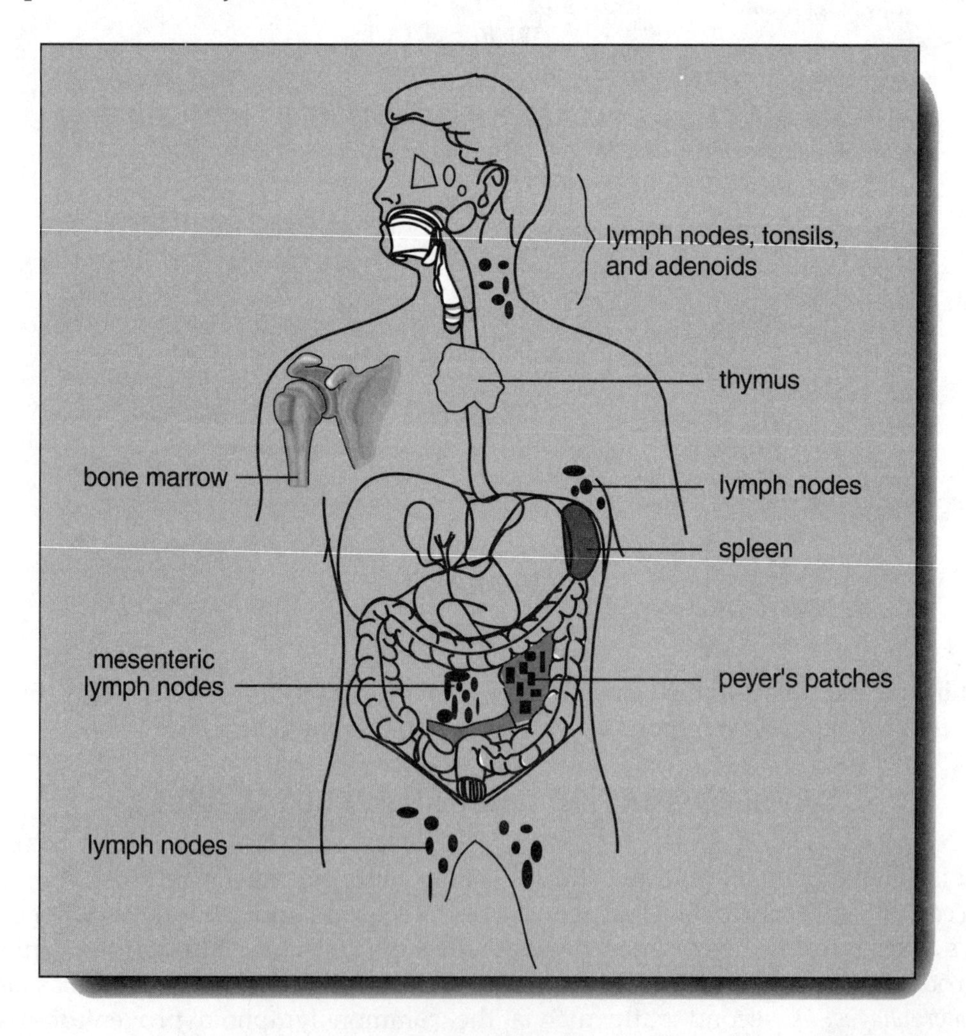

FIGURE 10.2

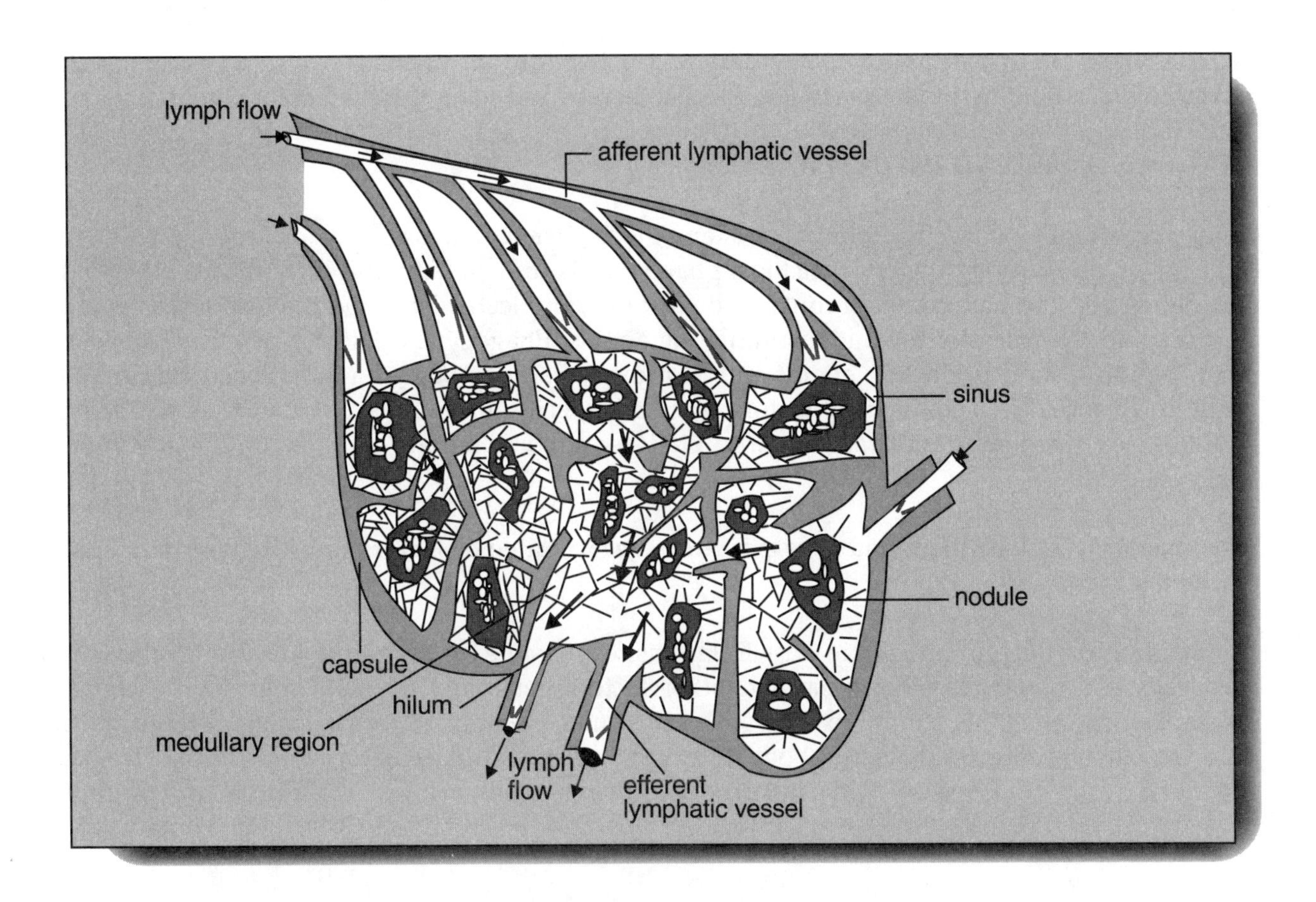

FIGURE 10.3

NONSPECIFIC IMMUNE RESPONSE

In humans and other vertebrates, the immune response can be subdivided into nonspecific immunity and specific or acquired immunity. Interactions between both types of immunity are absolutely necessary because they provide the body with a protective shield against microbial agents and foreign cells and substances that can lead to an alteration in normal homeostasis and disease.

All reactions, which are not directly dependent on antigen challenge, are classified as part of the nonspecific immune system. In addition, such nonspecific immune responses are not greater or more effective upon additional challenges after the initial first challenge. Thus no immunological memory exists for nonspecific response mechanisms.

Included in the list of nonspecific immune mechanisms are the mechanical barrier of the skin which prevents many bacterial agents from entering the body. In addition, chemical secretions of the skin create an acidic PH that inhibits the growth of many organisms, as does the presence of natural flora both on the skin and within the GI tract. The presence of lysozyme in body fluids, such as tears, saliva and blood, acts to inhibit the growth of gram positive organisms, while mucus traps particles and

bacteria which can then be removed by ciliary action. In addition, mucus may also act as a nonspecific barrier to block binding of viruses to target cells, thereby reducing their infective potential.

PHAGOCYTOSIS AND INFLAMMATION

Other important nonspecific immune system mechanisms that assist in destroying pathogenic bacteria, or other foreign agents, include phagocytosis and inflammation. Both are classified as nonspecific because both can be stimulated by a variety of mechanisms and the responses generated are the same regardless of the initial stimulating agents. Phagocytic cells such as macrophages and neutrophils will nonspecifically engulf and destroy bacteria or other foreign objects **(Figure 10.6)**. While this process is facilitated by factors from the specific immune system such as antibody, or lymphokines, these substances are not absolutely required to initiate or maintain the action of the phagocytic cells.

Like phagocytosis, the inflammatory response can be facilitated by factors generated from the specific immune system.

One such factor, the immunoglobin class E (IgE), is responsible for triggering histamine release from mast cells and basophile. Histamine and other factors generated by mast cells such as heparin, prostaglandins, enzymes, thromboxanes, leukotrines and chemotactic factors, then promote multiple actions on diverse targets all over the body. Furthermore, other inflammatory mediators such as those generated by the Kinin system, while not produced by mast cells, are also of importance in promoting

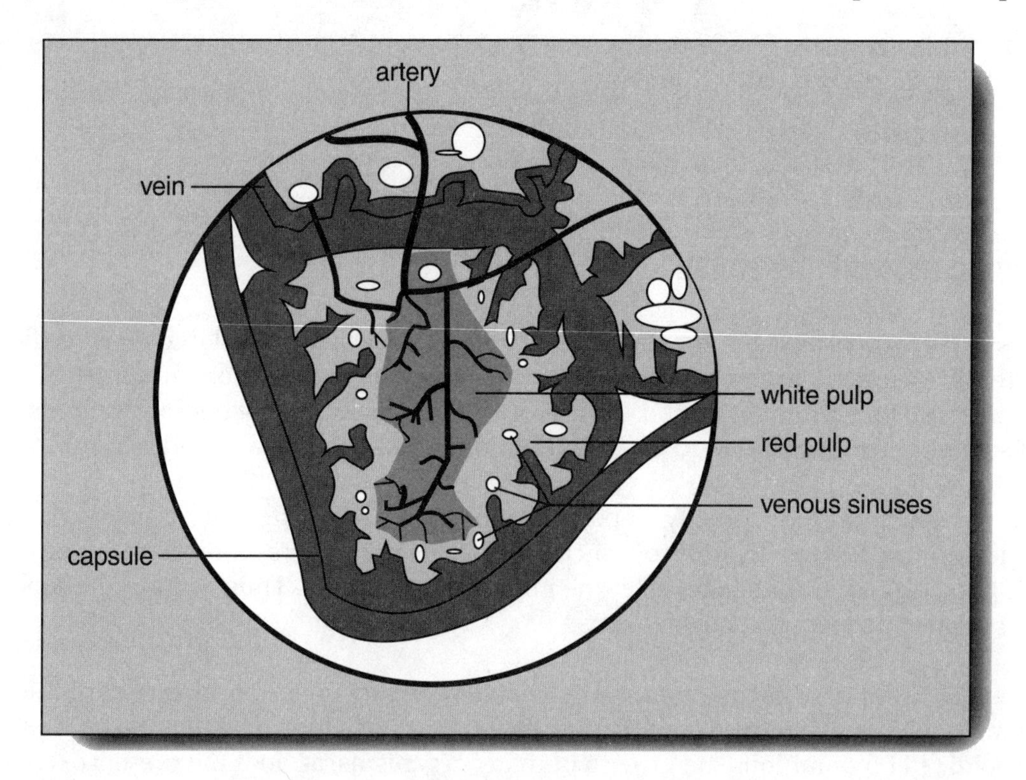

FIGURE 10.4

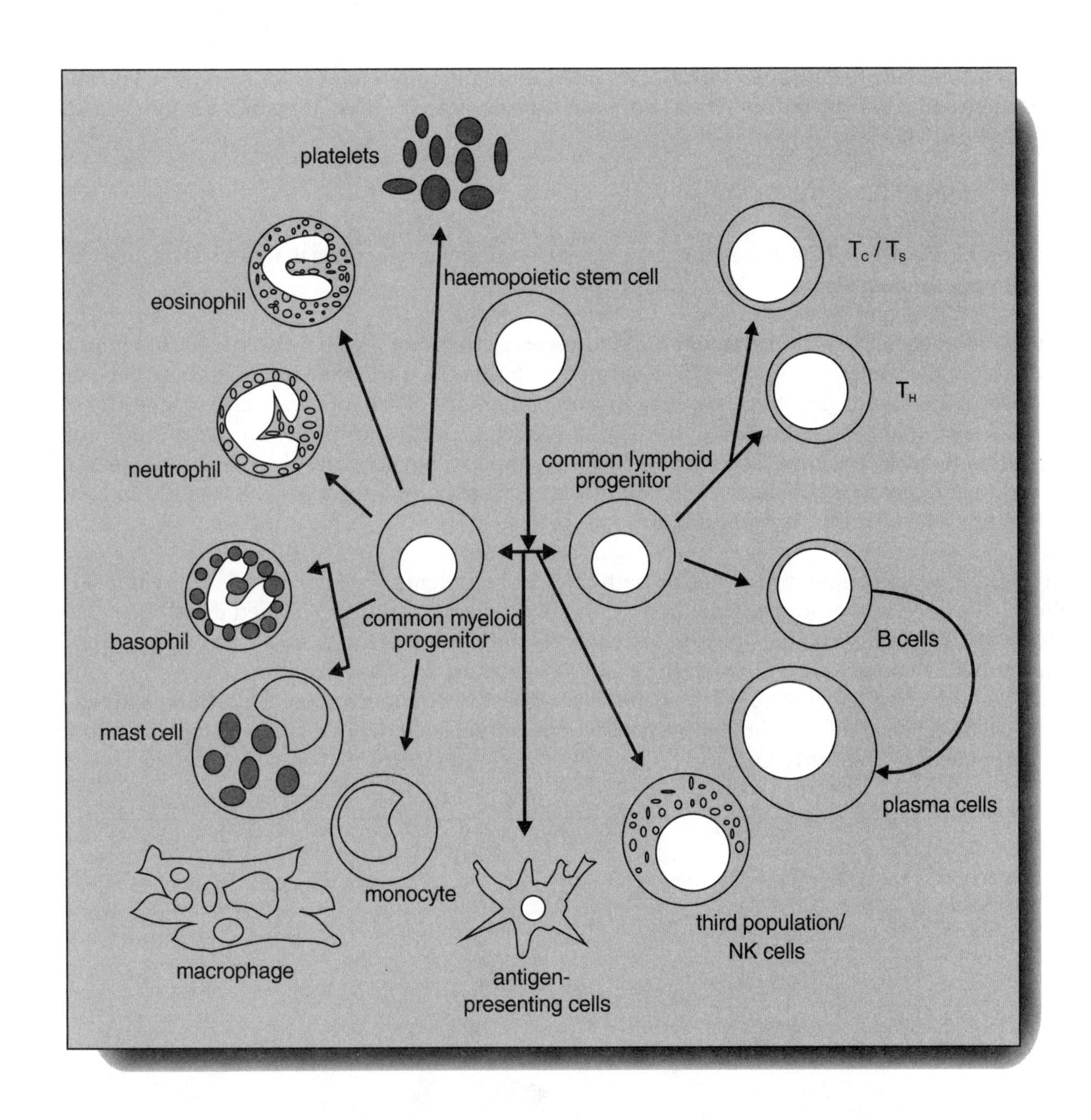

FIGURE 10.5

inflammation. Overall the inflammatory effects include increased capillary permeability and vasodilation leading to swelling (edema), bronchoconstriction, chemotaxis of eosinophils and neutrophils, platelet aggregation and activation of the complement system.

If these inflammatory effects are localized to the site of an ongoing infection, they result in the destruction of bacterial agents because they stimulate elements of the immune system including attraction and activation of phagocytic cells, and increased access of antibody to the infective site. Local increases in blood flow also act to increase available nutrients and O_2 and the removal of waste products

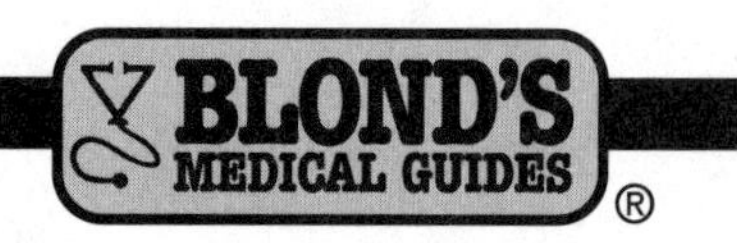

to promote healing of the tissue. On the other hand, if inflammatory mediators are released throughout the whole body, the outcome will be a sudden drastic decrease in circulating blood volume leading to anaphylactic shock.

THE COMPLEMENT SYSTEM

The complement system is yet another example of a series of interactions that can be triggered by both specific and non-specific switches.

Furthermore, complement activation leads to diverse effects on a variety of targets. One important activator of the complement system is generated by mast cells during the inflammatory response and consists of the enzyme tryptase; however, two other methods of activation are also of importance. The first takes place when antibodies of the G or M class (IgG, IgM) specifically bind to antigen, forming an antibody-antigen complex **(Figure 10.7)**. This complex then activates the first components of the complement cascade (there are nine complement proteins, C1 through C9) and this method of activation is termed the *classical pathway*.

In addition to activation by the classical pathway, the complement system can also be activated upon the release of certain products generated by microorganisms, and this is termed the *alternate pathway*. However, regardless of the method used, once the complement cascade has commenced, the physiological consequences will destroy microorganisms by opsonization (coating the antigenic particle with complement protein to increase phagocytosis by macrophages); cellular activation (of phagocytic cells); and lysis (direct destruction of the target bacteria by the final complement proteins in the sequence) **(Figure 10.8)**.

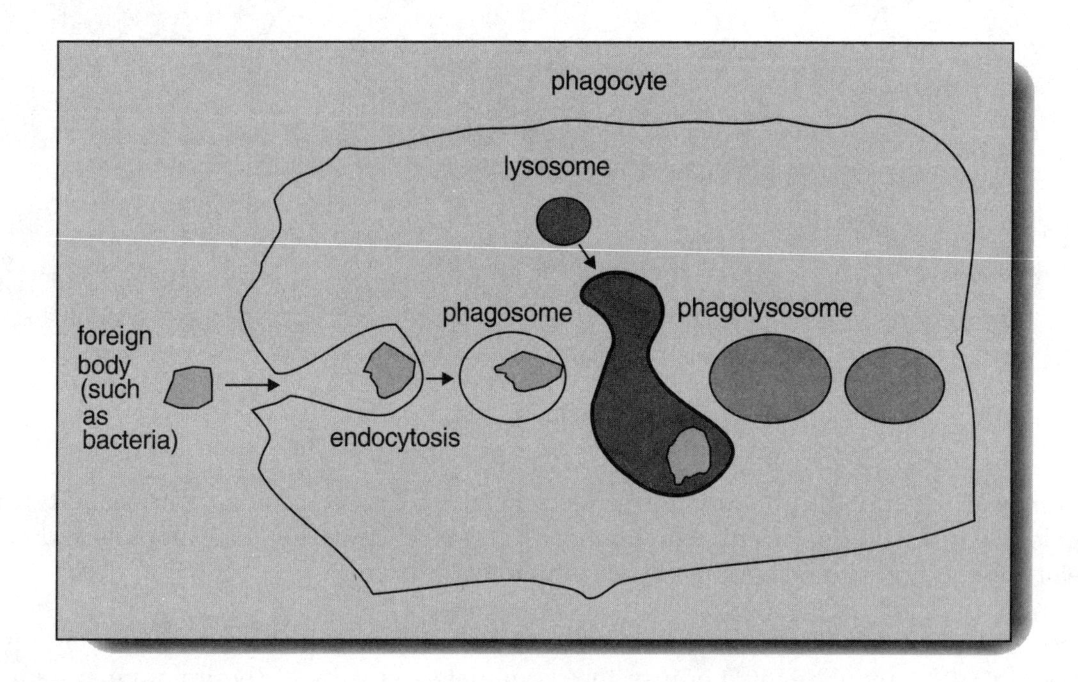

FIGURE 10.6

INTERFERONS

Interferons make up a class of proteins (see cytokine section) that nonspecifically inhibit viral replication inside host cells. Interferons are secreted by several different host cells including T-lymphocytes, epithelial cells and macrophages.

Secretion occurs in response to viral infection and after secretion, the interferon then binds through receptor interactions to the plasma membrane of the adjacent cells, regardless of whether they are infected by the virus or not. Binding of the interferon to the plasma membrane then triggers the synthesis of inter-cellular enzymes which act to block the synthesis of various viral proteins required by the virus for replication. Thus, the virus cannot replicate intra-cellularly and is inactivated. Most viruses induce interferon synthesis; however, those that do not do so will proceed to replicate and cause further infections and in some cases eventual death of the host.

THE SPECIFIC IMMUNE RESPONSE

The specific immune response is classified as such because each initiating antigen must bind to individual and specific antigen receptors located on the responding lymphocyte. Furthermore, subsequent contacts with the same antigen will induce a more extensive response within the effector lymphocytes because a memory of the initial event is maintained. Thus, while initial exposure to a pathogen may result in active disease, subsequent exposures to the same pathogen will not produce clinical symptoms because the individual is now protected (or immunized).

Functionally, the specific immune system can be further subdivided into cell mediated and Humoral immune components.

Cell mediated immunity consists of those immune responses that are mediated through lymphocyte action directly at body sites where foreign antigen on the cell membranes has been identified. Cell mediated immunity is especially effective in protecting the body against infections produced by intra-cellular pathogens such as viruses or facultative intra-cellular bacteria (tuberculosis), cancer, and parasitic infections (Protozoa, Helminths).

Humoral immunity consists of those immune responses that result from the action of the circulating *humoral factors* now known as *antibodies*.

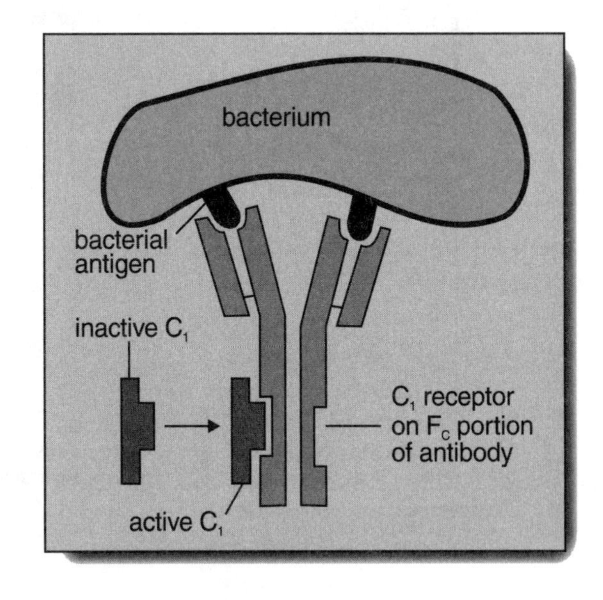

FIGURE 10.7

These antibodies, upon reaching the site of the infection, complex with the antigen containing structures and through a series of actions involving phagocytic cells or the complement system then destroy or inactivate the structure to which the

antigen is attached. Humoral immunity is especially effective in keeping the body free from infections by pathogenic bacteria, although it too may provide some protection against certain viruses and against some protozoan pathogens. It should also be noted that although the B and T lymphocytes (sometimes termed *effector cells*) responsible for these two components of specific immunity are derived from different subpopulations, both respond specifically to antigen stimulation and undergo clonal formation.

ANTIGEN

The presence of antigen stimulates the formation of antibodies. Indeed, the very term *antigen* (anti-gen) derives from "antibody generator."

Antigens can be any substance that will elicit a response from either the humoral or cell mediated immune systems. Antigen molecules are commonly located on the surface of larger structures such as bacteria and, as has been stated, a particular antigen will act specifically to induce the formation of the antibodies which are able to bind to it.

Antibodies do not bind to the entire antigen structure, but only to a small section termed an *antigenic determinant* or *epitope*. Thus it is quite possible that a single antigen molecule with many epitopes can elicit the production of many different antibodies, all of which bind to different determinants of the same antigen.

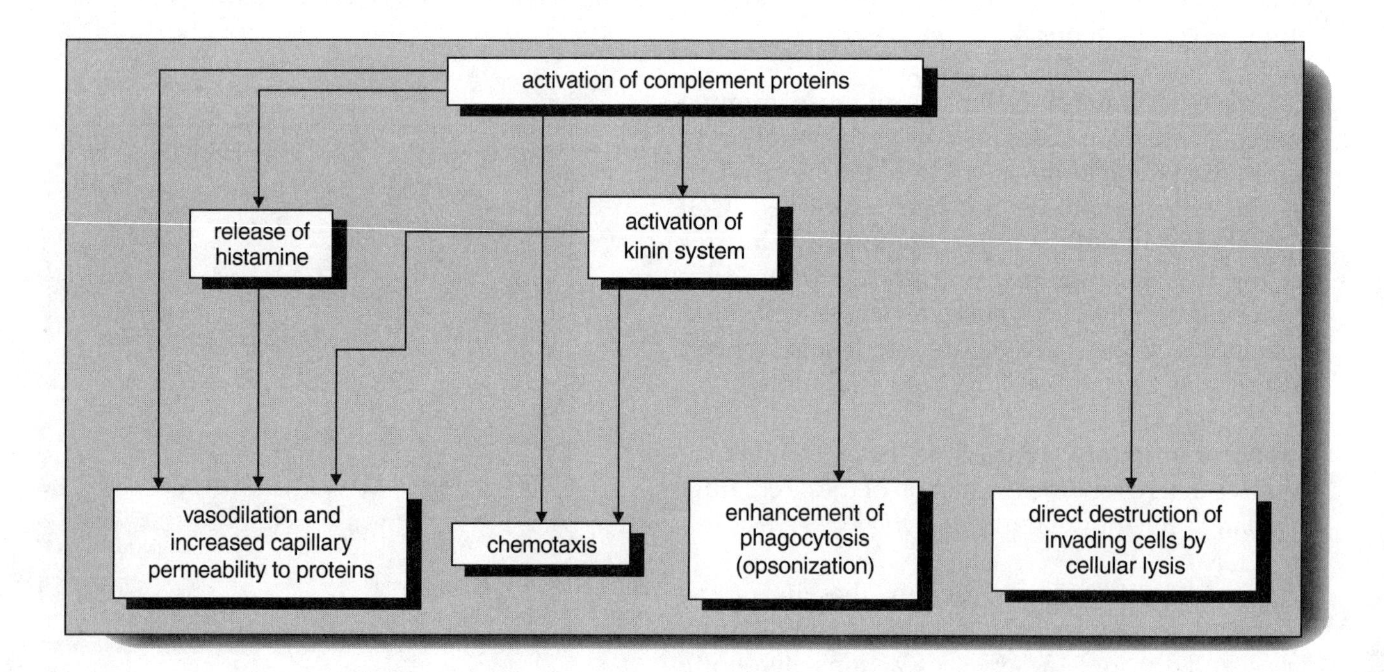

FIGURE 10.8

CLONAL SELECTION MODEL

Specific or adaptive immunity means that the immune system retains a "memory" of earlier contacts with an antigen.

If the immune system is later re-exposed to the same antigen, the response generated is much greater then at the time of initial contact. The response after the initial contact is termed the *primary immune response* while the response generated after subsequent contacts is termed the *secondary immune response*. For many years immunologists attempted to explain the mechanism that could account for

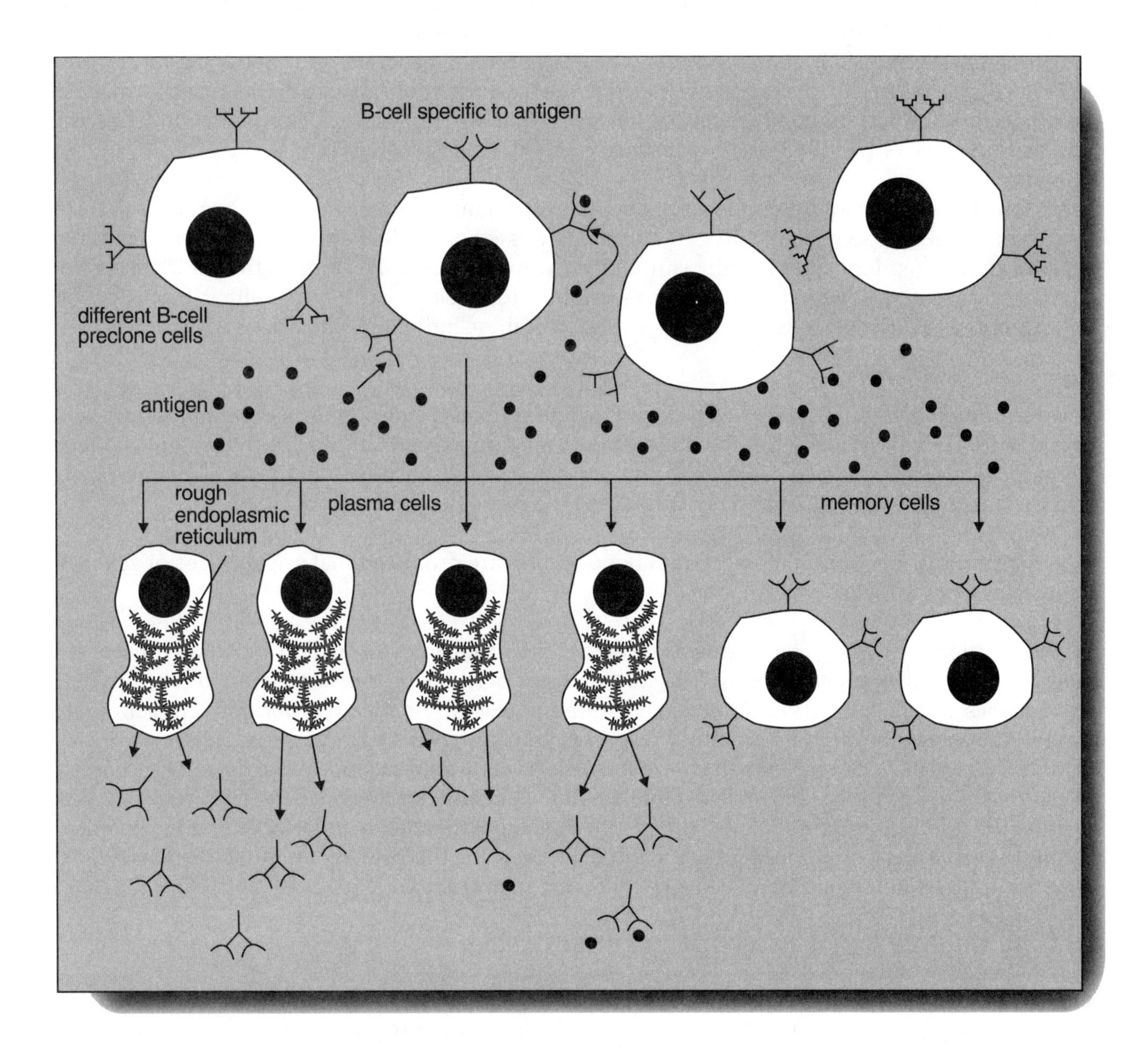

FIGURE 10.9

the specificity of millions of antibodies generated to a wide range of both natural occurring and artificially manufactured antigens. Investigators also attempted to explain how it was possible for the immune system to remember earlier contacts with the same antigens and generate a greatly amplified secondary immune response.

Finally, in 1959 Burnet proposed the Clonal Selection Model to explain the mechanism. He subsequently received the Nobel Prize for his work.

According to the Clonal Selection Model **(Figure 10.9)** initial contact with an antigen will stimulate only specific lymphocytes that possess receptors able to bind to this antigenic epitope. Such *pre-clone* cells can be either T-lymphocytes or B-lymphocytes, but in either case the cell that is stimulated in this way then undergoes DNA synthesis or *blastogenic transformation* and generates multiple copies or *clones* of itself. Furthermore, during the process of blastogenic transformation, genetic diversity is produced in these cells through specific gene activation, somatic mutations and somatic recombination. Thus, the final mature clone of lymphocytes (now termed *effector cells*) is generated.

If the clone is composed of B-lymphocytes, the final step in maturation is to form *plasma cells* that are capable of secreting specific antibody. This antibody possesses binding sites designed to specifically bind the antigen (epitope) that initially stimulated the formation of the clone. If the clone consists of T-lymphocytes, the cells generated will be a T-lymphocyte subclass which can be either T-helper cells, or T-suppressor/cytotoxic cells. The function of each of these individual subclasses will be discussed later.

However, unlike B-lymphocytes, T-cells do not secrete antibody. Instead they secrete *lymphokines* and certain types (T-cytotoxic cells) can specifically bind to and directly attack and destroy target cells. Such T-lymphocyte subclasses possess antigen receptors on their surface membranes which specifically bind the antigenic epitope that initially stimulated the formation of the clone.

As a byproduct of clonal formation, certain lymphocytes are produced that are not effector cells but are instead *memory cells.*

It is these lymphocytes that retain the memory of the initial antigenic stimulation and upon subsequent stimulation with the same antigen will rapidly undergo blastogenic transformation to produce new active clones of effector cells. The secondary immune response generated not only appears more rapidly than does the primary immune response, but the level of response is greater than that produced for the primary response. In **Figure 10.10** the levels of antibody produced during the primary and secondary response are shown. Note that the time after antigen challenge to generate peak levels of antibody is shorter for the secondary response. Also note that the majority of antibody generated during the secondary immune response is IgG, whereas for the primary immune response equal amounts of IgG and IgM are produced (see following material).

CHARACTERISTICS OF T-LYMPHOCYTE SUBCLASSES

As mentioned, different functional subclasses of T-lymphocytes have been described. While there is no apparent gross histological difference in appearance between different T-cell subclasses (and even between B-cells and T-cells), using monoclonal antibodies as markers, surface antigens can be shown to be different. Accordingly, those T-lymphocyte subclasses identified as T-helper cells possess CD4 antigen, while those T-lymphocyte subclasses identified as T-suppressor/cytotoxic cells possess CD8 antigens. Currently, at least 45 different CD antigens have been identified on all forms of leukocytes.

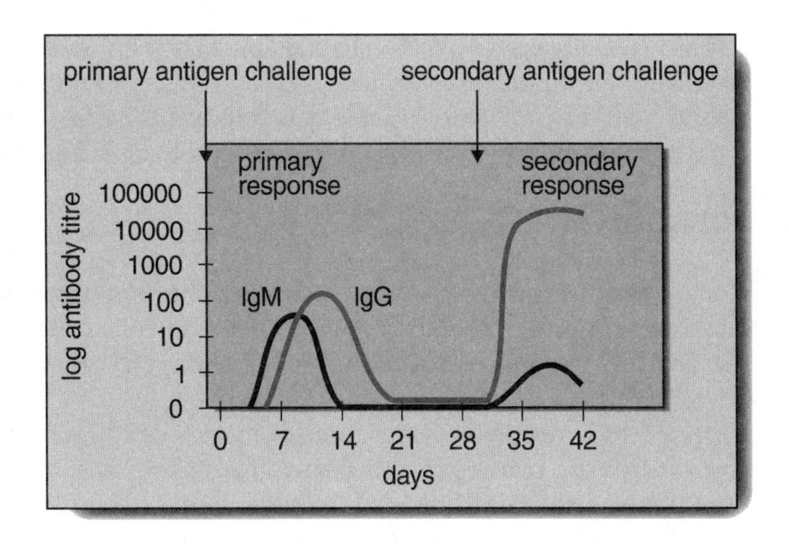

FIGURE 10.10

Functionally, T-helper (Th) cells assist other lymphocytes in fulfilling their functions.

For example, Th-cells assist B-cells to mature and to secrete antibodies. Th-cells also assist T-cytotoxic (Tc) cells to mature, after which the Tc-cells can then specifically attack and kill target cells.

On the other hand, T-suppressor (Ts) cells act to down regulate the function of other lymphocytes. For example, Ts-cells are probably involved in inhibiting over-production of antibody by B-cells after they have been activated (although other mechanisms are also of importance), and Ts-cells are also necessary to maintain the tolerant state (see following material).

THIRD POPULATION CELLS

A discussion of the important immune effector lymphocytes would not be complete without a description of what are sometimes referred to as *third population cells.*

Third population cells, or *natural killer* (NK) cells, also mediate a form of cellular immunity. Like T-cytotoxic cells, NK-cells are also able to lyse target cells. However, while Tc-cells will function only after prior sensitization via antigen stimulated clonal formation, NK-cells impart natural immunity because they function without such prior sensitization. NK-cell activity has been shown to be of importance in destruction of tumor cells. It has also been reported that activation of NK-cells takes place after they have been exposed to either *Interferon gamma* (IFNg) or *Interleukin-2* (IL-2). After such exposure, NK-cells are capable of killing target cells. Such activated NK-cells may also be transformed into *Lymphokine-Activated Killer* (LAK) cells, which are also effective in killing target cells.

Some confusion also exists regarding the K-cell. In actuality, K-cell refers to a cellular activity and not to one specific type of cell. Accordingly, K-cells are any effector cell that possesses receptor (*Fc Receptors*) sites for IgG antibody. Thus, target cells (tumor cells) coated with IgG can now bind with K-cells which will then kill the tumor cell by cellular lysis.

Certain Tc-cells, as well as NK cells, are able to lyse targets in this way. This IgG mediated lysis of targets by effector lymphocytes is termed *K-cell activity*. It should also be noted in passing that Tc-cells and NK-cells are not the only ones capable of lysing tumor cells. Both neutrophils and activated macrophages can also destroy tumor cells by releasing cytotoxic substances extra-cellularly.

TOLERANCE

A state of tolerance has been demonstrated to appear early in human development. Tolerance is necessary because the body's own tissues and cells possess surface antigens (*autoantigens*) that are theoretically capable of activating *autoimmune* responses.

If, in fact, such autoimmune reactions were to take place, the body's own immune system would attack and destroy its own cells and tissues (as is the case in such autoimmune diseases as lupus and rheumatoid arthritis). Thus, early in development, autoantigens tolerize the immune effector cells and thus the cells and tissues of the body are not attacked and destroyed. Furthermore, this state of tolerance must be continually restimulated after birth.

There are a number of pathways leading to tolerance of both T- and B-cells. Early in development, autoantigen appears to promote deletion of certain clones of T- and B-cells, while after birth autoantigen may promote mature Th- and Tc-lymphocytes to undergo functional deletions. Finally Ts-cells are an important component in the maintenance of the tolerant state after birth because they functionally depress the action of other effector lymphocytes (Th-, Tc- and B-cells).

CHARACTERISTICS, FUNCTIONS OF ANTIBODY PRODUCED BY B-CELLS

This section concerns B-cells—*antigen forming cells* (*AFC*) also know as *plasma cells*.

The structure of a basic "Y" shaped antibody molecule is shown in **Figure 10.11**. As can be seen, there are two identical heavy chains and two identical light chains, sometimes referred to as one $[LH]_2$ subunit. The light chains exist in two distinct forms (Kappa and Lambda). Either of the light chains can combine with the heavy ones. However, in a single antibody molecule, both light chains must be of similar type (both Kappa or both Lambda). The protein structure of both chains contains constant and variable domains. Constant heavy chain regions are formed from amino acid sequences that are common to all antibodies of a particular class. Thus, the amino acid sequences present in the constant heavy chain region (C_H), which is further subdivided into domains 1,2 and 3, will determine the specific class of antibody (immunoglobin). Five major classes of immunoglobin exist:

IgG
IgM
IgA
IgD
IgE

The constant light chain amino acid sequence (C_L) determines if the light chain is a Kappa or Lambda. Finally, the amino acid sequences of the light and heavy chain variable regions (V_L, V_H) are tailor-made to specifically fit individual antigens, thus forming two identical antigen binding site located on the arms of the "Y." The "stem" portion of the Y, which is formed from only the two heavy chain constant regions, is termed the Fc portion of the antibody and confers specific functions to each class of antibodies.

While all antibody molecules are designed to bind antigen, the Fc portion of individual classes confers specific functions to each class. For example, the Fc portion of IgE molecules can bind to specialized Fc receptors on mast cells, while the Fc portion of IgG molecules can bind to Fc receptors on macrophages, neutrophils and NK cells. Furthermore, the Fc portion of certain subclasses of IgG

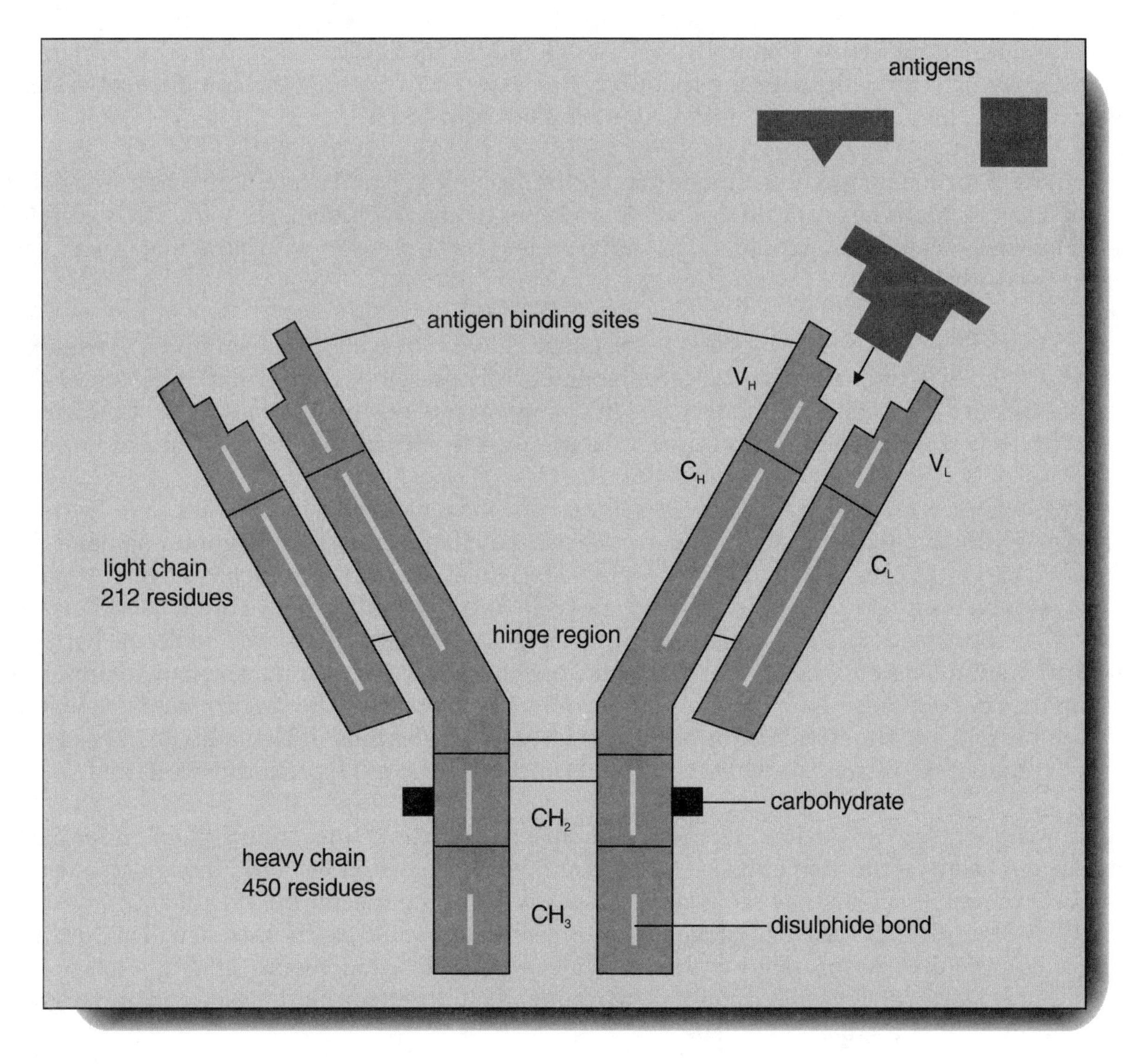

FIGURE 10.11

molecules—as well as the Fc portion of IgM molecules—can activate the complement system cascade (as described).

However, IgM and IgG differ in a number of important features. First, the heavy chain (constant region) is termed *Mu* for IgM and *Gamma* for IgG. Second, during the primary immune response, levels of IgM peak (approximately by day 8) before those of IgG **(Figure 10.10)**, although by day 12 the peak of IgG is greater. However, during the secondary immune response, peak levels of IgG are much increased above those of IgM and the time from antigen challenge to peak levels is shorter than for the primary immune response. IgM accounts for about 10% of the total immunoglobin pool, while that of IgG accounts for about 75%. IgM is found circulating in the blood in the form of 5 $[LH]_2$ subunits linked together with a "J" chain, while IgG is found only as a single $[LH]_2$ subunit, and IgM is the antibody responsible for transfusion reactions in the ABO blood group system. However, IgM is incapable of crossing membranes, while IgG can cross the placenta and is also present in milk. Thus, it is IgG that initially confers natural passive immunity to babies. Both IgM and IgG function to destroy bacteria or other antigenic structures through the following processes: activation of the complement system, opsonization, agglutination and precipitation (see following material).

Regarding the functions of IgA, this antibody accounts for about 15% of the total serum immunoglobin pool and it has an Alpha heavy chain. Unlike other classes of immunoglobins, IgA is actively secreted through mucous membranes in conjunction with a protein called *secretory component* that is manufactured by the epithelial cells.

This is an example of an unusual cooperative response between the cells of the immune system (that manufacture the IgA) and epithelial cells (that manufacture secretory component). IgA is probably important because it functions as the first line of defense at mucous membranes. Here it inhibits or destroys bacteria and possibly some viruses. Like IgG, IgA is also present in colostrum and milk.

Immunoglobin IgE is found in trace amounts in the serum and it has an Epsilon heavy chain. The main function of IgE is to act as antigen receptors on mast cells (and basophile). Like all immunoglobins, IgE is manufactured by plasma cells that were generated by clonal stimulation with an antigen. After the IgE is secreted by the plasma cell, it circulates to mast cells and basophile, where it binds to Fc receptors on the cell membrane. If antigen specific for the binding site of the particular IgE molecule happens to attach, this stimulates the mast cells or basophile to immediately release preformed mediators such as histamine, heparin, proteolytic enzymes and chemotactic factors. Furthermore, after a delay other newly synthesized mediators will also be released such as prostaglandins and leukotrienes. The overall effect is to produce an inflammatory reaction and this is referred to as Hypersensitivity-Type I.

The immunoglobin IgD is only found in trace amounts in the serum. This immunoglobin has a Delta heavy chain. The major function of IgD is to act as antigen receptors on B-cells (although on some B-cells, IgG or monomeric IgM, may act as receptors also). Thus, the specific binding of antigen to a B-clone cell will stimulate this cell to blastogenically transform into multiple copies that will mature into plasma cells. Individual plasma cells will then secrete one class of immunoglobin (IgM, IgG, IgA or IgE, but never IgD). However, regardless of the class of antibody, the antigen binding site will be designed to bind the antigen that initially stimulated the original clone cells.

For example, assume that Dinitrophenol (DNP) linked to Bovine Serum Albumin was used as the antigenic stimulator. Then the antibodies generated to the DNP epitope will all possess antigen receptors that can bind to DNP, regardless of whether they are IgG, IgM, IgA or IgE. It has also been demonstrated that a particular antibody secreting cell may initially synthesize one class of immunoglobin (for example, IgM), but may then switch and continue to produce a different class of immunoglobin (for example, IgG).

Thus, although the constant region of the old heavy chain will stop being synthesized (i.e., mu) and a new constant region will begin to be synthesized (i.e., gamma), the variable region remains the same. Therefore, while the class of immunoglobin changes, the antigen binding site remains the same.

WHY IS ANTIBODY EFFECTIVE IN FIGHTING INFECTION?

Antibody functioning alone would only be minimally effective in protecting the individual from infection by microorganisms. This is because simply binding the antibody to the microorganisms would do very little to inhibit or destroy them.

One effective way to destroy microorganisms is through *opsonization* by antibody. Here the antibody first must bind to the bacteria, after which the antibody coated organisms will adhere to the surface of neutrophils or macrophages through Fc receptor binding. The microorganism can then be easily phagocytized and destroyed **(Figure 10.12)**. Complement proteins activated by antibody-antigen binding can also act to opsonize bacteria. In addition, binding of antibody to bacteria may result in the formation of a large complex of cross-linked organisms that are immobilized, reducing their infective potential. Cross-linked bacteria are easy to phagocytize and this process is termed *agglutinization*. Soluble proteinatious substances, such as toxins, can be inactivated by antibody binding and may also be cross-linked, resulting in *precipitation*. Precipitated antigen can be removed by phagocytosis or the complexes may slowly solubilize. However, such precipitated antibody-antigen complexes may also collect within tissues or on blood vessel walls and produce *Immune Complex Hypersensitivity Type III* reactions (see following material).

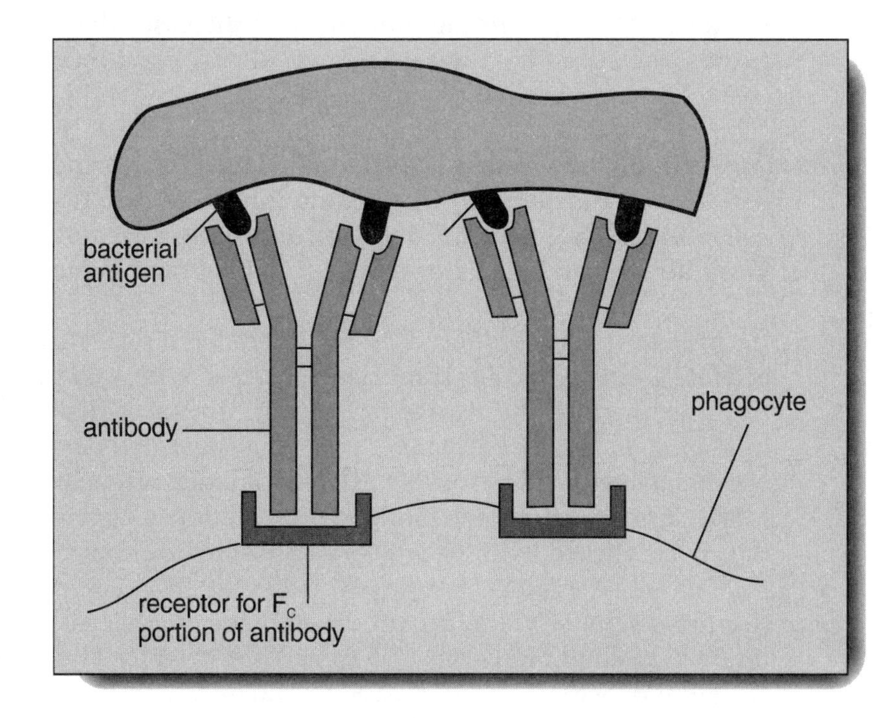

FIGURE 10.12

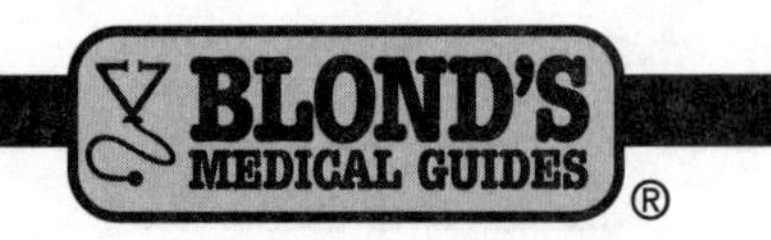

Finally, lymphocytes called *killer cells* (third population cells) possess Fc receptors on their cell membranes that bind antibodies. These antibodies can also bind to target cells through their antigen binding sites and thus the killer cell and the target cell are linked together through specific antibody binding. The killer lymphocyte can then destroy the target cell by release of cytotoxic substances.

FUNCTIONS OF CYTOKINES IN THE IMMUNE SYSTEM

Cytokines are factors generated by effector immune cells and include the *lymphokines* and the *monokines*. Lymphokines are produced by lymphocytes, while monokines are generated by monocytes/macrophages. Because cytokines may be generated at one site—but have an effect on target cells distant from the site of production—they may be considered hormones. While there are a wide variety of cytokines now known or suspected (for example, for the *interleukins* alone currently 11 have been identified and these constitute only a small fraction of all the cytokines), for the sake of clarity and brevity only a few of the more important ones will be discussed.

IL-1. It is generated by macrophages and thus it can be considered a monokine. IL-1 has many functions on diverse targets in the immune system. For example, it stimulates proliferation and differentiation of B-cells into antigen forming (plasma) cells; it increases the cytocidal activity of NK cells; it stimulates chemotaxis and metabolic activation of neutrophils; it stimulates the production of lymphokines (for example, IL-2) and proliferation and expression of IL-2 receptors by T-helper cells; it chemotactically autostimulates macrophages and increases their cytocidal activity and prostaglandin production; and finally, it also may have diverse effects on tissue cells. Furthermore, IL-1 has also been reported to systemically act on the hypothalamus causing the release of *Corticoreleasing Hormone* (CRH) which then releases *Adrenocorticotropic Hormone* (ACTH) from the pituitary and increases the release of *Cortisol* from the adrenal gland). This suggests a complex Immuno-Endocrine down-regulation of the immune response (see following material).

Interleukin-2 (IL-2). Produced by activated T-helper cells, it is therefore classified as a lymphokine. It has diverse functions which include autocrine action on other T-helper cells, monocyte activation and NK cell activation. It also activates and promotes B-cell division, and it promotes T-cell division and mediator release (i.e., interferon gamma).

Interferon gamma (IFNg). Produced by activated T-lymphocytes and by NK cells, IFNg is able to activate T-cytotoxic cells, stimulate NK cell activity, and promote B-cell differentiation. IFNg can also increase the expression of *class II Major Histocompatibility* (*MHC*) *Antigens* (see following material) on *antigen presenting cells* (*APC*), such as monocytes, macrophages, endothelial cells and astrocytes. Effective antigen presentation by APCs requires expression of MHC class II antigen. Since antigen presentation is a prerequisite for lymphocyte stimulation, the effects elicited by IFNg are very important to proper immune function.

Tumor Necrosis Factor (TNF). When produced by macrophages, it is called **TNF alpha (TNFa)** and is thus a monokine. However, TNF can also be produced by lymphocytes and is called *TNF beta* (*TNFb*) and then it is classified as a lymphokine. Both TNFa and TNFb have the same functions. Also some effects elicited by the TNFs many also be produced by other cytokines. For example, TNF activates eosinophils, and NK cells, promotes leukocyte adhesion, enhances MHC expression and antigen presentation, enhances macrophage and neutrophil cytotoxicity, activates B- and T-cells and promotes synthesis of such substances as IL-1, prostaglandin and platelet activating factor by endothelium. TNF has also been found to systemically regulate pituitary hormones and to induce fever and sleep by acting on the brain.

INTERACTIONS BETWEEN EFFECTOR CELLS

Complex interactions must take place between effector immune cells before an active immune response can be generated. One very important initial step is the processing of the antigen.

Initially, antigen present on foreign structures, such as bacteria, enters the body, and during the course of an infection, bacteria or pieces with antigen may find their way into the lymphatic system. This antigen may then pass into the lymph nodes where *antigen presenting cells* (*APC*), such as sinus macrophages, engulf this antigen. Alternately this antigen may be bound to or phagocytized by the APCs in other locations.

For example, APCs in the skin are *Langerhans cells*, in the liver they are *Kupffer cells*, in the brain they are *microglia*, in the lungs they are alveolar *macrophages*, in the thymus they are *interdigitating cells*, and in the blood they are monocyte/macrophages.

Having bound or engulfed the structure associated with this antigen, the next series of steps may vary slightly depending on the type of APC involved.

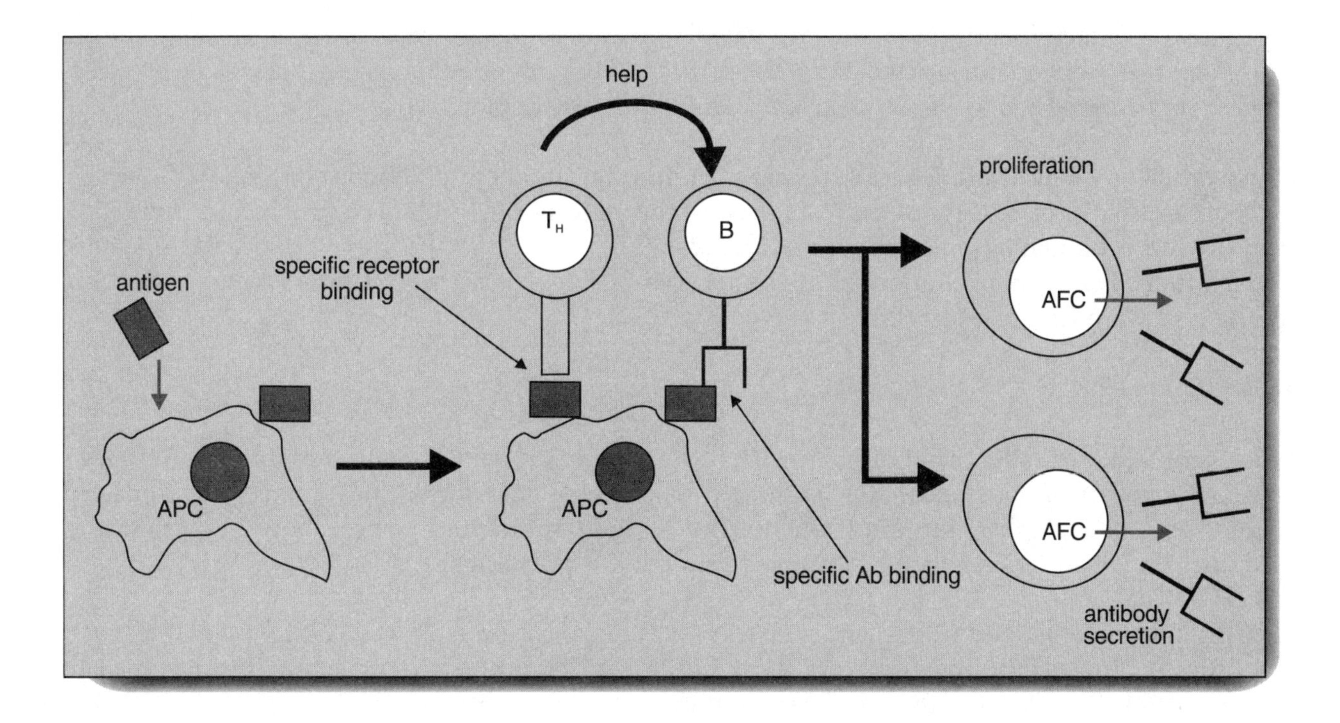

FIGURE 10.13

For example, certain APCs, such as macrophages, will partially digest the antigen, leaving only portions of the antigenic epitopes. However, regardless of whether the antigen is partially digested to epitopes or remains relatively intact, the APC will then present it to the pre B clone or pre T clone cell located within the lymphatic tissue (i.e., lymph node, spleen). If the antigens are digested into epitopes, they are then exposed on the cell surface to allow for effective presentation. Furthermore, the antigenic epitopes must be presented by the APC in conjunction with MHC autoantigens if the effective cell to cell interaction is to take place (see following material). Furthermore, it is now known that B-cells may actively present antigen to other effector cells such as T-helper cells to promote activation of the clone.

Once the antigen is available for presentation by the APC, T- and B-cells which have the appropriate specificity, receptors will bind to the antigen and B-cell clonal formation will then lead to antibody production by the humoral immune system.

Furthermore, to activate B-cells, not only is it necessary for the B-cell to bind the presented antigen via surface immunoglobin receptors, but help in the form of IL-2 must be supplied by the Th-cell which is also activated by binding to the antigen **(Figure 10.13)**. Also, while the same antigen can act to stimulate both T-cells and B-cells, different parts of the antigen (different epitopes) will be bound by the receptors on these two different types of lymphocytes.

T-Dependent V. T-Independent Antigens. Although the majority of antigen (i.e., *T-dependent antigens*) are only able to stimulate a B-cell response if Th-cell help is also present, there is a class of antigens that do not require the presence of Th-cells.

This group of *T-independent antigens* are polyvalent and apparently activate the B-cell by cross linking the antigen receptors on the surface of the B-cell. Some T-independent antigens, such as *Lipopolysaccharide* from bacterial cell walls, are also *mitogens* (see following material).

For activation of T-cells leading to cell-mediated immunity, similar interactions between APCs and T-cell clones must take place **(Figure 10.14)**. Here, however, additional levels of control include the action of suppressor cells to modulate the Th-cell and also the B-cell. Since the Th-cell plays an important pivotal role in both humoral and cell mediated immune response, Ts-cell interactions have wide ranging effects on immune response.

MITOGENS

In the event that the reader is not familiar with the term *mitogen*, a short definition follows: Mitogens are a large group of substances that stimulate blastogenic transformation of multiple clones of lymphocytes. Thus, unlike the effect elicited by antigen stimulation where only a specific and individual target clone is activated, mitogen stimulation is polyclonal.

Many mitogens are derived from plant extracts (e.g., Concanavalin A, Pokeweed mitogen, Phytohemagglutinin). However, mitogens do not stimulate clones of lymphocytes by acting through antigen receptors, but instead bind to these cells at separate mitogen receptors.

MHC

Located on chromosome 6 in humans and on chromosome 17 in the mouse are a series of genes that code for surface structures present on many different types of cells.

In the mouse, these genes are called the *major histocompatibility complex H-2* and in humans, they are called the *major histocompatibility complex HLA.*

In both the mouse and the human, these MHC genes code for three distinct classes of cellular surface MHC antigens (Class I, II & III). Furthermore, the probability is small that two unrelated individuals will have the same MHC genes and the same MHC antigens on their cells. For this reason, the success of tissue or organ transplants is very limited if the donor and recipient are unrelated. The closer the genetic relationship between the donor and recipient, the better the chances of a good MHC match and a successful transplant.

Without going into tedious detail, Class I MHC antigens located on target cells are recognized by Tc-cells. If the target cell also happens to be virally infected, then viral antigens will also be expressed on the surface of the target cell.

In order for the Tc-cell to attack and destroy the virally infected target cell, it must recognize both the viral antigen and the Class I MHC antigen at the same time. If this happens, the Tc-cell will kill the target cell. Thus, Tc-cells obtained from subject X will only kill virally infected target cells from subject X because they recognize the MHC Class I antigens on subject X's target cells plus the viral antigen co expressed on subject X's target cells. Tc-cells from subject X will not attack virally infected target cells from subject Y because, while the viral antigen on the surface of the virally infected target cell will be the same (as that on the virally infected target cells of subject X), the MHC class I antigens will not be recognized. Furthermore, if the Tc-cells from subject X are exposed to uninfected cells from subject X, although the MHC Class I antigens are recognized by the Tc-cell, it will not attack and kill the uninfected cell because no viral antigen is co-expressed.

Identification and killing of neoplastic cells is also accomplished by Tc-lymphocytes in much the same way that they kill virally infected cells. Again the neoplastic cell must not only express the correct MHC Class II antigens but also abnormal antigens on the surface. This process is termed *immunological surveillance.*

While Class I MHC antigens allow Tc-cells to recognize and destroy target cells, MHC Class II antigen is necessary to allow interactions to take place between various effector cells in the immune system. Thus, in the above analogy, APC cells from subject X will be able to present antigen to Th-cells or to B-cells from subject X because the APC recognizes the MHC class II antigens on the surface of the Th- and B-cells.

APC cells from subject X will not be able to present antigen to Th-cells from subject Y because they do not recognize the MHC Class II antigens on the Th-cells from subject Y. This characteristic is termed *genetic restriction.*

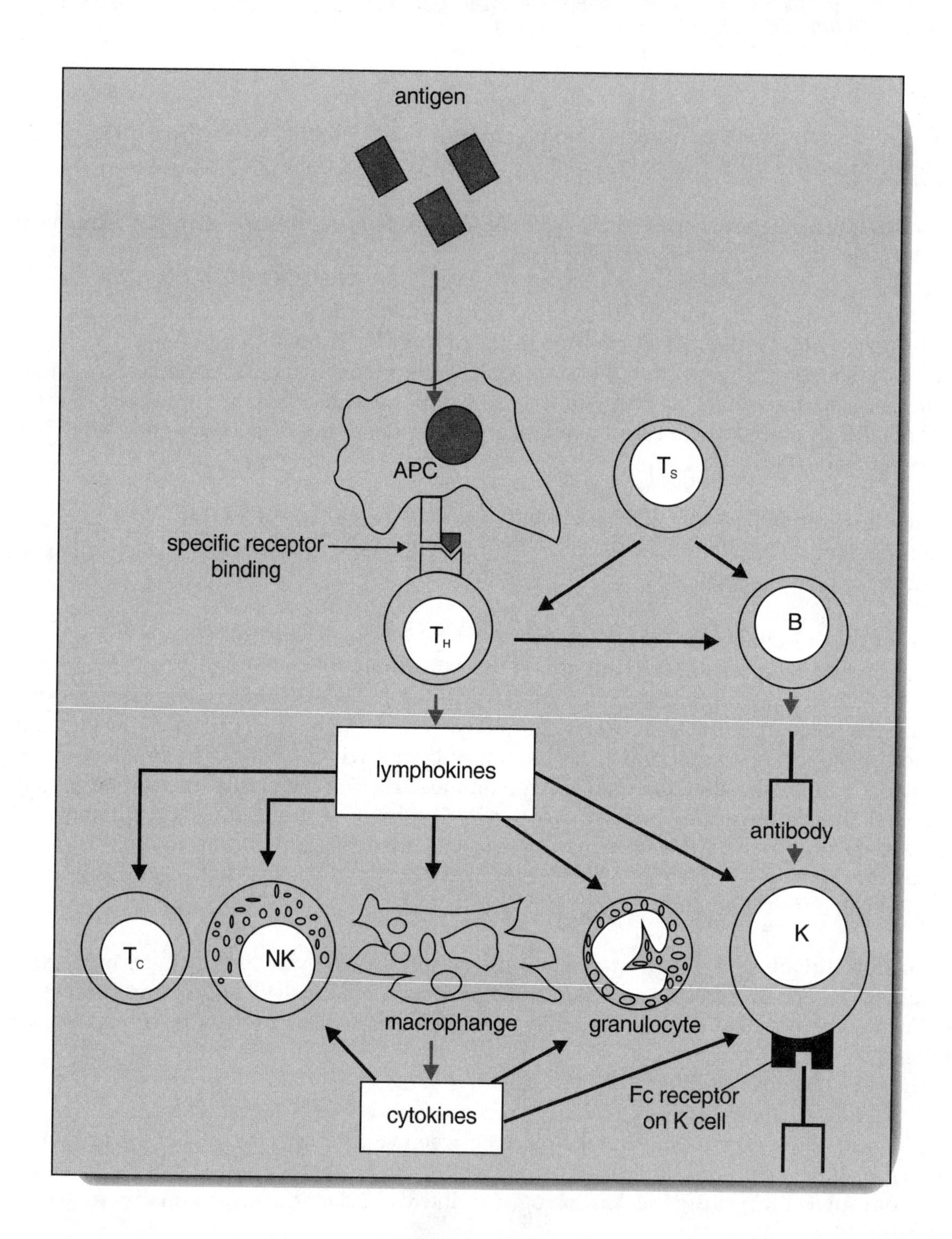
antigen
APC
specific receptor binding
T_S
T_H
B
lymphokines
antibody
T_C
NK
K
macrophange
granulocyte
cytokines
Fc receptor on K cell

FIGURE 10.14

DEVELOPMENT OF LYMPHOCYTES

As outlined in **Figure 10.5,** a common lymphoid progenitor cell is the source of all lymphocytes. This progenitor cells is believed to be located in the bone marrow.

However, during the time of fetal development, those progenitor cells destined to become T-lymphocytes migrated into the thymus (a *primary lymphoid organ*) where they continued to develop. In this thymic microenvironment they may have been exposed to local hormones which regulate this process.

When development is complete these *thymic* lymphocytes (or T-lymphocytes, or T-cells) then leave the thymus and migrate to the lymph nodes and spleen (which are called *secondary lymphoid organs*) where they colonize the *thymic dependent areas*. Apparently, once this colonization is completed prior to birth, the thymus can be removed without effecting the cell mediated immune response. On the other hand, since the thymus is the source of thymic hormones, removal may result in alteration in some aspects of immune system regulation.

While the steps leading to development of T-lymphocytes are fairly well understood, the process of B-cell development is not as clear.

In birds it is known that progenitor cells that are destined to become B-cells migrate from the bone marrow to an intestinal organ called the Bursa of Fabricius where they mature. After this process is complete, the *bursal* lymphocytes (or B-lymphocytes, or B-cells) then migrate to the secondary lymphoid organs where they colonize the *thymic independent areas*. However, in animals there is no Bursa, but B-cells are still present, thus some other structure (a *Bursal analogue*) must take the place of the Bursa. Organs that have been proposed as this Bursal analogue are the liver, the GALT (or *gut associated lymphatic tissue*), the appendix and the bone marrow itself. Currently the Bone Marrow seems to be the strongest contender as the source of B-cell development. (If correct this is convenient since B can now stand for bone marrow-derived lymphocytes).

OTHER REGULATORY PATHWAYS

Complex regulatory pathways exist that are designed to limit excessive immune response. For example, idiotypic and anti-idiotypic pathways (described first in Jerne's Network Hypothesis), where primary antibodies stimulate the production of secondary (idiotypic) antibodies, these idiotypic antibodies stimulate tertiary or anti-idiotypic antibodies, and so on. The ultimate purpose is to down-regulate the system. T-Helper and T-Suppressor pathways, functioning something like Jerne's Network, where different cell populations regulate each other, have also been proposed as additional levels of control **(Figure 10.15)**.

Endocrine hormones have also been implicated in important regulatory interactions with the immune system.

Hormones from the adrenals (cortisol), gonads (estrogen, androgen, progesterone), pituitary (growth hormone, prolactin), thymus (thymosin, thymulin) and from effector immune cells themselves (IL-1, Glucocorticoid Increasing Factor, possibly ACTH) are involved in interactions between the immune

and endocrine systems. Furthermore, the difference in immune response between males and females, termed immunological sexual dimorphism, is probably based in part on the differences in the hormonal milieu present both during development and after birth in both sexes.

HYPERSENSITIVITY REACTIONS AND AUTOIMMUNITY

Normal adaptive immunity is a very important component of the immune response. However, if an individual is excessively sensitized to an antigen the response generated may cause extensive damage to the normal body cells and tissues. This is termed hypersensitivity and there are four types of hypersensitivity reactions.

Type I or Immediate Hypersensitivity. Here mast cells bind IgE through Fc receptor interactions. Then upon antigen binding to the IgE the mast cells release vasoactive products such as histamine which generate inflammatory responses.

Type II or Antibody-Dependent Cytotoxicity. Here autoantibody binds to antigen on normal body cells. This antibody then mediates attack by K-cells and complement.

Type III or Immune Complex Disease. Here formation of antibody-antigen complexes precipitate in the tissues. These complexes then mediate attack by complement or polymorphonuclear neutrophils.

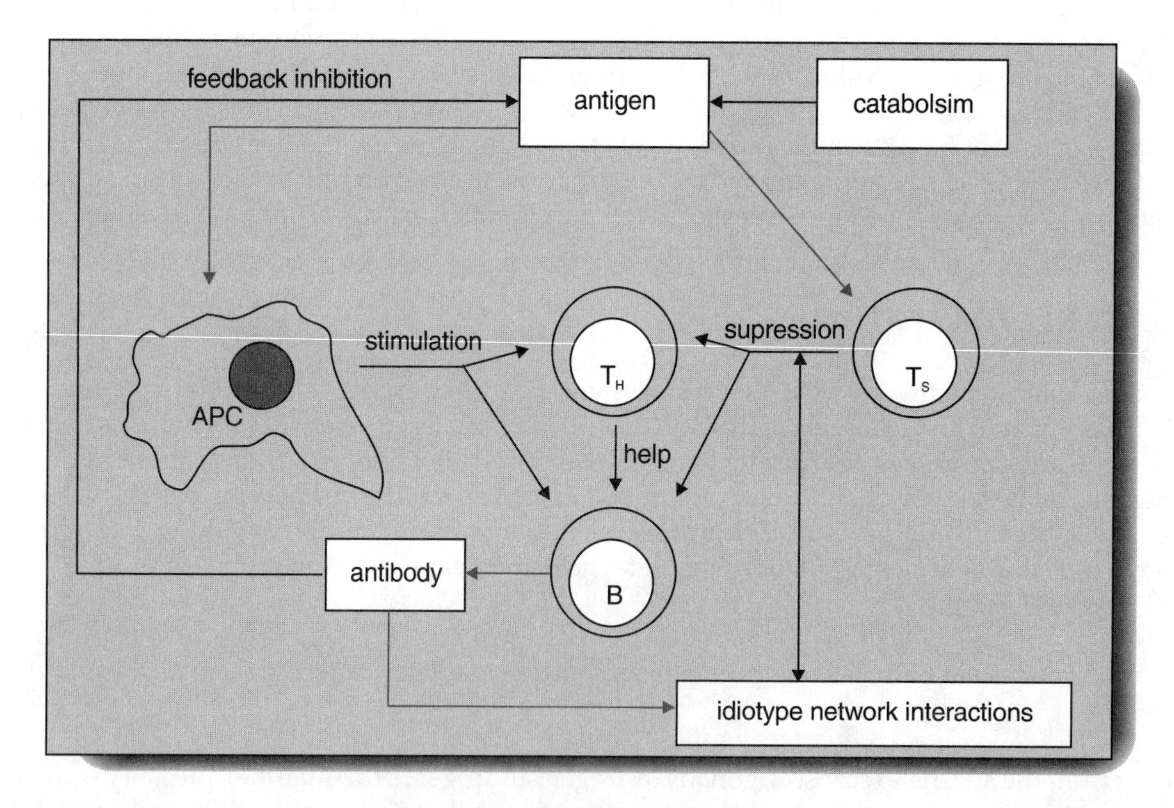

FIGURE 10.15

Type IV or Delayed Hypersensitivity. Here T-lymphocytes activated by antigen release lymphokines that stimulate release of inflammatory mediators. After the antigen challenge, the response may take hours or even days to develop, which is why this is classified as "delayed" hypersensitivity.

It is not always easy to separate hypersensitivity from autoimmunity. While hypersensitivity requires antigen presensitization, overt antigen presensitization may not always be obvious in autoimmune diseases. Also autoimmunity seems to be genetically predisposed where certain specific MHC genotypes are more susceptible to certain autoimmune diseases.

Of interest is the fact that many autoimmune diseases are more commonly expressed in one sex or the other. For example, lupus, rheumatoid arthritis and hyperthyroidism are more common in females then in males and these diseases are MHC Class II associated. On the other hand, ankylosing spondylitis and Reiter's Syndrome are MHC Class I associated and are more common in males. These differences in expression of autoimmunity between males and females may be based on the differences in immunological sexual dimorphism, but the underlying pathways responsible are still under investigation. However, it is known that altering the sex hormone environment during early development can alter the expression of autoimmune disease in the adult.

Chapter 11
RESPIRATORY SYSTEM

INTERNAL AND EXTERNAL RESPIRATION

The respiratory system is involved in the exchange of gases at both the lung and tissue levels. Various processes involved with gas exchange at the lung level are termed *external respiration,* while the utilization of O_2 and generation of CO_2 at the cellular level is described as *internal respiration.*

STRUCTURAL ELEMENTS OF THE RESPIRATORY SYSTEM

The respiratory system is basically a series of tubes that conduct atmospheric gases to the alveoli **(Figure 11.1)**. Beginning in the nasal and oral cavities, the air passes into the pharynx (nasopharynx, oropharynx, laryngopharynx), then into the trachea, to left and right primary bronchi, to secondary bronchi, to tertiary or segmental bronchi, to bronchioles, terminal bronchioles, to respiratory bronchioles, and then finally to alveolar ducts. From there, it enters the alveoli where gas exchange takes place. No gas exchange takes place anywhere in the system of transport tubes and thus this region is termed the *anatomical dead space* (see following material).

MEMBRANES LINING THE RESPIRATORY SYSTEM

Mucous membranes that line the respiratory system begin in the nose. Cells that comprise these membranes are the ciliated epithelial cells, and they secrete mucous from goblet cells, also located there. Movement of the cilia push dust particles trapped in the mucous upwards in the system towards the back of the pharynx where the material is then either swallowed or coughed out.

Thus, the membranes lining the respiratory system are self-cleaning, and when this ability is coupled with the phagocytic action of alveolar macrophages, the process is very efficient in the removal of dirt and other foreign matter.

NOSE AND NASAL CAVITY

The two openings to the nose are termed the *nostrils* or *external nares.* Internal hairs at this location limit the entrance of larger particles into the respiratory system. The nasal cavity is the hollow space behind the nose and is divided into left and right sides by the *nasal septum.* Protruding into the nasal cavity are the *superior, middle and inferior nasal conchae* covered with mucous membranes. As air passes into the nasal cavity and flows over the membranes of the nasal concha, it is warmed and humidified. (This is especially important during the winter when cold, dry air from outside may possibly damage the delicate alveolar membranes.) Also located in the upper nasal cavity are the patches of olfactory membrane responsible for the sense of smell.

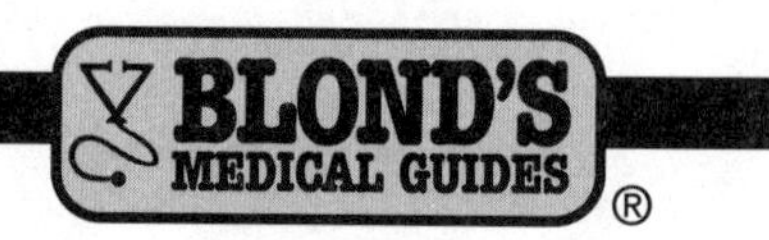

SINUSES

Sinuses are air spaces located in the maxillary, frontal, ethmoid and sphenoid bones of the skull. They are connected to the nasal cavity and, like the rest of the respiratory system, are lined with mucous membranes. Thus, inflammation, swelling and increased secretions from these membranes can result in a build-up of pressure in these spaces, leading to pain and headaches. Swelling of these membranes may be due to allergic reactions resulting from allergins such as plant pollens, mold spores and animal dander. Respirator infections can also promote similar swelling and drainage.

PHARYNX, LARYNX, TRACHEA AND BRONCHI

Located behind the nasal and oral cavities, the *pharynx* also extends to the top of the *larynx*. It provides a shared function carrying air into the larynx and food into the *esophagus*. It is composed mainly of a

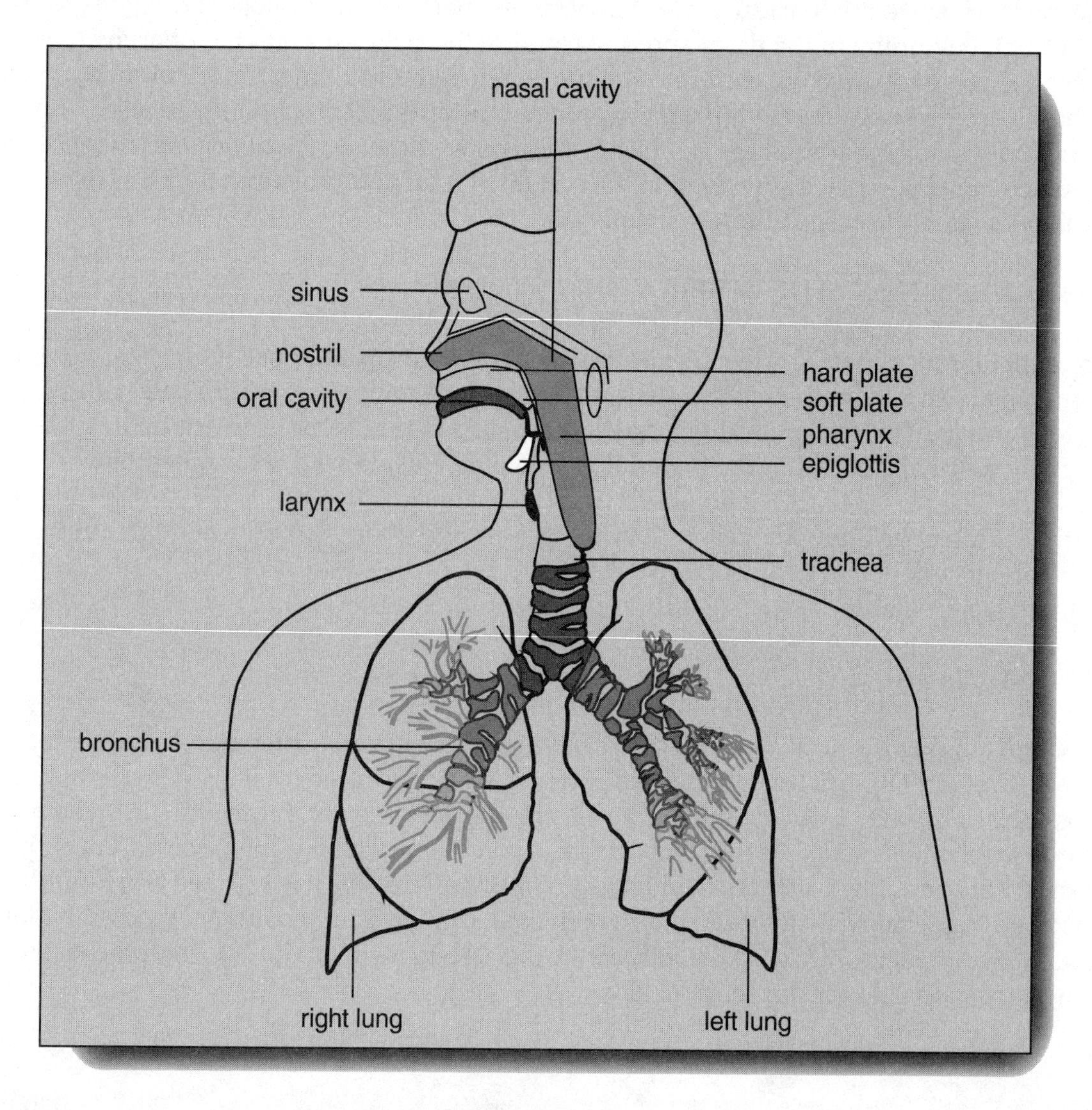

FIGURE 11.1

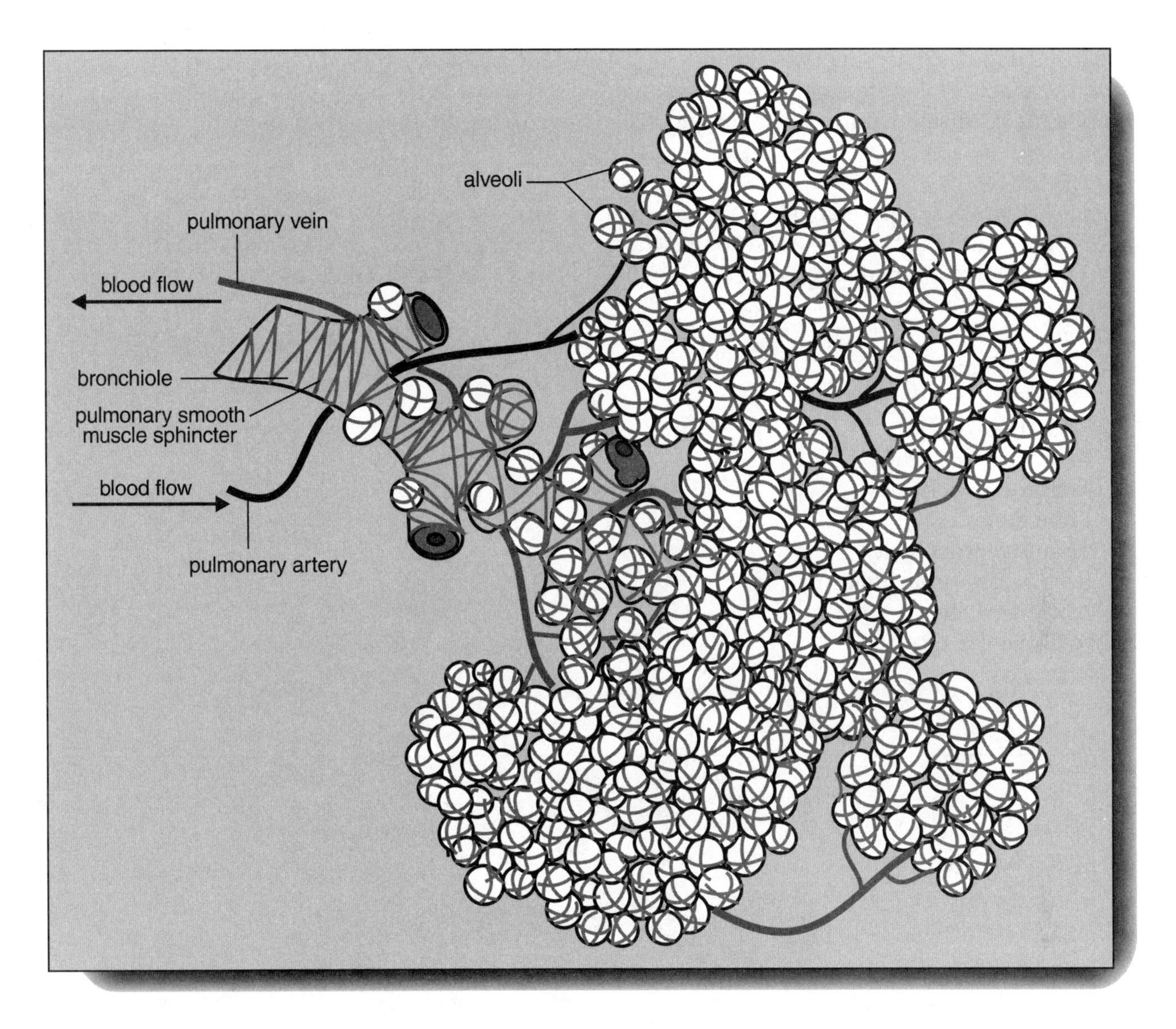

FIGURE 11.2

skeletal muscle but these muscles are not under conscious control and during swallowing undergo automatic and reflexive contraction.

The *larynx* is an enlargement of the airway at the top of the *trachea* and carries air from the pharynx into the trachea. In addition, it contains the *vocal chords*. The opening into the larynx can be closed by a flap-like structure (called the *epiglottis*) during swallowing. This prevents food or liquids from entering the larynx.

The *trachea* is 2.5 cm in diameter and about 12.5 cm long and is comprised of about 20 C-shaped hyaline cartilages. The open portion of the C-shape is directed posteriorly and is closed by smooth muscle and connective tissue. Pressing against this portion of the trachea is the esophagus, which is normally

collapsed. However, during swallowing, the bolus of food distends the esophagus as it passes down to the stomach. This distension also presses into the soft tissue at the back of the trachea allowing the esophagus to distend during swallowing. The inner lining of the trachea is made up of *ciliated epithelium*.

A *bronchial tree* is composed of all of the branched tubes that begin at the trachea and become smaller and smaller until they reach the level of the *alveoli*. Accordingly, the trachea branches into the left and right *primary bronchi* that supply the two lungs. The total number of tubes and the branches in both lungs are as follows:

Name	Number of tubes
Primary bronchi	2
Secondary bronchi	4
Tertiary bronchi	8
Bronchioles	16
Terminal bronchioles	$6x10^4$
Respiratory bronchioles	$5x10^5$
Alveolar ducts/ Alveolar sacs	$8x10^6$

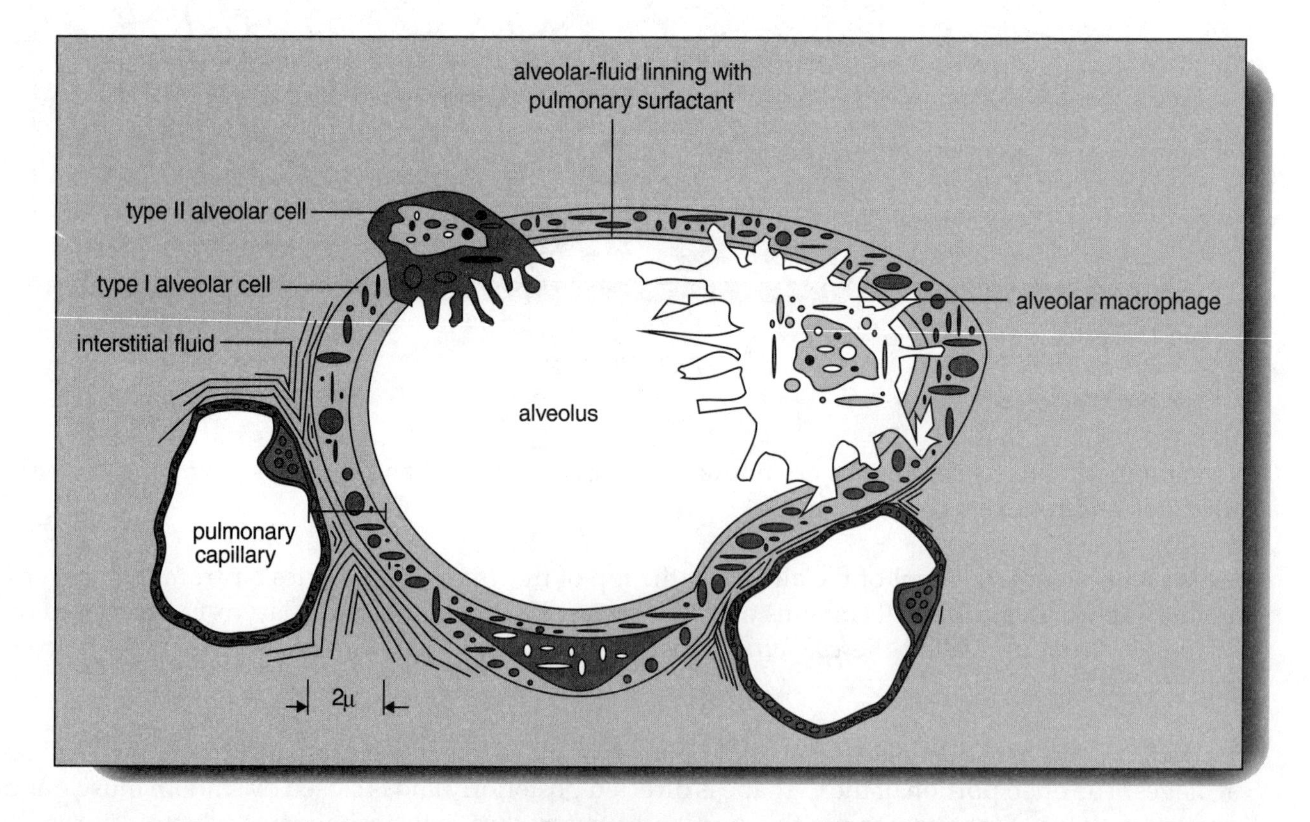

FIGURE 11.3

The *alveoli* appear as clusters of thin-walled, inflatable microscopic sacs closely enclosed in a dense network of pulmonary capillaries **(Figure 11.2)**. Alveolar walls are composed of a single layer of *simple squamous epithelium*; however, for gases to exchange with the capillary blood the capillary wall must also be traversed. Termed the *respiratory membrane*, this structure is normally only 0.2 um thick and this very thin structure facilitates gas exchange. The total respiratory membrane that supports gas exchange is structured as follows: **(Figure 11.3)**:

1. A single layer of simple squamous epithelium.
2. The basement membrane of the squamous epithelium.
3. The interstitium, (or interstitial space).
4. The basement membrane of the endothelial cells.
5. The single layer of endothelial cells in the capillary wall.

In the lungs, there are approximately 300 million alveoli whose walls are composed of flattened simple squamous epithelium called *Type I Alveolar Cells*. Given the remarkably dense network of pulmonary capillaries associated with these alveoli, the total surface area available for gas exchange in the lungs is 75 mm^2 (or about the size of a tennis court). Since the rate of diffusion is proportional to the surface area, this arrangement is well suited to maximize gas exchange.

FLICK'S LAW

From *Flick's Law of Diffusion* we know that the molecular diffusion rate (or transfer of molecules through a membrane) is:

1. Inversely proportional to the distance traveled; thus the thicker the membrane the smaller the rate of diffusion.
2. Directly proportional to the surface area; thus as the surface area increases, so does the rate of diffusion.
3. Directly proportional to the concentration gradient; thus, as the concentration gradient increases so does the rate of diffusion.
4. Directly proportional to the permeability of the membrane; thus as the membrane permeability increases so does the rate of diffusion.
5. Inversely proportional to the molecular weight of the diffusing molecules; thus, as the molecular weight increases, the rate of diffusion decreases. Since O_2 and CO_2 are low mass molecules, they can more effectively transfer across the respiratory membrane.

Type II Alveolar Cells, also present within the alveoli, are responsible for the secretion of *pulmonary surfactant*. This substance is composed of a phospholipid-protein complex whose purpose is to reduce surface tension in the alveoli and facilitate lung expansion (see following material).

Alveolar Macrophages are also found in the alveoli and these cells are phagocytic in function. They are classified as fixed macrophages of the *Reticuloendothelial system*, and their purpose is to remove foreign particles from the alveolar spaces.

THORACIC CAVITY AND THE LUNGS

The thoracic cavity is a closed compartment whose walls are composed of the ribs and associated skeletal muscles (external intercostals, internal intercostals), and whose floor consists of the diaphragm. Each lung is cone shaped with the pointed top of the cone facing up; each is connected to a

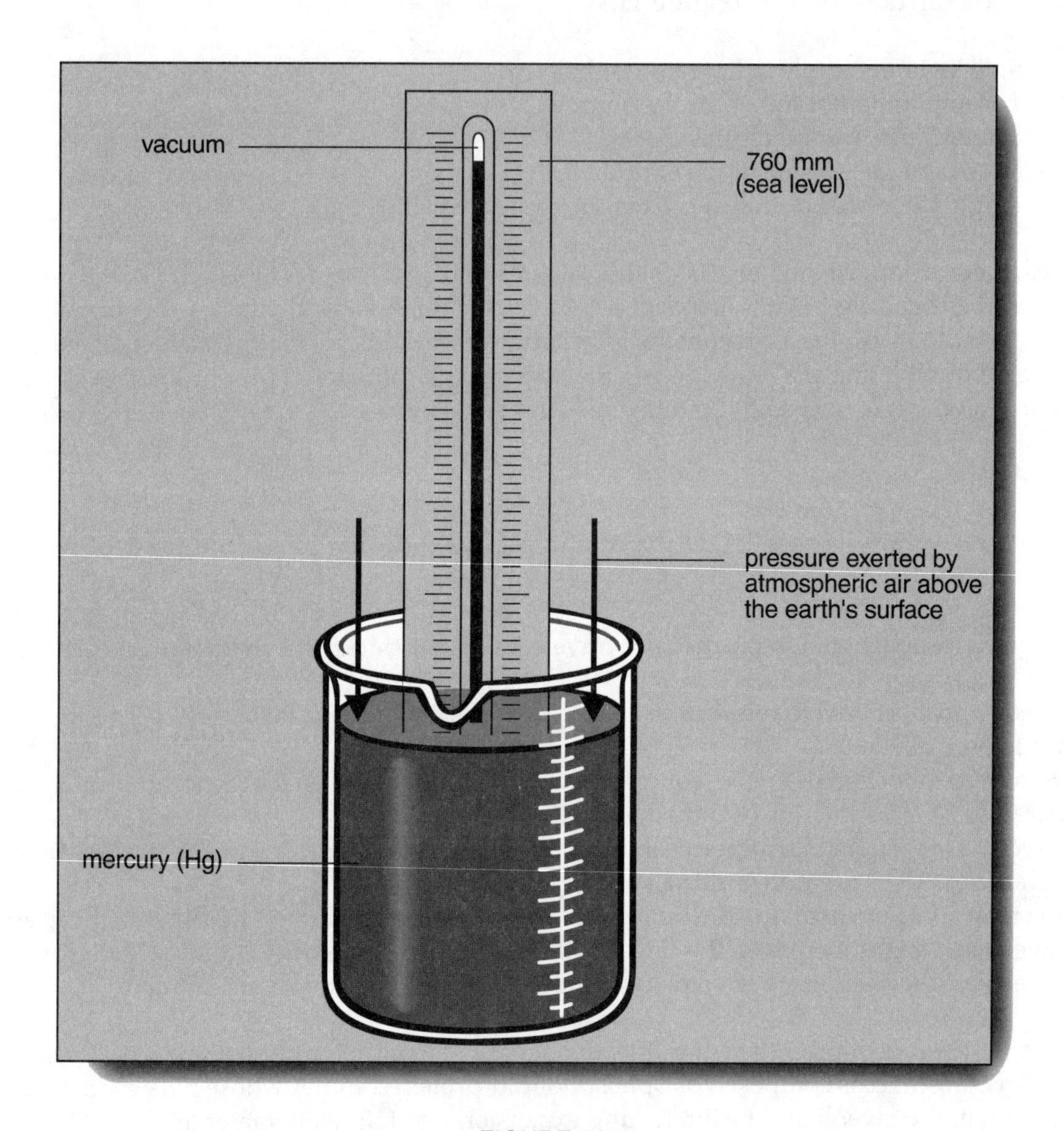

FIGURE 11.4

primary bronchus. The connection is located at the hilus where the blood vessels also enter and leave, as do the nerves and lymphatics. Thus, each lung is suspended from a single point (the *hilus*) within the thoracic compartment. Each lung is also separated from the other by the central mediastinum containing the heart and some other structures (the thymus gland, and a portion of the esophagus and trachea).

Lining the inner walls of the thoracic cavity is a membrane composed of serous epithelial cells called the *Parietal Pleura*. A similar membrane called *Visceral Pleura* covers the lungs. The space between these two membranes is the *Interpleural Space* and is filled with serous fluid. This arrangement effectively reduces friction as the lungs expand and contract against the thoracic walls.

MECHANICS OF RESPIRATION

In order for air to flow into—and out of—the lungs, flow must be driven by a pressure gradient. Given the relationship F = ΠP/R here are the various components that affect flow:

Atmospheric (Barometric) Pressure. Atmospheric pressure results from the gravitational attraction of air molecules (which have mass) and thus, can be measured as weight. Consider a column of air molecules one square inch in size that extends from the earth's surface all the way to the fringes of the earth's atmosphere. At the surface, this square inch column of air molecules will weigh 15 lbs. Thus at sea level every square inch of the earth's surface has 15 lbs of air pressing down on it from above (i.e., the atmospheric pressure is 15 lbs/inch^2).

In addition, atmospheric pressure can be measured in the form of *Atmospheres* so that 1 Atmosphere = 15 lbs/inch^2. It can also be measured in *mm of Mercury (Hg)*. Such a form of pressure measurement is commonly used in both physiology and medicine and derives as follows.

The weight of a column of air pressing down from above can hold up a column of mercury in a glass tube (a mercury barometer) that is 760 mm high **(Figure 11.4)**. Thus at sea level, the atmospheric pressure is equivalent to 760 mm Hg. Note also that as you ascend from sea level to higher altitudes, the atmospheric pressure falls because there is less air on top of you. Thus, at the top of a high mountain, the atmospheric pressure may only be 240 mm Hg.

Intra-Alveolar Pressure. This pressure (also called *intra-pulmonary pressure*) is the pressure of gas inside the alveoli of the lungs. Because the alveolar spaces are connected to the outside through a network of tubes, air flow can pass from the atmosphere to the alveoli and vice versa, but this depends on the pressure gradient. The intra-alveolar pressure varies, but during normal quiet respiration it ranges from about 761 to 759 mm Hg.

Inter-Pleural Pressure. This is the pressure present within the pleural cavity between the visceral and parietal pleura. It is also known as the *intrathoracic pressure* and is usually less than atmospheric pressure. Interpleural pressure ranges from about 756 to 754 mm Hg during normal quite respiration. In addition, normally it never will equilibrate with either the atmospheric or intra-alveolar pressures because the pleural cavity is a closed compartment.

Surface Tension and Transmural Pressure. During human development, the thoracic cavity grows more rapidly than do the lungs. Thus, after birth the thoracic cavity is larger than the unstretched lungs. However, the pleural membranes lining the lungs and the thoracic wall are held together by *surface tension* and by the *transmural pressure gradient*.

Surface tension results from the interaction of water molecules present on the surfaces of the pleural membranes. It holds the lungs and thoracic walls closely together and is partly responsible for changes in the lung volume as the thoracic cavity expands and contracts during respiration.

Transmural pressure gradient develops across the lung wall and across the thoracic wall due to the differences in pressure on the two sides of these walls (i.e., the alveolar pressure and the interthoracic pressure are different). Because the thoracic cavity is larger than the lungs, the thoracic wall is trying to expand outward. However, it cannot completely do so because as it tries to expand it moves away from the lung wall on the inside leaving a space called the *interpleural space*. Because the interpleural cavity is closed to the outside, as the volume of this space enlarges the pressure of the air trapped within decreases **(See Boyle's law)**. The result is that the interpleural pressure is lower than atmospheric pressure by about 4 mm Hg (i.e., it is 756 mm Hg if atmospheric pressure is 760 mm Hg). However, the lungs are connected to the outside atmosphere by way of the respiratory airways. Thus, within the lungs the inter-alveolar pressure of air is equal to the atmospheric pressure (i.e., 760 mm Hg). Because the pressure in the lungs is greater than in the interpleural space, the lungs expand in an attempt to fill the thoracic space. This pressure difference across the surface of the lungs is the *transmural pressure gradient across the lung wall* and it keeps the lungs partially inflated. A similar *transmural pressure gradient across the thoracic wall* is also present and this keeps the thoracic cavity partially compressed.

PNEUMOTHORAX AND ATELECTASIS

In the event that a hole develops either in the thoracic wall or in one lung, the transmural pressure gradient on the side with the puncture is abolished. A puncture in the thoracic wall is termed a *traumatic pneumothorax*, while a hole in the lung is called a *spontaneous pneumothorax*. Because the two lungs are separated by the mediastinum, only the lung on the side with the hole will collapse (atelectasis) and at the same time, the chest wall on that side will spring outward.

BOYLE'S LAW

Boyle's law states that at a constant temperature, the pressure exerted by a gas is inversely proportional to the volume of the gas (P is proportional to 1/V). For example, if the pressure of O_2 in a closed container is 50 mm Hg and the volume of the container is reduced by half, then the pressure should increase to 100 mm Hg. Indeed it is such changes in the volume leading to changes in the pressure gradient that promote air flow into and out of the lungs, as outlined later.

FLOW DEPENDS ON A PRESSURE GRADIENT

In order to cause air to flow into the respiratory system, the pressure outside (the atmospheric pressure) must be greater than the pressure inside the lungs. This will provide a pressure gradient that is directed inward. Since the atmospheric pressure, P_{atm} = 760 mm Hg, the pressure within the lungs (in the alveoli, P_{alv}) must be less than 760 mm Hg (for example, 759 mm Hg).

Given: $\Pi P = P_{atm} - P_{alv}$; and P_{atm} = 760 mm Hg, P_{alv} = 759 mm Hg, *then* ΠP = 760 - 759 = 1 mm Hg, and this small inwardly directed pressure gradient will cause air to flow into the lungs. If, on the other hand,

P_{alv} = 761 mm Hg, and P_{atm} = 760 mm Hg, then the gradient will again be 1 mm Hg but the direction of flow will be from inside to outside resulting in the outflow of air from the lungs during exhalation **(Figure 11.5)**.

Since the atmospheric pressure is constant for all intent and purposes, the only way to promote air flow into and out of the lungs is by increasing or reducing the intra-alveolar pressure above or below atmospheric. If this can be accomplished, the resulting gradient will allow the lungs to inflate and deflate. The problem then is how to change intra-alveolar pressure. To accomplish this, the volume of the thoracic cavity can be changed by the action of the inspiratory muscles of the diaphragm and external intercostals. Stimulation of the phrenic nerve to the diaphragm causes it to contract and descend downward, enlarging the thoracic cavity. Additional contraction of the external intercostal muscles, upon stimulation by the intercostal nerves, causes these muscles to elevate the thoracic cavity upwards and

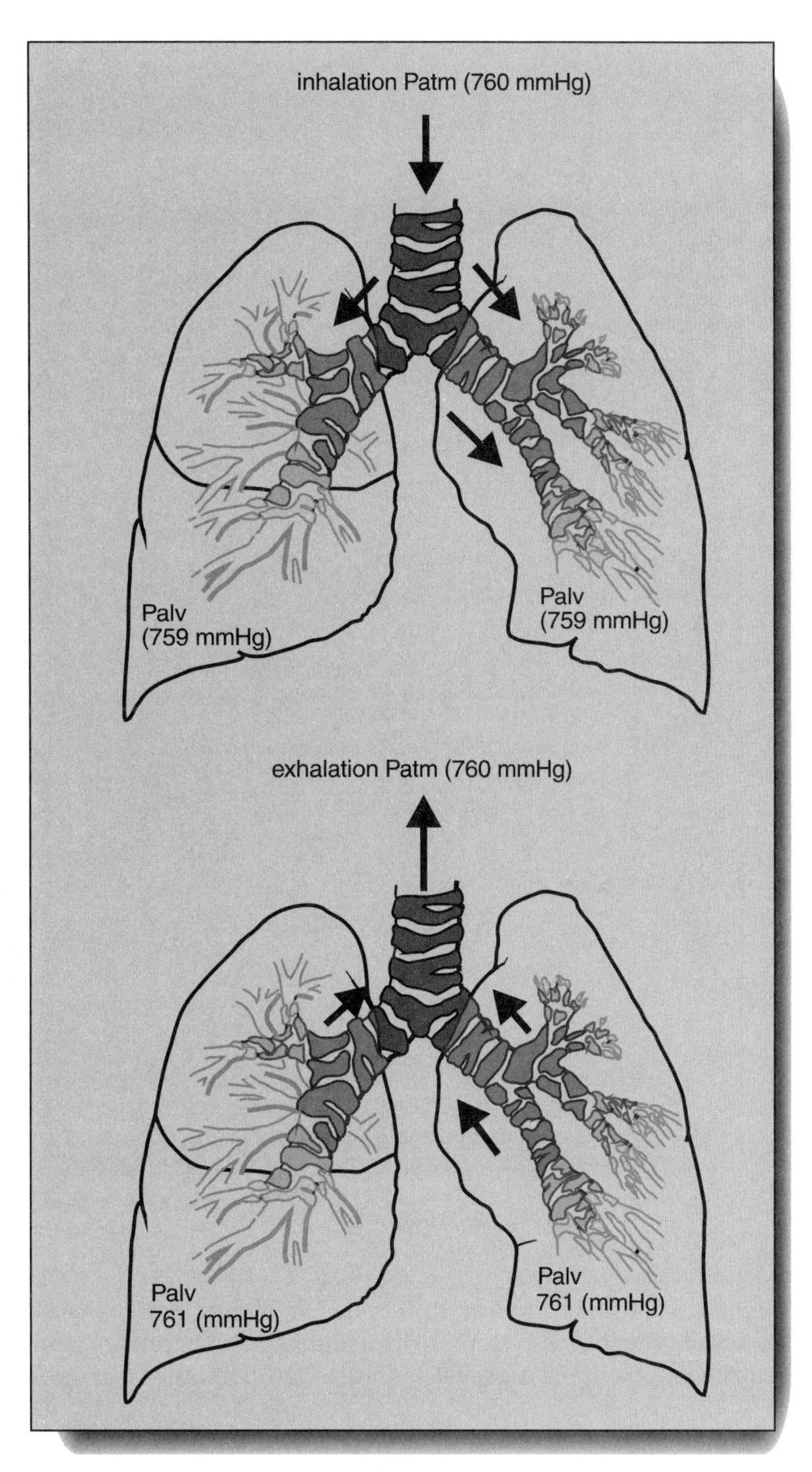

FIGURE 11.5

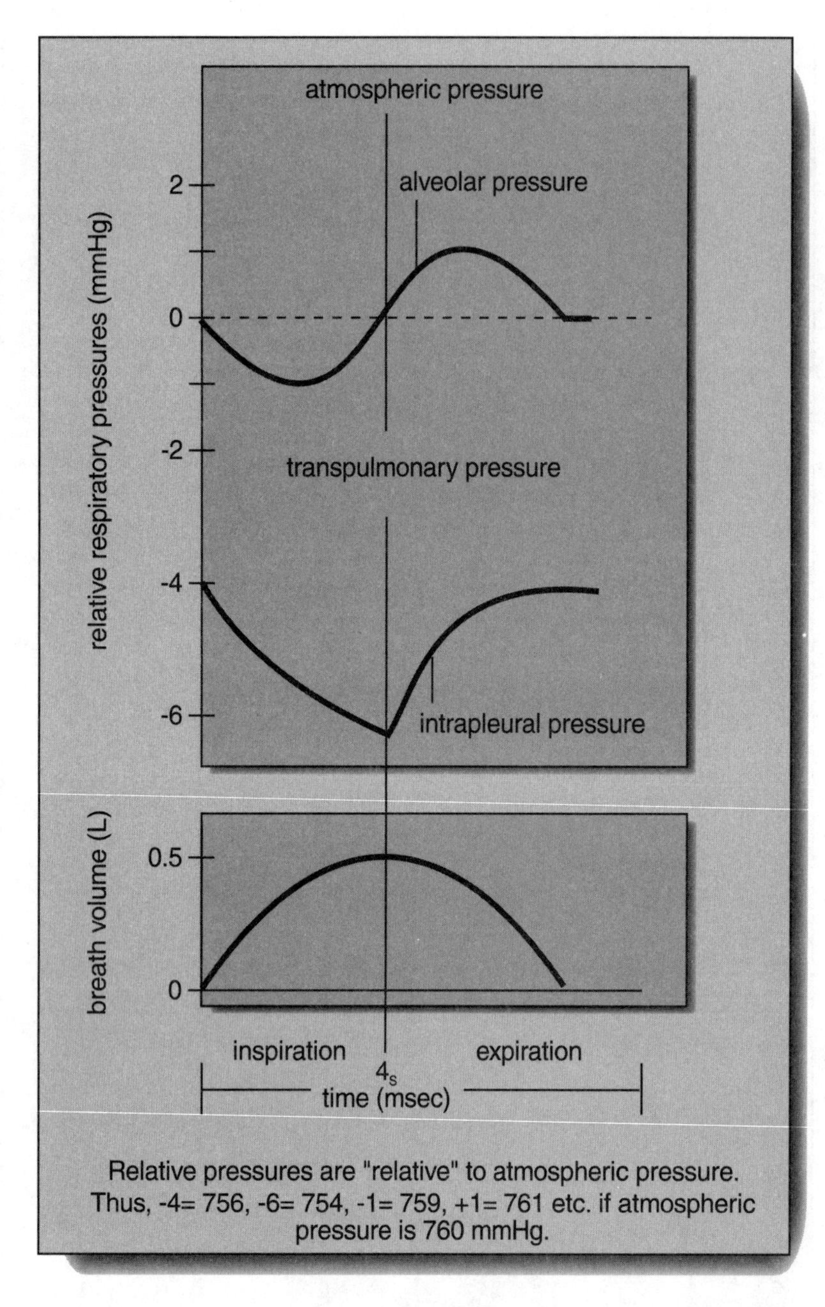

FIGURE 11.6

outwards. This combination causes the volume of the thoracic cavity to enlarge.

INFLATION OF THE LUNGS

Upon enlargement of the thoracic cavity, the interthoracic pressure decreases from 756 mm Hg (-4 mm Hg with respect to atmospheric) to 754 mm Hg (-6 mm Hg with respect to atmospheric). This reduction in interthoracic pressure from -4 to -6 means that the transmural pressure gradient will have changed by -2 mm Hg, and this causes the intra-alveolar pressure to fall to 759 mm Hg. Accordingly, there is now a pressure gradient of -1 mm Hg (atmospheric is 760 mm Hg and interalveolar is 759 mm Hg) directed inwards which causes air to flow into the lungs resulting in inflation. Inflation continues until about 500 ml of additional air have entered the lungs **(Figure 11.6)**.

As the lungs inflate, the elastic recoil of the lung wall increases. Eventually the outward force due to the pressure exerted by the air within the alveoli is balanced by the inward forces of the air pressure in the interthoracic space, plus the elastic recoil of the wall. Thus, the transmural pressure gradient is again reduced and the interalveolar pressure returns to 760 mm Hg. The inwardly directed pressure gradient disappears and flow of air into the lungs stops. Keep in mind that the pressure gradient driving air into the lungs must return to zero or the lungs will continue to fill until they explode.

DEFLATION OF THE LUNGS

Deflation of the lungs is basically the reverse of the processes leading to inflation. Thus, upon relaxation of the inspiratory muscles, the thoracic cavity *passively*, elastically rebounds to the original smaller volume causing the interthoracic pressure to increase to 756 mm Hg (again becoming -4 with respect to the atmospheric pressure of 760 mm Hg). This increase in interthoracic pressure, from -6 to -4, means that the transmural pressure gradient will have changed by +2 mm Hg and this causes the intra-alveolar pressure to rise to 761 mm Hg. Accordingly there is now a pressure gradient of +1 mm Hg directed outward causing 500 ml of air to flow out of the lungs.

As the lungs deflate, the outward force due to the pressure exerted by the air within the alveoli is balanced by the inward forces of the air pressure in the interthoracic space, plus the reduced elastic recoil of the wall. Thus, the interalveolar pressure returns to 760 mm Hg, the outwardly directed pressure gradient disappears, and flow of air out of the lungs stops.

AIRWAY RESISTANCE

Because $F = \Pi P/R$, as airway resistance increases flow decreases. Airway resistance is regulated by bronchial smooth muscle constriction or dilation which changes the radius of the bronchioles. The relationship of resistance and radius is the same as in blood vessels, in that $R = r^4$.

Bronchiolar smooth muscle is under two levels of control: systemic and local. *Systemic control* consists of autonomic nerve innervation with branches from both the *sympathetic and parasympathetic divisions*. Upon *sympathetic stimulation* (and release of the neurotransmitter norepinephrine), bronchodilation leads to increased air flow as a result of reduced resistance. Additionally circulating epinephrine (adrenalin from the adrenal cortex) will also bring about significant bronchodilation. On the other hand, stimulation of the *parasympathetic division* causes bronchoconstriction due to the decreased radius and increased resistance of the bronchioles.

Various factors can also act to exert local control on bronchiolar smooth muscles. For example, decreasing levels of CO_2 will cause bronchoconstriction and decreased air flow to these local alveoli, while increased levels of CO_2 will cause bronchodilation and increased air flow to these local alveoli (see below). Additionally, during disease processes, bronchoconstriction may take place due to allergic reactions because cells involved in immune response release slow reactive substances, resulting in anaphylaxis, and histamine release. Narrowing and/or blockage of the airways during disease may also result from increased mucus secretions, or edema in the walls which also will reduce air flow.

MATCHING AIR AND BLOOD FLOW

In addition to local control of the air way diameter, local control of blood flow also takes place in blood vessels that supply alveoli. Since gasses of the air within the alveoli must exchange with blood supplied to the alveoli, there is an optimum level of air flow and blood flow to each alveolus that maximizes this exchange process. As explained, air flow can be regulated by local CO_2, but it is also important to add that blood flow can be regulated by the local concentration of O_2. Increased levels of O_2 will result in dilation of the arteriolar smooth muscle supplying the capillaries of these local alveoli and blood flow

in this region will increase. Decreased levels of O_2 will result in constriction of these arterioles and decreased blood flow to these local areas.

Example No. 1. In an area of the lungs where blood flow is greater than air flow:

1) Initially the small air flow increases CO_2 levels locally, which relaxes local bronchial smooth muscle, decreases resistance and increases air flow.

2) Initially the large blood flow decreases levels of dissolved O_2 locally, which constricts arteriolar smooth muscle, increases resistance and decreases blood flow.

3) The outcome thus allows the air and blood flow to match each other more effectively leading to a more efficient exchange of gases between the local alveoli and the local capillaries.

Example No. 2. In an area of the lungs where air flow is greater than blood flow:

1) Initially the large air flow decreases CO_2 levels locally, which constricts local bronchial smooth muscle, increasing resistance and decreasing air flow.

2) Initially the small blood flow increases levels of dissolved O_2 locally, which dilates arteriolar smooth muscle, decreases resistance and increases blood flow.

3) The outcome thus allows the air and blood flow to match each other more effectively leading to a more efficient exchange of gases between the local alveoli and the local capillaries.

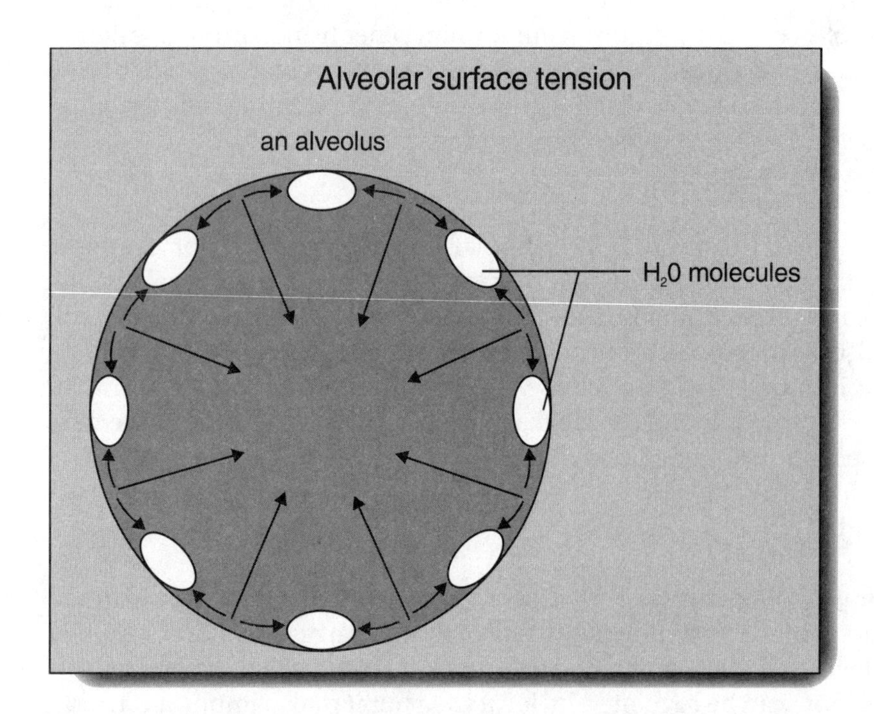

FIGURE 11.7

ELASTIC RECOIL AND COMPLIANCE OF THE LUNGS

Changes in the transmural pressure gradient are responsible for the ability of the lungs to inflate and deflate. Remember that the transmural pressure gradient results from the difference between the alveolar pressure and interthoracic pressure acting across the lung wall. However, elastic recoil also plays an important role in the process of deflation. Elastic recoil makes the lungs rebound to their preinspiratory volume when the inspiratory muscles relax.

Take a look at the forces acting on the lungs:

1. Outwardly directed alveolar pressure causes the lungs to inflate.
2. Inwardly directed interthoracic pressure acts against inflation.
3. Inwardly directed elastic recoil acts against inflation.
4. Surface tension acts against inflation.

Naturally, if the outward force is greater than the sum of the inward forces the lungs will inflate, while if the inward forces are greater than the outward force, the lungs will deflate.

Elastic recoil is due to the physical construction of the lung tissue which functions much like a balloon trying to rebound to a smaller volume. *Compliance* is a measure of the elasticity of this lung "balloon" and is calculated as:

$$C = \pi V / \pi P$$

where: C = compliance

πV = change in volume

πP = change in pressure

As the lungs inflate, the elastic recoil increases in the same way that a balloon does as it is inflated; thus it becomes more difficult to blow up as the volume increases. Because the elastic recoil increases upon inflation, the inwardly acting force increases as the lung expands. Initially then the outward force is

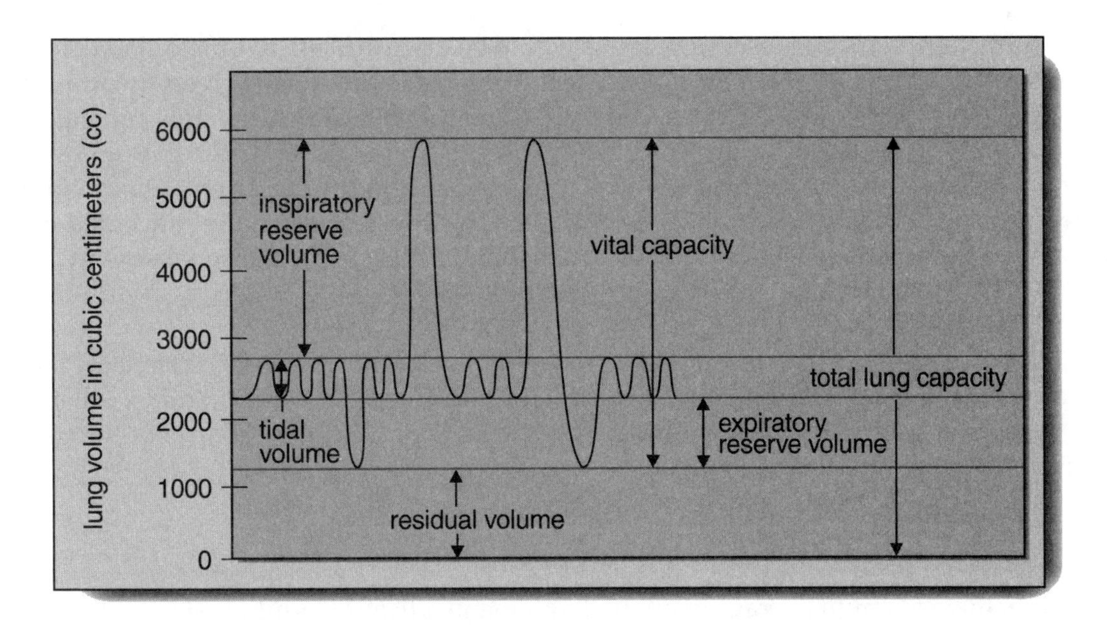

FIGURE 11.8

greater than the inward force and the lungs begin to inflate. However, as the volume in the lungs increases, the elastic recoil also increases. Eventually, the inwardly directed forces become equal to the outward force and the lungs stop filling with air.

ALVEOLAR SURFACE TENSION AND SURFACTANT

As has been mentioned, surface tension result from the cohesive forces exerted between adjacent water molecules at an air -water interface. Since each alveoli is lined with water and is filled with air, the huge surface area within the lungs would be expected to produce an enormous surface tension effectively preventing the lungs from expanding. Such surface tension will not only act to prevent the lungs from expanding, but will also attempt to cause the alveoli to collapse to their smallest volume **(Figure 11.7)**. Thus, the force produced by the surface tension can be considered as a component of lung compliance. In other words, as surface tension increases, elastic recoil also increases and compliance decreases.

To reduce surface tension to a level which will allow the lungs to expand easily, *Type II Alveolar cells* in the walls of the alveoli secrete the substance *Pulmonary Surfactant*. This mixture of lipoproteins act to reduce surface tension by reducing the attractive forces between the adjacent water molecules. Surfactant acts to:

1. Reduce elastic recoil and this prevents lung collapse.

2. Increase pulmonary compliance. Using an analogy of the lungs as a balloon, then as compliance increases the balloon becomes easier to blow up, thus it takes less work to inflate the lungs when surfactant is present.

ALVEOLAR STABILITY - FACTORS THAT PREVENT COLLAPSE

Obviously, if surfactant were not present, the lungs would be more difficult to inflate, and they would also be unstable and tend to collapse. This is because the collapsing pressure is inversely proportional to the radius, and therefore the smaller the alveoli, the greater the collapsing pressure generated. Thus, if two alveoli, one with a smaller radius than the other, are connected together through the same terminal bronchiole, the smaller alveolus will collapse and expand the larger one (this condition is described by the *Law of LaPlace*). This unstable condition, however, is normally prevented because surfactant reduces the surface tension sufficiently such that the collapsing pressure in the two alveoli is equal.

Alveolar stability is also maintained through the process of *Alveolar Interdependence*. Here neighboring alveoli are all connected together by connective tissues and if one alveolus starts to collapse, the neighboring ones are stretched. The surrounding alveoli now prevent the collapse of the neighbor because they resist being stretched and recoil.

LUNG VOLUMES

During inflation, the maximum average combined volume in a healthy adult, that can fill the lungs, is approximately 5,700 ml. During normal breathing the combined volume at the end of normal inspiration is about 2,700 ml, and at the end of normal expiration is 2,200 ml (which is a difference of

500 ml of *tidal volume*). Finally, the absolutely smallest volume that can be attained upon forced exhalation is 1,200 ml, and this is known as the *residual volume*. These lung volumes can be measured using a spirometer and are presented in **Figure 11.8**. A complete listing of lung volumes and their calculation is presented here.

Tidal Volume (TV) — is the volume of air that enters and leaves the lungs during a single breath, measured when the person is resting. TV = 500 ml.

Inspiratory Reserve Volume (IRV) — is the maximum amount of additional air that can be inhaled above the TV. IRV = 3,000 ml.

Expiratory Reserve Volume (ERV) — is the absolutely greatest volume of air that can be exhaled using maximal contraction of expiratory muscles. ERV = 1,000 ml.

Residual Volume (RV) — is the air that remains in the lungs after maximal expiration. This air cannot be removed unless the lungs collapse during a pneumothorax. RV = 1,200 ml.

Inspiratory Capacity (IC) — is the maximal volume of air that can be inhaled if the person first exhales normally and then inhales maximally. IC = 3,500 ml (IC = IRV + TV; IC = 3,000 ml + 500 ml).

Functional Residual Capacity (FRC) — is the volume of air remaining in the lungs after a normal relaxed expiration is completed. FRC = 2,200 ml (FRC = ERV + RV; FRC = 1,000 ml + 1,200 ml).

Vital Capacity (VC) — is the maximal volume of air moved in a single breath, that a person has control over. Thus, if you breath in as much as you can and then breath out as much as you can, the volume of air that leaves your lungs would be equivalent to the VC. VC = 4,500 ml (VC = IRV + TV + ERV; VC = 3,000 ml + 500 ml + 1,000 ml).

Total Lung Capacity (TLC) — is the maximum volume of air that is present in the lungs after maximal inspiration. TLC = 5,700 ml (TLC = VC + RV; TLC = 4,500 ml + 1,200 ml).

PULMONARY VENTILATION

Pulmonary ventilation *(PV)*, which is also known as the *minute respiratory volume*, is a measure of the total amount of air moved into and out of the lungs in one

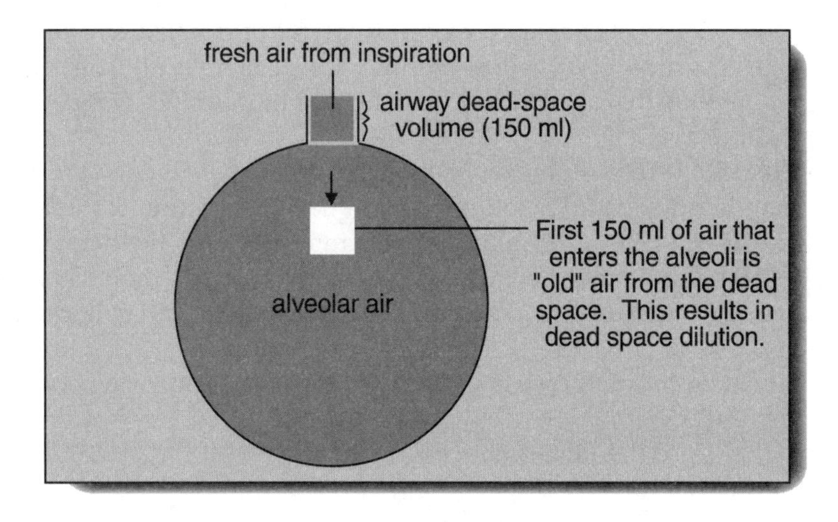

FIGURE 11.9

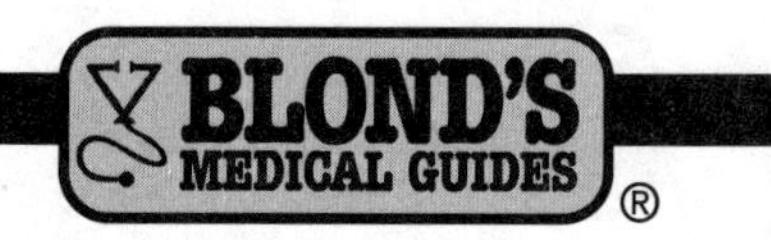

minute. You can calculate PV if you know the tidal volume (TV) and the respiratory rate (RR), thus:

PV (ml/min) = TV (ml) x RR (breaths/*min)*

If this reminds you of the calculation for cardiac output you are correct, because it is measuring a similar kind of parameter; thus: CO (ml/min) = SV (ml) x HR (beats/min).

However, knowledge of PV is not enough to determine if air from outside is effectively entering the alveoli where it can exchange with the blood. To ascertain if the outside air is ventilating the alveoli, calculate *Alveolar Ventilation*. To do this, you must also know the volume of *dead space* in the lungs.

DEAD SPACE

In order for air from outside to reach the alveoli, it must first pass through the conducting airways. But gas exchange does not take place here as it does in the alveoli. In an average adult, the volume of air that fills these conducting airways is 150 ml, and since this volume is built into the human organism, it is termed the *anatomical dead space*. In addition to the anatomical dead space present in all people, some individuals also have *alveolar dead space*, if they have regions of alveoli in their lungs that are non-functional.

Alveolar ventilation is a measure of the actual amount of gas that enters and leaves the alveoli and is also directly related to the amount of gas exchange taking place in the alveoli. To calculate alveolar ventilation, both the pulmonary ventilation and the dead space volume must be known. In the simplest of examples, if a person takes a single 500 ml breath, all of this volume does not enter the alveoli. Instead, the first air that enters the alveoli will be 150 ml of old, used air present in the dead space from the previous breath. Following this dead space air, 350 ml of new fresh air will also enter the alveoli and the dead space will now be filled with 150 ml of new fresh air. Thus, the outside air is diluted by the air in the dead space. Upon exhalation, the first air that leaves the airway is the 150 ml of fresh air in the dead space. Since dead space air does not exchange, this first 150 ml of exhaled air will be high in 0_2 and low in CO_2 (like the atmosphere). Following this will be 350 ml of air from the alveoli that has undergone exchange; thus this air will be low in O_2 and high in CO_2 **(Figure 11.9)**.

To calculate the volume of fresh air that enters the alveoli you must, therefore, subtract the dead space volume from the tidal volume. This equals the volume of fresh air actually ventilating the alveoli (500 ml - 150 ml = 350 ml). Taking this example one step further, if a person were to breath a tidal volume of 150 ml and have a dead space volume of 150 ml, then 150 ml - 150 ml = 0 ml of fresh air entering alveoli. Under these conditions, the individual would collapse for lack of O_2.

However, to correctly calculate alveolar ventilation, also consider the number of breaths, thus:

AV = (TV - DS) x RR

where:

AV = Alveolar Ventilation
TV = Tidal Volume

effect of different breathing patterns on alveolar ventilation

breathing pattern	TV	RR	DV	$\dot{V}$ = (TV) (RR)	$\dot{A}$ = (TV - DV) RR
normal, quiet breathing	500	12	150	6000	4200
deep, slow breathing	1200	5	150	6000	5250
shallow, rapid breathing	150	40	150	6000	0

TV = Tidal Volume (ml/min)

RR = Respiratory Rate (breaths/min)

DV = Dead - Space Volume (ml)

$\dot{V}$ = Pulmonary Ventillation (ml/min)

$\dot{A}$ = Alveolar Ventillation (ml/min)

DS = Dead Space Volume (Usually 150 ml)
RR = Respiratory Rate

An example of a standard calculation for AV follows:

given:

TV = 500 ml
DS = 150 ml
RR = 12 breaths/min

then: AV = (500 ml - 150 ml) x 12 breaths/min
AV = 4,200 ml/min

The effects of different breathing patterns on alveolar ventilation is presented in the **Table**. Clearly, breathing slowly and deeply significantly increases the efficiency of alveolar ventilation, while shallow rapid breathing reduces alveolar ventilation. (This physiological truth has probably been known to serious athletes for centuries.)

It should also be pointed out that as alveolar dead space increases (with disease), total dead space will become greater than 150 ml, and this will decrease the efficiency of alveolar ventilation. For example, in the calculation above if TV = 500 ml, Total dead space = 500 ml, and RR = 12 breaths/min, then AV = 0 (and not 4,200 ml/min). Clearly, such a patient would have to breath a tidal volume much greater than 500 ml (for example, 800 ml or 900 ml) to effectively ventilate the alveoli. This significantly increases the work of breathing and seriously weakens the patients.

GAS AND PARTIAL PRESSURE

As described by *Dalton's law*, in a mixture of gases each gas is has its own partial pressure; however, the sum of all the individual partial pressures is the total pressure of the mixture. For example, supposing that in a closed container the pressure of O_2 alone was 40 mm Hg, and N_2 is added at a pressure of 60 mm Hg, and CO_2 at a pressure of 70 mm Hg. The partial pressure of O_2 in the mixture remains 40 mm Hg (P_{O2} = 40 mm Hg) and similarly PN_2 = 60 mm Hg, and PCO_2 = 70 mm Hg. The sum of these three partial pressures is equal to the total pressure of 170 mm Hg, or:

$$P_{Total} = P_{O2} + P_{N2} + P_{CO2}$$
$$P_T = 40 \text{ mm Hg} + 60 \text{ mm Hg} + 70 \text{ mm Hg}$$
$$P_T = 170 \text{ mm Hg}$$

In addition, since P_T is the sum of all the partial pressures of the gases in the mixture it is equal to 100%, and thus each gas in the mixture comprises a percent of the total; so if P_{O2} = 40 mm Hg it must be equivalent to 23.5% of the total gas pressure in the mixture:

$$100\%/170 \text{ mm Hg} = X\%/40 \text{ mm Hg}$$
$$(40)(100\%) = 170X$$
$$X = 23.52\%$$

A similar calculation can be made for the other gases in the mixture, thus:

N_2 = 35.29%, CO_2 = 41.17%,

and: $\%O_2 + \%N_2 + \%CO_2$ = % total, or:

23.52 + 35.29% + 41.17% = 99.98%

This is indeed very close to the expected 100%.

Given the fact that the atmosphere is composed of a mixture of gases, a similar calculation of partial pressures and percents can also be made:

Thus, the total atmospheric pressure (P_T) is composed of a mixture of N_2, O_2, CO_2, as well as trace gases (helium, argon, neon, xenon, etc) which can be described as P_{other}:

$P_T = P_{O2} + P_{N2} + P_{CO2} + P_{other}$, and

P_T = 760 mm Hg (at sea level)

P_{O2} = 159.6 mm Hg

P_{N2} = 592.8 mm Hg

P_{CO2} = 0.3 mm Hg

P_{trace} = 7.3 mm Hg

Therefore, the partial pressures of the individual gases must be equal to the total atmospheric pressure:

152 mm Hg + 592.8 mm Hg + 0.3 mm Hg + 14.9 mm Hg = 760 mm Hg

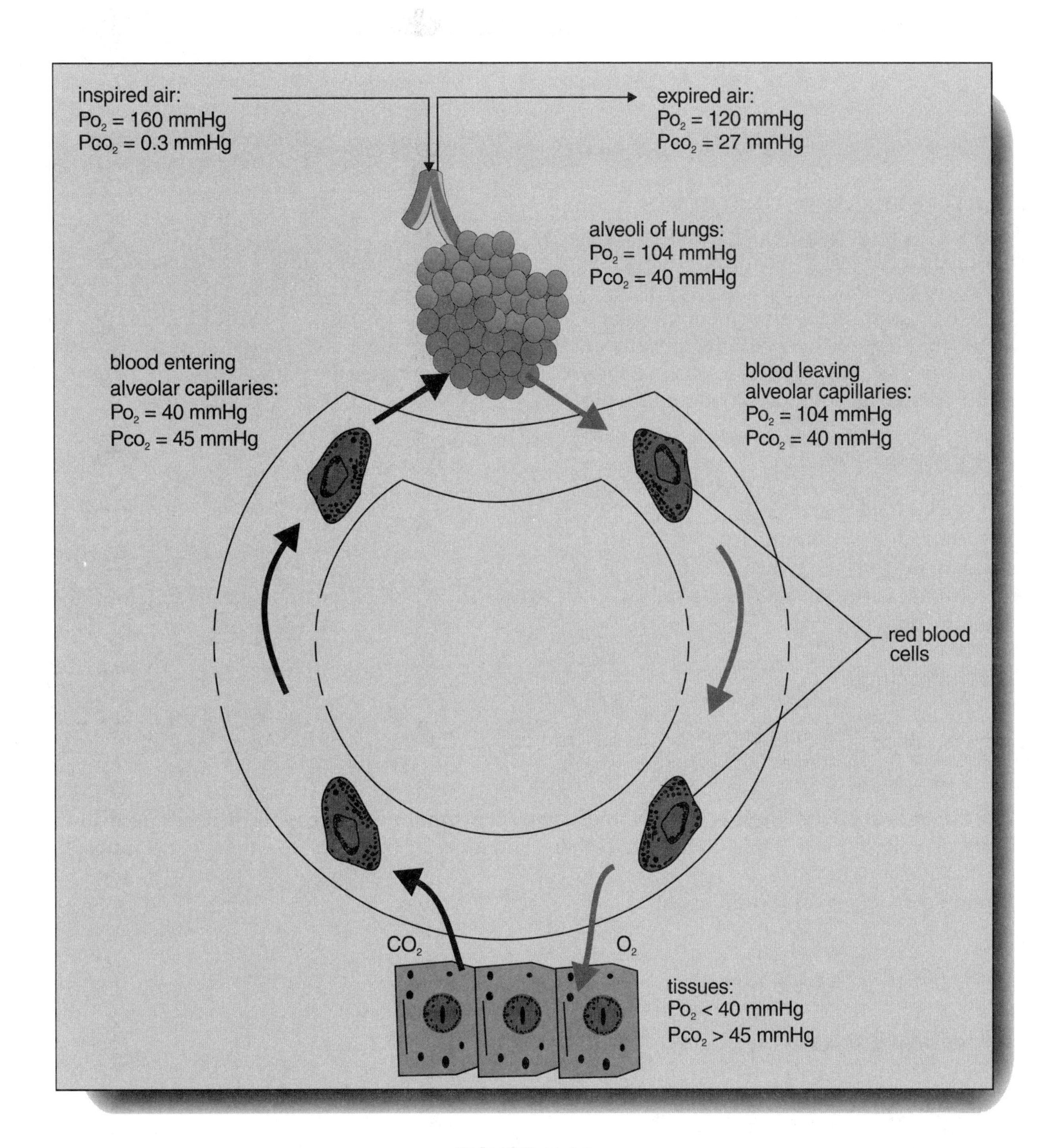

FIGURE 11.10

Since the partial pressure of each gas in the atmosphere is a component of the total pressure, if $P_T = 760$ mm Hg and is 100%, then the percent of the other gases in the atmosphere are:

O_2 = 21%, N_2 = 78%, CO_2 = 0.039%, trace gases = 0.96%

and:

21% + 78% + 0.039% + 0.96% = 99.999%, or 100% as expected.

PARTIAL PRESSURE GRADIENTS

Gases always move down their partial pressure gradients from higher to lower partial pressures. Thus, if the pressure of O_2 in the alveoli is 100 mm Hg, and the pressure of O_2 in the blood is 40 mm Hg, then O_2 will move from the alveoli into the blood until the partial pressure of O_2 in the blood is equivalent to 100 mm Hg. At this point the movement will stop because there is no longer a pressure gradient. To clarify this concept further, observe **Figure 11.10** which outlines the partial pressures of the gases in the atmosphere, lungs, blood and tissues.

Inspired air:

P_{O2} = 160 mm Hg
P_{CO2} = 0.3 mm Hg

These values are calculated as described above, where P_{O2} is 21% of 760 mm Hg, and P_{CO2} is 0.039% of 760 mm Hg.

Alveoli of lungs:

P_{O2} = 104 mm Hg
P_{CO2} = 40 mm Hg

The difference between these values and those present in the atmosphere is due to dead space dilution. Thus, alveolar P_{O2} is less than in the atmosphere, and alveolar P_{CO2} is greater than in the atmosphere.

Blood entering alveolar capillaries:

P_{O2} = 40 mm Hg
P_{CO2} = 45 mm Hg

Blood leaving alveolar capillaries:

P_{O2} = 104 mm Hg
P_{CO2} = 40 mm Hg

Alveolar P_{O2} = 104 mm Hg, but the P_{O2} in the blood (from the venous side) entering the alveolar capillaries is less (P_{O2}= 40 mm Hg). The resulting gradient drives O_2 from the lungs into the blood. This movement continues until the blood P_{O2} equilibrate with the alveolar P_{O2} and the movement stops. Thus, the blood leaving the lungs in the alveolar capillaries has a P_{O2} level similar to that of the alveoli.

A similar situation exists for P_{CO2}, but here the gradient is in the direction that drives the gas from the blood into the lungs. Thus, the blood entering the lungs has a P_{CO2} concentration of 45 mm Hg (similar

to the venous P_{CO2} level), but the P_{CO2} in the alveoli is 40 mm Hg. Thus, the gas moves out of the blood and eventually the blood levels equilibrate with the alveolar concentration of 40 mm Hg which then leaves the lungs.

Note that the gas concentration in the alveoli never changes because the lungs are being ventilated and are in contact with the atmosphere.

Tissues:

P_{O2} = <40 mm Hg
P_{CO2} = >45 mm Hg

Gas levels in the tissues result from metabolism, thus O_2 is being used up and CO_2 is being generated. Since the level of P_{O2} in the arterial blood from the lungs is 104 mm Hg, O_2 moves from the blood into the tissues until the P_{O2} level is equivalent to the tissues. Venous blood leaving the tissues, therefore, has a P_{O2} level that is similar to that in the tissues (40 mm Hg). Additionally, CO_2 moves out of the tissues (where it is generated) into the blood until the blood level is similar to that of the tissues (P_{CO2} = 45 mm Hg). The blood in the venous side than travels back to the lungs to begin the whole process over again.

FACTORS THAT INFLUENCE EXCHANGE OF GASES

Obviously one very important factor that affects gas exchange is the magnitude of the partial pressure gradient. However, various other factor can also influence gas exchange, as follows:

1) The surface area available for gas exchange from alveoli to capillary. During exercise this surface area increases because the increased blood pressure forces open additional pulmonary capillaries that normally are closed during resting conditions.

2) Thickness of the exchange membrane.

3) The *diffusion coefficient* (D) of the gas that is undergoing the exchange. D is related to the solubility of the gas that is undergoing the exchange. This value is 20 fold greater for CO_2 than for O_2, and thus the rate of the diffusion of CO_2 across the membrane is 20 times greater than O_2. However, because the P_{O2} gradient is significantly greater than P_{CO2} gradient, the driving

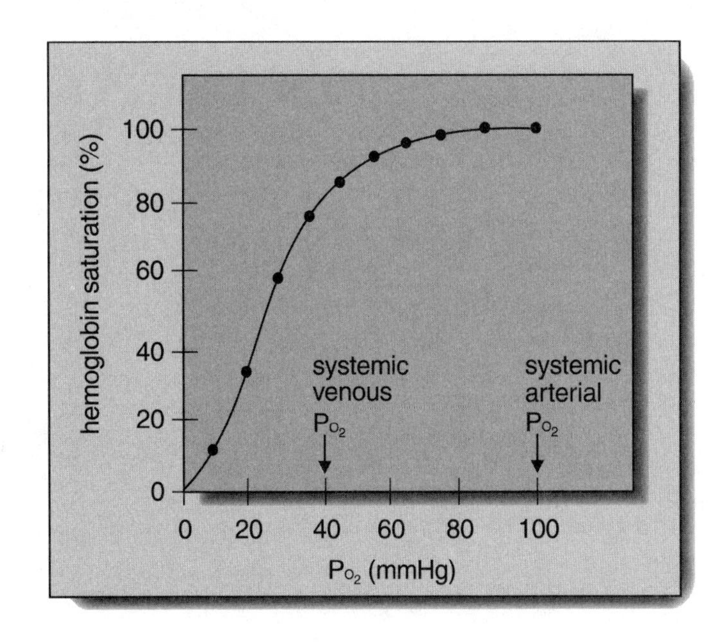

FIGURE 11.11

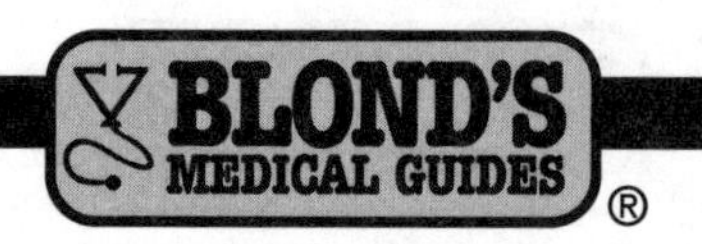

force moving O_2 across the membrane is greater than it is for CO_2. Therefore, the increased D for CO_2 is offset by the decreased P_{CO2} gradient, and the decreased D for O_2 is offset by the increased P_{O2} gradient. The outcome is that under normal conditions approximately equal amounts of O_2 and CO_2 are exchanged.

TRANSPORT OF O_2 FROM THE LUNGS TO THE TISSUES

There are two methods by which O_2 is transported in the blood from the lungs to the tissues. It can be dissolved in the water of the plasma. However, since O_2 is only slightly soluble in water, 100 ml of blood contains about 0.3 ml of this gas. Blood that contains this volume of dissolved O_2 will have a P_{O2} of 100 mm Hg. The majority of transported O_2 is not dissolved in the plasma water, but is instead physically bound to hemoglobin in the red blood cell. Thus, in 100 ml of blood approximately 20.1 ml of O_2 can be transported bound to hemoglobin. Since the total O_2 transported (dissolved + bound) is 20.4 ml/100 ml blood, then 1.5% is transported dissolved in the plasma and 98.5% is transported bound to hemoglobin. *Remember that only the dissolved gas has a partial pressure.*

CHARACTERISTICS OF O_2 - HEMOGLOBIN BINDING

Oxygen binds to the iron containing protein called hemoglobin present in erythrocytes. This binding process is reversible:

$$Hb + O_2 \leftrightarrow HbO_2$$

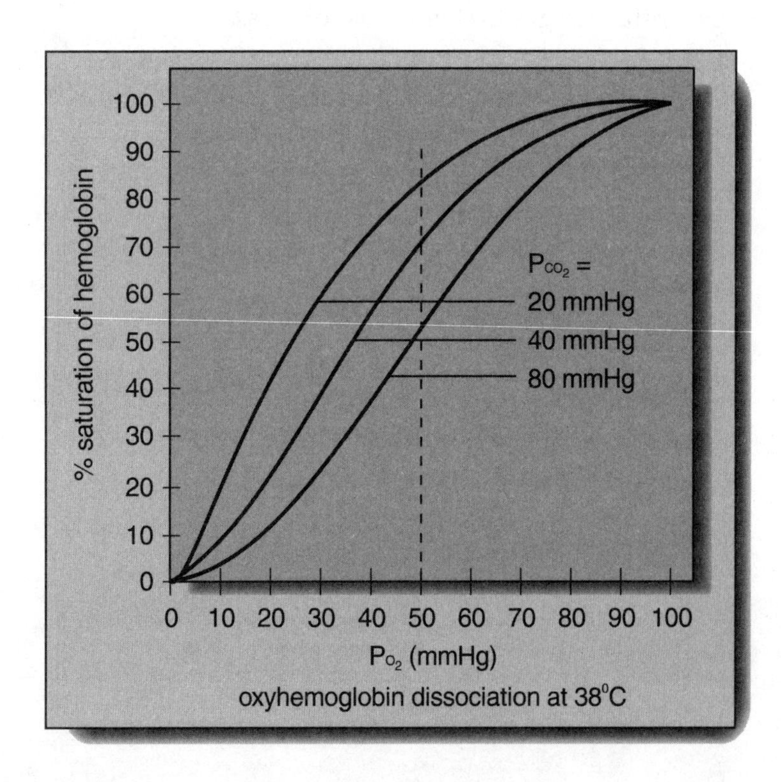

FIGURE 11.12

Here, Hb is the *reduced hemoglobin* and HbO_2 is called *oxyhemoglobin*. This reaction is driven by the *law of mass action* because if the concentration of one of the reactants in a reversible chemical reaction is increased, this drives the reaction towards the opposite side until a new equilibrium is obtained.

The importance of Hb as the major O_2 transporter is demonstrated if we consider that in a healthy adult (with a normal hematocrit) there is:

About 15 gm Hb/100 ml blood, and
The resting cardiac output is 5 lt/min, and therefore:
About 1,000 ml of O_2/min can be transported in the blood bound to the Hb in erythrocytes;
As compared to only 15 ml/min

that can be transported as dissolved gas in the plasma water.

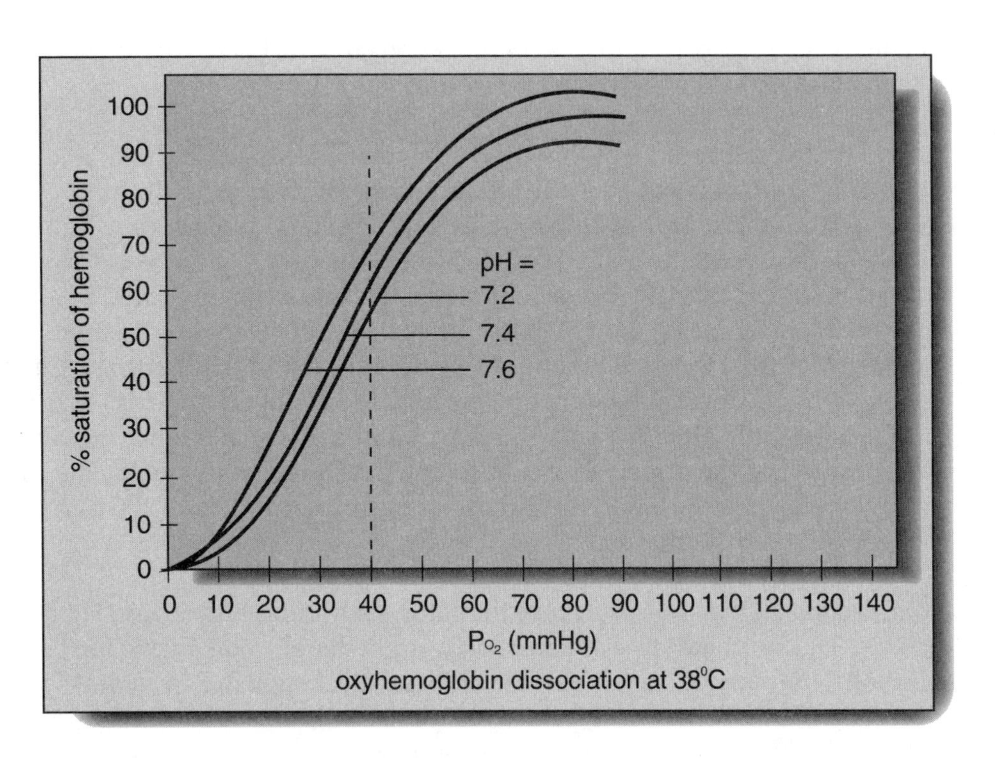

FIGURE 11.13

The binding process that takes place between the reduced form of Hb and O_2 is primarily driven by the partial pressure of the dissolved O_2. Thus, as P_{O2} increases, the amount of oxyhemoglobin formed also increases. A diagram of this is found in **Figure 11.11**. Note that the relationship between the dissolved P_{O2} (on the X-axis) and the percent hemoglobin saturation (on the Y-axis) is not linear, but is instead sigmoidal.

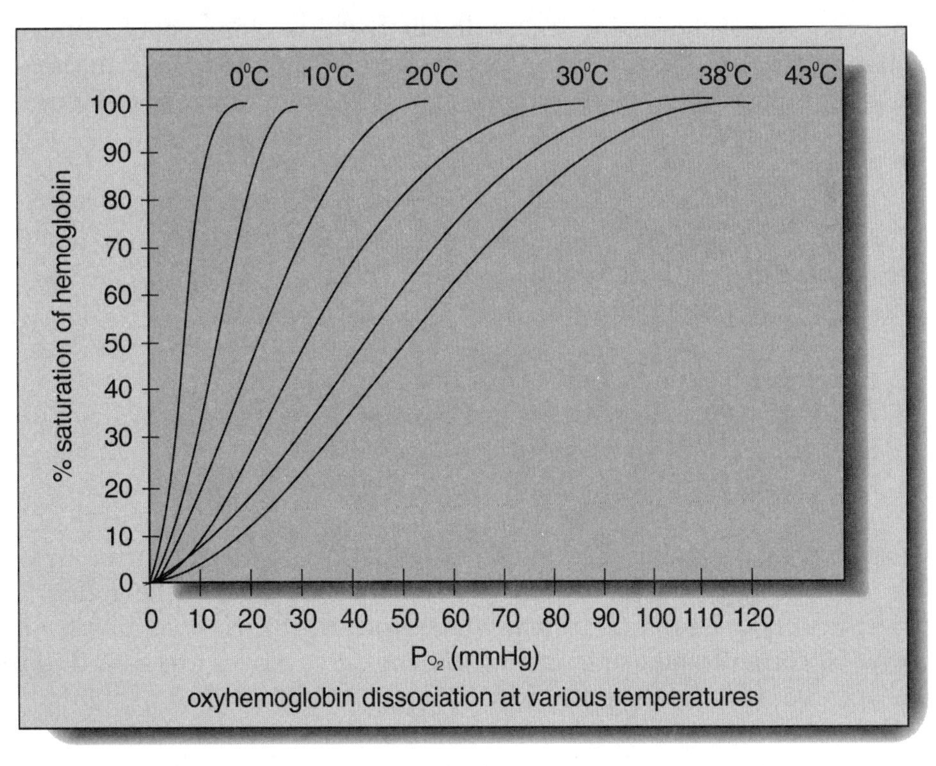

FIGURE 11.14

The Percent Hb Saturation. It is a measure of the amount of Hb that has combined with O_2. Thus, at 50% saturation, 50% of the available Hb is bound with O_2 and 50% of the Hb is unbound (free); at 30% Hb saturation, 30% of Hb is bound and 70% is free, and so on.

Clearly as P_{O2} increases, so does saturation. However, a plateau re-

gion in the binding exists at P_{O2} levels between 60 mm Hg and 100 mm Hg, while there is a steep slope at P_{O2} values between 0 mm Hg and 60 mm Hg.

To explain the importance of this sigmoidal relationship, consider that at the lung level the P_{O2} in the blood is 100 mm Hg, and thus the Hb saturation is close to 100%, but even if the blood P_{O2} falls well below 100 mm Hg, (for example, during pulmonary disease or when the person goes to higher altitudes), Hb continues to carry almost a full load of O_2. For example, at a P_{O2} of 60 mm Hg Hb saturation is 90%, and at a P_{O2} of 40 mm Hg, Hb saturation is about 75%. This mechanism builds in a safety net (so to speak) for the body because relatively wide fluctuations in arterial P_{O2} will not significantly affect the total amount of O_2 transported to the tissues (i.e., the carrying capacity). This also explains why breathing pure O_2 (resulting in a P_{O2} of 600 mm Hg) only increases the Hb saturation by 10% (since it is already nearly maximized at a P_{O2} of 100 mm Hg), although it does increase the concentration of the dissolved gas in the blood. Functionally, then, the tissues receive an adequate supply of O_2 when the blood P_{O2} is in the range of 60 mm Hg to 100 mm Hg because the Hb remains between 90% - 99% saturated.

Note also that at the lower ranges of the curve (between 0 and 60 mm Hg P_{O2}), a small drop in systemic capillary P_{O2} can make large amounts of O_2 immediately available to the tissues. This is because within this range, the curve is steeply sloping and a small change in P_{O2} results in a large change in the percent saturation of Hb.

EFFECTS OF H^+, P_{CO2} AND TEMPERATURE ON O_2-HB BINDING

The processes of association of O_2 with Hb, and dissociation of O_2 from Hb, driven by the law of mass action, is not only effected by the concentration of the reactants in the mixture, but is also regulated by certain physical parameters—namely, H^+, P_{CO2} and temperature **(Figures 11.12, 11.13 and 11.14)**.

H^+ AND P_{CO2}

H^+ and P_{CO2} are actually "two sides of the same coin." That is to say that they are intimately related to each other through the law of mass action. This relationship will be described in detail below. Suffice it to say at this point that:

Increasing P_{CO2}, increases H^+ (pH falls; more acidic)
Decreasing P_{CO2}, decreases H^+ (pH rises; more basic)
Increasing H^+ (pH falls, more acid), increases P_{CO2}
Decreasing H^+ (pH rises, more basic), decreases P_{CO2}

Furthermore, the ability of Hb to associate/dissociate with O_2 turns out to be directly effected by P_{CO2}-H^+. Accordingly, with an increasing P_{CO2} concentration, (or the greater the H^+ concentration, or the lower the pH) the saturation curve moves towards the *right*, and with a lower P_{CO2} concentration, (or lower H^+ concentration, or increased pH) the curve moves to the *left* **(Figures 11.12 and 11.13)** because the affinity of Hb for O_2 is altered.

To clarify this point, look at **Figure 11.12 or 11.13** and consider what takes place at a P_{O2} of 40 mm Hg (although other P_{O2} values could be used just as effectively in this example). Draw a line up from 40

mm Hg P_{O2} value on the X-axis until it intersects the control curve (in **Figure 11.12** this is the curve at 40 mm Hg P_{CO2}; in **Figure 11.13** this is the pH 7.4 curve). Now cross to the Y-axis and note the percentage of saturation. For both curves, this is about 75% saturation. Thus, at the control levels of P_{CO2} and pH about 75% of the Hb is saturated (and thus 25% is free). Now repeat this process, but note the intersection point on the lower curves (in **Figure 11.12,** this is the 80 mm Hg P_{CO2} curve, and in **Figure 11.13** this is the pH 7.2 curve). Again draw a line from the intersection points on these curves to the Y-axis and note that now saturation is about 60% (40% is free). Similarly if you perform this with upper curves (20 mm Hg P_{CO2}, pH 7.6), the saturation is about 80% (20% free). Clearly shifting the curves left or right changes the percent of saturation of Hb and the amount of free Hb.

Temperature Effects. As shown in **Figure 14** a similar effect can be elicited with temperature. Thus, the curves shift to the right as the temperature increases and, and to the left as temperature decreases.

The Effect of Diphosphoglycerate. The curve will also be shifted to the right in the presence of the substance 2,3,- diphosphoglycerate (DPG). This substance, produced by erythrocytes, also reduced the affinity of Hb causing increased O_2 unloading. Since production of DPG is inhibited by the presence of HbO_2, under conditions when Hb in the arterial blood is under-saturated (for example at high altitudes) increased levels of DPG would be present. This would then assist

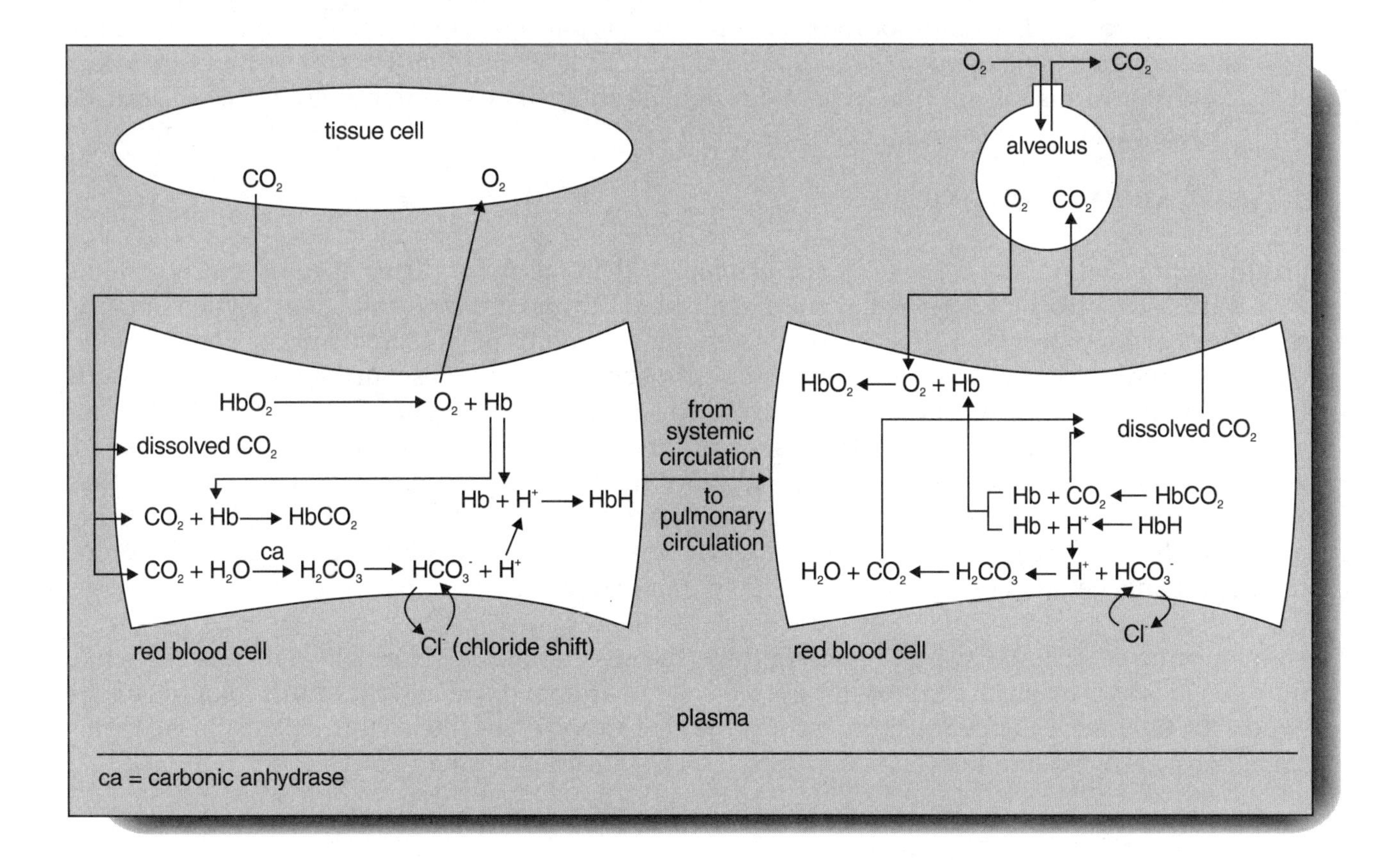

FIGURE 11.15

in increasing O_2 availability in the tissues by increasing unloading of O_2. Elevated levels of DPG are also present in pregnant women and this may assist in the transfer of O_2 from the mother to the fetus. While DPG assists in increasing O_2 availability at the tissue level, it also reduces the ability of Hb to bind O_2 at the lungs. This is unlike the effects elicited by temperature and P_{CO2}-H^+, which increase O_2, unloading at the tissues but not the lungs.

Shifting the Curve to the Right. The Hb saturation curve shifts to the right in the presence of increased levels of P_{CO2} and increased H^+ (more acid). When the curve shifts to the right as a result of these particular parameters (P_{CO2}, H^+), this is known as the *Bohr effect*. Additionally, increased temperature or the presence of DPG can also cause the curve to shift to the right. The outcome of shifting the curve to the right is that Hb holds less O_2 (more is free, less is bound). To put it another way, more O_2 is unloaded from Hb because the affinity of Hb is reduced under these conditions. The arrangement is eminently suited to supply O_2 to the tissues (where it is needed to support metabolism) because at the tissue level there is increased P_{CO2} - H^+ and elevated temperature.

Shifting the Curve to the Left. The Hb saturation curve shifts to the left when the levels of P_{CO2} - H^+ are reduced and when temperature is lower. By shifting the curve to the left, Hb affinity is increased and Hb binds more O_2 and releases less O_2. At the lungs P_{CO2} is being removed from the blood and transferred to the alveoli. Thus, at the lungs the concentration of P_{CO2} is reduced (H^+ levels are also reduced). In addition, blood temperature is lower (as compared to the tissues) because heat is being transferred to the air that is exhaled. Thus, at the lungs Hb affinity for O_2 increases and the formation of HbO_2 is maximized, meaning that effective volumes of O_2 can be transported to the tissues.

Hb AND CARBON MONOXIDE

Hb affinity for Carbon Monoxide (CO) is much higher than it is for O_2. Thus, if equal concentrations of O_2 and CO are present, the Hb will preferentially bind CO and not O_2, forming not oxyhemoglobin, but instead *carboxyhemoglobin* (*HbCO*). Thus, very small amounts of CO can disproportionately bind to large amounts of Hb and therefore prevent O_2 transport to the tissues. Furthermore, HbCO shifts the saturation curve to the left and therefore reduces tissue unloading of whatever O_2 still remains bound to the Hb. The outcome is that the tissues rapidly become anoxic, which if not quickly reversed, can result in death. Additionally, an individual suffering from CO poisoning has no warning because the O_2 receptors that signal an increase in respiration are not effected by CO poisoning (see following material). The outcome is that in CO poisoning, victims do not feel breathless, and suddenly become dizzy and faints before they can get out of the contaminated area.

An interesting adjunct in CO poisoning relates to the color of HbCO. Normally, HbO_2 is red while reduced Hb is bluish, which accounts for the red color of arterial blood and the bluish color of venous blood. HbCO, however, appears cherry red and thus in CO poisoning, the victim appears flushed. This can be most easily seen in the nail beds which appear much pinker than normal.

TRANSPORT OF CO_2 FROM THE TISSUES TO THE LUNGS

There are three methods by which CO_2 is transported from the tissues to the lungs. Namely, about 7% as P_{CO2} dissolved in the plasma water, about 23% which is physically bound to the Hb protein and is called *carbamino hemoglobin*; and about 70% which is transported through the *chloride shift* mechanism in the form of Bicarbonate Ions.

Dissolved P_{CO2}. CO_2 dissolved in the plasma water has a partial pressure (P_{CO2}) that can be measured in mm Hg. Thus, in the venous blood the P_{CO2} is about 46 mm Hg, and in the arterial blood the P_{CO2} is about 40 mm Hg. This dissolved CO_2 is also capable of stimulating the respiratory chemoreceptors (see following material and **Figure 11.15**).

Carbamino Hemoglobin. CO^2 can be physically bound to the globin protein of the Hb as $HbCO_2$. Note that the CO_2 binds at a different location (the globin) than does O_2 which binds to the iron of the heme portion of the Hb. Theoretically, both O_2 and CO_2 could be transported at the same time on a molecule of Hb, since they bind at different locations. However, in practice O_2 is transported from the lungs to the tissues where it is unloaded, while CO_2 is transported from the tissues to the lungs. This is facilitated because reduced Hb has a greater affinity for CO_2 than oxidized Hb. Thus, as O_2 is unloaded from the Hb at the tissue level (due to the Bohr effect, and increased temperature), CO_2 can be bound to the Hb more effectively **(Figure 11.15)**.

The Chloride Shift Reaction. The majority of CO_2, generated at the tissues as a byproduct of metabolism, is transported by chemically converting it into another form. This chemical conversion takes place in the red blood cells (RBC) because they contain the essential enzyme called *carbonic anhydrase* (*CA*). The reaction is as follows:

$$CO_2 + H_20 \rightarrow CA \rightarrow H_2CO_3 \rightarrow H^+ + HCO_3^-$$

Where:

- CA = Carbonic Anhydrase
- H_2CO_3 = Carbonic Acid
- HCO_3^- = Bicarbonate Ion
- CO_2 = Carbon Dioxide
- H_20 = water
- H^+ = Hydrogen Ion

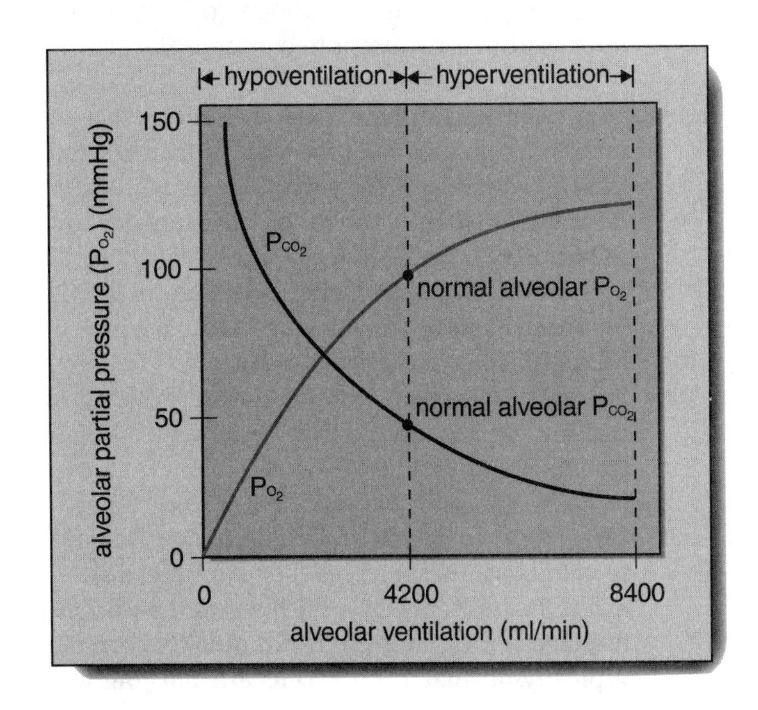

FIGURE 11.16

As can be seen from this equation (and as has been mentioned previously), as a result of the law of mass action, CO_2 and H^+ are intimately related since this is a reversible equation. Accordingly, dissolved CO_2 reacts with water in the RBC and is converted into carbonic acid, and

then into bicarbonate ions and hydrogen ions **(Figure 11.15)**. Additionally note that as HCO_3^- is generated from this reaction it passes out of the RBC into the plasma and Cl^- enters the RBC to balance electric charges. This accounts for the name *chloride shift reaction*. Furthermore, the H^+ generated can either pass into the plasma or be buffered on the hemoglobin protein ($Hb^- + H^+ \rightarrow HHb$). The increased concentration of H^+ in the venous blood accounts for the lower pH in the venous blood than in the arterial blood.

At the lung level, the steps in this reaction are reversed and CO_2, dissolved in the plasma, attached to the Hb protein, and generated from the reverse chloride shift reaction then passes into the alveoli to be exhaled from the body **(Figure 11.15)**. Note also that the release of CO_2 from the Hb protein promotes O_2 binding.

HYPOXIA

When the level of O_2 in the body falls, this is referred to as *hypoxia*. In other words, hypoxia takes place if the levels of O_2 being supplied to the tissues are insufficient to support the tissue needs. Different conditions that can all result in hypoxia:

Hypoxic Hypoxia. Takes place when the level of total arterial O_2 is reduced. Thus, both P_{O2} and HbO_2 are abnormally low. Such an effect can be brought about because the levels of O_2 in the inspired air is reduced (as it would be at high altitudes), or the respiratory membrane is damaged and O_2 cannot easily traverse it from the alveoli to the capillary blood, or the airway is blocked and air cannot enter the lungs, and so on. In this condition, the reduced levels of PO_2 will be sensed by the O_2 chemoreceptors (see following material) and respiration rate and depth will be increased in an attempt to compensate.

Anemic Hypoxia. Takes place when the levels of dissolved O_2 remain normal (arterial P_{O2} can be 100 mm Hg or greater), but the total O_2 transported is reduced because the amount attached to Hb is reduced (HbO_2 is reduced). The usual reason is that the individual is suffering from anemia, either because the number of RBCs are abnormally low (reduced hematocrit), or the amount of Hb/RBC is reduced (or both). Under these conditions, the individual will not be stimulated to increase respiration because as far as the O_2 chemoreceptors are concerned (see following material), the P_{O2} levels are quite normal. (*These receptors only monitor the dissolved gas concentration, since bound gas has no partial pressure.*) Such anemic individual will usually feel lethargic, but in an extreme situation this could become life threatening. For example, CO poising, as described above, generates a kind of acute anemic hypoxia because O_2 cannot be transported by the Hb.

Circulatory Hypoxia (or Ischemic Hypoxia). Takes place when blood flow to all tissues or only to selected tissues is reduced. For example, during a myocardial infarction, the drop in blood flow to the heart creates a local hypoxic condition in the heart muscle which can do permanent damage if the condition is not quickly corrected. During ischemic hypoxia, the arterial P_{O2} is typically normal, but the cells are starving for O_2.

Histotoxic Hypoxia. Takes place when O_2 cannot be utilized in the metabolic pathways at the cellular level. O_2 must act as the final electron acceptor in the electron transport chain present in the

mitochondria. If the enzymes in this chain are blocked by the action of a chemical poison such as ***cyanide***, O_2 cannot perform its function. The resulting significant loss in ATP production can kill the cell. However, while histotoxic hypoxia blocks the processes of internal respiration, arterial P_{O2} levels will be in the normal range.

CONTROL OF RESPIRATION

Changes in the rate and depth of respiration will alter the levels of the gases (P_{O2}, P_{CO2}) dissolved in the plasma water. As can be seen in **Figure 11.16**, at the optimum alveolar ventilation of 4,200 ml/min, arteriolar P_{O2} is 100 mm Hg and arteriolar P_{CO2} is 40 mm Hg. As the respiration rate increases (hyperventilation), P_{O2} levels rise, while P_{CO2} levels fall. Also as expected during reduced respiration (hypoventilation), P_{O2} levels fall while P_{CO2} levels rise. Note, however, that during exercise the increased metabolic demands of the tissues will generate excess CO_2 and use up O_2. This will lead to increased respiration and this is termed *hyperpnea*.

The regulation of respiratory rate and depth in conjunction with demands by the tissues is accomplished through a series of chemical receptors, coupled with specific regulatory areas in central nervous system. Within the brain stem there are three such centers: the *Pneumotaxic center*, the *Apneustic center* and the *Medullary respiratory center*.

Pneumotaxic and Apneustic Centers. These two centers are located in the Pons. The neurons in these regions exert control on the Dorsal Respiratory group (DRG) in the Medullary respiratory center. These pontine centers regulate the duration of activity generated by the DRG and in so doing, control depth of inspiration by the respiratory muscles. The Pneumotaxic center control the Apneustic center and also switches off the DRG to limit the duration of inspiration (see following material). In certain forms of brain damage involving the Pneumotaxic center, *apneusis* is present. The outcome is prolonged inspiration interrupted by very short periods of expiration.

Medullary Respiratory Center. Within the medulla there are two regions (or groups of neurons) called the *dorsal respiratory group* (*DRG*) and the *ventral respiratory group* (*VRG*).

1) The DRG consists of inspiratory neurons whose axons terminate on the motor neurons that control the respiratory muscles. Neurons of the DRG possess pacemaker-like activity —turning themselves first on, and then off again. Accordingly, activation of the DRG neurons results in inhalation, while exhalation takes place when the DRG neurons turn themselves off again. The basic rate of firing by the DRG inspiratory neurons is also influenced by inputs from the

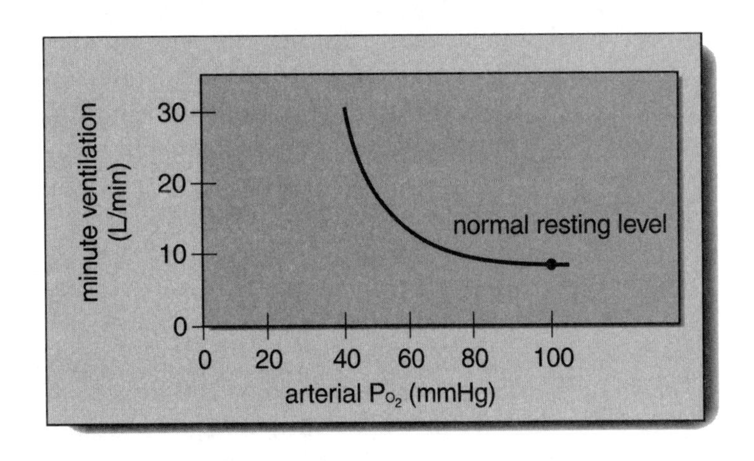

FIGURE 11.17

Pneumotaxic and Apneustic centers in the Pons, certain reflexes, chemoreceptors and stimuli from higher brain areas.

2) The VRG is comprised of both inspiratory and expiratory neurons and is believed to remain inactive during normal, quiet breathing. However, under conditions of exercise or emotional stress when increased demands are placed on the respiratory system, expiratory neurons in the VRG function to promote active expiration. Thus, the main function of the VRG is to activate expiratory muscles to cause active exhalation. Note that during normal quiet breathing, however, expiration is accomplished passively and the expiratory muscles are not involved. Other functions attributed to neurons in the VRG include: assisting the inspiratory neurons of the DRG to cause inspiration during periods of increased ventilatory demands; and promoting a negative feedback between neurons of the VRG and DRG to cause expiration. Thus, when the inspiratory neurons of the DRG fire, they also stimulate the expiratory neurons of the VRG. These expiratory neurons then inhibit the inspiratory neurons of the VRG and terminate inspiration.

The Hering-Breuer Reflex (Inflation Reflex). This reflex, originally identified in dogs, is actually a stretch reflex originating with stretch receptors located in the bronchiolar smooth muscle tissue, visceral pleura and alveoli. Accordingly, as the lungs inflate the stretch receptors are activated and send stimuli through the vagus nerve to the DRG. As the lungs inflate, the increasing stimuli acts to inhibit the DRG and thus terminate inflation of the lungs. While this reflex is highly sensitive in dogs, in humans the higher threshold means that this reflex is only of importance during periods of increased ventilation.

Other Reflex Controls Passively (or actively) moving the limbs can also increase ventilation, probably acting through pathways originating in the proprioceptive receptors in the joints, muscles and tendons that lead to the higher centers of the brain.

OTHER FACTORS THAT AFFECT RESPIRATION

The breathing pattern can also be altered by inputs from the higher centers generated by such emotions as fear, pain and also through an increase in body temperature. Such stimuli generally cause hyperventilation by acting through the sympathetic nervous system. Additionally, conscious control can be exerted to over-ride the medullary pattern of breathing, allowing active inhalation or active exhalation. Furthermore, prior to the onset of exercise, respiration increases, apparently due to the cerebral recognition of the impending exercise.

RESPIRATORY CHEMORECEPTORS

Chemoreceptors to control the rate of respiration are located both in the periphery and in the central nervous system.

The Peripheral chemoreceptors. They are located in the two carotid sinuses and in the aortic arch (carotid and aortic bodies), and nerve tracks from these receptors enter the brain stem and synapse in the respiratory centers. These peripheral chemoreceptors are sensitive to arterial blood concentrations of P_{CO2}, H^+ and P_{O2}. As pointed out, changes in P_{CO2} and H^+ are related by the law of mass action. An increase in either will stimulate an increase in the rate and depth of respiration, but the overall effect is weak as compared to the effect elicited by the *central chemoreceptors*. Peripheral chemoreceptors that monitor arterial P_{O2} are more effective in eliciting a response in the rate of respiration, but this effect does not take place until the blood levels of P_{O2} fall below 60 mm Hg **(Figure 11.17)**. Thus, the P_{O2} chemoreceptors are not important for minute to minute regulation of respiration. Instead, they function only during periods of hypoxic hypoxia when the dissolved arterial O_2 concentration fall significantly below the normal range.

Remember that these chemoreceptors do not monitor the concentration of any gas bound to Hb, but are only designed to monitor the partial pressure of the dissolved gas. Bound gases do not have a partial pressure.

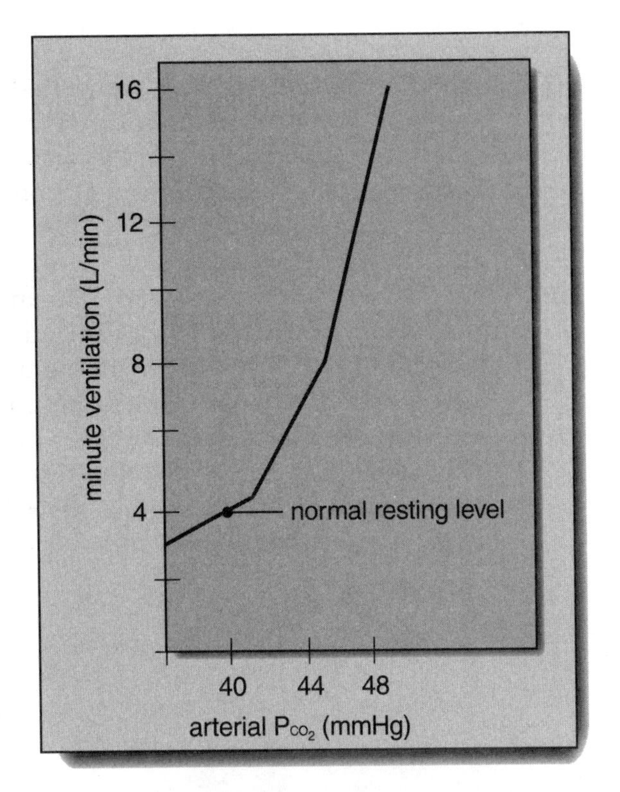

FIGURE 11.18

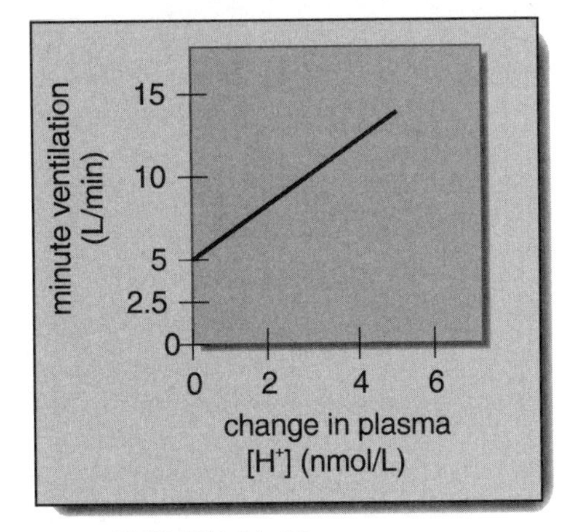

FIGURE 11.19

Central Chemoreceptors. These are located in the medulla and monitor the cerebral spinal fluid (CSF) space. They are not responsive to dissolved O_2, and only indirectly react to arterial P_{CO2} because P_{CO2} can be converted to H^+. Accordingly, as the levels of P_{CO2} rise in the arterial blood entering the brain, this gas passes by diffusion from the extracellular fluid (ECF) compartment into the CSF where it is then converted to H^+. As the H^+ concentration increase, this stimulates the central chemoreceptors and increases respiration. These central receptors are responsible for minute to minute control of respiration because they are very sensitive to small change in P_{CO2}-H^+ concentration **(Figure 11.18)**.

Although elevated levels of P_{CO2}-H^+ acting through the central receptors will increase respiration, very high levels of P_{CO2} can directly depress the entire brain. Thus, abnormally high levels of P_{CO2} (70 - 80 mm Hg) can actually depress respiration and also lead to severe respiratory acidosis which is life threatening (**see acid-base regulation described in Chapter 13**).

H+ Ions and Respiration. H^+ can be generated from CO_2 or can be produced during metabolism from other substances (such as lactic acid). However, unlike P_{CO2}, elevated blood levels of H^+ cannot penetrate the blood brain barrier and it cannot stimulate the central chemoreceptors that monitor the CSF. Instead, such elevated peripheral H^+ concentrations stimulate the peripheral chemoreceptors to increase respiration **(Figure 11.19)**.

HCO_3^- and the CSF. H^+ present in the CSF can be buffered (neutralized) by HCO_3^- ions that cross the blood brain barrier from the ECF. Under conditions where a patient is suffering from acute and long-standing CO_2 retention (during prolonged hypoventilation, for example), the presence of increased body H^+ may result in large amounts of HCO_3^- crossing into the CSF which then will neutralize the elevated H+ in the CSF compartment. Thus, in these patients, the elevated level of blood P_{CO2} is not stimulating respiration. Instead, the stimulus to breath is the hypoxic hypoxia that also accompanies their condition. Therefore, if you give these patients O_2 to breath, it may depress this stimulus and they will stop breathing completely.

Chapter 12
THE URINARY SYSTEM AND RENAL PHYSIOLOGY

FUNCTIONS OF THE URINARY SYSTEM

To maintain a stable internal environment, it is absolutely imperative that fluid and electrolyte balance be maintained, and that waste products be removed from the body. These tasks are performed by the kidneys whose functions are summarized as follows:

1. Regulation of water and electrolyte balance.

2. Removal of metabolic waste products from the blood and excretion into the urine.

3. Removal of foreign chemicals from the blood and excretion into the urine.

4. Metabolism or inactivation of some foreign chemicals or other substances from the blood or tubular fluid.

5. Indirect control of blood pressure by angiotensin II acting through release of the hormone renin.

6. Regulation of the $N^+/K^+/H^+$ blood levels by aldosterone, acting through angiotensin II and the release of the hormone renin.

7. Control of Ca^{++} balance in the body through the production of 1,25-dihydroxyvitamine D_3.

8. Stimulation of erythrocyte production through the release of the hormone erythropoietin.

THE KIDNEYS

The two kidneys are located *retroperitoneally*. Each kidney has a convex and concave side, and the single hilum is located in this concave region. It is here that the renal artery and vein are connected, along with nerves and lymphatic drainage. In addition, the *ureter* exits from this location to connect with the urinary bladder **(Figure 12.1)**. From the bladder the single *urethra* drains stored urine to the outside.

Each kidney is covered by a capsule and directly under this can be found the outer granular area called the *cortex*. Below the cortex is the *medulla* that contains the *renal pyramids*, as well as the cortical structures called *renal columns*. These renal columns project between the pyramids and each contains a branch of the *Interlobar artery and vein*. Deep within the center of each kidney is a hollow region called the *renal pelvis*. Along the margins, the renal pelvis narrows into a number of smaller funnel-like structures, each called a *Calyx*. It is here that urine, produced by the *nephrons* (see following material),

enters the pelvis from the *collecting ducts*. Once in the pelvis the urine will then be directed into the ureter and thence by peristaltic action, pumped into the bladder.

THE NEPHRON

Here is where filtration of blood takes place and where urine is manufactured. There are approximately one million nephrons in each adult kidney. The basic structure of a nephron consists of the *Renal corpuscle* (*Malpighian corpuscle*) and the *renal tubules*. In addition, each nephron connects to a collecting duct which is shared with other nephrons. These collecting ducts extend through the medulla and end at the Calyx of the renal pelvis **(Figure 12.2)**. Many collecting ducts lying together form the medullary structures, called the *renal pyramids*.

On examination **(Figure 12.2)**, it can been seen that the renal corpuscle has an outer covering called *Bowman's capsule* surrounding Bowman's space, and that in the center is the tuft of *glomerular capillaries*. Blood enters the glomerulus by way of the *afferent arteriole*, and leaves through the *efferent arteriole*.

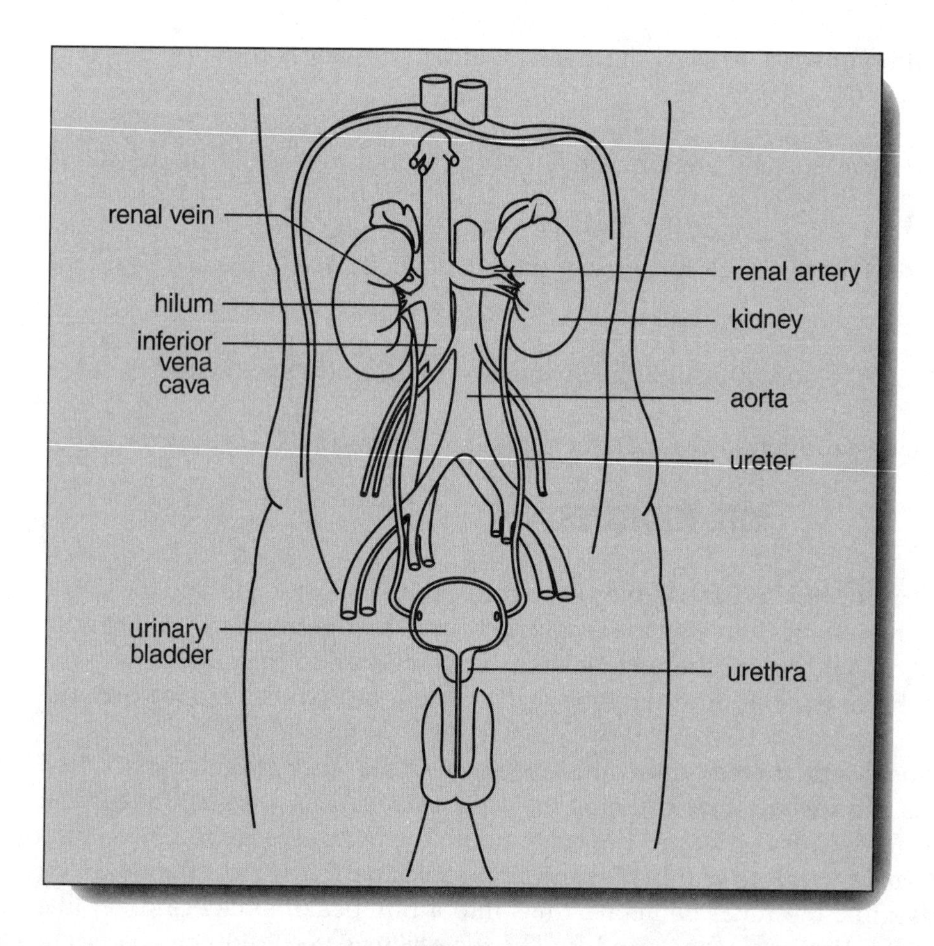

FIGURE 12.1

Bowman's capsule itself is comprised of an inner visceral layer of squamous epithelium that closely cover the glomerulus, and an outer parietal layer of squamous epithelium that is continuous with the *visceral layer* and the wall of the renal tubule. The renal tubules in turn are connected to Bowman's capsule and conduct the ultra-filtrate, generated by glomerular filtration, to the collecting ducts.

The order of the various structures of the renal tubule and the arrangement of the vascular components are presented below.

<u>TUBULAR COMPONENTS</u>

Bowman's Capsule
Proximal Convoluted Tubule
Descending Loop of Henle
Ascending Loop of Henle

Distal Convoluted Tubule
Collecting Duct

VASCULAR COMPONENT

Afferent Arteriole
Glomerular Capillaries
Efferent Arteriole
Peritubular Capillaries

The *Arcuate Artery and Vein* delineate the division between the renal cortex and the medulla. The Renal Corpuscle, and the Proximal and Distal Convoluted Tubules, lie within the cortex while the loop of Henle extends into the medulla.

Two distinct types of nephrons can be identified in the human kidneys: *Cortical nephrons* and *Juxtamedullary nephrons* **(Figure 12.2)**.

Cortical nephrons have glomeruli that lie close to the surface of the kidney and in addition, they have short loops of Henle that do not extend very far into the medulla.

On the other hand, the Juxtamedullary nephrons possess glomeruli that are deep in the cortex, close to the medulla, and have long loops of Henle that extend deep into the medulla. Between one-third and one-fifth of the nephrons in the adult kidney are Juxtamedullary nephrons. It is this type of nephron that is responsible for water reabsorption through the *Counter Current Mechanism* (see following material).

MECHANISMS INVOLVED IN THE GENERATION OF URINE.

GLOMERULAR CAPILLARY STRUCTURE

The structure of the glomerular capillary wall is responsible for regulating the composition of the *ultrafiltrate* passing into Bowman's space. Like other vessels of the circulatory system, the inner lining of the glomerular capillaries contains endothelial cells.

However, the glomerular endothelium is pierced by small holes called *fenestrae*, which make it many times more permeable than are other capillaries in the body. The size of these fenestrae allow small molecules such as salts, water, amino acids, glucose, and the like to pass through, but larger substances, such as blood cells, cannot exit from the glomerular capillaries.

In close contact with the endothelium—but separated by a *basement membrane*—is the *visceral squamous epithelium* of Bowman's capsule. The cells of this layer are highly modified, and are called *podocytes*. Each Podocyte possesses primary processes that extend from the cell body and then branch into *pedicles*. Because the pedicles interdigitate with each other, they form spaces with adjacent pedicles called *slit pores*.

pedicles. Because the pedicles interdigitate with each other, they form spaces with adjacent pedicles called *slit pores*.

It has been suggested that the pedicles can control the size of the slit pores and further regulate the composition of the filtrate. Thus, in order for the filtrate to move from the glomerular capillaries into Bowman's space, it must pass thorough fenestrae and also through the slit pores. In addition, the filtration of certain substances, such as the protein albumin, is further impeded because the basement membrane contains negatively charged proteoglycans.

Forces Involved in Glomerular Filtration. As in other capillaries, outward forces favor filtration, while inward forces oppose filtration and may even promote reabsorption. In the case of the Glomerular capillaries the outward force favoring filtration is due to *hydrostatic pressure* or P_{GC}

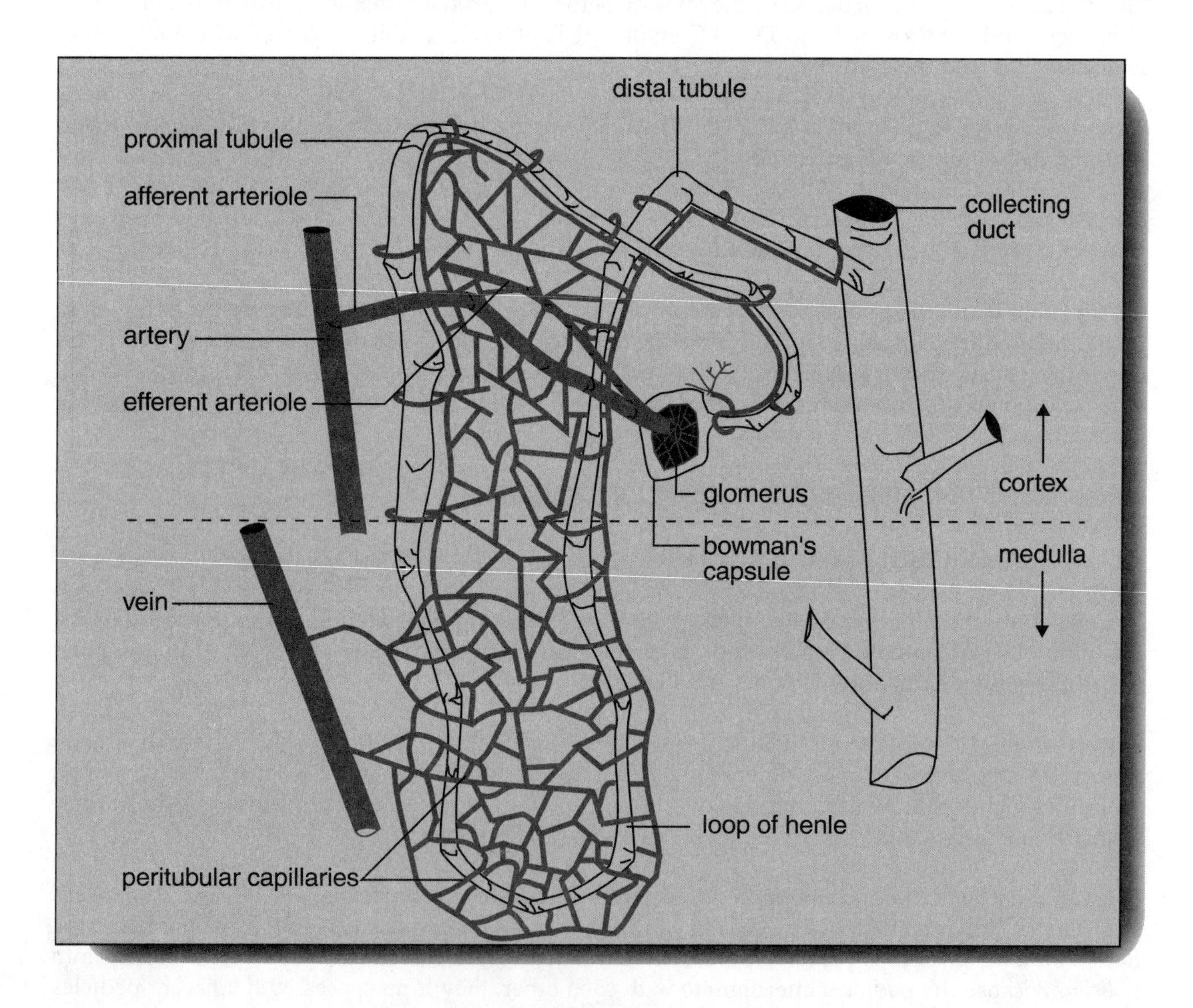

FIGURE 12.2

(55 mmHg), while the inward forces that oppose filtration are the *osmotic pressure* of plasma protein or $ð_{GC}$ (30 mmHg) and hydrostatic pressure in Bowman's space or P_{BC} (15 mmHg). Now, to calculate *net filtration pressure* (P_{NF}) the following formula is used:

$$P_{NF} = P_{GC} - (ð_{GC} + P_{BC})$$

Clearly, P_{NF} = 55 - (30 + 15) or 10 mmHg. Thus, assuming that the outward force is greater than the inward forces, filtration will take place. Naturally, if these forces are altered, for example, by constricting or dilating the afferent or efferent arterioles (see following material), or by altering the concentration of plasma protein, this will alter net filtration pressure. Although we have calculated net filtration pressure, this is not glomerular filtration rate (GFR). Glomerular filtration rate (GFR) actually depends not only on the P_{NF} but also on glomerular surface area and permeability of the glomerular membrane. This property of the membrane is referred to as the filtration coefficient (K_f).

thus: $GFR = K_f \times P_{NF}$

Approximately 20% of the plasma that enters the glomerulus is filtered at a P_{NF} of 10 mmHg. Combining the filtrate generated by all glomeruli of both kidneys results in a GFR of 180 Lt/day of filtrate, or an average GFR of 125 ml/min in males. In females the GFR is 160 Lt/day or about 115 ml/min.

Calculation of GFR. A commonly used method to estimate the GFR in a patient is to infuse a marker substance into the circulation and measure its concentration in the urine over a period of time. One commonly utilized substance is the carbohydrate inulin. Inulin is utilized because it is filtered, (and not reabsorped or secreted), it is not metabolized by the body or changed chemically in the kidney and it is easily measured in the urine. To facilitate ease of calculation, one method used is to maintain a constant plasma level of inulin through intravenous infusion of the marker. The following formula is then utilized:

$$GFR = (U)(V)/P$$

Where:

U = the Inulin concentration in the total volume of urine in g/L
V = the total volume of urine collected in a set period of time (for example L/24 hr)
P = the plasma concentration of Inulin in mg/ml.

Using the above formula the following hypothetical calculation can be made.

Given:

V = 3 L/24hr
U = 60 g Inulin/L
P = 1 mg/ml

Thus:

$$GFR = (U)(V)/P = (60\ g/L)(3\ L/24\ hr)/1\ mg/ml$$

Calculation of Filtered Load. It is possible to measure the total amount of any substance that is filtered at the glomerulus. To accomplish this, multiply GFR x plasma concentration of the substance. The value obtained is the filtered load of that particular substance.

For example, if:

$$GFR = 180\ L/day$$
$$Plasma\ glucose\ concentration = 1\ g/L$$

$$Filtered\ Load\ of\ Glucose = (180\ L/day)(1\ g/L)$$
$$Filtered\ Load\ of\ Glucose = 180\ g/day$$

Note that meaningful information can only be obtained from the above calculation if the substance chosen *is actually able to be filtered across the glomerulus*. For example, this calculation could have been made for protein (since the plasma concentration is a known value), but the results would still be meaningless since little if any protein is filtered at the glomerulus under normal conditions.

COMPOSITION OF THE GLOMERULAR FILTRATE

Because the glomerulus is highly permeable, the composition of the glomerular filtrate (or ultrafiltrate) will be strikingly similar to that of the plasma. However, little or no protein will be present under normal physiological conditions. Proteinuria is a pathological condition and implies inflammation within the glomerular wall structure (nephritis).

Nevertheless, it is interesting to note that in athletic individuals after very strenuous exercise, the presence of protein in the urine (athletic pseudonephritis) is quite normal and reversible. The concentration of certain major substances within the plasma, glomerular filtrate (filtered load), and the urine is presented below. Keep in mind that the difference between the urine concentration of these substances and the glomerular filtrate is due to the actions of reabsorption and secretion taking place in the peritubular capillaries.

	Substance Concentration in mEq/L		
	Plasma	**Glomerular Filtrate**	**Urine**
Sodium	142	142	128
Potassium	5	5	60
Calcium	4	4	5
Magnesium	3	3	15
Chloride	103	103	134
Bicarbonate	27	27	14
Sulfate	1	1	33
Phosphate	2	2	40

	Concentration in Mg/100ml		
	Plasma	**Glomerular Filtrate**	**Urine**
Glucose	100	100	0
Urea	26	26	1820
Uric Acid	4	4	53
Creatine	1	1	196

FACTORS THAT REGULATE GFR

GFR is most commonly regulated by changes in glomerular capillary hydrostatic pressure (P_{GC}). To alter $P_{GC,}$ both the afferent and efferent arterioles can be constricted or dilated. For example, if the afferent arteriole is constricted this will increase resistance to blood flow into the glomerulus reducing P_{GC} and GFR will decrease (the opposite will happen if the afferent arteriole is dilated). On the other hand, if the efferent arteriole is constricted this will increase resistance to blood flow out of the glomerulus increasing P_{GC} and GFR will increase. Constriction and dilation of the afferent and efferent arterioles can be regulated locally by the Juxtaglomerular Apparatus (see following material).

Such regulation is termed autoregulation and is designed to prevent, or minimize, spontaneous changes in GFR as systemic blood pressure fluctuates. In addition to this *autoregulation*, extrinsic sympathetic control of the afferent arteriolar smooth muscle supplies long-term regulation of arterial blood pressure.

AUTOREGULATION OF GFR

Fluctuations in systemic arterial blood pressure would, under normal circumstances, affect the hydrostatic pressure (P_{GC}) of the glomerular capillary and thus alter GFR. To maintain a relatively constant GFR in the face of changing systemic arteriolar blood pressure, both the afferent and efferent arteriole can undergo dilation or constriction. Such changes in the arteriolar radius are believed to be controlled by two interrenal processes which are: (1) The myogenic mechanism, and (2) a tubulo-glomerular feedback.

1) The myogenic mechanism results from the action of stretch on smooth muscle. Accordingly, the greater pressure present in the afferent arteriole, the greater will be the stretch on the smooth muscle in the arteriolar wall. In response the smooth muscle will then constrict and reduce the radius of the afferent arteriole. Naturally as radius decreases resistance to flow increases, and P_{GC} in the glomerular capillary decreases.

remember that: $F = (\Delta P)/R$

where: F = flow
$\Delta P = (P_2 - P_1)$ = pressure gradient
R = resistance

and the most important element of the resistance is the radius of the vessel (r):

where: $R = 1/r^4$

rewriting: $F = (\Delta P)(r^4)$ or $\Delta P = F/r^4$

And thus changing the radius significantly alters the pressure gradient and the flow which will, in turn effect GFR.

2) Tubulo-glomerular feedback mechanism involves the *juxtaglomerular apparatus* (or JG). This structure **(Figure 12.3)** is located at the point where the distal convoluted tubule loops back and touches the afferent arteriole. At this junction the cells lining the arteriolar wall form specialized *juxtaglomerular* (*or granular*) cells, and the cells lining the distal convoluted tubule wall are the *macula densa*. Macula densa cells monitor either changes in rate of fluid flow through the efferent arteriole, possible alterations in Cl^- concentration in this region.

Assume that there is a drop in systemic blood pressure and thus a drop in GFR. This, in turn, results in a decrease in the fluid passing through the distal convoluted tubule.

The drop in blood pressure in the afferent arteriole causes the smooth muscle of the wall to relax and dilates this vessel (*myogenic mechanism*). In addition, the reduction in the Cl^- concentration in the fluid of the efferent arteriole is then sensed by the Macula densa which stimulates the juxtaglomerular cells to release the substance *renin*. Renin then acts to convert the plasma protein *Angeotensinogen* into *Angiotensin I*. Angiotensin I is then converted to *Angiotensin II* by an enzyme in the lungs. Angiotensin II then causes the constriction of the efferent arteriole. In this example, the dilation of the afferent arteriole coupled with the constriction of the efferent arteriole act together to increase P_{GC} and thus increase GFR (correcting the initial decrease). It should also be understood that in addition to the

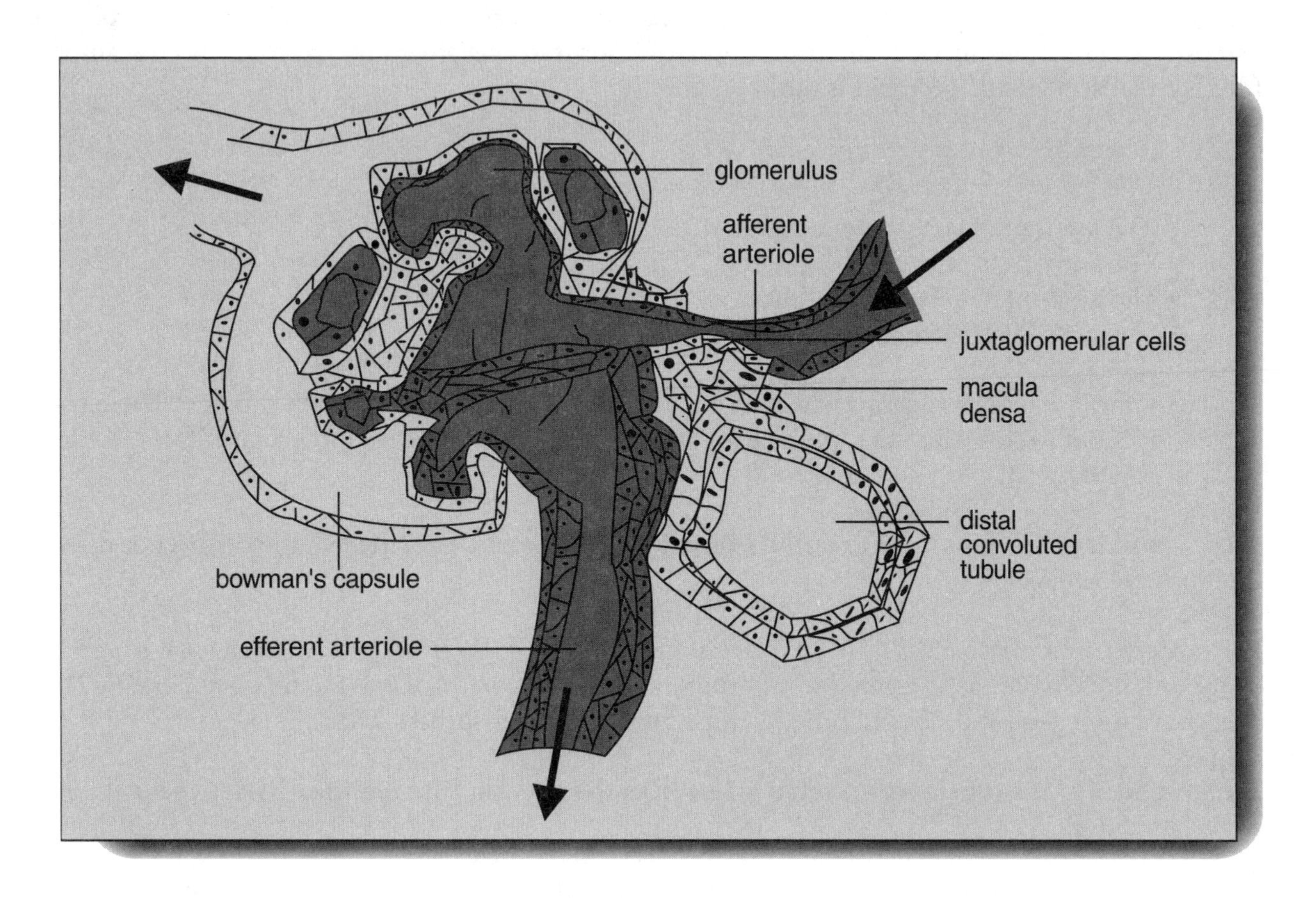

FIGURE 12.3

vasoconstrictive effects elicited by Angiotensin II on the smooth muscle of the efferent arteriole, it is also a potent vasoconstrictor of the arterioles throughout the body.

This hypertensive effect will increase systemic blood pressure. In point of fact, one form of hypertension is termed *Renal Hypertension* and is believed to result from elevated levels of renin leading to elevated Angiotensin II and high blood pressure.

Angiotensin II is also involved in releasing the *mineralocorticoid Aldosterone* from the adrenal cortex. Aldosterone increases the reabsorption of Na^+ in the distal convoluted tubule while increasing K^+/H^+ secretion. This effect will, however, be described later.

TUBULAR REABSORPTION AND SECRETION

Since nondiscriminant filtration of all plasma constituents except protein takes place at the glomerulus, it is necessary to salvage valuable substances from the tubular fluid and transfer them back to the peritubular capillary blood by the process of reabsorption In addition, certain substances are not filterable or not completely cleared by filtration and, if they are to be removed from the body, they must be transferred from the peritubular capillary blood back to the tubular fluid by the process of secretion. Movement of substances across the tubular epithelial cells during secretion and reabsorption is accomplished either by passive diffusion or active transport. Movement of water takes place by osmosis. The relationship between the amount of a substance excreted in the urine and the amounts filtered, reabsorbed and secreted is outlined below:

$$E_X = F_X + S_X - R_X$$

where:

E_X = The amount of substance X excreted
F_X = The amount of substance X filtered
S_X = The amount of substance X secreted
R_X = The amount of substance X reabsorbed

To clarify the above relationship, return to the **Table** and note that the concentration of the excreted substance is not necessarily the same as the concentration found in the glomerular filtrate. It should now be obvious that:

If the amount of the substance excreted is greater than the amount filtered, the difference must be due to secretion.

For example, in the case of Cl^-, 103 mEq/L are filtered, 134 mEq/L are present in the urine, and thus 31 mEq/L must have been secreted. On the other hand, 142 mEq/L of Na^+ are filtered, 128 mEq/L are present in the urine, and thus 14 mEq/L must have been reabsorbed. Thus:

If the amount of the substance excreted is less than the amount filtered, the difference must be due to reabsorption.

A substance is *reabsorbed* if it moves from the tubular fluid into the peritubular capillary blood. A substance is *secreted* if it moves from the peritubular capillary blood into the tubular fluid **(Figure 12.4)**.

1) *Tubular reabsorption*—this process functions very efficiently and, therefore for important substances that must be retained by the body, only very small amounts of these plasma constituents appear in the urine. On the other hand, waste products are poorly reabsorbed, and thus large amounts are present in the urine. During tubular reabsorption the substances to be saved cross the tubular epithelial cells entering the interstitium and then pass into the peritubular capillary blood. Such substances must pass through the epithelial cell membranes since adjacent cells are connected via tight junctions. The steps involved in this transepithelial transport process resulting in tubular reabsorption are outlined below:

1. The substance crosses the tubular epithelial cell membrane from the luminal side.

2. The substance passes through the tubular epithelial cell cytoplasm.

3. The substance crosses the basolateral membrane of the tubular epithelial cell.

4. The substance diffuses through the interstitial fluid.

5. The substance moves by diffusion across the endothelial cell wall of the peritubular capillary and enters the blood plasma.

If passive diffusion is the driving force for this transepithelial transport, all the steps outlined above are also considered to be passive. Thus, the reabsorption of the substance results from the movement of the substance down its concentration gradient. On the other hand, if any one of these steps require the expenditure of energy and involves transport pumps, the entire process of reabsorption for this substance would be classified as active transport. Keep in mind that with active transport, the net movement of the substance is against an electrochemical gradient. Examples of

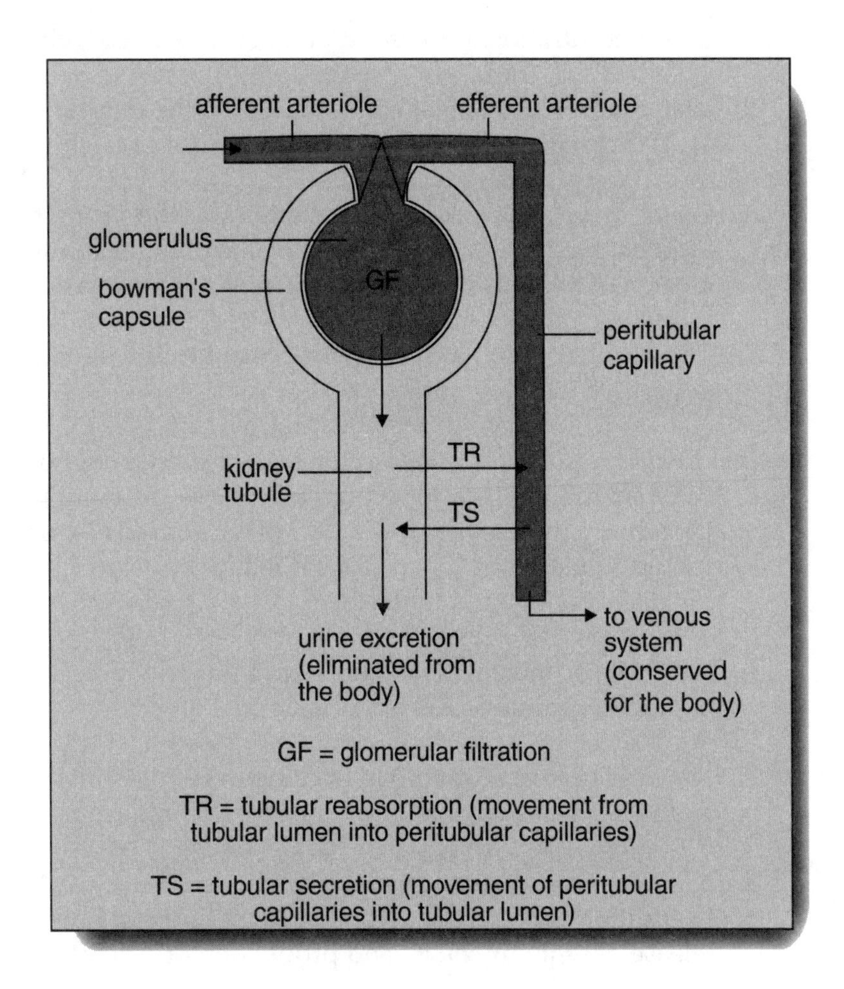

FIGURE 12.4

some of the many substances that are reabsorbed by active transport in the proximal convoluted tubule include: glucose, amino acids, and electrolytes such as Na^+, Ca^{++} and $PO4^{--}$.

Cl^- is reabsorbed in the proximal convoluted tubule by passive diffusion due to the electrical attraction between the negatively charged Cl^- and the positively charged and actively transported Na^+. Movement of water out of the proximal convoluted tubule is by the process of osmosis. To understand the process of osmosis you must recognize that water passively follows Na^+ and the other substances as they leave the tubule either by active transport or passive diffusion. The steps below will help to clarify the process of osmosis:

1. Na^+ ions are reabsorbed by active transport.

2. Negatively charged ions are attracted to positively charged Na^+ ions (considered passive transport due to electrical attraction).

3. Other substances are also removed by active transport or, in some cases, by passive diffusion.

4. The concentration of the solution decreases in the tubular fluid and increases in the plasma.

5. Water is then at a higher concentration in the tubular fluid than in the plasma (because water is all that remains in the tubular fluid, since everything else has already been removed).

6. Water moves by osmosis from the tubular fluid into the plasma down its own concentration gradient (because it moves from an area of higher water concentration in the tubular fluid, to an area of lower water concentration in the plasma).

7. The volume of tubular fluid in the proximal convoluted tubule decreases significantly (and the volume of the plasma increases).

Sodium Reabsorption. Na^+ reabsorption takes place throughout the entire tubular system. However, different mechanisms to account for this Na^+ reabsorption operate in the different segments of the tubule. The majority (over 60%) of the filtered Na^+ is reabsorbed in the proximal convoluted tubule, while about 25% is reabsorbed in the loop of Henle and less than 10% is reabsorbed in the distal convoluted tubule.

1. Na+ reabsorption in the proximal convoluted tubule is also linked with the active reabsorption of glucose, amino acids, water and Cl^-.

2. Na^+ (and Cl^- reabsorption) in the loop of Henle is involved in the counter current mechanism, and water reabsorption under the control of hormone ADH.

3. Na+ reabsorption in the distal convoluted tubule involves the linked Na+/K+ ATPase pump. In addition, this process is controlled by the hormone Aldosterone and is further linked to the release of renin by the JG and production of Angiotensin II (as described above).

Generally the Na^+ transport requires an energy-dependent Na^+/K^+ ATPase carrier protein that is located on the basolateral (interstitial side) of the tubular epithelial cell. While movement of the Na^+ from the tubular lumen into the epithelial cells is through Na^+ channels in the membrane, the movement out of the epithelial cells is via the active transport carrier located on the interstitial side of the cell. As this pump removes the Na^+ from the cellular cytoplasm, it also brings in a K^+ since it is a linked pump and utilizes ATP as the energy source. The Na^+ concentration builds up in the interstitial fluid, and then this ion passes by diffusion across the endothelial cell membrane of the capillary and into the plasma.

In the proximal convoluted tubule, the reabsorption of nutrient substances such as glucose and amino acids depend on a co-transport system and is linked with Na^+ reabsorption. Such an effect depends on the Na^+ concentration gradient which is maintained by the action of the Na^+/K^+ ATPase pump. In the distal convoluted tubule, the reabsorption of Na^+ also depends on the Na^+/K^+ ATPase pump but, as was pointed out earlier, it is also regulated by the hormone Aldosterone.

Active Reabsorption of Substances and the T_M. For the active reabsorption of all substances with the exception of Na^+, carrier molecules, specific for each substance, are required.

There are a limited number of these carrier molecules in the tubular epithelial cells. Thus, if excess amounts of a particular substance are filtered, it is possible that the carriers will become saturated and some of this material will be spilled in the urine. The T_m or transport maximum is the saturating concentration of the particular substance, above which the excess is lost in the urine and not reabsorbed.

An example of this effect is shown in **Figure 12.5** for the substance glucose. Here, as the plasma concentration of glucose increases, so does the filtered load. However, because the glucose carriers are capable of reabsorbing all of the glucose filtered, no glucose will be excreted in the urine. This is because all of the filtered glucose is being transported back into the peritubular capillaries. Here the amount of glucose reabsorbed also increases, and thus no glucose is excreted in the urine.

For example, at a plasma concentration of 100 mg/ml, a GFR of 125 ml/min and a filtered load of 125 mg/min, the amount of glucose reabsorbed is 125 mg/min, and thus the amount of glucose excreted is 0. If the plasma concentration of glucose increases to 300 mg/ml, with a GFR of 125 ml/min, the filtered load of glucose now increases to 375 mg/min and the amount transported also increases to 375 mg/min. Thus, again no glucose is excreted in the urine. However, should the plasma glucose increase above 300 mg/ml, (as, for example, in a diabetic) to 500 mg/ml, then the amount filtered will be 625 mg/min. But the amount reabsorbed will remain at 375 mg/min and thus 250 mg/min (625 mg/min - 375 mg/min) will be lost in the urine. In this example, the levels of 375 mg/min are the T_M for glucose.

Each substance (with the exception of Na^+) has a T_M if its transport is *carrier mediated*.

Assuming that this T_M is close to the normal plasma concentration, then even a slight increase in the plasma concentration above normal levels will saturate the carriers and the filtered excess will be lost in the urine. This will return the plasma concentration to the normal range. However, for substances like glucose, because this T_M is significantly greater then the normal levels of plasma glucose, this will

not serve to regulate the plasma levels. In the case of glucose, other factors (i.e., the hormones insulin and glucagon) regulate the plasma levels.

2. *Tubular secretion*—tubular secretion provides an alternate pathway (bypassing the filtration mechanisms) by which the body can expel substances. Secretion is utilized when substances, which cannot be filtered, must still be cleared from the circulation. (For example, penicillin which is too large to be filtered at the glomerulus is secreted because the body considers it to be a foreign chemical.) In addition, a substance may be both filtered and secreted. In this way, more of the substance will be removed in one pass through the kidney than if the substance was only filtered, or only secreted.

Examples of some important substances that are secreted include H^+ and K^+ ions, organic ions and various drugs.

H^+ ions are secreted in the proximal and distal convoluted tubules and collecting ducts. The amount of H^+ ion secretion depends on the concentration present in the body fluids, with an increased concentration (acidosis) leading to an increased secretion and vice versa.

K^+ ions also undergo secretion but, in addition, these ions are also filtered and reabsorbed. Specifically, K^+ ions are filtered with other substances at the glomerulus and then completely reabsorbed in the proximal convoluted tubule. In the distal convoluted tubule, Na^+ transport from the tubular lumen is linked with K^+ transport from the interstitial fluid via an ATP driven pump. Furthermore, this process is also stimulated by the hormone *aldosterone*. Thus, for every Na^+ reabsorbed, a K^+ or H^+ is secreted as a result of the linked pump action and also electrical repulsion. Although aldosterone secretion from the adrenal cortex is under the control of the renin-angiotensin system, an alternate pathway via direct stimulation of the adrenal cortex by elevated levels of blood K^+ is also of importance **(Figure 12.6)**.

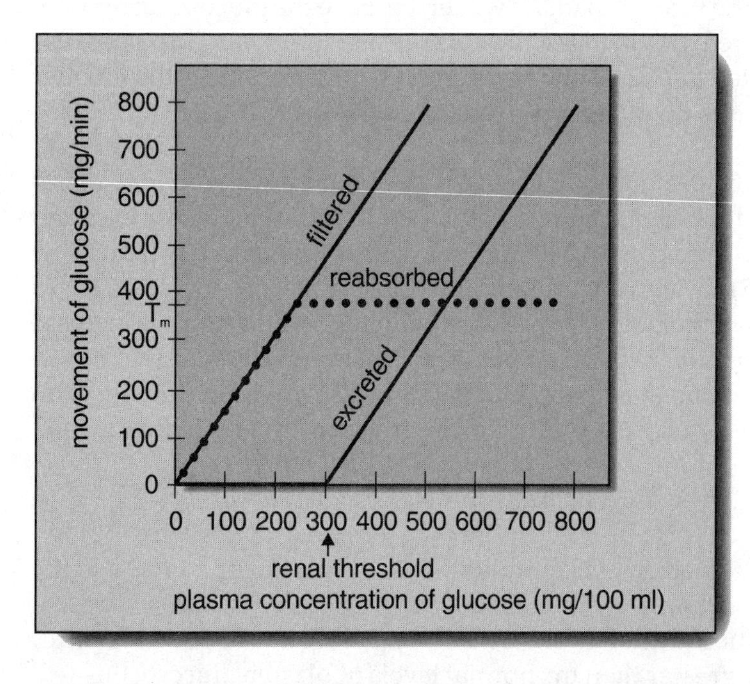

FIGURE 12.5

THE IMPORTANCE OF OSMOLARITY

Under normal circumstances more than 99% of the glomerular filtrate is reabsorbed as it passes through the tubular system. While the majority of this filtrate is reabsorbed in the proximal convoluted tubule (termed *obligatory reabsorption*) in response to Na^+ active transport, some fluid remains to be processed farther along the tubules. It is the fate of this remaining fluid which we will now consider. Although this volume is quite small, it plays a major role in maintenance of body fluid homeostasis.

The fluid in the extra-cellular compartment is Isotonic which actually means that it has an osmolarity of 300 mOsm. Any solution that has an osmolarity which is greater then 300 mOsm is considered to be Hypertonic, while a solution that has an osmolarity that is less than 300 mOsm is considered to be Hypotonic.

Terminology	Osmolarity	Composition
Hypotonic	less than 300 mOsm	More water and less salt than Isotonic.
Isotonic	300 mOsm	Equivalent to body fluids and plasma.
Hypertonic	greater than 300 mOsm	More salt and less water than Isotonic.

1) A RBC placed in a Hypertonic solution will lose water and will shrink since the cytoplasm will be 300 mOsm, but the outside will be greater than 300 mOsm. In this case, there will be more water inside the cell than outside, and the osmotic gradient will cause the water to move out of the cell.

2) A RBC placed in a Hypotonic solution will gain water and will rupture since the cytoplasm will be a 300 mOsm, but the outside will be less than 300 mOsm. In this case, there will be more water outside the cell than inside, and the osmotic gradient will cause the water to move into the cell.

3) A RBC placed in an Isotonic solution will experience no net water movement either in or out because the osmotic pressure will be equal on both sides of the membrane (i.e., 300 mOsm).

THE IMPORTANCE OF UREA

Active Na^+ reabsorption drives the passive reabsorption of both Cl^- and water, and in addition, it is also responsible for the passive reabsorption of urea. Although urea is actually a waste product, a certain amount of this substance is also reabsorbed as follows:

1) Active Na^+ reabsorption in proximal convoluted tubule induces water to be reabsorbed out of the proximal convoluted tubule by osmosis.

Note: About 180 lt/day (or 125 ml/min) are filtered at the glomerulus. About 117 lt/day (or 81 ml/min) are reabsorbed in the proximal convoluted tubule, secondary to Na^+ reabsorption. Thus, about 63 lt/day (or 44 ml/min) of fluid remain in the proximal convoluted tubule, and it is this volume that enters the descending Loop of Henle. The remaining 63 lt/day (or 44 ml/min) will be further reduced by the counter current mechanism (see following material).

2. As water is reabsorbed it leaves urea behind in the lumen of the proximal convoluted tubule. Initially the concentration of urea filtered at the glomerulus is the same as the concentration in the plasma. However, the urea concentration in proximal convoluted tubule increases (as water

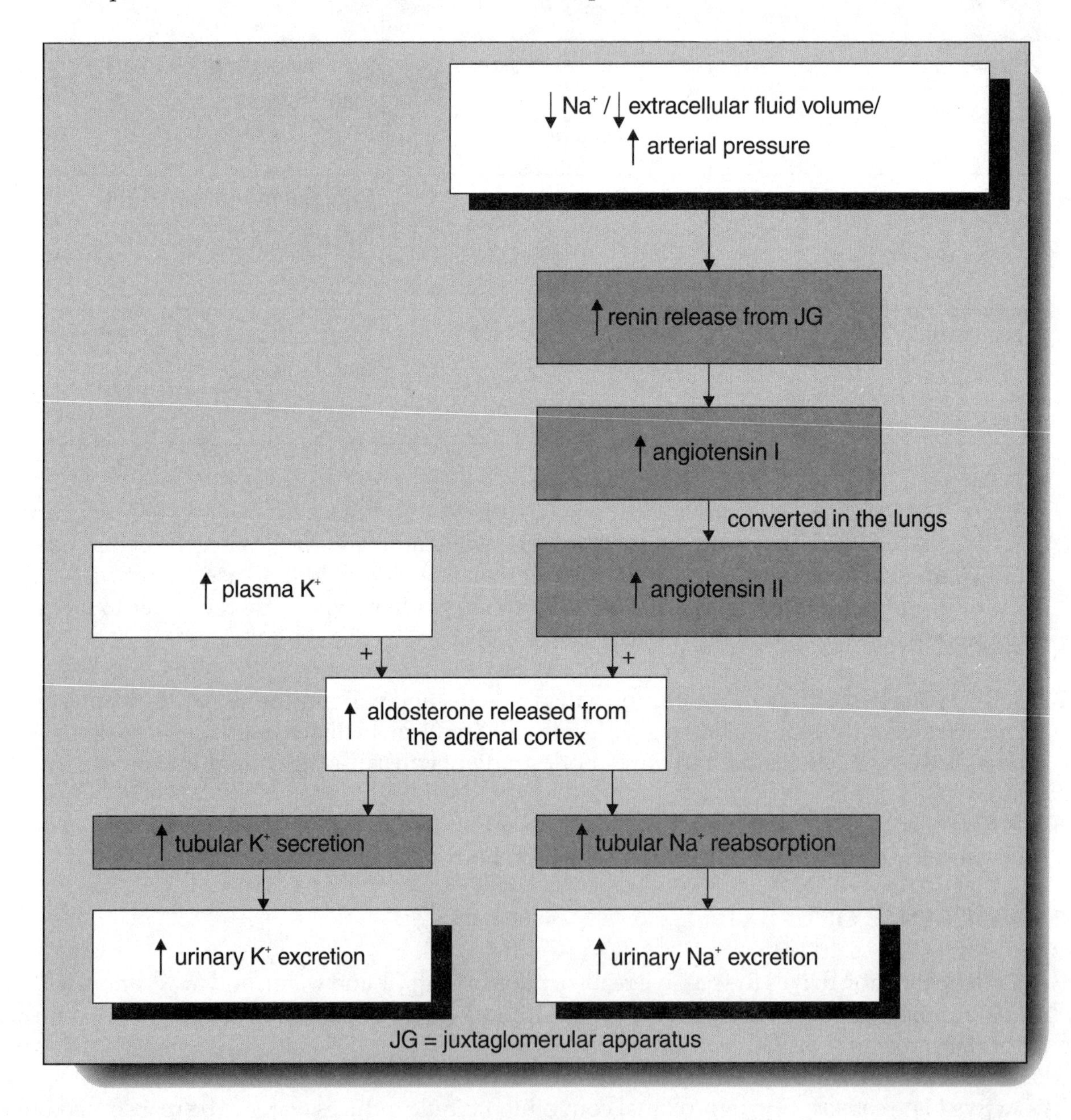

FIGURE 12.6

is removed and the volume decreases) until it is about threefold greater than in the plasma.

3) Because the urea concentration is now greater than in the plasma, urea now passively diffuses from the tubular lumen into the plasma of the peritubular capillaries. Only about 50% of the urea, present in the proximal convoluted tubule, diffuses into the peritubular capillary because the walls of the tubule are only partially permeable to urea. Thus, 50% of the urea is passively reabsorbed, and 50% is cleared from the body with each pass of the blood through the kidneys.

THE COUNTER-CURRENT MECHANISM

The *counter-current mechanism* (also termed the *counter-current multiplier*) operates only in those nephrons termed *Juxtamedullary;* Cortical nephrons do not play a role in this process.

To understand the processes involved here, consider the volume and concentration of the ultrafiltrate in Bowman's capsule: 125 ml/min at 300 mOsm (like the plasma). As this now moves through the Proximal Convoluted tubule, obligatory reabsorption (driven by Na^+ active transport) will significantly reduce this volume (to 44 ml/min) but since salts, glucose, amino acids and urea are also removed, the smaller volume that now enters the loop of Henle remains at 300 mOsm. Assume that this fluid now enters the loop of Henle, passes down the descending limb, and then back up the ascending limb. Up to this point the fluid has remained at 300 mOsm. Now, as this fluid moves into the ascending limb of the loop of Henle, active transport pumps transfer the Na^+ and Cl^- into the interstitium, but the wall of the tube is impervious to water.

Thus, the fluid in the ascending limb of the loop of Henle becomes Hypotonic (100 mOsm) and this hypotonic fluid now enters the Distal Convoluted tubule **(Figure 12.7)**.

Meanwhile the salt removed in the ascending limb of the loop of Henle concentrates in the interstitium of the medulla and moves by diffusion back into the fluid in the descending limb of the loop of Henle. Furthermore, water moves out of the descending limb of the loop of Henle into the interstitium by osmosis.

Thus, the fluid in the descending limb of the loop of Henle now becomes Hypertonic. This hypertonic fluid moves into the ascending limb of the loop of Henle where the salt is pumped out and it becomes hypotonic. The salt thus fluxes down the loop of Henle, back up the ascending limb, and crosses into the interstitium to reenter the descending limb again. This "counter current" movement generates a vertical osmotic salt gradient in the renal medulla that begins at the level of the cortex at 300 mOsm continuing to increase with depth into the medulla; reaching a maximum of 1,200 mOsm near the level of the renal pelvis. Maintenance of this salt gradient is assisted by loops of the peritubular capillaries called the *Vasa Recta* that carry blood containing the salt into the medulla and then back up again (see following material).

Nephron Component	Fluid Osmolarity
Bowman's Capsule	300 mOsm (Isotonic like plasma)
Proximal convoluted tubule	300 mOsm (Isotonic but volume is significantly reduced)
Descending limb of loop of Henle	Increasing to 1,200 mOsm (Hypertonic due to salt entering and water leaving)
Ascending limb of loop of Henle	100 mOsm (Hypotonic, salt is pumped but water cannot leave tubule)
Distal convoluted tubule	Variable (from 100 - 300 mOsm depending on the concentration of ADH present)
Collecting duct	Variable (from 300 - 1,200 mOsm depending on the concentration of ADH present)

To utilize this vertical salt gradient, consider the permeability of the tubular walls to water.

If the proximal convoluted tubule is permeable to water (but not salt), water will move *OUT* into the cortex because there is more water in the tubular fluid (100 mOsm) than in the cortex (300 mOsm, like the blood). Similarly, if the walls of the collecting duct are permeable to water, it will move as a result of the osmotic pressure generated by the salt gradient in the medulla. For example, if the tubular fluid is 300 mOsm and the interstitial gradient is 400 mOsm, water moves out of the collecting duct. As the fluid moves farther down the collecting duct, the concentration of the fluid now becomes 400 mOsm, but the interstitial salt gradient deeper in the medulla is now 500 mOsm, so water moves out, and so on. This process continues with water moving out by osmosis, due to the increasing salt gradient deeper in the medulla. Since only water can leave the collecting duct, the salts in the fluid are being concentrated. The final volume of hypertonic urine excreted may be as small as 200 ml/day with an osmolarity of 1,200 mOsm.

The Vasa Recta. This is a loop of capillaries originating from the peritubular capillary system that dips down into the medulla. It helps to maintain the medullary salt gradient generated by the loop of Henle, acting as an additional counter current multiplier because salt diffusing into the blood flowing slowly down in the Vasa Recta is then carried back up towards the renal cortex where it then diffuses back out again (**Figure 12.8**).

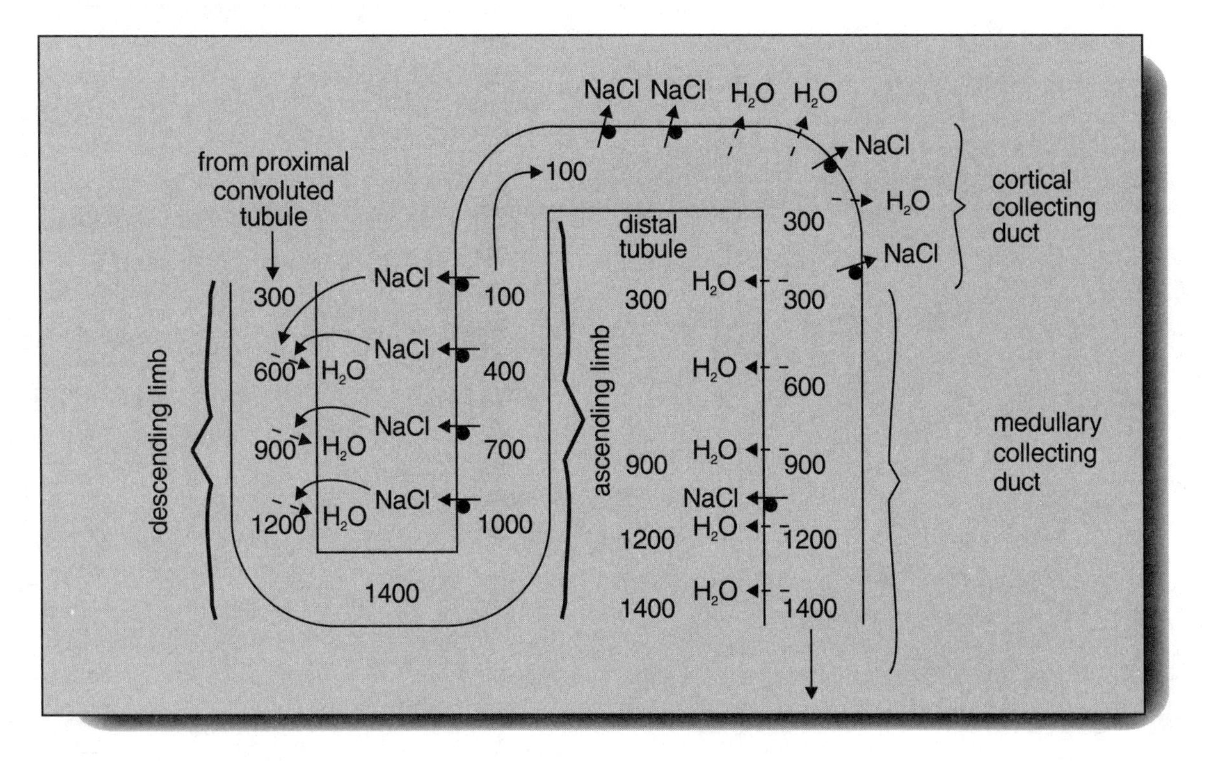

FIGURE 12.7

ADH (VASOPRESSIN) REGULATION OF URINE VOLUME AND OSMOLARITY

It is important to realize that the movement of water out of the distal convoluted tubule and collecting duct will only take place if *Antidiuretic Hormone* (*ADH*, also called *Vasopressin*) is present. When ADH binds to the ADH receptors on in the tubular cell membrane, it increases levels of intra-cellular *Cyclic AMP*. The tubular cell wall then becomes more permeable to water. As ADH levels increase, permeability also increases and water reabsorption also increases. Basal (or base-line) levels of ADH are normally present and thus some water is normally reabsorbed. However, as the osmolarity of the body fluids change, ADH levels will increase or decrease (above or below the basal level) to maintain fluid homeostasis **(Figure 12.9)**.

FLUID OSMOLARITY AND CONTROL OF ADH SECRETION.

ADH is synthesized in the *Supraoptic Nucleus of the Hypothalamus* and carried down nerve tracks into the *Posterior Pituitary* (*Neurohypophysis*), where it is released into the blood.

The *Osmoreceptors* are also located in the area of the Supraoptic nucleus, and it is these osmoreceptors that monitor the osmolarity of the blood. If the blood osmolarity increases above 300 mOsm (hypertonic), the osmoreceptors will increase the release of ADH from the posterior pituitary. This will act on the kidney to increase water reabsorption and dilute the hypertonic blood. On the other hand, if the blood osmolarity decreases below 300 mOsm (Hypotonic), the osmoreceptors will reduce the

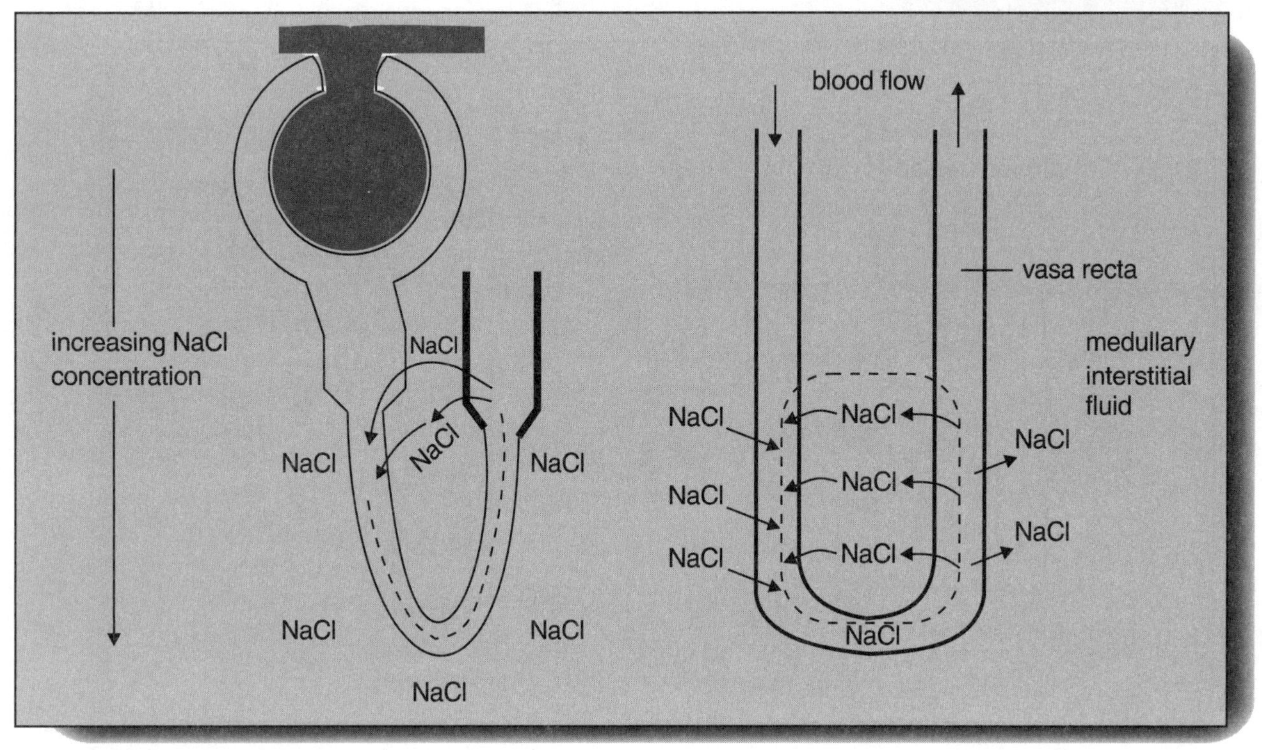

FIGURE 12.8

amount of ADH released from the posterior pituitary. This will then act on the kidney to decrease the amount of water reabsorbed and increase the amount of fluid excreted. Thus, water will be lost from the body and blood osmolarity will return to normal (Isotonic).

BARORECEPTOR REFLEX AND ADH SECRETION

When arterial blood pressure decreases, this stimulates the baroreceptor reflex and activates corrective responses designed to normalize the blood pressure. One of the effects elicited during this reflex is the stimulation of the sympathetic nervous system. Sympathetic nervous system stimulation has multiple effects (i.e., increasing heart rate, increasing stroke volume, constricting the veins, increasing total peripheral resistance, increasing respiration) and, in addition, it also leads to increased release of ADH.

ADH mediates two important responses related to arterial blood pressure regulation. First, it constricts the peripheral arterioles which is why it is also called *Vasopressin*. Such vasoconstriction assists in increasing blood pressure. Second, it increases water reabsorption in the kidney and thus expands the blood volume. The increased blood volume, by increasing venous return and cardiac output will assist in elevating the blood pressure.

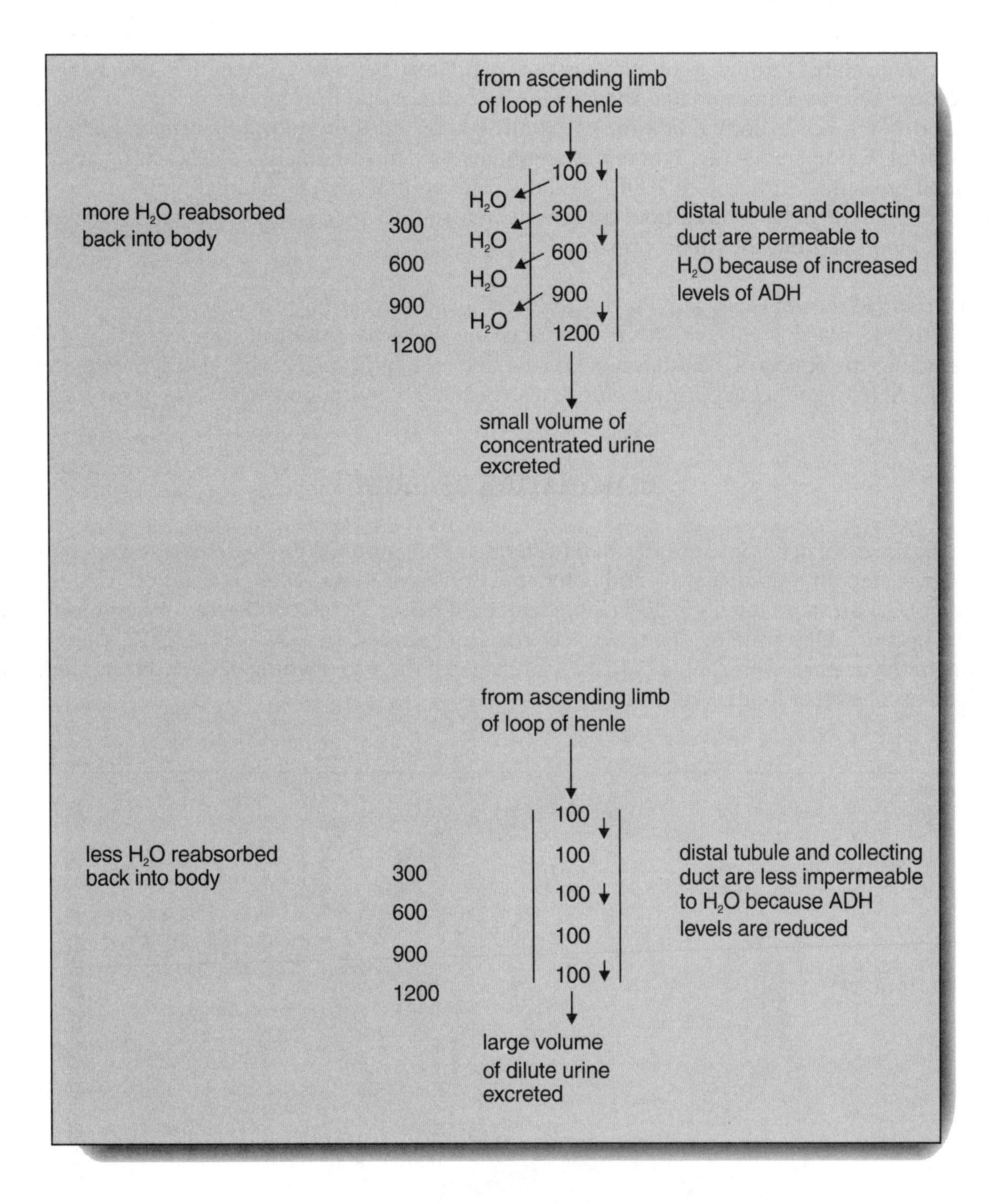

FIGURE 12.9

INTERACTIONS BETWEEN GFR, NA+ LEVELS, ADH AND ATRIAL NATRIURETIC FACTOR IN THE REGULATION OF FLUID VOLUME

In should now be apparent that body osmolarity, fluid volume, blood pressure and blood volume are

all closely interrelated through a complex series of regulatory pathways. Changing GFR changes body fluid volume, blood volume and Na^+ levels in the body fluids and thus blood pressure as well. On the other hand, Na^+ levels in body fluids are also controlled by aldosterone and this effects body water as well. Control of aldosterone involves renin/angiotensin II, and angiotensin itself acts on arterioles to alter blood pressure. Furthermore, ADH regulates fluid volume and, thus blood pressure. In addition to the above interactions another hormone produced from the Atria of the heart is also involved in control of blood volume and blood pressure.

Called *Atrial Natriuretic Factor* (*ANF*), it is released when the cardiac atria are distended due to an increase in the plasma volume. ANF then acts on the renal tubule to increase Na^+ excretion (counteracting the effects of aldosterone). Since a decrease in N^+ levels will also promote increased water loss, ANF by reducing plasma volume, also reduces blood volume and thus blood pressure as well.

ELIMINATION OF URINE

Urine exits the renal pelvis into the attached ureter. It is then pumped through the ureter by peristaltic action of the smooth muscular wall, and enters the urinary bladder where it is stored until released. The wall of the urinary bladder is distensible and, on the inner side contains the structure called the *trigone* **(Figure 12.10)** consisting of an inverted triangular shaped area. The point of the triangle is the opening to the urethra, while at the base there are located the two opening of the ureters. The wall of the bladder consists of four layers:

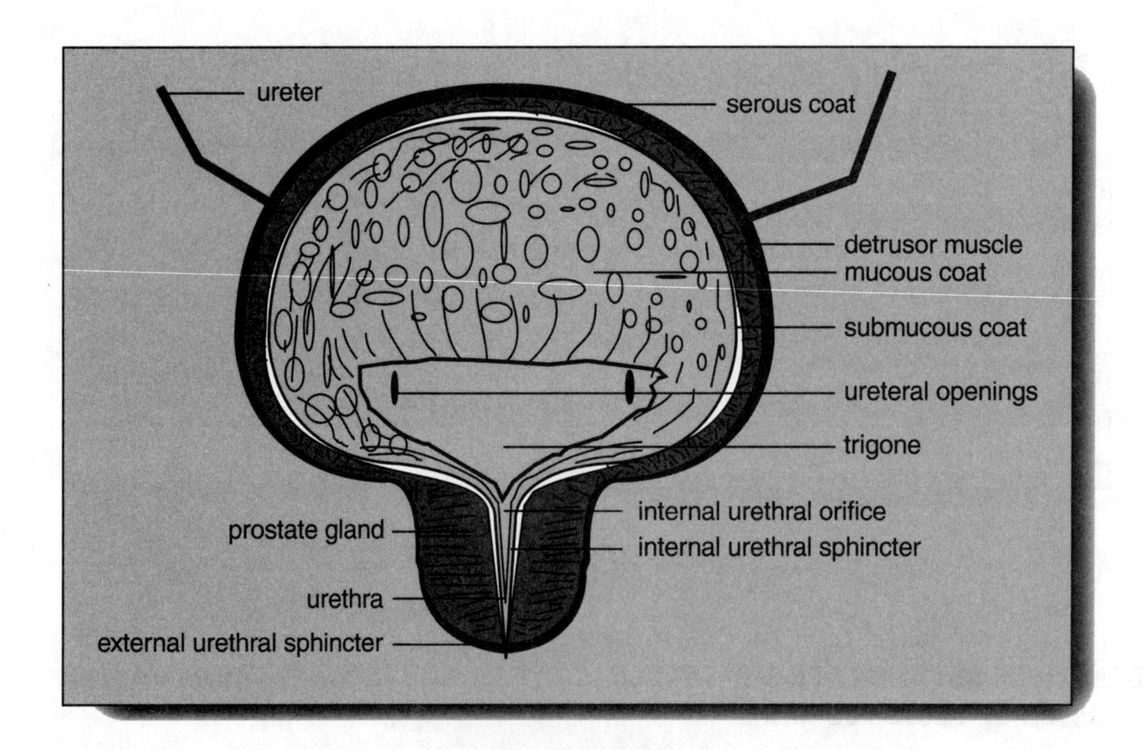

FIGURE 12.10

1. Inner Mucous Coat
2. Submucous Coat
3. Muscular
4. Serous Coat

The muscular coat (or layer) is composed of smooth muscle fibers in interlaced bundles and is called the *detrusor muscle*. Where the detrusor muscle narrows to form the opening of the ureter it forms the *internal urethral sphincter*.

The *urethra* is the tube that carries the urine from the bladder to the outside of the body. In females the urethra is about 2.5 cm long, and it ends between the labia minor, posterior to the clitoris. The opening of the urethra is called the *external urethral orifice* (*or urethral meatus*).

The urethra in the male is about 19.5 cm long and is divided into three segments **(Figure 12.11)**.

1) The *prostatic urethra*, which is 2.5 cm long, and passes from the bladder through the prostate gland (located directly below the bladder). Secretions from the prostate gland enter the urethra at this location, and in addition the ejaculatory ducts are also open into the urethra at this location (see the chapter on Reproduction).

2) The *membranous urethra*, which is about 2 cm long, and passes through the *urogenital diaphragm*. Skeletal muscle fibers form the external urethral sphincter that surrounds the urogenital diaphragm.

3) The *penile urethra*, which is 15 cm long, passes through the corpus spongiosum of the penis. The opening of the urethra is called the *external urethral orifice* (or *urinary meatus*).

MICTURITION REFLEX

Micturition (or urination) is the process by which urine is excreted from the body. The reflex that causes this to take place is initially activated when the bladder is distended with about 150 ml of urine. A stronger stimulation is produced when the bladder is filled with 300 ml of urine (maximum volume that can be contained is about 600 ml). Steps involved in micturition are as follows:

1. Bladder becomes distended.

2. Stretch receptors in the wall are stimulated

3. Impulses are transmitted to the micturition center in the sacral spinal cord.

4. Nerve impulses are then transmitted back to the detrusor muscle through the parasympathetic nervous system.

5. The detrusor muscle contracts in a rhythmic fashion.

6. The impending need to urinate is sensed by the cerebral cortex.

7. To prevent urination the individual must voluntarily keep the external urethral sphincter contracted. The reflex is also inhibited by impulses originating from the midbrain and cerebral cortex.

8. If the decision to urinate is made, then the external urethral sphincter is allowed to relax. The reflex is further facilitated by impulses originating from the pons and hypothalamus.

9. The internal urethral sphincter relaxes involuntarily, as contraction of the detrusor muscle proceeds. Urine is then expelled from the body.

10. As the neurons in the micturition center of the cord become fatigued, the detrusor muscle relaxes and the reflex ends.

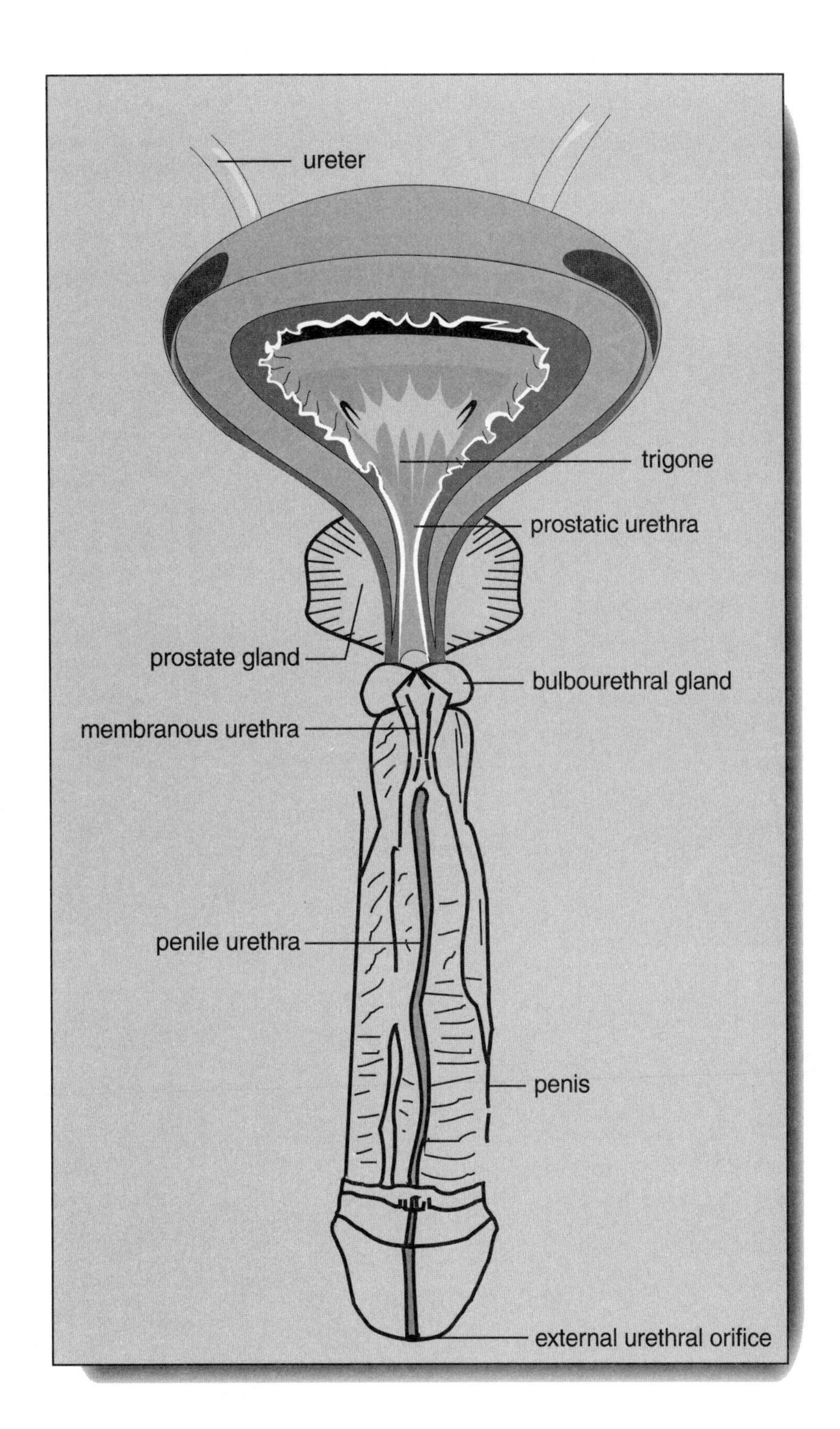
ureter
trigone
prostatic urethra
prostate gland
bulbourethral gland
membranous urethra
penile urethra
penis
external urethral orifice

FIGURE 12.11

Chapter 13
FLUID, ELECTROLYTE AND ACID-BASE BALANCE

THE INTRA-CELLULAR AND EXTRA-CELLULAR COMPARTMENTS

In an average adult, the body contains about 40 liters of water (63% by weight). Dissolved in this water are a variety of electrolytes as well as many other substances, and the fluid is distributed in two major compartments (i.e., the extra-cellular and the intra-cellular compartment). Of the total 40 liters of water, 37% is present in the extra-cellular compartment, while the remaining 63% is found inside of cells (intra-cellular compartment). Composition of the fluid in these two locations is not the same, primarily because the plasma membrane that separates the intra-cellular space from the extra-cellular space is semipermeable.

COMPOSITION OF INTRA-CELLULAR AND EXTRA-CELLULAR FLUID

The composition of both fluids is quite different as can be seen in **Figure 13.1**. Specifically, the majority of Na^+ is present in the extra-cellular compartment along with Cl^- and HCO_3^-, while the majority of K^+ is present intracellularly along with Mg^{+2} and PO_4^{+3}. Remember that these differences are due to the action of the semipermeable membrane that can control ion flux through specific membrane channel proteins.

DIVISIONS OF THE EXTRA-CELLULAR COMPARTMENT

Although the extra-cellular compartment contains 37% of the total body water, it can be further subdivided into four other regions containing the following fluids:

1 Interstitial fluid space — consists of the fluid present within the connective tissue spaces between the cells.
2 Plasma — comprised of the blood within the various vessels.
3 Lymphatic fluid — present within the various components of the lymphatic system (e.g., lymphatic ducts, nodes,).
4 Transcellular fluid — which includes the cerebral spinal fluid in the central nervous system, aqueous and vitreous humors of the eyes, the synovial fluid within synovial joints, and glandular secretions.

MOVEMENT OF FLUID BETWEEN VARIOUS SPACES

Fluid movement between the various spaces is predicated on two mechanisms. If the fluid is moving across semipermeable membranes, the process is termed *equilibrium*. If the fluid passes through

membranes in which there are pore-like openings the driving force is dependent on hydrostatic and osmotic pressures **(Figure 13.2)**.

Accordingly, movement of fluid from the plasma or vascular space into the interstitium is dependent on the hydrostatic pressure. The complete process involved here has been previously described as the Starling Mechanism of the Capillary.

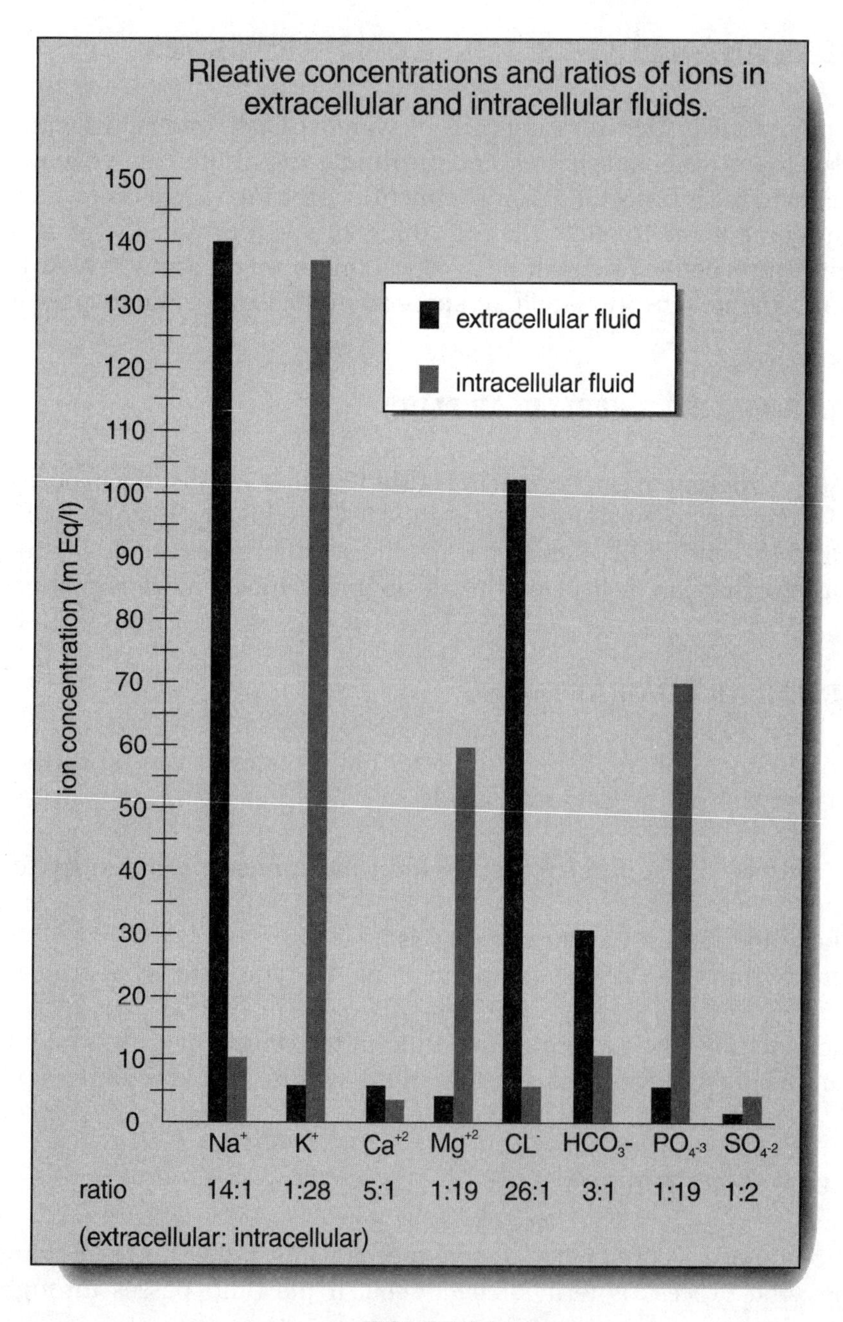

FIGURE 13.1

This movement of fluid from the interstitial space into the lymphatic system is dependent on the osmotic pressure of protein which acts as a suction force to drain the fluid from the interstitium (reducing edema).

Movement of fluid across the semipermeable membranes, separating the transcellular compartment from the extra-cellular compartment, and the inter-cellular from the extra-cellular compartment, is dependent on equilibrium. This process is specifically related to the osmosis of water. Note that osmotic water movement is affected by the salt concentration, because areas of higher salt concentration contain less water. To clarify this point consider the following hypothetical example.

1) Assume that the concentration of salt and water intra-cellularly is exactly equal to the concentration extra-cellularly (it is isotonic). Thus, no net movement of water will take place.

2) Suppose that water is lost from the extra-cellular compartment (but salt is not lost). Such an effect might take place during profuse sweating, for example. Thus, the concentration of salt would increase in the extra-cellular compartment and the concentration of water would decrease (thus the extra-cellular compartment would become hypertonic, but the intra-cellular compartment would at least initially remain isotonic). Because there is more water in the intra-cellular compartment (it is isotonic) than the extra-cellular one (it is hypertonic), the concentration gradient for *water* would drive the water out of the cells. The intercellular compartment would shrink and become hypertonic. (Note here that salt will not move because the semipermeable membrane will prevent it from passing into the cells.)

The point of the above example is that water moves by *osmosis* down its own concentration gradient depending on the salt/water concentration on the two sides of the membrane. Obviously changing the salt concentration in the extra-cellular v. intra-cellular compartment will also effect water movement.

DEHYDRATION AND WATER INTOXICATION

When the extra-cellular compartment loses water (as outlined in the above example) this is termed *dehydration*. It is a deficiency condition and may develop when water intake is restricted, accompanied by excessive sweating. Because dehydration eventually leads to loss of water from the intra-

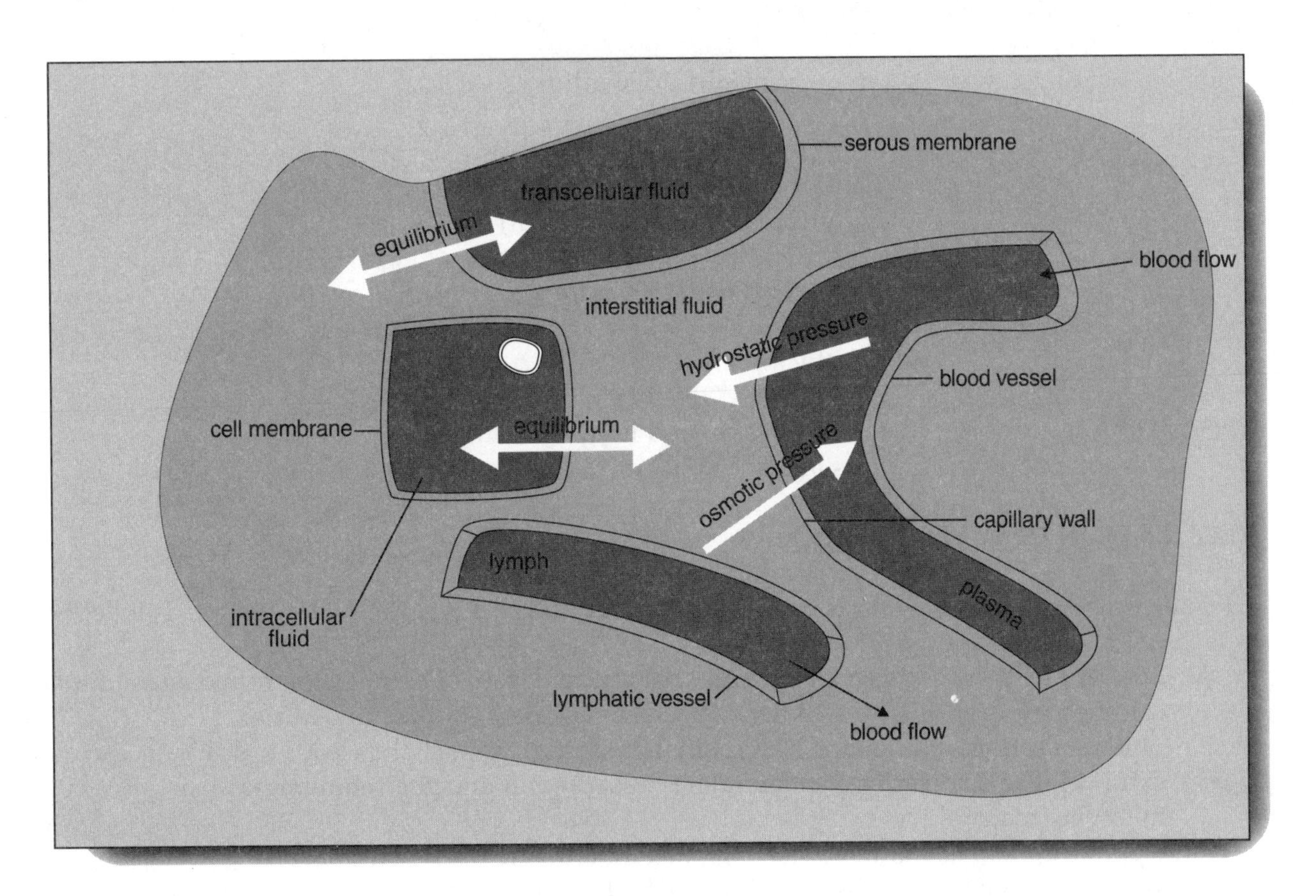

FIGURE 13.2

cellular compartment, it causes cerebral disturbances, mental confusion, delirium and coma; this is a result of changes in the salt and water concentration intra-cellularly and the accumulation of metabolic waste products.

Treatment requires the replacement of both the water and salt to maintain isotonic concentration of the fluid in both the extra- and intra-cellular spaces.

In the event that excessive water (but not salt) is replaced, this can lead to fluid shifts into the intra-cellular compartment, swelling the cells, diluting their cytoplasm and leading to *water intoxication*. Here because the fluid in the extra-cellular compartment becomes hypotonic, the osmotic driving force moves water into the intra-cellular compartment. Such an effect could be generated by drinking large volumes of pure water without accompanying salt. Symptoms of this condition include intense muscle cramping, convulsions, confusion and coma, associated with increased edema of the brain tissue primarily related to hyponatremia (low extra-cellular Na^+ levels).

GAIN AND LOSS OF WATER

The average daily water intake and loss must be equal, otherwise this will lead to either an excess or a deficiency which will alter homeostasis. Given an average daily intake and loss of 2500 ml, the sources and losses of body water are as follows:

Gain (2500 ml/day)

1. Gain from metabolism (250 ml)
2. Gain from moist food (750 ml)
3. Gain from beverages (1500 ml)

Loss (2500 ml/day)

1. Loss by sweating (150 ml)
2. Loss in feces (150 ml)
3. Loss from skin and lungs (700 ml)
4. Loss in urine (1500 ml)

REGULATION OF WATER GAIN

The major method to gain water is by drinking fluids, and this is controlled by thirst, as follows:

1. Loss of 1 to 2% of body water increases osmotic pressure of extra-cellular fluid and stimulates osmoreceptors in the thirst center which activate the hypothalamic nuclei.
2. Thirst sensation is stimulated by hypothalamic activity which then leads to drinking. Increased extra-cellular osmolarity also stimulates vasopressin and the renin-angiotensin system (see following material).
3. Upon drinking, distention of the stomach inhibits the thirst sensation.
4. Water in the GI track is absorbed into the body and decreases extra-cellular osmotic pressure.

REGULATION OF WATER LOSS

Water loss can only be effectively regulated in the kidney. Loss from other sites (e.g., skin, lungs, GI track) is not under body control. Certainly it is true that loss of water through the skin is affected by body temperature, and thus increased on hot days. However, this response is necessitated only because body temperature must be maintained within physiological limits and this is not predicated on water conservation. Mechanisms responsible for water regulation in the kidney are as follows:

Antidiuretic Hormone, ADH, (Vasopressin) — this hormone is released by the posterior pituitary and controls the permeability of the distal convoluted tubule and collecting duct in the nephron. Increasing ADH increases water reabsorption from the tubules back into the interstitial compartment. The result is the excretion of a small volume of hypertonic urine. Stimuli that release ADH include an increase in the osmolarity of the tissue fluids and activation of the Baroreceptor reflex.

Aldosterone — this hormone is elaborated from the zona glomerulosa of the adrenal cortex and is classified as a mineralocorticoid. Release is stimulated by increased levels of extra-cellular K^+, and by the Renin-Angiotensin pathway (see following material). Elevated levels of aldosterone act upon the distal convoluted tubule causing Na^+ (and under certain conditions H^+) to be reabsorbed back into the interstitial compartment and the blood. In exchange, K^+ is secreted by electrical repulsion. Since aldosterone causes increased Na^+ reabsorption back into the body, it indirectly also results in increased water retention in the body, since water follows Na^+ (by osmosis).

GFR — factors that influence GFR will also affect the amount of water lost from the body. These include: changes in the radius of the afferent and efferent arterioles, changes in systemic blood pressure, sympathetic nerves that control smooth muscle constriction of the afferent arterioles, plasma protein osmotic pressure, and Bowman's capsule osmotic and hydrostatic pressures.

ELECTROLYTE BALANCE

The average daily intake and loss of individual electrolytes must be equal otherwise this will lead to either an excess or a deficiency altering homeostasis. The processes involved in gain or loss of these substances is under the control of various mechanisms. Like water, electrolyte gain can occur in the body by eating these substances in foods, drinking them in fluids, or obtaining them from metabolic reactions. They can be lost from the body in the sweat, in the feces and in the urine.

Imbalances in Na^+ and K^+ levels can result in serious disorders in body function as follows:

1) **Hyponatremia (low Na^+ concentration)** — causes include severe and prolonged sweating, vomiting, diarrhea, renal disease and Addison's disease (of the adrenal cortex). Outcome is hypotonic extra-cellular fluid, and swelling of the intra-cellular compartment (water intoxication); symptoms include muscle spasms, convulsions, confusion and coma.

2) **Hypernatremia (elevated Na+ concentration)** — causes include loss of water from severe uncorrected Diabetes Insipidus and Diabetes Mellitus. Alterations in central nervous system function can result in confusion and coma.

3) **Hypokalemia (low K^+ concentration)** — causes include Cushing's disease (of the adrenal cortex) leading to increased release of aldosterone and loss of K^+ from the body. The effects of reduced K^+ can include muscle weakness or paralysis, atrial or ventricular arrhythmias, and the inability of the respiratory muscles to function correctly.

4) **Hyperkalemia (elevated K^+ concentration)** — causes include Addison's disease (of the adrenal cortex), use of drugs that promote K^+ retention in the kidney, renal disease associated with decreased excretion of K^+ by the kidney, or the presence of increased H^+ (acidosis) causing movement of K^+ from the intra-cellular to the extra-cellular compartment. Symptoms include severe skeletal and cardiac muscle alterations in function. Cardiac arrest can result from elevated levels of extra-cellular K^+.

ELECTROLYTE REGULATORY PATHWAYS

Calcium. Ca^{++} blood levels are controlled by two hormones: *Calcitonin and Parathyroid hormones (Parathormone)*. Calcitonin reduces blood Ca^{++} levels that are above the normal range, while Parathormone has the opposite effect.

EFFECTS OF CALCITONIN

1. Elevated level of blood Ca^{++}.
2. Calcitonin released from the thyroid "C" cells.
3. Stimulation of osteocytes in the bone matrix.
4. Activated osteocytes put Ca^{++} into the mineral matrix of the bone; (the Ca^{++} is removed from the blood).
5. Ca^{++} blood levels fall.

EFFECTS OF PARATHORMONE
SEE FIGURE 13.3

1. Blood levels of Ca^{++} are lower than normal.
2. Parathormone is released from the Parathyroid glands (four small glands embedded in the thyroid tissue, but not actually part of the thyroid).
3. Parathormone stimulates the following:
   a. Increase in the absorption of Ca^{++} by the intestine.
   b. Increase in the absorption of Ca^{++} by the kidney tubules.
   c. Increase in the excretion of phosphate by the kidney tubules.
   d. Increased activity of the osteocytes/osteoclasts in the bone matrix. Increasing reabsorption of mineral matrix releasing Ca^{++} and phosphate into the blood.
4. Due to the loss of phosphate in the kidney, coupled with the action of the intestine and release of stored minerals from the bone, blood levels of Ca^{++} increase while phosphate levels are maintained unchanged.

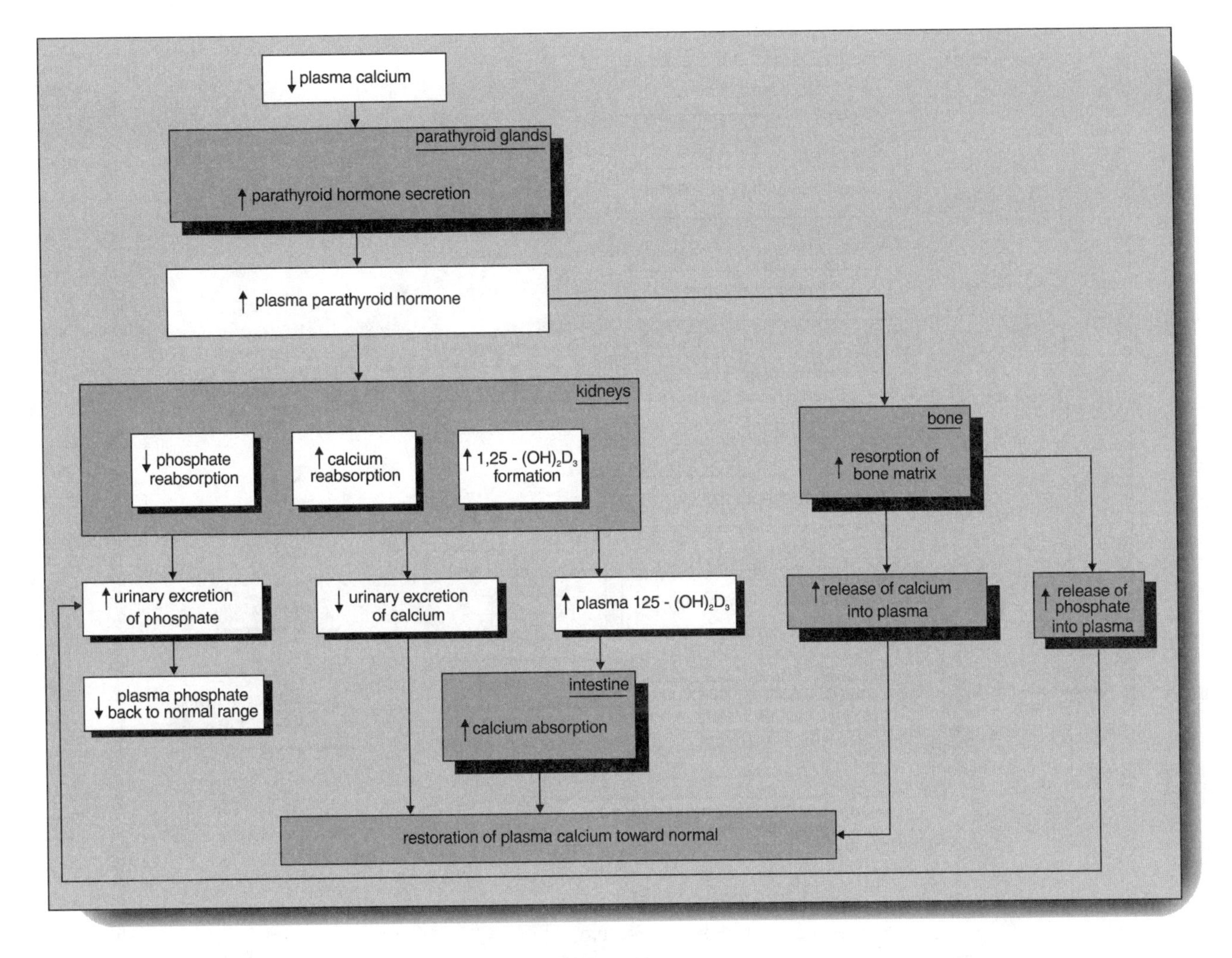

FIGURE 13.3

5. Note that one of the requirements for absorption of calcium in the GI track is that Vitamin D ($1,25\text{-}(OH)_2D_3$) be present. Vitamin D can be obtained in the diet, or can be synthesized in the skin in the presence of sunlight.

Sodium. Regulation of body Na^+ is dependent on a complex series of interactions.

1) Changes in NaCl affect *blood volume* which alter systemic blood pressure (operating through venous return). This in turn will effect GFR and loss of water in the kidney **(Figure 13.4)**.

2) Changes in NaCl will affect the release of *angiotensin* and *aldosterone*. This will in turn affect constriction of arterioles in many locations (effecting BP), and will also affect the reabsorption of Na^+ and secretion of K^+ (see following material) **(Figure 13.5)**.

3) Changes in NaCl will affect *osmolarity* of the extra-cellular fluid. Since osmolarity in this compartment is monitored by the central osmoreceptors, alterations in NaCl concentration

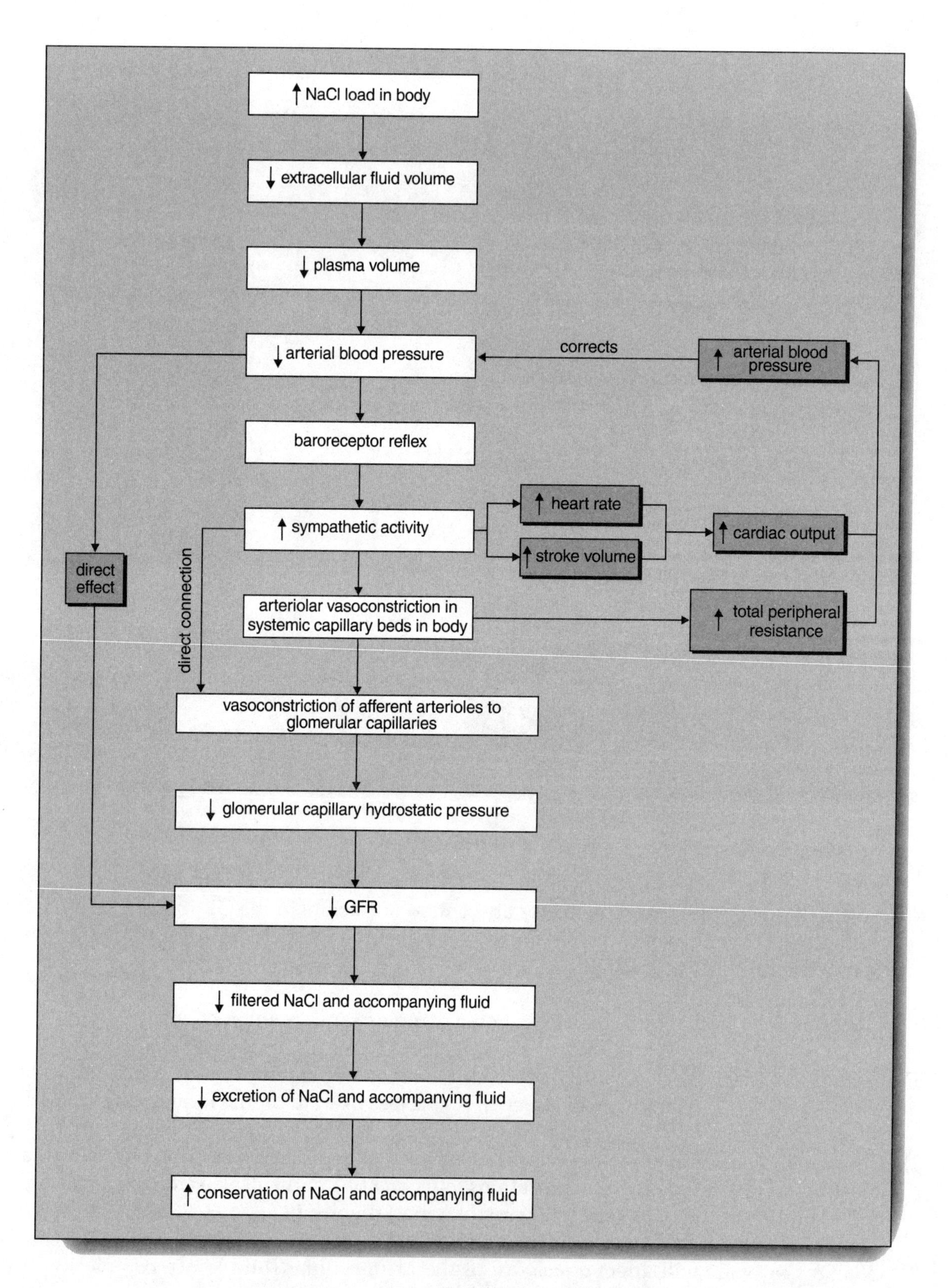
↑ NaCl load in body
↓ extracellular fluid volume
↓ plasma volume
↓ arterial blood pressure
corrects
↑ arterial blood pressure
baroreceptor reflex
↑ sympathetic activity
↑ heart rate
↑ stroke volume
↑ cardiac output
direct effect
direct connection
arteriolar vasoconstriction in systemic capillary beds in body
↑ total peripheral resistance
vasoconstriction of afferent arterioles to glomerular capillaries
↓ glomerular capillary hydrostatic pressure
↓ GFR
↓ filtered NaCl and accompanying fluid
↓ excretion of NaCl and accompanying fluid
↑ conservation of NaCl and accompanying fluid

FIGURE 13.4

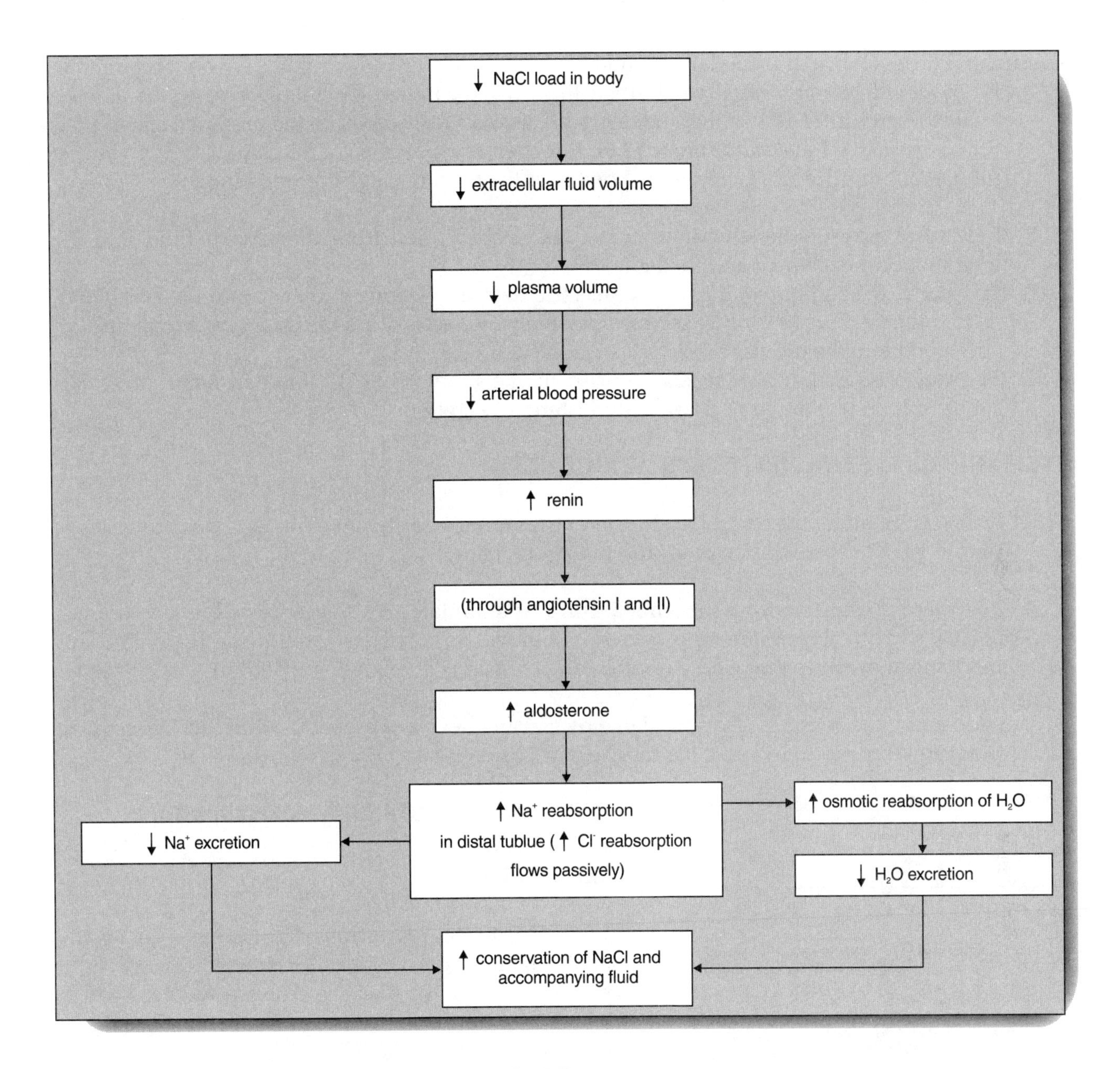

FIGURE 13.5

will affect the release of *vasopressin* (*ADH*) which will then alter the amount of fluid reabsorbed by the kidney.

4) Na^+ reabsorption within the tubules of the kidney is inhibited by *Atrial Natriuretic Peptide* (*ANF*) released from the atria of the heart when blood volume is expanded. When Na^+ reabsorption is reduced, this reduces blood volume and thus BP is also decreased. In addition, ANF also directly lowers BP by promoting arteriolar dilation.

Potassium. The levels of extra-cellular K^+ (and H^+) are inversely linked with those of Na^+ through the hormone aldosterone. Initially K^+ is freely filtered in the kidney and this process is thus affected by factors that alter GFR. All of the filtered K^+ is then reabsorbed in the proximal convoluted tubule by active transport **(Figure 13.6)**. It is then resecreted back into the distal convoluted tubule, but this process is under the control of the hormone aldosterone as follows:

1. Release of aldosterone stimulates active Na^+ reabsorption from the tubular fluid into the interstitial compartment and the blood.
2. To balance the build up of the positive electric charge collecting in the interstitium and blood, K^+ (or in some cases H^+) will be secreted passively by electrical repulsion, from the interstitium and blood into the tubular fluid.
3. The outcome is that Na^+ levels increase in the body while K^+ (and in some cases H^+) concentration will fall in response to aldosterone secretion.

MECHANISMS RESPONSIBLE FOR K^+ REGULATION

1. K^+ levels are under the control of aldosterone (as described above). Thus, as aldosterone levels rise, blood K^+ concentration falls (and blood Na^+ rises).

2. Aldosterone secretion from the adrenal cortex is stimulated by a decrease in blood volume, a decrease in blood pressure, or a decrease in blood Na^+ levels (see following material). This mechanism involves the release of renin-angiotensin (see following material).

3. Aldosterone secretion is also stimulated by elevated levels of blood K^+ which acts directly on the adrenal cortex to increase the release of aldosterone.

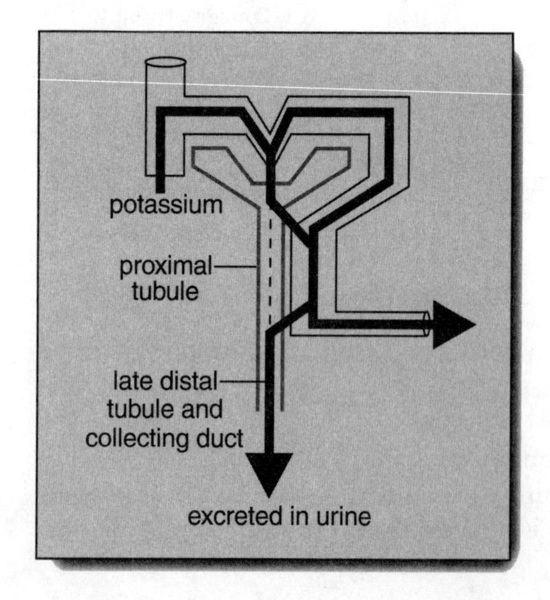

FIGURE 13.6

THE RENIN-ANGIOTENSIN-ALDOSTERONE MECHANISM
SEE FIGURE 13.7

1. The Juxtaglomerular Apparatus (JG) of the nephron monitors the NaCl concentration of fluid passing through the distal convoluted tubule. A fall in blood pressure, blood volume or salt will cause Renin to be released from the JG.

2. Renin acts as an enzyme to convert the plasma protein Angeotensinogen to Angiotensin I. (Angeotensinogen itself is synthesized and released from the liver to circulate in the blood.)

3. Angiotensin I is then converted to Angiotensin II by a converting enzyme present in the lungs.

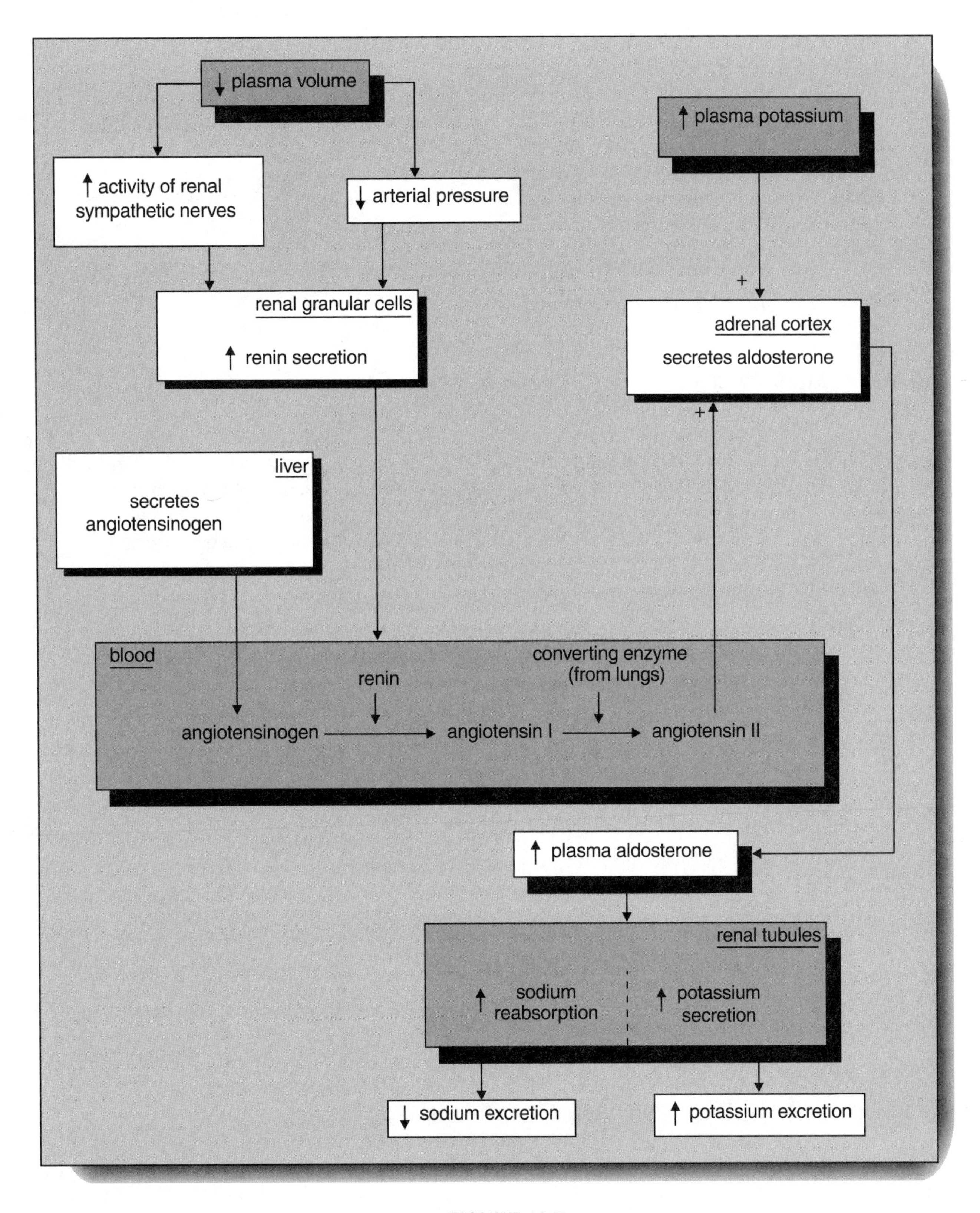

↓ plasma volume
↑ activity of renal sympathetic nerves
↓ arterial pressure
↑ plasma potassium
+
renal granular cells
↑ renin secretion
adrenal cortex
secretes aldosterone
+
liver
secretes angiotensinogen
blood
converting enzyme (from lungs)
renin
angiotensinogen
angiotensin I
angiotensin II
↑ plasma aldosterone
renal tubules
↑ sodium reabsorption
↑ potassium secretion
↓ sodium excretion
↑ potassium excretion

FIGURE 13.7

4. Angiotensin II acts on the smooth muscle of the precapillary sphincters to cause constriction and in so doing increases systemic arterial BP.

5. In addition, Angiotensin II also acts at the level of the adrenal cortex (on the zona glomerulosa) to cause the release of Aldosterone.

6. Aldosterone acts on the cells of the distal convoluted tubule to increase the reabsorption of Na^+, coupled with the secretion of K^+ (and in some cases H^+).

7. The outcome is that blood Na^+ levels will increase and blood K^+ levels will decrease. The increase in blood Na^+ promotes an increase in blood volume, which increases venous return, and leads to an increase in systemic BP.

ACID BASE REGULATION

Maintenance of body homeostasis requires that the constant low level of free H^+ be maintained. The presence of low, but controlled, concentration of free H^+ is critical to enzyme function and cellular homeostasis. This is because protein conformation, including enzyme active site interactions, depends on the concentration of H^+ within the cytoplasm. Under normal conditions H^+ levels will increase as a result of acid build up from various sources. Thus, H^+ can accumulate in the body from:

1. Anaerobic respiration of glucose leading to an increase in the level of lactic acid.

2. Aerobic respiration of glucose leading to an increase in the level of carbonic acid.

3. Oxidation of sulfur containing amino acids leading to an increase in the level of sulfuric acid.

4. Incomplete oxidation of fatty acids leading to an increase in the level of acidic ketone bodies.

5. Hydrolysis of substances containing phosphorus (phosphoproteins and nucleoproteins) leading to an increase in the level of phosphoric acid.

MEASUREMENT OF H^+

The following is presented here in the (perhaps unlikely) event that the student needs a slight refresher in the meaning of the pH scale (measurement of H^+).

The level of free H^+ is measured as the negative log of the H^+ concentration, and thus a smaller number on the pH scale represents a larger concentration of H^+ **(Figure 13.8)**. Free H^+ is highly reactive because it is actually the nucleus of a H atom containing a single proton and lacking the electron. This free proton avidly strips an electron from practically any substance it comes in contact with. This accounts for the high reactivity of H^+, and the destructive effect elicited by acids.

To understand pH, it is also important to know that an oppositely charged species called a OH^-, upon reaction with the H^+, will produce water (H_2O). Neutrality in a chemical solution is therefore the point where an equal number of H^+ and OH^- exist together, forming pure water. At neutrality a very small number of free H^+ (balanced by free OH^-) are present, producing a pH of 7.0. If the number of free H^+ is greater than the free OH^- this constitutes acid. Such acid solutions can range in pH from just below 7.0 to 0 depending on the number of free H^+, and they are acid solutions. If the number of free H^+ in a solution is less than the free OH^-, this solution is said to be alkaline or basic. Solutions that range from above pH 7.0 up to pH 14 are said to be basic.

Note that each unit of pH is indicative of a ten fold change in the number of free H^+, (thus a pH 5 solution has 10 x more free H^+ than a pH 6 solution, a pH 3 solution has 100 x more free H^+ than a pH 5 solution, and a pH 10 solution has 1000 x more H^+ than a pH 13 solution, and so forth).

MECHANISMS THAT CONTROL CONCENTRATION OF FREE H^+

Body pH should remain in the range of 7.35 to 7.45 with a mean of 7.40. To maintain the pH in this normal range two lines of defense exist that are designed to provide protection and limit pH changes. The first line of defense is supplied by chemical buffer systems that react rapidly to correct possible fluctuations. If buffer action is not sufficient for this purpose then the second line of defence kicks in and here respiratory mechanisms also respond rapidly. Finally mechanisms in the kidney are also available if buffers coupled with respiratory mechanisms are unable to bring the pH back into the normal range.

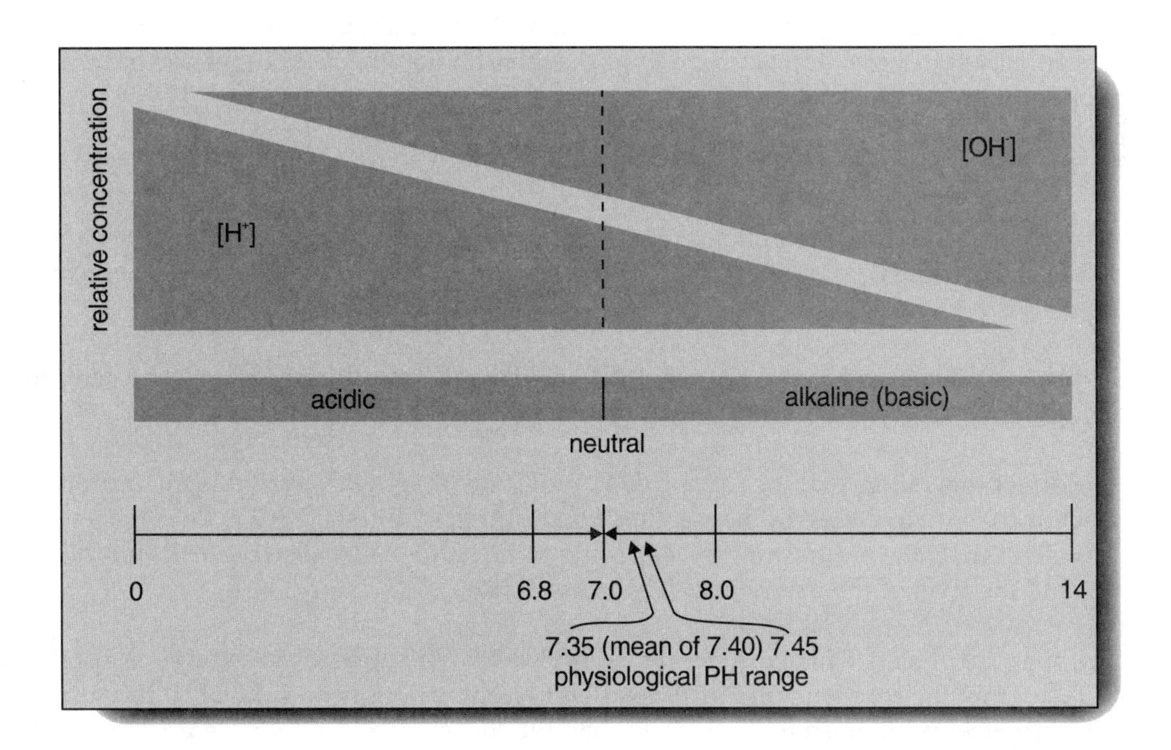

FIGURE 13.8

BUFFERS

Buffers constitute chemical reactions that minimize changes in the number of free H^+ in a solution when an acid or base is added. A buffer system consists of two compounds that react in a reversible process to supply, or remove H^+ and thereby maintain a constant pH. To understand how this is possible let us first consider what happens if you were to add HCl to water.

$$HCl \rightarrow H^+ + Cl^-$$

Under these conditions all of the original, parent acid (HCl) will dissociate into free H^+ and Cl^-. A similar situation will take place with sulfuric acid, thus,

$$H_2SO_4 \rightarrow 2H^+ + SO_4^{-2}$$

and here again the original parent acid (H_2SO_4) will be completely dissociated and $2H^+$ will be generated. Acids that completely dissociate when dissolved are said to be *strong* acids. Strong acids are very poor buffers.

On the other hand, certain types of acids do not completely dissociate in solution. Consider for example what happens when carbonic acid is placed in solution:

$$H_2CO_3 \rightarrow H^+ + HCO_3^-$$

Here some of the H_2CO_3 forms $H^+ + HCO_3^-$, but some of the original parent acid compound still remains in solution unchanged. Such a weak acid can act as a buffer because the reaction is reversible (it can move from either left to right, or from right to left) depending on the need to release or remove free H^+ from the solution.

To clarify this point , assume that:

1. This reaction is taking place in a beaker of water, where there would be some of the parent acid, H_2CO_3, some free H^+, and some HCO_3^- salt in equilibrium and the pH in the beaker (due to the free H^+) is 7.40.

2. If we were now to add to this beaker additional free H^+ (by adding some strong acid) the pH would fall to 7.1, because extra free H^+ would be present in the mixture.

3. To correct this fall in pH, the extra free H^+ will now react with available HCO_3^- forcing the reaction to move to the left by mass action, and forming additional H_2CO_3. Thus the level of free H^+ will return to the normal range (7.40), some of the HCO_3^- salt (called buffer base) will be used up, and the level of the parent acid (H_2CO_3) will rise.

4. The opposite process will take place if the pH rises above 7.40 because some H^+ is lost from the beaker. Thus, the reaction will move from left to right to supply additional H^+ to cause the pH to return to the normal range. As this happens, the concentration of the parent acid (H_2CO_3) will fall, and the concentration of the salt (HCO_3^-) will also rise.

BUFFER SYSTEMS

Remember that buffer systems are usually comprised of two (or more) sets of chemical substances. Specifically, there are three major buffer systems acting in the body.

Bicarbonate Buffer System. It is present in the intra-cellular and extra-cellular body fluids. It consists of the parent carbonic acid (H_2CO_3), and the sodium bicarbonate salt ($NaHCO_3$). If a strong acid such as HCl, is present it reacts with the sodium bicarbonate to produce the weaker carbonic acid and sodium chloride:

$$HCl + NaHCO_3 \rightarrow H_2CO_3 + NaCl$$

If a strong base like NaOH is present, it reacts with the carbonic acid to produce the weak base sodium bicarbonate plus water:

$$NaOH + H_2CO_3 \rightarrow NaHCO_3 + H_2O$$

Phosphate Buffer System. This is found in the bone matrix due to the presence here of Calcium - Phosphate salts. It consists of the compounds sodium monohydrogen phosphate (Na_2HPO_4) and sodium dihydrogen phosphate (NaH_2PO_4). If a strong acid is present it reacts with the sodium monohydrogen phosphate to form the weaker acid, sodium dihydrogen phosphate and sodium chloride:

$$HCl + Na_2HPO_4 \rightarrow NaH_2PO_4 + NaCl$$

If a strong base is present, it will react with the sodium dihydrogen phosphate to form the weak base, sodium monohydrogen phosphate and water:

$$NaOH + NaH_2PO_4 \rightarrow Na_2HPO_4 + H_2O$$

Protein Buffer System. It is found in all cells and additionally consists of the various plasma proteins of the blood as well as the hemoglobin in the RBCs. Chemical groups on the amino acids that compose the protein molecules are capable of both releasing and binding H^+. Thus, carboxyl groups can release H^+; ($COOH \rightarrow COO^- + H^+$), while amino groups can bind free H^+; ($NH_2 + H^+ \rightarrow NH_3^+$). Since proteins can both release and bind H^+, this makes them effective buffers, and an important component in the regulation of body pH, especially in the intra-cellular compartment.

RESPIRATORY REGULATION OF PH

The respiratory system is intimately involved in pH regulation because:

$$CO_2 + H_2O \leftrightarrow H_2CO_3 \leftrightarrow H^+ + HCO_3^-$$

What this formula demonstrates is that carbon dioxide can combine with water to form carbonic acid (a weak acid). The carbonic acid can then further dissociate into a hydrogen ion and bicarbonate salt. If this seems familiar consider that

$$H_2CO_3 \leftrightarrow H^+ + HCO_3^-$$

is the bicarbonate buffer system. Clearly through the above relationship respiratory system function is linked to buffer action. This relationship may also be familiar to the student given the fact that it is involved in the Chloride Shift reaction that takes place during CO_2 transport in the RBC. Remember that for this reaction to operate the conversion of $CO_2 + H_2O\ H_2CO_3$ depends on the presence of carbonic anhydrase (CA) to catalyze this reaction, and CA is present in the RBC.

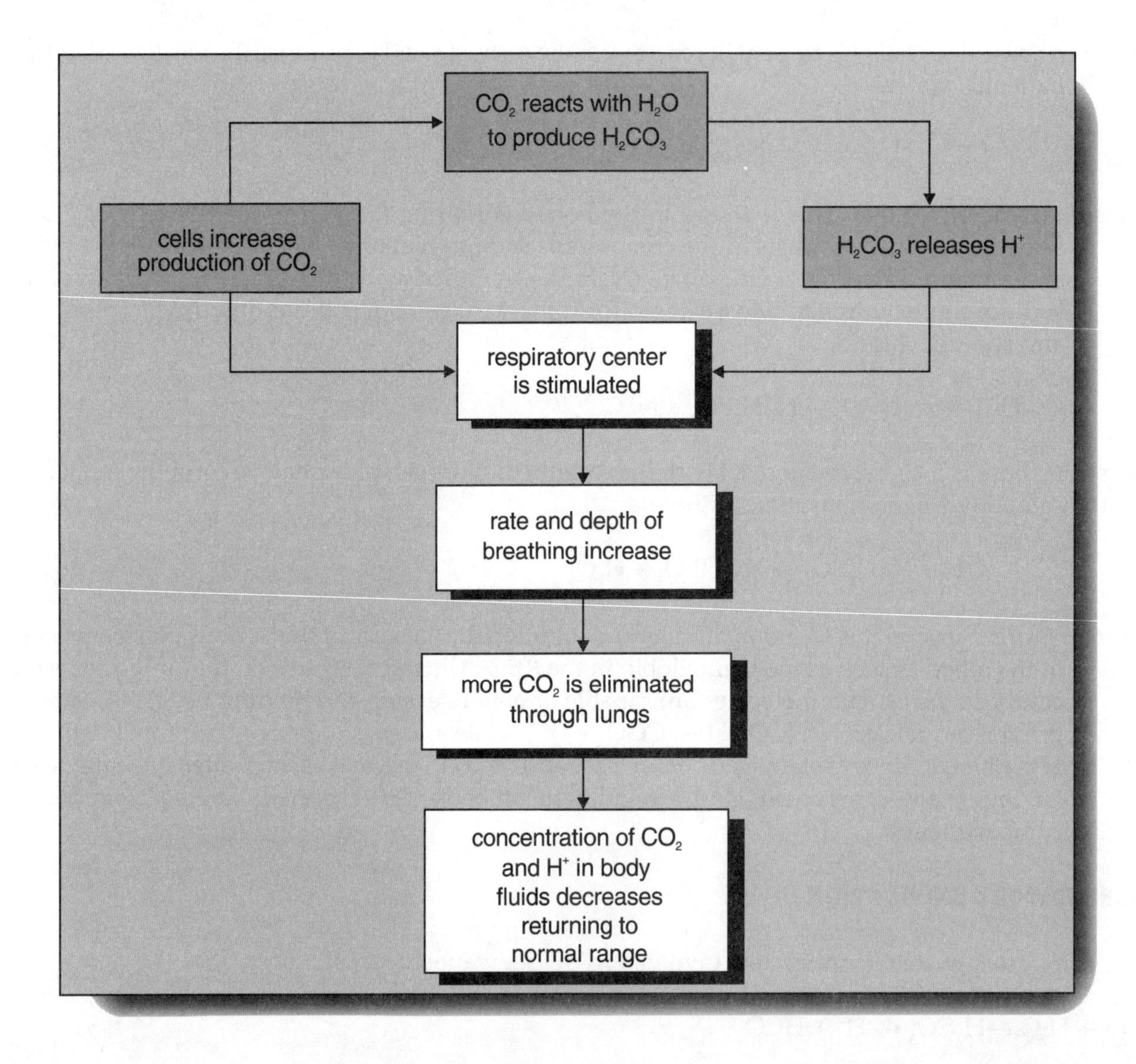

FIGURE 13.9

Consider then the connection between respiration and body pH. If excess H^+ was to build up and reduce the pH below the normal range (an acidotic condition), it would then react with bicarbonate to form carbonic acid. The carbonic acid would then, in turn, be converted into CO_2 and H_2O. Since elevated levels of H^+, as well as elevated levels of pCO_2 both act through the chemoreceptors to stimulate an increase in respiration, the outcome would be that the elevated levels of CO_2 would be blown off, accompanied by a rise in the pH back into the normal range.

Since by mass action chemical reactions operate in both directions (from left to right, or form right to left), changes in the level of pCO_2 will also alter levels of H^+, and visa versa. It is for this reason that the concentration of CO_2 and H^+ are really two sides of the same coin (so to speak) **(Figure 13.9)**. Thus, the respiratory system plays a central role in regulation of body pH in conjunction with the bicarbonate buffer system.

KIDNEY REGULATION OF PH

In the event that neither the buffer reactions nor the respiratory system is capable of maintaining a normal body pH, the kidneys are capable of secreting H^+ directly into the urine. Secretion by the epithelial cells of the nephrons takes place in the proximal and distal convoluted tubules, and in the collecting ducts **(Figure 13.10)**. H^+ secreted into the kidney tubules is then buffered by the phosphate buffer system present in the tubular fluid, and in addition buffering also takes place through the action of ammonia which easily diffuses from the cells of the renal tubules into the urine as follows:

H^+ + NH_3 NH_4^+

Note that while the NH_3 is highly permeable across cell membranes and easily enters the urine in the tubules, upon formation into NH_4^+ it becomes impermeable and must, therefore, remain within the tubules where is then excreted from the body.

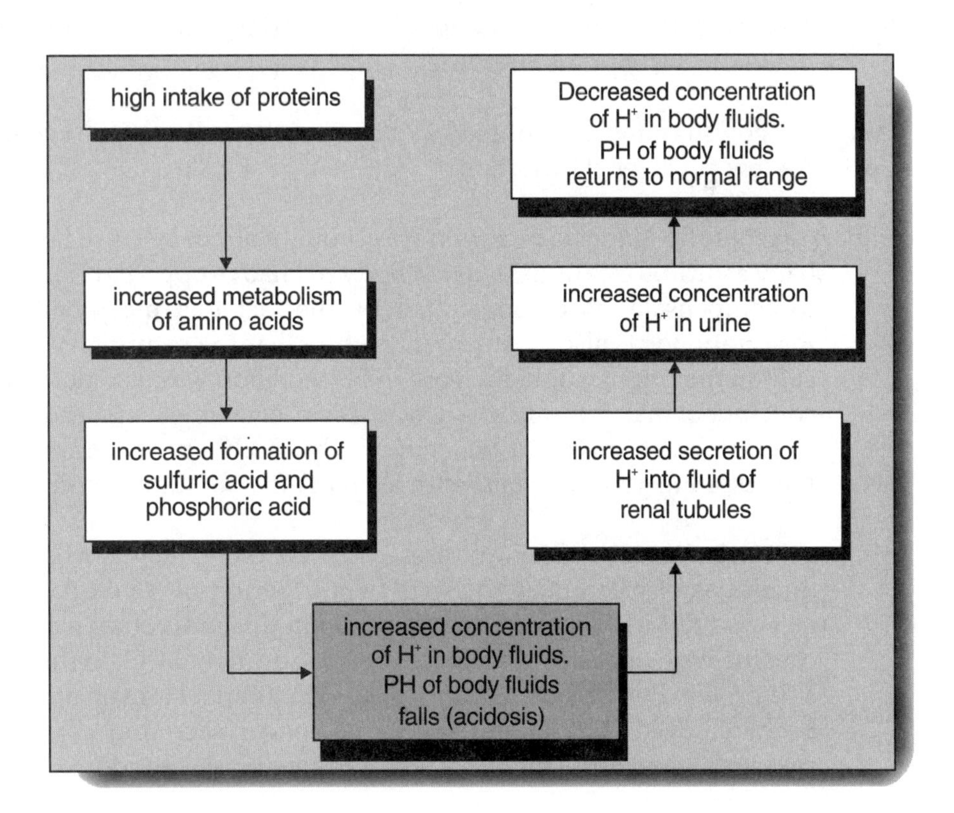

FIGURE 13.10

ACIDOSIS, ALKALOSIS AND COMPENSATIONS

To further clarify the mechanisms involved in maintaining homeostasis of the body pH, the following information is also of importance:

1. Body pH should remain within the normal range which is between pH 7.35 to 7.45; (mean pH 7.40). If the pH falls below 7.35 this is said to be acidosis, and if it rises above 7.45 this is alkalosis. Shifts in pH out of the normal range can alter homeostasis so significantly that they can be life-threatening. Death usually occurs if the pH falls to 7.0 or rises to 7.8, because such changes in the levels of free H^+ significantly effect enzyme action which critically disrupts metabolism of cells.

2. Perturbations in body pH out of the normal range can be caused by problems arising from metabolic events or respiratory malfunctions. Thus, if alterations in pH take place which lead to acidosis, and the cause is due to problems arising from metabolism, this is said to be *Primary Metabolic Acidosis*. Similarly, if alterations in pH take place which lead to alkalosis and the cause is due to problems arising from metabolism, this is said to be *Primary Metabolic Alkalosis*. If, on the other hand, the cause of the difficulty is due to a problem in the respiratory system, depending on the pH (above or below normal), this is said to be either *Primary Respiratory Acidosis*, or *Primary Respiratory Alkalosis*.

3. To correct for abnormal changes in pH, various body systems (buffers, respiratory, kidney), make the appropriate responses. These responses are said to be *compensation* effects.

Using the above terminology coupled with an understanding of the events involved in pH regulation, consider a few cases for clarification:

Primary Metabolic Acidosis in a child was brought about because he had consumed three bottles of baby aspirin. Tests indicated that body pH had dropped to 7.32. Buffer systems had attempted to compensate (as demonstrated by the fact that bicarbonate buffer base was reduced). Respiration was also elevated, indicating that the respiratory system was blowing off excess carbon dioxide. A slight increase in H^+ secretion was also noted in the urine. It was therefore concluded that the initial event (aspirin consumption) leading to the Primary Metabolic Acidosis had been compensated for fairly effectively by the action of the buffers, coupled with the respiratory system, and with a slight effect of kidney secretion of additional H^+.

Primary Metabolic Alkalosis was present in an old man who had repeatedly consumed large quantities of bicarbonate of soda over a period of seven days. Body pH was significantly elevated to 7.61. The respiratory system in this individual was attempting to compensate by causing hypoventilation and retention of additional CO_2 in the body. Bicarbonate, phosphate and protein buffers had acted to release additional H^+ in an attempt to replace the H^+ that had been removed. The kidneys were no longer secreting any H+ into the urine and were reabsorbing any H^+ that was present in the tubular fluids.

Primary Respiratory Acidosis leading to a pH of 7.21 was observed in person with pulmonary edema because the exchange of gases across the alveoli was hampered by the fluid in the interstitial compartment of the lung tissues. Elevated pCO_2 was present in the blood, and pO_2 levels were reduced (as expected). The patient was hyperventilating (in an attempt to blow off the excess pCO_2 and increase the low levels of pO_2, but since the cause of the disorder was the respiratory system itself, the hyperventilation was not effective). Since the bicarbonate buffer system was promoting the increase in H+ (through the mass action relationship with CO_2 and H_2O), the remaining phosphate and protein buffer systems had attempted to compensate, but due to the long standing acidotic condition, were no longer effective because the chemical reactants had been used up. The kidney was actively secreting H^+ into the urine and this was the only remaining line of defense to prevent the fall of body pH into the lethal range.

Primary Respiratory Alkalosis leading to a pH of 7.63 was observed in a patient who had taken a drug overdose that stimulated the sympathetic nervous system. The patient was hyperventilating and blowing off excess amounts of CO_2 which had resulted in the reduction in the levels of free H^+ in the body. The bicarbonate buffer system was involved in the problem because it was actively converting H^+ into CO_2. The phosphate buffer system and protein buffer systems had released most of their available H^+ in an attempt to compensate but the pH was still way above the normal range. The kidney was actively reabsorbing H^+ from the urine.

FACTORS LEADING TO ACIDOSIS AND ALKALOSIS

1. **Metabolic Acidosis** can be caused by kidney disease and retention of H^+, prolonged vomiting where the alkaline contents of the duodenum are lost, prolonged diarrhea, Diabetes Mellitus, or intake of acids from the GI track (like aspirin).

2. **Metabolic Alkalosis** can be caused be gastric lavage, prolonged vomiting in which only the contents of the stomach are lost, or intake of excess antacids.

3. **Respiratory Acidosis** can be caused by injury to the respiratory center of the brain stem leading to hypoventilation, obstruction of the airways, diseases that affect the respiratory membrane and reduced exchange of gases.

4. **Respiratory Alkalosis** can be caused by hyperventilation due to anxiety, high fever, poisoning from aspirin (although this can also cause metabolic acidosis) or other drugs, going to high altitudes where the low levels of pO_2 stimulate increased ventilation.

Chapter 14
THE DIGESTIVE SYSTEM

FUNCTIONS OF THE DIGESTIVE SYSTEM

DIGESTION

The primary function of the digestive system (gastro-intestinal, or GI tract, alimentary canal) is to promote the digestion of food substances. During digestion, the larger substances (proteins, carbohydrates, lipids, nucleic acids) are broken down into smaller components (amino acids, monosaccharides, fatty acids, nucleotides) and absorbed into the body to be used in metabolism.

In addition to the digestion, the GI tract also is involved in motility, secretion and absorption. These additional functions are all necessary in order for the GI tract to process the food and make the nutritional components available to body cells.

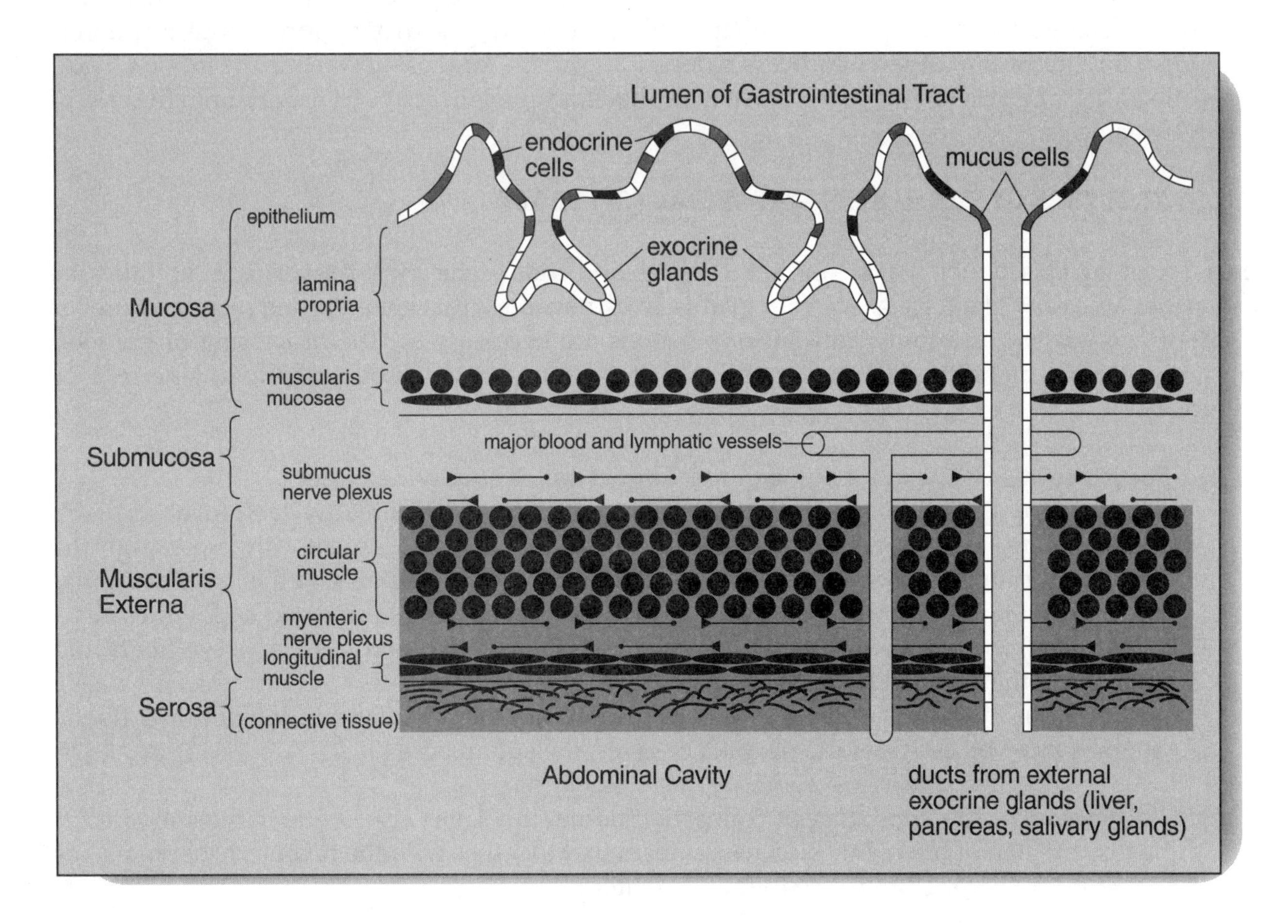

FIGURE 14.1

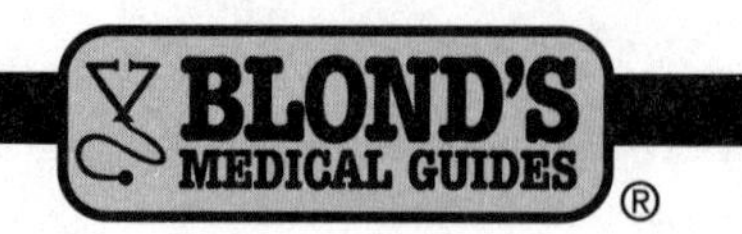

MOTILITY

Motility describes the various types of movements brought about by muscular contraction (primarily smooth muscle) that mix the food within the lumen of the stomach and intestines and propel the chyme (digested food mixed with enzymes and acid) down the lumen and eventually expel it from the digestive system.

SECRETION

Cells of the GI tract are involved in *secretion* of various substances (such as enzymes, acids and hormones) into the stomach and the lumen of the intestines. Exocrine glands involved in secretion of acid and enzymes are located in the intestines, liver, pancreas and walls of the stomach. In addition, hormones secreted by endocrine glands, located in the walls of the stomach, intestine and pancreas, are involved in regulation of digestive system functions.

ABSORPTION

Absorption is the process whereby products of digestion move from the lumen of the stomach and intestines back into the interstitium, lymphatic fluid and blood. The driving force for absorption is dependent on a number of different processes depending on the kind of substance to be moved. Thus, absorption may take place by passive diffusion, facilitated diffusion, active transport, or in the case of water, by osmosis.

WALL STRUCTURE OF THE GASTRO-INTESTINAL (GI) TRACT

The GI tract is a closed tube that begins at the mouth and ends at the anus. Associated with this tube are various accessory glands (e.g., salivary glands, liver, pancreas) that generate and secrete a number of important substances into the tube. The secretions are necessary for the processing of the food substances during digestion. Specifically the wall of the tube is comprised of the following regions **(Figure 14.1)**:

1) **The Mucosa** — is the inner lining of the GI tract and is subdivided into:
   a) Epithelium — this highly convoluted layer increases the surface area available for absorption. Cells here are linked together by tight junctions; thus, substances can only cross from the luminal side into the blood side by passing through the epithelial cells. In this layer are found mucus secreting cells (goblet cells) as well as various types of exocrine and endocrine cells.
   b) Lamina propria — comprised of connective tissue, small blood vessels, nerve fibers and branches of the lymphatic system.
   c) Muscularis mucosa — this thin layer of smooth muscle (circular and longitudinal fibers) separates the Mucosa from the next major layer, the submucosa.

2) **The Submucosa** — consisting of connective tissue, this layer also contains the submucosal plexus, a region of interconnected nerve fibers that together with the myenteric nerve plexus constitute the Enteric Brain (see following material).

3) **The Muscularis Externa** — comprised of an inner layer of circular smooth muscle fibers and an outer layer of longitudinal smooth muscle fibers, this layer is responsible for the movements (mixing, peristalsis) of the tube.

a) Sandwiched between the circular and longitudinal layer of smooth muscle is the myenteric nerve plexus, which like the submucosal nerve plexus is composed of nerve fibers of the autonomic nervous system involved in regulation of GI tract function. Called the *enteric brain*, this region is responsible for control of motility and secretion of the various regions of the GI tract. Higher nerve centers in the medulla are not required for minute to minute regulation of GI tract function since this process is overseen by the enteric brain. However, higher center inputs can override the local control exerted by the enteric brain when necessary. (For example, during stress, sympathetic stimulation to the GI tract reduces motility, secretions and blood flow effectively shutting down all the processes involved in digestion.)

4. **The Serosa** — composed of a thin layer of connective tissue, that covers the outside of the tube. It also connects the tube to the abdominal wall.

ORGANIZATION AND FUNCTIONS OF THE GI TRACT

Although there are local variations in the structure of the GI tract, the primary layers (as outlined above) are found in the wall of the tube throughout its entire length. The functions (and a limited consideration of their anatomy) of the specific regions of the GI tract are outlined here:

The Mouth is the beginning of the alimentary canal and is involved in chewing the food, mixing it with salivary secretions and initiating the swallowing process. Additionally, it is involved in speech.

The Cheeks. They limit the movement of the food inside the oral cavity during the process of chewing (mastication). In addition, they assist in keeping the bolus (the mass of food particles mixed with salivary secretions) lined up with grinding surfaces of the teeth. The inner lining of the cheeks is stratified squamous epithelium.

The Lips. These mark the external boarders of the mucous membranes lining the GI tract. They contain sensory receptors that determine the temperature and consistency of the food entering the oral cavity.

The Tongue. Thick layers of skeletal muscle are covered by mucous membrane. Rough projections on the tongue's surface, called papillae, allow the tongue to exert friction and more effectively control the movement of the bolus during chewing and swallowing. Located along the sides of the papillae are the gustatory (or taste) receptors. The tongue is also important in control of the speech process because it assists in changing the shape of the oral cavity.

Pharynx. It is located behind both the nasal and oral cavities. This muscular tube is lined with mucous membrane and divided into three regions: the *nasal pharynx* (behind the nasal cavity), the *oral pharynx* (behind the oral cavity) and the *laryngopharynx* (connected to the larynx).

The Tonsils. They are found in the three locations that open into the pharynx. The *pharyngeal* tonsils (also known as the *adenoids*) are located in the upper back of the Nasopharynx. A pair of *palatine* tonsils are located at the rear of the oral cavity on either side of the opening to the oropharynx. Finally the *lingual* tonsils are found at the rear of the tongue, also located opposite to the opening of the oropharynx.

Tonsils are associated with the immune system and are composed of lymphatic tissue. Their purpose is probably to guard the openings to the GI tract and to prevent the entrance of infective agents. How efficiently they perform is a matter of conjecture, but it is clear that they may, themselves, become a source of bacterial infection. With repeated infection, they may have to be removed through a *tonsillectomy*. Tonsillectomies can be performed to remove the Pharyngeal and/or Palatine tonsils, but although the Lingual tonsils can also be removed, this is only infrequently done because the process is extremely painful to the patient.

The Palate. It forms the roof of the oral cavity. The front portion is the *hard palate*, formed by the palatine processes of the maxillary bones in the front and the palatine bones in the back. Behind the hard palate is the soft palate, composed of muscle tissue which extends downwards to end in the small projection called the *uvula*.

The Teeth. Teeth are anchored in sockets called *alveoli* located in the upper and lower jaw (termed the *maxillary and mandible*). Their function is to break up large chunks of food into smaller pieces that can be more effectively mixed with salivary secretions, formed into a *bolus* and then swallowed. The structural shape of the tooth determines the particular function it performs. Thus, teeth can be designed for cutting (*incisors*), grasping and tearing (*cuspids*) or grinding (*bicuspids, molars*).

Between the age of 6 months to 4 years, the first set of *primary* (or *deciduous*) teeth erupt from the gums (*gingiva*) in children. During this time period 20 teeth, of the following types, will erupt:

4 Central Incisors
4 Lateral Incisors
4 Cuspids
4 First Molars
4 Second Molars

Primary teeth are lost beginning at the age of about 6 years because the root is reabsorbed into the jaw bone. Since this weakens their connection in the alveoli, the developing *secondary teeth* can now push the primary teeth out of the jaw. There are a total of 32 secondary (or adult) teeth, of the following types:

4 Central Incisors
4 Lateral Incisors
4 Cuspids
4 First Bicuspids
4 Second Bicuspids
4 First Molars
4 Second Molars
4 Third Molars

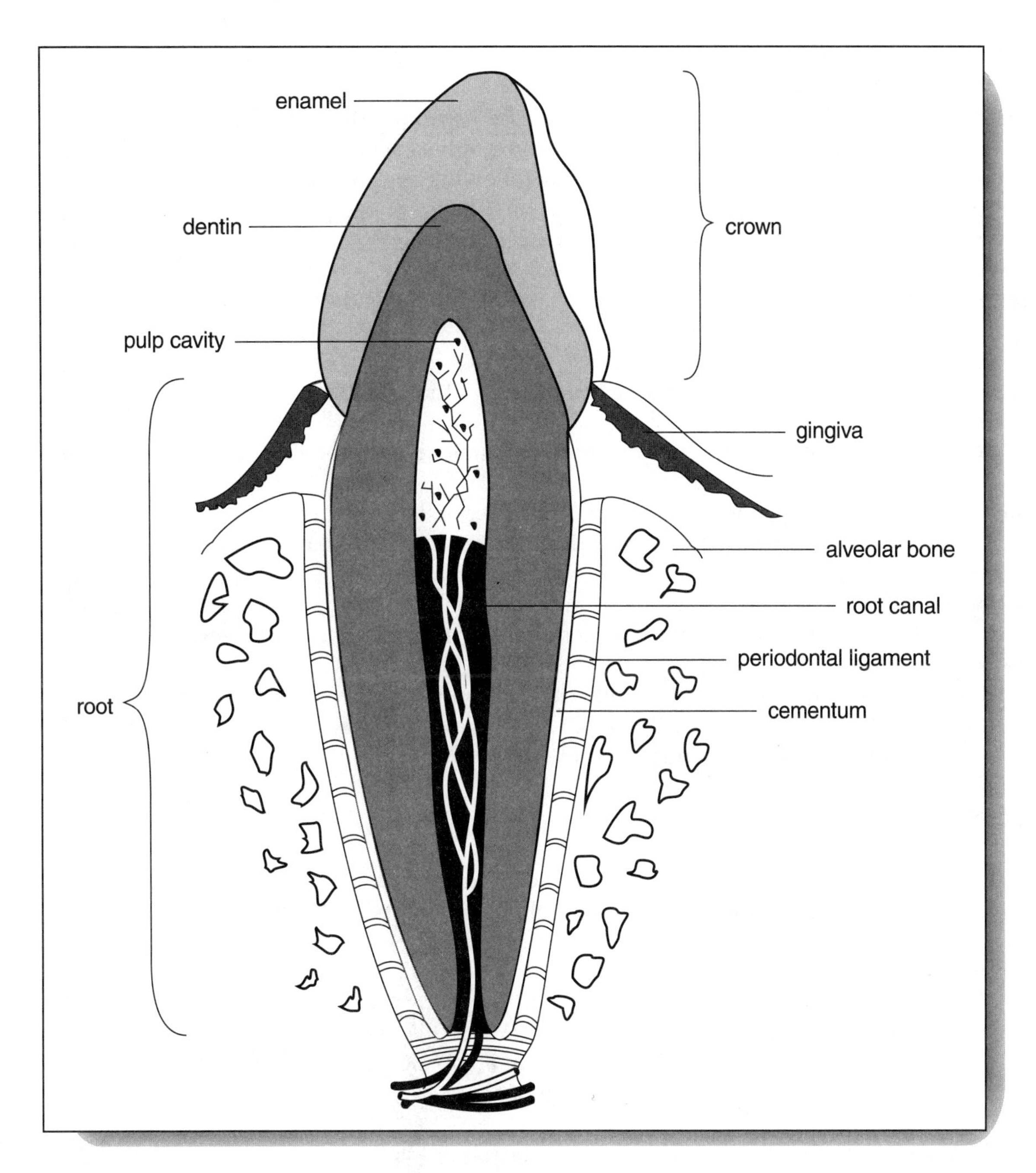

FIGURE 14.2

Third molars (also called *wisdom teeth*) may not erupt until around the age of 17 to 20 years. Eruption of these third molars may crowd and displace the other adult teeth that are already present in the jaw. In addition, these teeth may not erupt straight up through the gingiva, but may attempt to exit at an angle and become wedged against other teeth. If this happens, the wisdom teeth will have to be extracted or removed by an oral surgeon.

TOOTH ANATOMY

Anatomy of a tooth is seen in **Figure 14.2**. The tooth is subdivided into the region above the gingiva (gum), called the *crown*, and the region below the gingiva called the *root*. The root is firmly anchored in the alveolar bone of the jaw. In addition the following is also found in a tooth:

Enamel — the extremely hard mineralized matrix covering the crown.

Dentin — the inner hard mineralized matrix of the tooth. Dentin is not as hard as enamel, but it is harder than common bone.

Neck — the region where the crown and root meet at the gum line.

Pulp Cavity — located in the center of the tooth, it contains blood vessels, nerves, and connective tissue.

Root Canal — is a tubular channel opening at the base of the root and extending up to the pulp cavity, though which the blood vessels and nerves pass to reach the pulp cavity.

Cementum — a thin layer of mineralized, bone-like material that is secreted during the formation of the tooth, and which anchors the dentin of the root into the periodontal ligament.

Periodontal Ligament — a membrane that contains bundles of thick collagenous fibers, which binds the root of the tooth to the alveolar bone of the jaw. It is attached to the dentin through the layer of cementum that surrounds the root.

SALIVARY GLANDS

The salivary glands are located in the mouth and secrete saliva. Salivary secretion binds food particles together to form a bolus, and moistens the bolus to facilitate swallowing. In addition, the enzyme amylase, present in these secretions, begins the process of starch digestion. Salivary secretions are comprised of the following:

1) **Serous fluid** containing *bicarbonate* ions (HCO_3^-) is used to neutralize acid generated in the mouth by bacterial action. The presence of an alkaline pH in the mouth reduces tooth decay, and facilitates the action of salivary amylase.

2) **Mucus secretions** that assist in keeping the food particles bound together, thus lubricating the bolus for swallowing.

3) **Amylase**, an enzyme that converts starch into sugar, and which optimally functions at the alkaline pH present in the mouth. Once the bolus reaches the stomach, the acid pH present here inactivates the amylase in the bolus. However, other amylase enzymes present in intestinal and pancreatic secretions continue the process of starch digestion.

Neural Control of Salivary Secretions. The salivary glands are innervated by branches of both the parasympathetic and sympathetic nervous systems. Parasympathetic stimulation elicits copious

secretions containing serous fluid high in amylase and bicarbonate, while sympathetic stimulation produces a small volume of viscous saliva primarily containing mucus.

Anatomy of the Salivary Glands. Salivary glands are found in three locations in the mouth as follows:

1) A pair of *parotid glands* are located in the regions of the ears on the left and right sides of the face. The parotid glands are the largest of the major salivary glands and secrete their saliva through the parotid (Stensen's) ducts which open opposite the upper second molar on either side of the jaw. Histologically, the parotid glands contain many cells that manufacture serous fluid and amylase. Thus, the large volume of saliva secreted from this gland (under parasympathetic control) is very watery and contains bicarbonate and amylase.

2) A pair of *submandibular glands* (also known as the *submaxillary glands*) are located on the left and right inside of the lower jaw. The submandibular glands secrete their product through the submandibular (Wharton's) ducts which open under the tongue, close to where the tongue is fixed to the floor of the mouth (at the *frenulum*). Histologically, the submandibular gland contains cells that primarily manufacture serous fluid, but some mucus-secreting cells are also present here. Thus, stimulation of the submandibular glands by the nervous system produces a saliva which is more viscous than the secretion produced by the parotid glands.

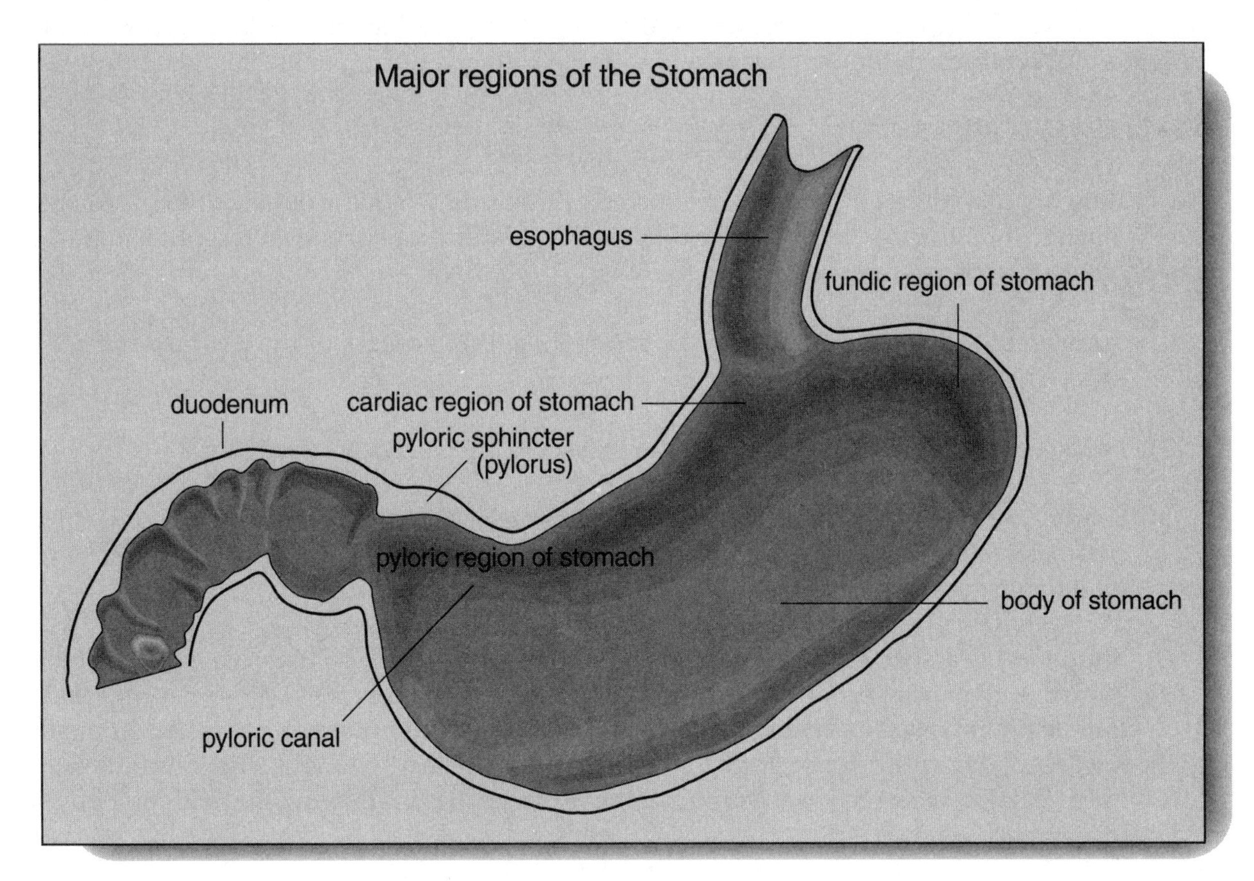

FIGURE 14.3

3) The *sublingual glands* are located under the tongue, in the floor of the mouth. A series of small ducts that open under the tongue carry the sublingual secretions into the mouth. Histologically, the sublingual gland contains mucus secreting cells, and thus the sublingual secretions appear thick and stringy.

ESOPHAGUS

A straight tube 25 cm long, the esophagus attaches to the oropharynx at the upper end and to the Cardiac region of the stomach at the lower end. At the point where the pharynx attaches to the esophagus, fibers of the inferior constrictor muscle act as a sphincter to prevent air from entering the esophagus. A second lower esophageal sphincter (the cardiac sphincter) is also present at the point where the esophagus attaches to the stomach. This cardiac sphincter is formed from a thickened region of circular smooth muscle fibers. These fibers also supply constriction to keep the contents of the stomach from refluxing up into the esophagus. During swallowing, the peristaltic waves temporarily open the sphincters, allowing the bolus of food to pass through.

The esophagus, which remains collapsed except when food moves through it, passes through the mediastinum and penetrates the diaphragm at the esophageal hiatus. Like other parts of the GI tract, the walls of the esophagus consist of an inner layer of mucosa, a middle layer of smooth muscle and an outer serosal layer. Smooth muscle within the walls is responsible for peristaltic action that moves the bolus of food through the esophagus and into the stomach during the swallowing process.

SWALLOWING REFLEX

The swallowing (deglutitation) process is initiated voluntarily when the bolus of food is forced back into the opening of the oral pharynx by the tongue. Stimulation of pharyngeal receptors in this region then promotes a series of reflex actions as follows:

1) To prevent food from re-entering the mouth, skeletal muscles of the pharynx contract. The tongue also blocks the opening of the pharynx.

2) Food is prevented from entering the nasopharynx by elevation of the uvula.

3) Food is prevented from entering the trachea because the larynx elevates. The vocal cords also contract to close off the glottis. The epiglottis also closes over the opening of glottis to further inhibit entrance of the bolus into the larynx.

4) The pharyngeal skeletal muscles contract reflexively to force the bolus down into the esophagus. Note that the upper esophageal sphincter (formed by the constriction of pharyngeal skeletal muscle fibers) relaxes as the bolus of food passes down into the esophagus. Additionally, the lower esophageal sphincter (formed by constriction of smooth muscle fibers at the terminus of the esophagus) also relaxes as the peristaltic wave reaches it, allowing the bolus of food to enter the stomach.

STOMACH

The stomach **(Figure 14.3)** wall has three layers of smooth muscle (*circular, oblique and longitudinal*) and is lined with mucous membrane. The stomach is hollow and J shaped (the inner side of the J comprises the *lesser curvature* and the outer side the *greater curvature*). Additionally, the stomach is divided into the *cardiac* region (where the esophagus enters), the *fundic* region (on the top), the *body* (or main portion), and the *pyloric* region (where the *duodenum* is attached at the *pyloric sphincter*).

Gastric Motility. This is mediated by the action of the smooth muscles in the wall of the stomach. It can be further subdivided into the following specific processes.

1) **Gastric filling** — takes place as food enters the stomach. Initially the stomach has a volume when empty of about 50 ml, but can stretch to hold a volume of about 1 lt when completely full. The wall tension during filling, however, remains constant because (unlike skeletal and cardiac muscle) smooth muscle fibers allow an increase in length to take place without a concomitant increase in tension. This process, known as *plasticity*, will only take place up to a certain level of stretch, after which the smooth muscle cells begin to contract in response. In addition, the stomach undergoes *receptive relaxation* where folds in the inner lining of the stomach wall relax as food enters the mouth, thus increasing the interior volume. This relaxation is mediated by vagal impulses.

2) **Gastric storage** — contraction/mixing movements of the stomach wall do not occur continuously, and only very weak mixing movements are present in the body of the stomach. Therefore, food may be stored within this stomach region for relatively long periods, only slowly moving into the antrum, and then into the pyloric region through the action of peristaltic waves.

3) **Gastric mixing** — pacemaker cells in the smooth muscle layers of the stomach generate rhythmic slow wave potentials. Termed *basic electrical rhythm* (*BER*), these slow wave potentials occur approximately three times per minute. This BER alone is not sufficient to cause the smooth muscle of the wall to undergo contraction unless the level of smooth muscle excitability is increased by additional electrical or chemical stimuli. If this happens, the contractions resulting during a burst of BER will result in peristaltic activity. As peristaltic contractions take place, chyme is forced towards the pyloric sphincter. Some of the more liquified chyme then passes through the partially relaxed sphincter into the duodenum by the increasing pressure build-up in this region. The sphincter then undergoes a stronger contraction when the depolarization wave reaches it, preventing the release of any additional chyme into the duodenum. Meanwhile, the thicker chyme will remain in the stomach where additional peristaltic activity continues to propel it towards the pyloric sphincter (termed *retropulsion*) and mix it with digestive juices.

4) **Gastric emptying** — takes place as a result of increased peristaltic contractions in the stomach. The strength of the peristaltic waves is dependent on the level of excitability of the smooth muscle and this is controlled by hormonal, chemical and neurological stimuli. In addition, BER can be affected by the rate of depolarization of the gastric pacemaker cells which are themselves under the control of the autonomic nervous system and gastric hormones.

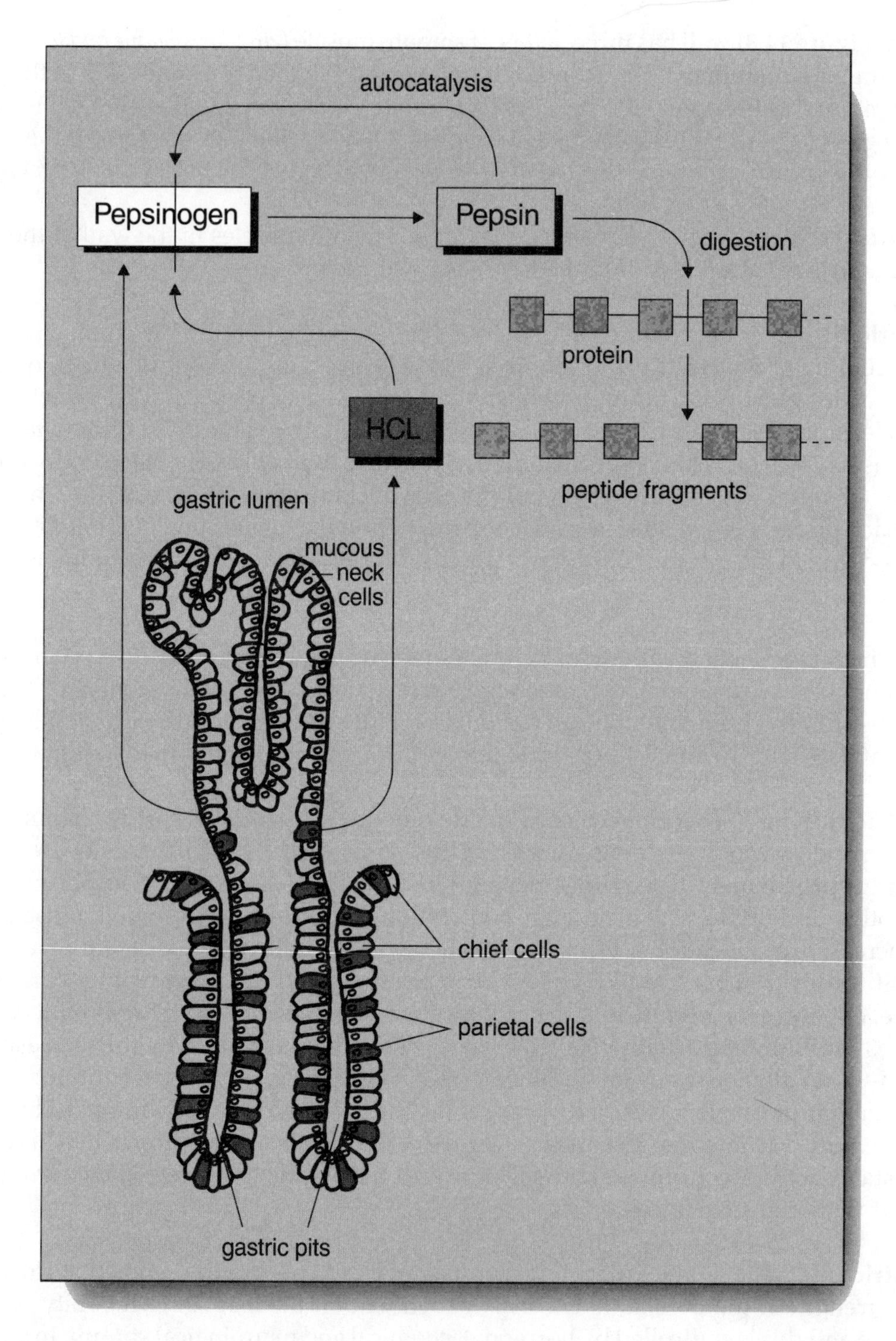
autocatalysis
Pepsinogen
Pepsin
digestion
protein
HCL
peptide fragments
gastric lumen
mucous
neck
cells
chief cells
parietal cells
gastric pits

FIGURE 14.4

Gastric Secretions. *Gastric glands* are present within the mucous membrane lining the inside of the stomach wall **(Figure 14.4)**. They open onto the lumen of the stomach through *gastric pits* in the mucous membrane. A number of cell types within the gastric glands manufacture and release secretions as follows:

1) **Goblet cells** — produce *mucus,* which functions to protect the inner lining of the stomach wall. Since mucus is comprised of a glycoprotein (carbohydrate-protein substance), it is not affected by the protein digesting enzymes present in the gastric juice. Thus, it protects the stomach wall from enzymatic degradation. Additionally, since mucus is alkaline (basic pH), it neutralizes acid present in the gastric juice and protects the wall from acid damage.

2) **Chief cells** — manufacture the substance *pepsinogen,* an inactive form of the protein-digesting enzyme *pepsin.* After pepsinogen is secreted and enters the lumen of the stomach, it is split by the action of *HCl* in the stomach juice into the active pepsin. Active pepsin then also acts *autocatalytically* on additional unsplit pepsinogen to promote the formation of more pepsin. Pepsin then functions to split large proteins into smaller polypeptide fragments. The activation of pepsin, only after it has entered the lumen of the stomach, is designed to protect the gastric glands from autodigestion by this enzyme while it is being secreted.

3) **Parietal cells** — manufacture and release HCl which is necessary for the activation of pepsin in the lumen of the stomach. Parietal cells manufacture HCl through a mechanism involving active transport of both Cl^- and H^+ to produce the H^+. CO_2 combines with H_2O to form H_2CO_3 and this reaction is catalyzed by carbonic anhydrase. A similar process has been previously described for the chloride shift reaction that takes place in RBCs. However, in the case of the parietal cell, the H^+ generated from the breakdown of the H_2CO_3 is actively pumped out of the

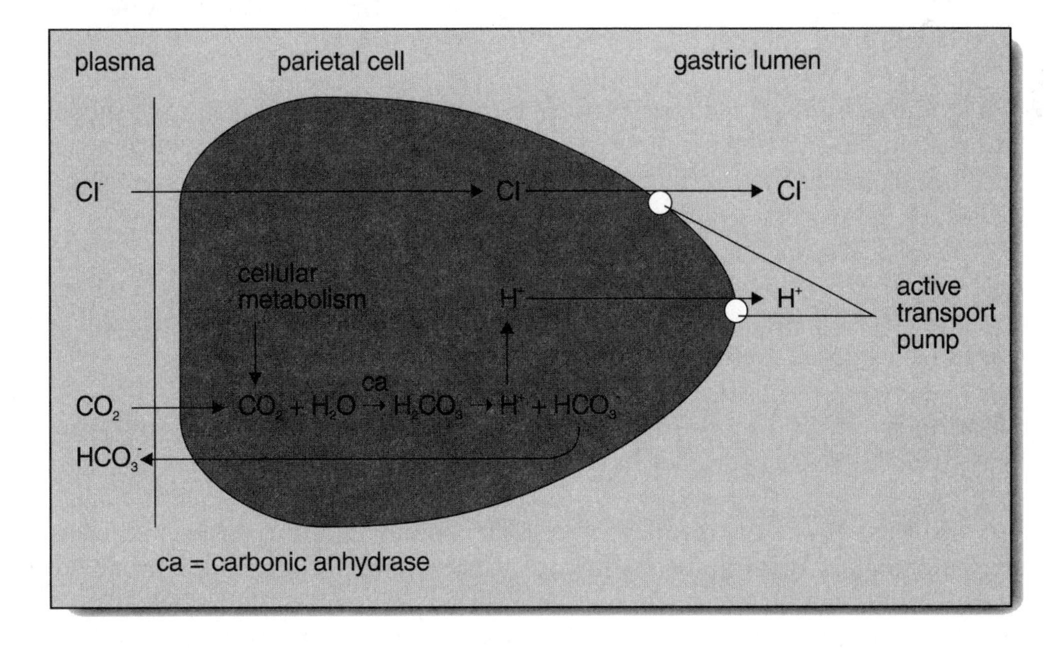

FIGURE 14.5

parietal cell and into the duct of the gastric gland. Additionally, Cl^-, which moves by passive diffusion from the blood into the parietal cell, is then actively pumped into the duct of the gland, where it works in combination with the H^+ to form the HCl released into the stomach. Bicarbonate salt (HCO_3^-), generated as a byproduct of the reaction in the parietal cell, then passes by passive diffusion back into the blood **(Figure 14.5)**. In addition to the HCl, parietal cells also release *intrinsic factor* which is necessary for the absorption of *vitamin* B_{12} in the duodenum.

Control of Gastric Secretion. Hormones released by cells within the gastric mucosa are involved in regulation of gastric secretion, motility and emptying. Specifically, the hormones involved are:

Somatostatin — this hormone is produced by specialized endocrine cells associated with the parietal cells within the gastric glands. A similar hormone is also generated by pancreatic D cells, and by the hypothalamus.

Gastric gland somatostatin acts locally (paracrin effect) to inhibit secretion of acid by the parietal cells. However, the release of acetylcholine by vagus stimulation (a branch of the parasympathetic nervous system) inhibits release of somatostatin and stimulates increased release of gastric juice.

Pancreatic somatostatin inhibits digestion of nutrients and inhibits nutrient absorption within the small intestine. This hormone is released by the pancreas in response to increased blood glucose and blood amino acids. Release of this hormone is believed to prevent excessive increase in plasma nutrients while a meal is in progress.

Hypothalamic somatostatin acts to inhibit the secretion of somatotropin (growth hormone, GH) and thyroid stimulating hormone (TSH) from the anterior pituitary. The actions of the hormone will be considered in detail in the chapter on the endocrine system.

Gastrin — is produced by G-cells located in the pyloric region of the stomach and functions on the cells of the gastric gland to increase their secretory activity, releasing gastric juice rich in enzyme and acid. Stimulation of the G-cells by the parasympathetic nervous system results in the release of gastrin.

Histamine — while not considered a classic hormone, this substance acts locally (a paracrin-like effect) to increase the release of gastric juice. Histamine is released from gastric mucosal mast cells when they are stimulated by parasympathetic impulses and by gastrin.

Phases of Gastric Secretion. To more clearly define the complex relationships involved in regulation of gastric activity, the processes have been subdivided into three phases of gastric secretion.

1) **Cephalic Phase** — as a result of the sight, smell, thought, taste and chewing of food, parasympathetic vagal impulses stimulate gastric juice secretion, accompanied by increased stomach motility. A classic example is Pavlov's experiment with the bell and salivating dogs. Vagal impulses also release gastrin which then increases the secretion of gastric juice.

2) **Gastric Phase** — the presence of food in the stomach mechanically stimulates the release of gastrin from the G cells. Gastrin then circulates in the blood back to the gastric glands where it stimulates the release of gastric juice rich in enzyme and acid. Proteins, peptide fragments, caffeine, alcohol, spices and distension of the stomach are all capable of initiating the gastric phase.

3) **Intestinal Phase** — the intestinal phase of gastric secretion can be subdivided into an *excitatory component* and an *inhibitory component*.

EXCITATORY COMPONENT

Protein fragments in the chyme that enter the duodenum of the small intestine initially will activate the excitatory component. Specifically these protein fragments stimulate cells in the wall of the intestine to release *intestinal gastrin*. It is then carried by the blood to the gastric glands of the stomach where it results in increased gastric secretions.

INHIBITORY COMPONENT

This component of the intestinal phase overrides the excitatory component. It is caused by the presence of fat, acid, and hypertonicity in the duodenum, or by duodenal distention. The inhibitory component, as its name implies, acts to inhibit gastric secretion, gastric motility and emptying through two pathways, namely the *enterogastric reflex* and hormones collectively called *enterogastrones*.

The Enterogastric Reflex. This reflex is mediated by intrinsic nerve plexuses (short reflex), and autonomic nerves (long reflex) that suppresses secretion by gastric glands and reduces stomach motility.

Enterogastrones. This name actually refers to a group of hormones secreted by cells of the duodenum. The enterogastrones currently identified are: secretin, cholecystokinin and gastric inhibitory peptide. Their functions are:

Secretin — stimulates the pancreas to secrete pancreatic juice high in bicarbonate.

Cholecystokinin — inhibits secretion by gastric glands. Inhibits gastric motility. Stimulates the pancreas to secrete pancreatic juice rich in digestive enzymes. Stimulates the gall bladder to contract and release liver bile.

Gastric inhibitory peptide — inhibits secretion by the gastric glands.

VOMITING

Vomiting, or *emesis*, constitutes a complex protective reflex activated as a result of noxious stimuli from the stomach, tactile stimuli from the back of the throat, or other inputs from nervous system and higher centers. Reverse peristalsis does not take place, and the stomach itself does not undergo active contractions during vomiting. However, the stomach, esophagus and pyloric sphincter undergo relaxation. Contraction of the diaphragm and respiratory muscles then forces the stomach contents to

be expelled. The vomit center located in the medulla controls this reflex. Nausea is believed to result from nervous system activation of the vomit center, which may not reach some necessary level required to promote the complete reflex response.

GASTRIC ABSORPTION

The only substances that are absorbed, to any appreciable degree through the gastric mucosa, are alcohol and certain weak acids (for example, aspirin). Other substances (carbohydrates, proteins and fats) are not absorbed in the stomach because they have not been degraded by the digestive process to the extent needed for absorption.

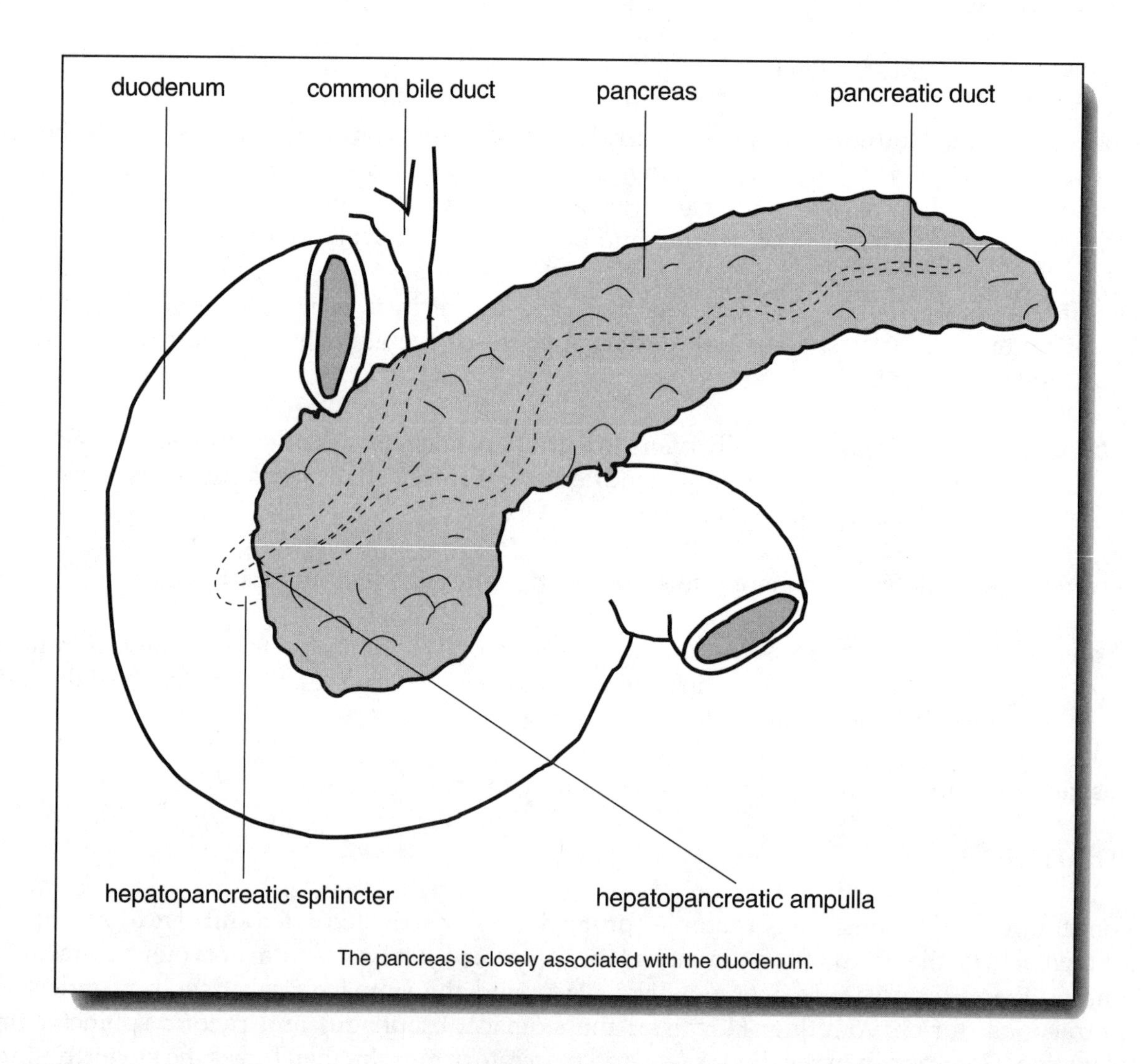

The pancreas is closely associated with the duodenum.

FIGURE 14.6

PROTEIN, CARBOHYDRATE DIGESTION AND PH

Carbohydrate digestion begins in the mouth when the salivary amylase are mixed with the food. Since salivary amylase will only function at pH 7.0, when the bolus reaches the stomach, the acid pH inactivates the amylase, but only on the surface of the bolus. Within the interior of the food mass, the pH remains at 7.0, and thus amylase continues to function. Meanwhile, the process of protein digestion begins on the surface of the bolus where the pepsin + HCl comes in contact with the food. Eventually, as the food moves into the antrum of the stomach, the increased peristaltic activity mixes the food completely with the pepsin and HCl, amylase activity is inhibited, and protein hydrolysis increases.

GASTRIC ULCERS

An intact stomach lining protects the underlying mucosa from damage by pepsin and HCl. This gastric mucosal barrier is comprised of an outer mucus layer, a luminal membrane impermeable to H^+, and the presence of tight junctions between the epithelial cells.

Additionally, since the entire stomach lining is replaced every three days, cells are removed before they are seriously damaged, which maintains an intact mucosal barrier. In the event that the barrier is breached, the wall is then injured by enzyme and acid action producing a *peptic ulcer*. The underlying causes leading to breakage of the barrier are not clearly known, but factors such as stress have been implicated.

A report has suggested that the organism called *helicobacter pylori* is a causative agent in ulcer formation, and treatment with an antibiotic is effective in curing the ulcer.

Regardless of the mechanism, after the barrier has been breached, gastric secretions can diffuse through mucosa and trigger release of histamine from the underlying mast cells within the connective tissue spaces. The increased release of histamine then acts on the parietal cells to increase secretion of HCl. The additional HCl then diffuses through the damaged mucosa, releasing more histamine, increasing HCl, and so on. The positive feedback thus created erodes the mucosa, further enlarging the ulcer and possibly producing extensive damage. Treatment with the antihistamine drug (H-2 receptor blocker) cimetidine is an effective method of blocking the histamine release and interrupting the positive feedback cycle.

PANCREAS

The pancreas has a curved head located in close proximity to the duodenum **(Figure 14.6)** and a long, straight tail. A *pancreatic duct* passing through the central core of the pancreas then exits from this organ to combine with the *common bile duct* at the *hepatopancreatic ampulla* (*ampulla of Vater*), which then enters the duodenum at the *hepatopancreatic sphincter* (*sphincter of Oddi*). The pancreas possesses both an endocrine and exocrine function.

ENDOCRINE PANCREAS

Present within the pancreatic tissue are isolated islands of endocrine cells called the *islets of Langerhans*. The islets are responsible for production and secretion of a number of endocrine hormones, the most

important of which are *insulin* and *glucagon*. Functions of these hormones will be considered in detail in the chapter on endocrinology.

EXOCRINE PANCREAS

Acinar cells are responsible for production and secretion of pancreatic juice. The enzymes secreted by these cells and their functions are as follows:

1) **Pancreatic amylase** which functions like salivary amylase to split starches and glycogen into disaccharides.

2) **Pancreatic lipase** which splits triglycerides into three fatty acid molecules and one molecule of glycerine.

3) **Trypsin, chymotrypsin and carboxypeptidase**, three proteolytic enzymes, break long protein molecules into shorter polypeptide chains, or into single amino acids. These three (plus the enzyme from the intestinal glands described below) must function together to promote complete digestion of proteins. This is because there are multiple combinations of amino acids present in proteins and a single enzyme is only specific for certain amino acid combinations.

4) **Pancreatic nuclease** which splits nucleic acids (DNA, RNA) into nucleotides.

Pancreatic (as well as intestinal enzymes) can only function in an alkaline environment. Thus, pancreatic juice also contains a high concentration of bicarbonate ions to neutralize the acid present in the chyme entering the duodenum from the stomach.

ACTIVATION OF PANCREATIC PROTEOLYTIC ENZYMES

Trypsin, chymotrypsin and carboxypeptidase are potent proteolytic enzymes. If they were released in a active form from the pancreatic acinar cells, they would digest the internal structures of the pancreas itself. Thus, like the enzyme pepsin in the stomach, these pancreatic enzymes are secreted in an inactive form and activated after they enter the duodenum. The mechanism is as follows:

1) Inactive *trypsinogen* is released from the pancreas and enters the intestine. *Enterokinase* (which is secreted by the intestinal mucosal cells) then splits the trypsinogen to form the active enzyme *trypsin*.

2) Inactive *chymotrypsinogen* is released from the pancreas and enters the intestine. Trypsin then acts on the chymotrypsinogen to form the active enzyme chymotrypsin.

3) Inactive *carboxypeptidase* is released from the pancreas and enters the intestine. Trypsin then acts on the inactive carboxypeptidase to form the active carboxypeptidase.

Regulation of Pancreatic Secretions. As described, entrogastrones are involved in regulation of secretions from the pancreas. In the presence of acid in the duodenum, the hormone *secretin* is released and stimulates the pancreas to secrete juice with a high bicarbonate content to neutralize the acid. On the other hand, when chyme that contains fat enters the duodenum, this will cause the release of *cholecystokinin* which stimulates the pancreas to produce juice rich in enzymes.

THE LIVER

Located in the upper right and central abdominal region, the human liver is comprised of two lobes (a large right lobe and a smaller left one) and receives a large blood supply. It is the single most important metabolic organ in the body and is responsible for a very large number of functions.

1) The liver processes all major categories of nutrients (proteins, carbohydrates, lipids). Lipid metabolism includes oxidation of fatty acids, synthesis of lipoproteins, phospholipids and cholesterol. The liver also converts carbohydrates into proteins and fat. An especially important function provided by the liver involves protein metabolism. Thus, the liver deaminates amino acids, forms urea, synthesizes a number of important blood proteins (including the clotting factors, blood transport proteins, and angeotensinogen), and converts various kinds of amino acids to other kinds of amino acids.

2) The liver stores a variety of substances, including fats, glycogen, copper, iron, and the vitamins A, D, and B_{12}. When there is an increase in the blood concentration of iron, it is stored in liver cells in the form of ferritin. It also activates vitamin D in conjunction with the kidneys.

3) The liver detoxifies or chemically degrades foreign substances, hormones, drugs and waste products to reduce their potentially poisonous effects on metabolism. Thus, alcohol metabolism is localized in the liver.

4) Through the action of the phagocytic cells (Kupffer cells) the liver removes foreign particulate matter (such as bacteria) and damaged erythrocytes as they circulate through the sinusoids. As part of this process these RBCs are then degraded, and the hemoglobin broken down into the pigments bilirubin and biliverdin.

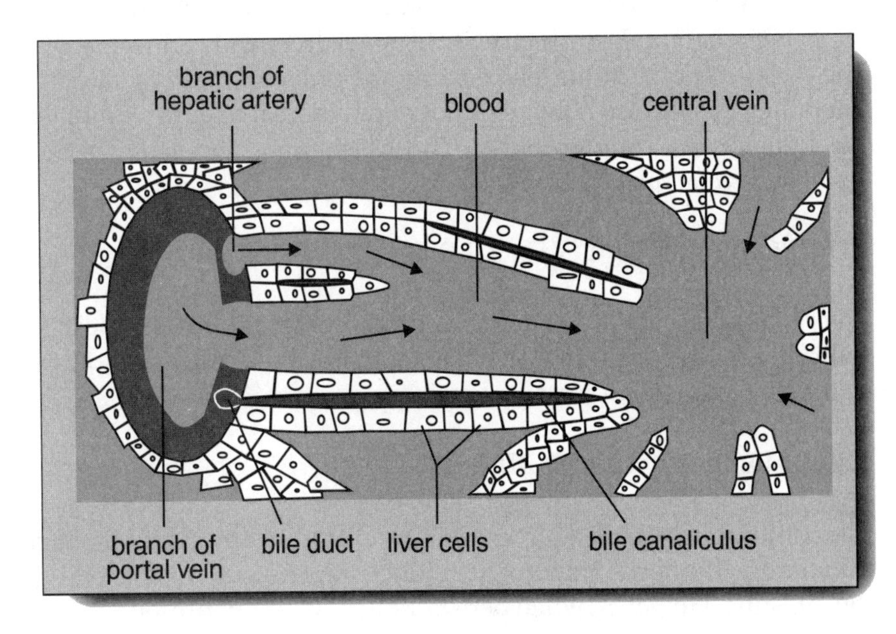

FIGURE 14.7

Iron from this process is saved and will be reused later for synthesis of new RBCs in the bone marrow.

5) The liver secretes liver bile which contains cholesterol, bilirubin and bile salts into the duodenum. The bile salts perform very important functions related to fat metabolism.

MICROSCOPIC ANATOMICAL FEATURES OF THE LIVER

The two lobes of the liver are subdivided into multiple hepatic lobules **(Figure 14.7)**. Each lobule is designed with a central vein in the middle and spoke-like sinusoids radiating outward. Hepatic cells are sandwiched between the sinusoids on one side and bile canals on the other. Thus, these cells have access to the blood passing through the sinusoids, but can also secrete bile into the bile canals. Around the outer margin of each lobule are a number of *portal triads*, each with a branch of the *portal vein*, a branch of the *hepatic artery* and a *bile duct*. The arrangement of the portal triad is such that blood from the portal system mixes with arterial blood and the mixture then passes into the sinusoids where the hepatic cells can process the blood as it passes by.

A *portal system* consists of two sets of capillary beds and two sets of veins. Therefore, the blood in the hepatic portal vessels has already drained through the capillary beds of the intestines, stomach and spleen. It thus contains a variety of digestion products not present in the systemic circulation and these substances are destined to be processed in the liver.

Once the blood has filtered through the sinusoids in the hepatic lobules where it is chemically processed, it drains into the central veins. Passing from here into the hepatic vein this blood then enters the systemic venus circulation and eventually returns to the heart.

The hepatic cells manufacture and secrete *liver bile* which is collected in the *bile canals*. These bile canals then drain into the *bile ducts* of the portal triads. Bile ducts from all over the liver then combine to form the *common hepatic duct* which exits from the liver and connects to the *cystic duct* from the gall bladder. The *common bile duct* thus formed after the union of these two ducts (common hepatic duct and cystic duct) then passes down to the duodenum where it usually combines with the pancreatic duct to form the hepatopancreatic ampulla. The *hepatopancreatic ampulla* then enters the duodenum at the *hepatopancreatic sphincter* (of *Oddi*).

COMPOSITION OF LIVER BILE

Liver bile is basically water and in addition contains bile salts (derived from cholesterol), bile pigments (bilirubin, biliverdin derived from hemoglobin breakdown), cholesterol (which has no special function in the bile) and electrolytes.

THE GALL BLADDER AND BILE RELEASE

Bile continuously secreted from the liver enters the common bile duct. During peristaltic activity in the duodenum, the hepatopancreatic sphincter opens to allow this bile (plus pancreatic juice) to squirt into the intestine. However, when peristaltic activity is not present, the sphincter remains closed. When the sphincter is closed, the bile then backs up into the cystic duct and instead enters the gall bladder where it is stored. While the bile is stored in the gall bladder, electrolytes and water are reabsorbed, causing the remaining bile salts, bile pigments and cholesterol to become concentrated. If cholesterol precipitates out of the solution, it can form solid crystals called gallstones. These can lead to *obstructive jaundice* and pain. Medical intervention is necessary to correct the problem.

During the digestive process, food in the duodenum stimulates the release of various regulatory hormones from the intestinal mucosa. One of these hormones (*cholecystokinin*) causes the gall bladder

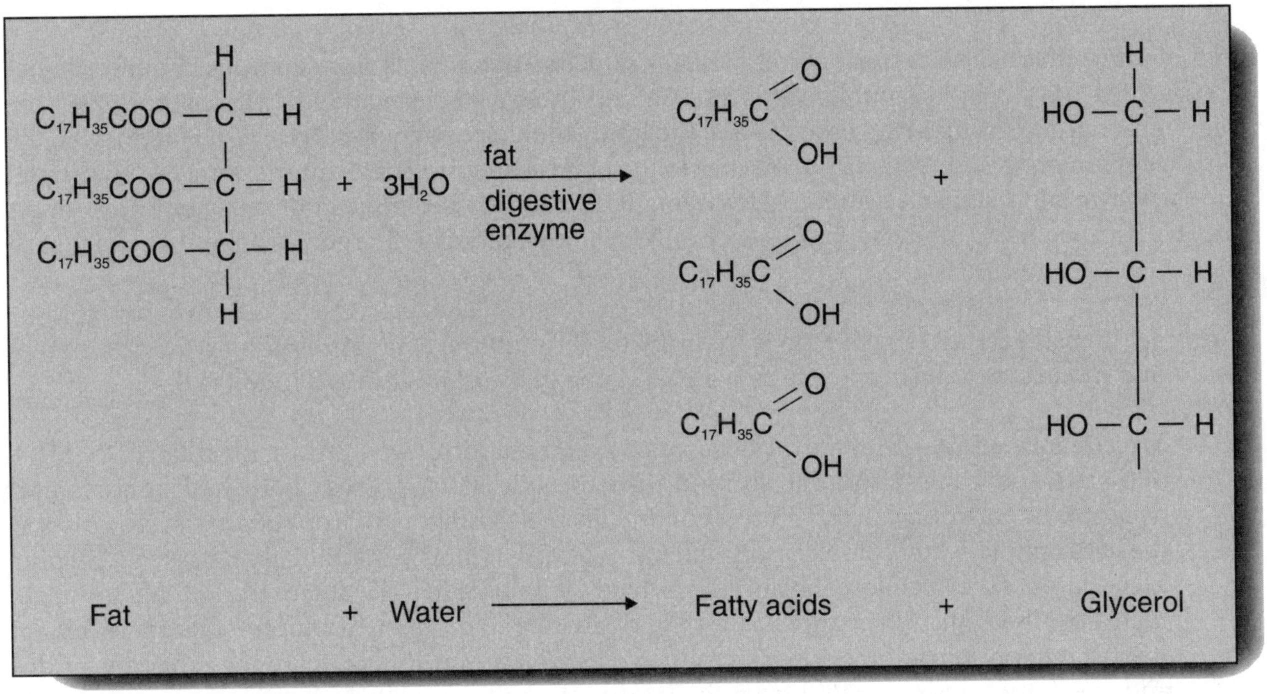

$$\begin{array}{c} C_{17}H_{35}COO-CH_2 \\ C_{17}H_{35}COO-CH \\ C_{17}H_{35}COO-CH_2 \end{array} + 3H_2O \xrightarrow[\text{digestive enzyme}]{\text{fat}} 3\,C_{17}H_{35}COOH + \begin{array}{c} HO-CH_2 \\ HO-CH \\ HO-CH_2 \end{array}$$

Fat + Water → Fatty acids + Glycerol

FIGURE 14.8

to contract, releasing liver bile. The bile is forced into the common bile duct and enters the intestine when the sphincter opens during peristaltic contractions. (Keep in mind that cholecystokinin has a wide range of additional actions in the digestive system that have been previously described.)

FUNCTION OF BILE SALTS IN FAT DIGESTION AND ABSORPTION

The most important component of liver bile are the bile salts, which are used during the process of fat digestion and absorption. Most of the bile salts that take part in these processes are not lost in the feces, but the majority (95%) are instead reabsorbed back into the body through an active transport mechanism located in the *terminal ileum*. The bile salts then return to the liver by means of the hepatic portal system where they are resecreted in the liver bile. This recycling is termed *enterohepatic circulation*.

Bile salts are important for two reasons. First, they possess a *detergent action*, and second, they are responsible for the formation of *micelles*.

1) **Detergent Action of Bile Salts** — bile salts are responsible for the emulsification of large droplets of fat into tiny fat droplets that are suspended in the aqueous chyme. By allowing these tiny droplets of fat to mix with the chyme, there is a massive increase in the surface area available to the enzyme lipase, which itself is water soluble. This increases the rate of fat digestion by this enzyme allowing it to more efficiently convert the triglycerides of the fat into fatty acids and glycerol.

Keep in mind that bile salts function, like other emulsification agents (for example, common soap), because they have a lipid soluble group at one end (a steroid derivative of cholesterol), and the negatively charged group at the other. While the lipid soluble portion dissolves in the fat droplet, the charged region projects outward and attaches to the polar water molecule, thus linking the fat droplet with the water and making it soluble. In addition, the small fat droplets cannot recombine into larger droplets because they are individually enclosed by the negatively charged groups that repel each other.

During fat digestion, each triglyceride is hydrolyzed through the action of lipase by the addition of three water molecules to form three fatty acids and one glycerol molecule **(Figure 14.8)**.

2) **Micelle formation** — they are formed from the aggregation of lecithin (a constituent of the bile), bile salts and cholesterol. The lecithin, like the bile salts, possess both hydrophobic and hydrophilic ends and can therefore form tiny droplets with the hydrophobic cores inward and the hydrophilic, water soluble groups facing outward. Water-insoluble substances can dissolve in the cores of the micelles. The micelles, which are soluble in the water of the luminal fluid, are then absorbed into the epithelial mucosa carrying the water-insoluble substances along. Important water-insoluble substances that are absorbed in this way are the monoglycerides and free fatty acids generated from the fat digestion, lipid soluble vitamins and cholesterol.

THE SMALL INTESTINE

The small intestine is subdivided into the *duodenum, jejunum and ileum*, and is anatomically 18 to 20 feet long (5.5 to 6.0 m). However, in a living individual, the muscular walls remain partially contracted (muscle tone) and therefore the length is reduced to about 10 feet. Next, the duodenum is attached to the pyloric region of the stomach and is about 25 cm long. Secretions from the liver and pancreas enter the duodenum and the mucosa contains intestinal glands that secrete various substances (see following material).

In addition, specialized mucus secreting glands (Brunners glands) are found in the submucosa of the proximal duodenum. These specialized glands are not present in any other region of the small intestine. The remainder of the small intestine is comprised of the jejunum and ileum, which cannot be clearly differentiated anatomically from each other. However, the jejunum has a slightly larger diameter, a thicker wall and demonstrates more contractile activity than the ileum.

In the small intestine, the inner lining must have a very great surface area to maximize the processes of absorption. This large surface area results from:

1) Numerous interior circular accordion-like folds (*plicae circulares*) in the wall of the small intestine which are most clearly defined in the lower duodenum and upper jejunum;

2) The entire inner surface contains very small structures called the *intestinal villi* that project into the intestinal lumen, and;

3) The outer covering of each individual villus is formed from a single layer of simple columnar epithelium. The lumen side of these epithelial cells contain brushlike boarders called *microvilli* that also greatly increase the available surface area.

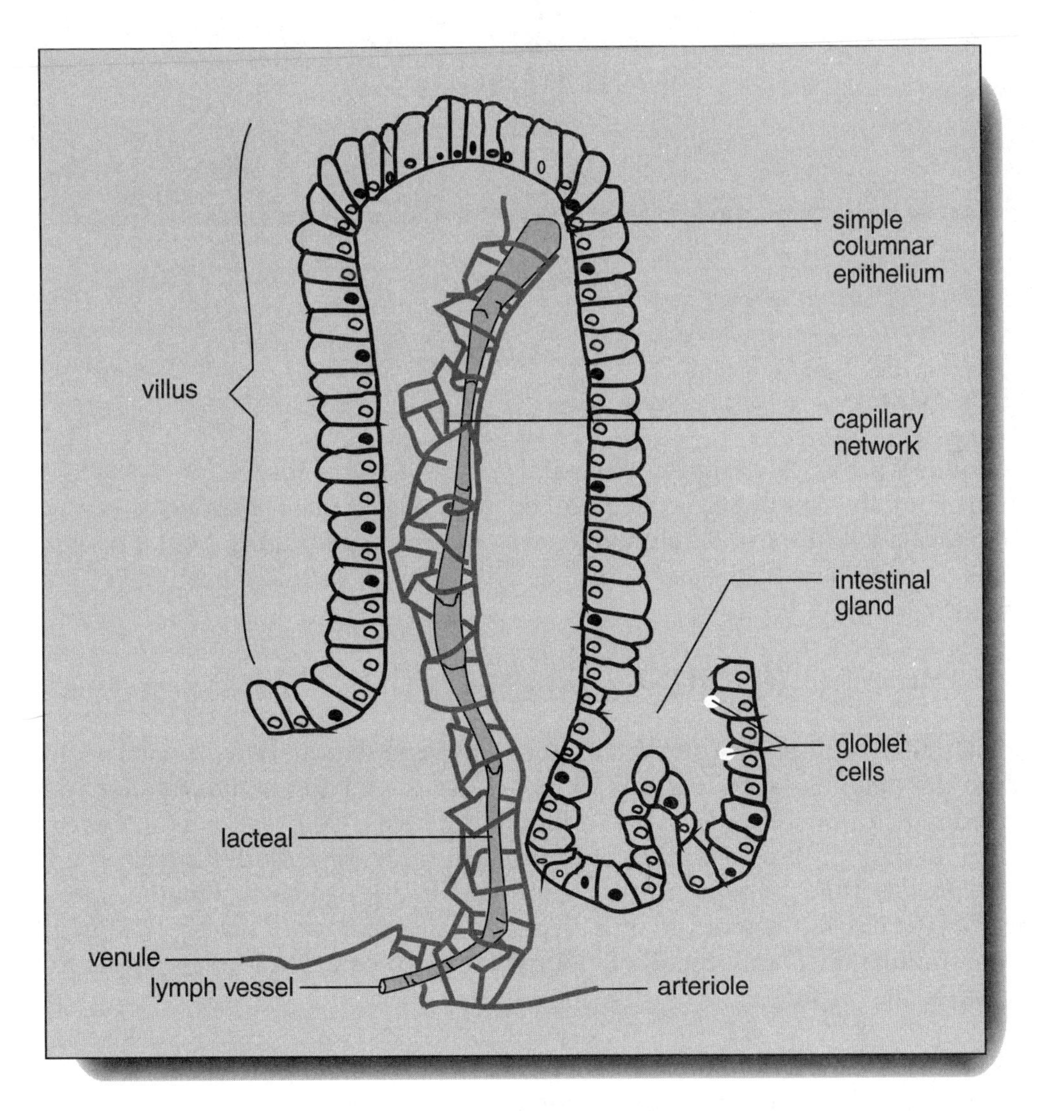

FIGURE 14.9

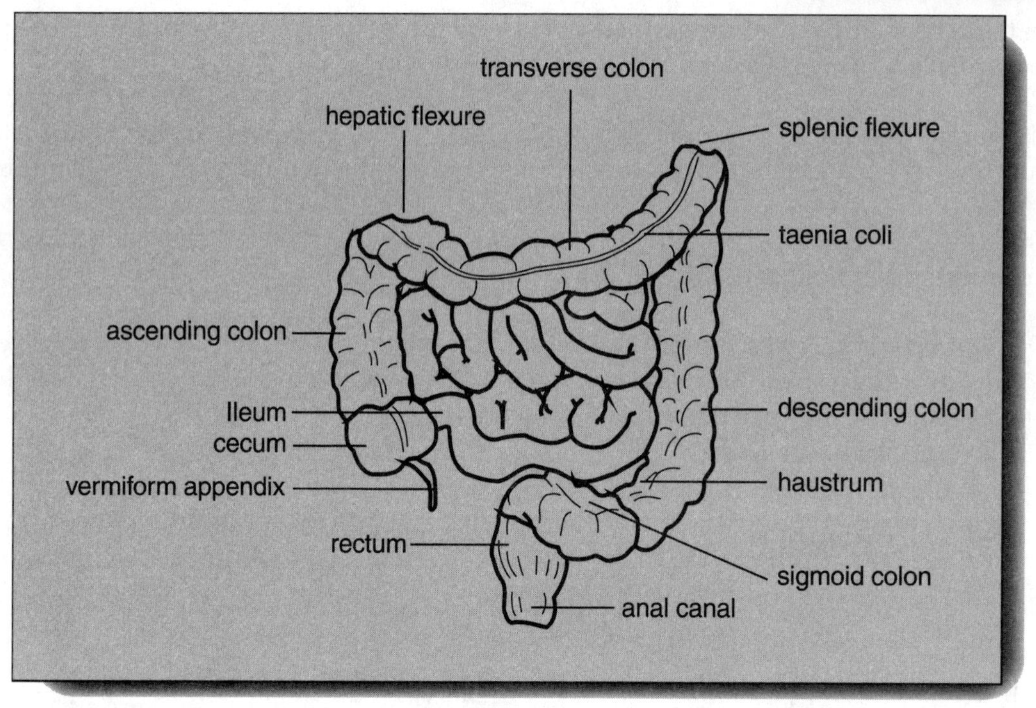

FIGURE 14.10

VILLUS STRUCTURE

An individual villus is primarily designed to maximize the process of absorption. Located within each villus is a branch of the lymphatic system called a *lacteal*, a *small arteriole*, a *small venule*, an interconnecting *capillary network*, and autonomic *nervous innervation* **(Figure 14.9)**. Present at the base of each villus is a *tubular intestinal gland* (*Crypts of Liberkuhn*), which is the source of the intestinal mucosal secretions.

MOTILITY OF THE SMALL INTESTINE

Within the small intestine, the most common form of motion is stationary segmental contraction and relaxation of the smooth muscle. During this process, there is a mixing of the chyme trapped in each segment coupled with a slow propulsion into adjacent segments. Contraction of localized regions of circular smooth muscle every few centimeters produces the segmental effect. Pacemaker cells produce a *basic electrical rhythm* (*BER*) similar to that generated in the stomach, but the intensity in the responsiveness of the circular smooth muscle depends on distention, coupled with extrinsic autonomic nervous stimulation. Parasympathetic stimulation enhances the effect while sympathetic stimulation reduces the response.

In addition to the segmental action, between meals weak peristaltic waves (called *migrating motility complexes*) proceed from the beginning of the duodenum, move a short distance and are extinguished. Additional waves then begin at the point where the first wave ended and continue farther down the intestine. The short bursts of peristaltic activity gradually clear the small intestine of its contents, emptying it into the large intestine.

As the contents of the ileum are expelled into the cecum of the large intestine, they must pass through the *ileocecal valve* **(Figure 14.10)**. This sphincter is closed normally to prevent the contents of the large intestine from refluxing back into the small intestine. However, when food enters and stretches the stomach during eating, this activates the *gastroileal reflex* (mediated by parasympathetic nerves) and increases peristaltic action in the ileum. The ileocecal valve opens and material from the ileum now enters the large intestine. This helps to clear the small intestine in preparation for chyme entering it from the stomach. Additionally reflex activity that assists in this process includes the *gastroenteric reflex*, and *duodenocolic reflex* (this last stimulates defecation).

If the intestinal wall is stimulated by over-distension or by inflammation, this results in a *peristaltic rush*. During a peristaltic rush, strong peristaltic contractions pass over the entire length of the small intestine sweeping all its contents rapidly into the large intestine. Since this rapid movement prevents adequate reabsorption of water and salts from the chyme, the result is a liquid chyme entering the large intestine which produces diarrhea.

SECRETIONS OF THE SMALL INTESTINE

The intestinal glands located at the base of the villi do not produce enzymatic secretions but instead are responsible for the production and release of *succus entericus*, an intestinal juice that is composed of aqueous salt and mucus. A relatively large amount of this secretion (1.5 lt/day) is produced in response to the presence of chyme in the intestine. Nonetheless, even though intestinal secretions do not contain any digestive enzymes, such enzyme proteins are physically embedded in the epithelial cells that compose the wall structure. Close examination indicates that these enzymes are part of the hair-like projections forming the *brush boarder* on the luminal surface of these epithelial cells. Enzyme action provided by the brush boarder is as follows:

1. **Peptidase** which acts on peptides to split them into amino acids. (Remember that these peptides were previously formed from proteins through the action of gastric and pancreatic enzyme.)

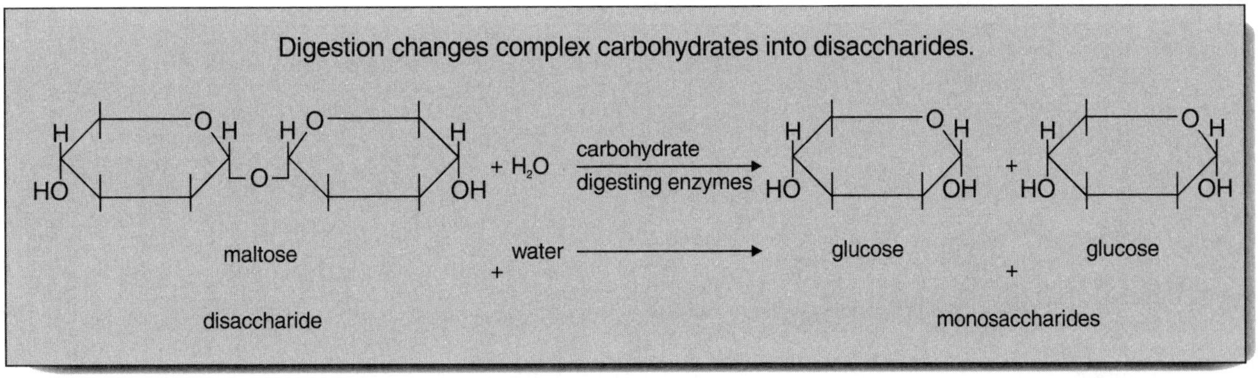

Digestion changes complex carbohydrates into disaccharides.

maltose + H_2O —carbohydrate digesting enzymes→ glucose + glucose

maltose (disaccharide) + water → glucose + glucose (monosaccharides)

FIGURE 14.11

2 **Sucrase, maltase and lactase** which act on the disaccharides, (sucrose, maltose and lactose) to form the monosaccharides, (glucose, fructose and galactose). Remember that these disaccharides were themselves formed by the action of salivary and pancreatic amylases.

3. **Intestinal lipase** which functions like the pancreatic lipase to form fatty acids and glycerol from fats.

4. **Enterokinase** is responsible for activation of pancreatic trypsinogen into trypsin. The active trypsin then activates the remainder of the cascade to form chymotrypsin and carboxypeptidase.

DIGESTION AND ABSORPTION IN THE SMALL INTESTINE

Within the small intestine, digestion of carbohydrates, proteins, and fats reaches its final phase when these compounds are broken down to their smallest constituents (monosaccharides, amino acids, fatty acids and glycerol) as follows:

1. Disaccharides are converted to monosaccharides through hydrolysis in the presence of the appropriate enzyme. For example, one molecule of maltose is converted into two molecules of glucose in the presence of maltase **(Figure 14.11)**.

2. Dipeptides are converted into amino acids by hydrolysis in the presence of the a peptidase **(Figure 14.12)**.

3. Fats are converted into three fatty acids and glycerol by hydrolysis in the presence of lipase **(Figure 14.8)**.

The products of this digestion process are then absorbed into the intestinal villi through special mechanisms individually tailored to maximize the process for each substance.

Sodium. It is absorbed both passively and by active transport.

The amino acids that result from protein digestion are absorbed by intestinal villi and enter the blood.

$$H-N(H)-C(H)(R)-C(=O)-N(H)-C(H)(R)-C(=O)-OH + H_2O \xrightarrow[\text{digestive enzyme}]{\text{protein}} H-N(H)-C(H)(R)-C(=O)-OH + H-N(H)-C(H)(R)-C(=O)-OH$$

peptide (portion of protein molecule) + water → amino acid + amino acid

FIGURE 14.12

Passive Na^+ transport. This occurs if there is an electrical gradient present. Here negatively charged substances on blood side of the epithelial cells attract the positively charged sodium to move from the luminal side of the epithelial cells. The Na^+, passing through leaky tight junctions, then collects in the lateral spaces on the blood side.

Active Na^+ transport. This results if the movement is energy dependent. In this case the Na^+ passively enters the epithelial cell on the luminal side because, either it is co-transported with glucose or amino acids, or because it enters by itself. However, in either case the Na^+ is then actively pumped out of the epithelial cell on the basolateral boarder (blood side) into the lateral spaces. This active pumping requires energy to be utilized. Once the Na^+ is in the lateral spaces it will then passively diffuse into the capillaries.

Water. Absorption depends on the active absorption of Na^+ into the lateral spaces. The presence of the increased concentration of the Na^+ in the lateral spaces generates an osmotic gradient that induces water to be absorbed between the epithelial cells through leaky tight junctions. A localized increase in the hydrostatic pressure in the lateral spaces then forces the water to move into the capillaries in the villi.

Carbohydrates. These are initially converted into the disaccharides by enzyme action, then further degraded into monosaccharides by the enzymes within the brush boarder of the epithelial cells. Glucose and galactose formed in this way are then both absorbed through a shared Na^+, energy-dependent, carrier mediated mechanism; while fructose is absorbed by a process involving facilitated diffusion.

Proteins. They must be converted into amino acids before they can be absorbed. Both ingested proteins from the food we eat, and *endogenous proteins*,are absorbed in the intestine.

Endogenous proteins. Produced by the body itself, they consist of enzymes secreted during the processes of digestion; proteins from the epithelial cells released during mucosal cell turnover; and small amounts of plasma proteins that leak out of capillaries into the intestinal lumen. All proteins are converted into amino acids (although some small peptides may also be absorbed intact), with the final steps taking place at the brush boarder through the action of the embedded aminopeptidases. Absorption is accomplished by secondary active transport where Na^+ is actively absorbed and the amino acids are co-transported at the same time. Different amino acid carriers are also apparently involved in this process.

Fats. They must first be converted into monoglycerides and free fatty acids. The water insoluble fatty end products of digestion then dissolve in the water soluble micelles which transport them to the epithelial cell luminal membranes. The monoglycerides and free fatty acids then passively diffuse out of the micelles passing through the epithelial cell membrane (which is itself lipid-based) and thus enter the interior of the epithelial cells.

Within the interior of the epithelial cells, triglycerides are reassembled from the monoglycerides and fatty acids. The triglycerides formed in this way then aggregate into large droplets, and are then coated with protein synthesized by the cell's endoplasmic reticulum. The coated droplets, called *chylomicrons*, are then actively extruded out of the lateral side of the epithelial cells by exocytosis, where they then

enter the lateral spaces. From here they pass into the lacteal (a branch of the lymphatic system) present in the center of each villus. (Note that chylomicrons cannot enter the blood capillaries in the villus because capillaries have a basement membrane which acts as a barrier.)

The majority of fats ingested in the diet are composed of long chains of fatty acids. This type of fat is processed as described above and will eventually enter the lacteals. However, fatty acids formed from fats with short or medium length carbon chains are not converted back into fats in the epithelial cells. These short and medium chain fatty acids can bypass this mechanism and can directly enter capillaries.

Since the majority of fat ingested in the diet passes into the lymphatic system as chylomicrons, it bypasses the portal system and enters the circulation unchanged. These chylomicrons are then transported in the blood to muscle and adipose tissues where they unload their contents. Specifically, the presence of *apoprotein* associated with the chylomicrons activates *lipoprotein lipase* present on the inner lining of capillaries within muscle and adipose tissue. This enzyme then causes the release of the monoglycerides and fatty acids from the chylomicrons, and these substances then enter the muscle and adipose cells where they are utilized for energy, or stored. The remnants of the chylomicrons then are transported in the blood to the liver. Here they bind to receptors on liver cells, enter the cells and are destroyed by lysosomal action.

Vitamins. Being wwater soluble they are absorbed by passive diffusion. Fat soluble vitamins are absorbed with the fat. Certain vitamins can also be absorbed by carrier mediated processes, and *vitamin* B_{12} must first be complexed with *intrinsic factor* (released by gastric glands) before being absorbed by a special transport mechanism present in the ileum.

Iron. It is actively transported from the lumenal contents into the epithelial cells, and then into the blood where the blood protein, *transferrin,* transports it to the bone marrow. The hormone erythropoietin (which is released by the kidney, and which stimulates RBC production in the bone marrow), also increases iron absorption from the lumen into the blood. In the event that iron is not needed, then it is instead stored in the epithelial cells in the form of *ferritin*. This stored iron is lost after three days because the epithelial cells containing the ferritin will slough off as new cells replace the old one. The sloughed cells containing the ferritin enter the feces, and if large quantities of iron are present, will give the feces a black color.

THE LARGE INTESTINE

The large intestine **(Figure 14.10)** is subdivided into the *cecum, colon* and *rectum.*

1. The cecum is connected to the ileum at the ileocecal valve. Attached to the cecum is the blind pouch call the *vermiform appendix* which is part of the *Gut Associated Lymphatic Tissue* (*GALT*) of the immune system.

2. The colon (or large intestine) is further subdivided anatomically into the *ascending colon, transverse colon*, and *descending colon.*

The *ascending colon* begins at the cecum, passes upwards in the abdominal cavity, and then turns to form the *transverse colon*, below the liver (at the *Hepatic flexure*, or *right colic flexure*).

The transverse colon passes horizontally through the abdominal cavity and then turns downward to form the descending colon at the spleen (at the *Splenic flexure*, or *left colic flexure*).

The descending colon passes downward through the abdominal cavity, attaches to the S-shaped sigmoid colon, which then attaches to the *rectum*.

3. The *rectum* attaches to the *anal canal*.

4 The *anal canal* has a lining of mucous membranes that are folded to form *anal columns*. The opening of the canal is the *anus* and this is closed by two sphincters. The *internal anal sphincter* consists of smooth muscle and is not under unconscious control, while the *external anal sphincter* (formed of skeletal muscle) is under conscious control.

STRUCTURE OF THE WALL OF THE LARGE INTESTINE

The wall of the large intestine has no villi and the longitudinal smooth muscle does not uniformly cover the wall as it does in the other parts of the Gastrointestinal tract. In the large intestine, longitudinal smooth muscle forms three distinct bands (called *taeniae coli*) that extend across the length of the colon and create pouches called *haustra*. Mucus-secreting tubular glands are present in the wall of large intestine, and mucus is the only substance secreted. The mucus protects the wall from abrasion and, in addition, holds the fecal particles together. Mucus is alkaline in pH, and also assists in neutralizing acid generated in the large intestine from the action of intestinal bacteria. Secretion of mucus by the tubular glands is stimulated when chyme contacts the wall of the large intestine, or through parasympathetic nerve activity.

ABSORPTION IN THE LARGE INTESTINE

Nothing is digested in the large intestine, and the only substances that are absorbed here are water and electrolytes. The absorption of electrolytes is mediated either by active transport or passive diffusion, and water is absorbed by osmosis. Absorption of salts and water usually only takes place in the proximal portion of the large intestine, and the osmotic process is so efficient that the majority of the water and salts initially present in the chyme that enters from the small intestine are absorbed and not lost in the feces.

FECES

Colon bacteria are present in the large intestine and utilize the indigestible material in the feces (for example, cellulose) for their own metabolic needs. Some of these bacteria may synthesize certain vitamins (K, B12, thiamine, riboflavin) which can be absorbed by the intestinal mucosa.

The presence of these bacteria (normal flora), however, is also very important because it prevents other more virulent pathogens from taking up residence here and causing disease.

It has also been suggested that these normal bacterial flora are responsible for sensitizing the immune system at an early age against blood group antigens, thus resulting in the presence of circulating ABO

blood group antibodies responsible for transfusion reactions. Bacterial metabolic action can also lead to generation of gas (termed *flatus*), and in addition, the products generated from their metabolism are responsible for the unpleasant odor of feces.

In addition to the indigestible matter present in feces, water, electrolytes, mucus and bacteria are also present. Even with water reabsorption in the large intestine, fecal material still contains about 75% water, and the color is primarily a result of bile pigments that have been altered by bacterial action.

DEFECATION REFLEX

Within the large intestine, mixing and peristalsis take place, but these actions are less active then in the small intestine. Mixing and segmentation movements assist in the reabsorption of water and salts. Peristaltic movement of the fecal material down the length of the large intestine takes place much less frequently than in the small intestine. This strong peristaltic *mass movement* occurs about 2 to 3 times a day forcing the contents of the large intestine to move to the *rectum* and activating the *defecation reflex*. Stimulation of the large intestine to produce such mass movements take place after eating when the duodenum is stretched as food enters it. The reflex that activates the large intestine is thus termed the *duodenocolic reflex* and is mediated by the parasympathetic nervous system.

The defecation reflex is initiated when the wall of the rectum is stretched as fecal matter enters it. Conscious initiation of the defecation reflex can also be accomplished by deep breath holding and contraction of the abdominal muscles (called the Valsalva maneuver) which will force fecal matter into the rectum stretching it. Rectal wall distention stimulates peristaltic waves in the descending colon and the opening of the internal anal sphincter (made of smooth muscle). If the external anal sphincter (made of skeletal muscle) is then consciously allowed to relax, the reflex continues and fecal pellets are excreted. On the other hand, if the external anal sphincter remains contracted, this will inhibit the defecation reflex.

The Valsalva Maneuver is interesting because not only does it promote defecation, but in addition it has an unexpected side effect on Venous Return (VR). Remember that VR is assisted by respiration, and by the negative pressure in the thoracic compartment. However, during the process of abdominal muscle contraction and breath-holding, the thoracic pressure rises significantly and this will reduce blood flow in the abdominal aorta as it passes from the abdominal to the thoracic compartments. With a drop in VR, Stroke Volume (SV) falls, Cardiac Output (CO) falls, and systemic blood pressure can be affected. In a person with a weak or compromised cardiovascular system, this may pose a real danger. In fact it is not unusual to learn that a person has died from a heart attack while in the process of defecation.

DISORDERS ASSOCIATED WITH THE LARGE INTESTINE

Diarrhea. When water is not removed from the chyme, diarrhe results. This can take place if peristaltic action within the intestine is increased abnormally. Inflammation of the intestine and colon will increase peristalsis and may be caused by ulceration of the intestinal wall, or from microbial action. Emotional stress may also increase peristaltic action. Hypersecretion of intestinal mucus (caused by bacterial toxins, certain dietary nutrients, or hormones) will cause an increase in the fluid content of the chyme and lead to diarrhea.

Constipation. If the water content of the feces is reduced, a the fecal material can become hard and dry. The usual cause is prolonged inhibition of the defecation reflex (which is a conscious decision), coupled with poor eating habits. In addition, lack of intestinal muscle tone or muscle spasms can also block the movement of material and these are much more serious and may need to be treated medically.

For those sufferers of constipation due to prolonged inhibition of the defecation reflex (certainly the majority in the USA and possibly the world), the simple solution is to resensitize them to the "call of nature." In addition, increasing the fiber content in the diet is helpful (eating less processed starches, for example), since indigestible fiber increases the osmotic pressure of the material in the intestine and holds water in the fecal matter. The use of over-the-counter laxatives—other than the fiber variety—is not recommended (unless prescribed by a physician) since the mechanism of action is to stimulate the intestinal mucosa, thereby increasing peristalsis abnormally. The use of mineral oil is a worse idea, since it is a hydrocarbon and certainly does not belong in the body.

Chapter 15
THE ENDOCRINE SYSTEM

The endocrine system is a body-wide communication system employing blood born hormones as messengers.

Such endocrine hormones exert wide-ranging controls on body homeostasis, and function in close association with the nervous system. Some of the most obvious functional similarities and differences between the endocrine and nervous systems are:

1) **Response time**: Communication between the nervous system and its associated end organs is practically instantaneous. Thus, stimulating a nerve produces an immediate response at the end organ. In comparison, upon stimulus of the endocrine gland, there is a delay (which may be in some cases hours in length) before the hormone affects the target tissue. Such a delay is expected since first, the hormone takes time to circulate to the target tissue, and second, metabolic changes in the target cells may, in some cases, be dependent on changes in protein synthesis.

In addition, the stimulus off-response of the nervous system is also nearly instantaneous. Thus, once the nerve is no longer stimulated, the effect at the end organ ceases almost immediately. However, in the case of the endocrine system, even if hormone release from the endocrine gland ceases, the hormonal effects exerted at the target tissue may continue for extended periods (in some cases hours, or even days). This is because the hormone continues to circulate in the blood and must be cleared by the liver and kidneys before the levels fall, and the target tissue is no longer affected.

2) **Extent of communication**: The nervous system is wide ranging. While nerve branches innervate many cells and tissues, they do not reach every single cell in the body. On the other hand, all cells in the body are in contact with the blood, or with substances carried by the blood. Thus, hormones of the endocrine system transported by the blood can potentially reach every cell in the body.

THE CONCEPT OF THE TARGET CELLS AND TARGET TISSUES

Although endocrine hormones can potentially reach all cells in the body via the blood, these hormones do not necessarily effect all cells. This is because the only cells that respond to these hormones are those which possess specific hormone receptors to bind the particular hormone in question.

Cells with such specific hormone receptors are said to be *target cells*, and obviously target cells make up *target tissues* for the particular hormone. Individual target cells can (and do) possess receptors for many different kinds of hormones.

Thus, a cell can be a target for growth hormone, prolactin, epinephrine and insulin—all at the same time—because this cell possesses individual receptors for each of these hormones. However, since this

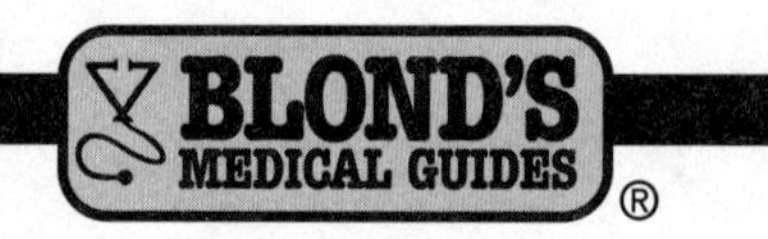

cell does not have receptors for Testosterone, it is not a target cell for this hormone, and will not respond to that hormone, even if the hormone reaches this cell from the blood.

EXOCRINE V. ENDOCRINE GLANDS

Exocrine glands secrete their products into ducts and these substances are then conducted to some location (for example, the mouth, the GI tract, the skin) where they serve some particular purpose. On the other hand, *endocrine glands* do not have ducts, and instead release their products into the interstitial fluids and from here products are picked up by the blood and carried to all regions of the body.

HORMONES OF THE ENDOCRINE SYSTEM

The classical description of an endocrine hormone is a substance that is transported by the blood and exerts its effects at target tissue distant from its origin. However, it is now known that while many hormones indeed function via blood transport others may function locally. Specifically, hormonal communication between cells can be defined in the following ways:

Paracrines. They are hormones secreted by an endocrine cell that acts locally on surrounding target cells to elicit a response. Here the paracrine secretion does not get transported by the blood and the substance is distributed by simple diffusion. Thus the effects elicited by paracrine actions are restricted to short distances.

Classical Hormones. They are released by endocrine cells and these substances are transported in the blood. The target tissue for these hormones can therefore be distant from the endocrine gland that secreted them.

Neurohormones. They are secreted by nerve cells, but unlike classical neurotransmitters, do not bind directly to target cell end organs (at synapses). Instead these neurohormones are released into the blood which transports them to target tissues that can be distant from the neuroendocrine cell that released the neurohormone.

CLASSES OF HORMONES

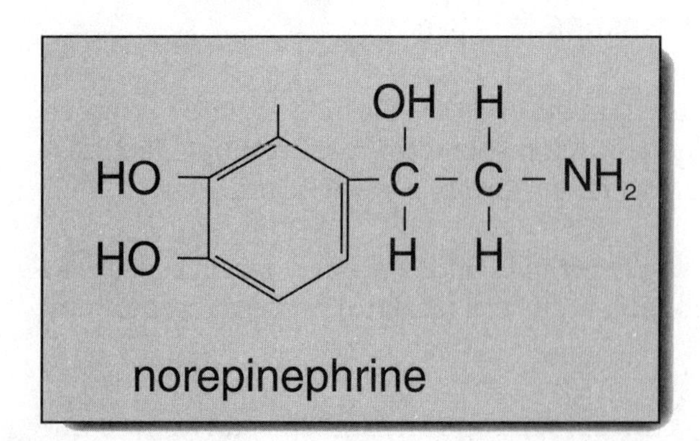

FIGURE 15.1

Classical hormones can be subdivided into three general groups according to their chemical structures. These are the amines, the protein/peptides, and the steroids.

Amines—contain nitrogen and are derived from the amino acid tyrosine. Included in this group are hormones from the thyroid (thyroxin, triiodothyronine) and the catecholamines from the adrenal medulla (epinephrine, norepinephrine) **(Figure 15.1)**.

Proteins/peptide hormones—are synthesized like

all proteins, and consist of chains of amino acids linked together. Longer chain hormones are termed *proteins*, whereas shorter chain hormones are classified as *peptides*; and the majority of all hormones of the endocrine system fall into the category of protein/peptides. The endocrine glands responsible for synthesis of this class of hormones are widespread and include (but are not limited to) the hypothalamus, anterior and posterior pituitary, liver, kidneys, heart, thymus, parathyroids and gastrointestinal tract.

Steroids—are synthesized from the common precursor *cholesterol* through various metabolic pathways as outlined in **Figure 15.2**. The majority of this cholesterol, utilized by endocrine cells during steroid synthesis, is derived from low density lipoprotein degradation within these cells. Since the synthesis of particular steroids depends on the presence of specific enzymes, it follows that all endocrine cells cannot synthesize all steroid hormones. Thus, cortisol can only be synthesized by the adrenal cortex because other steroidogenic organs lack the necessary enzyme for this synthetic pathway.

Since all steroids are initially derived from cholesterol, their structure mirrors to some degree that of this original parent compound. Thus although all steroids possess the fourring signature, minor differences in side groups and ring saturation can produce major differences in their biological functions **(Figure 15.3)**.

Prostaglandins—fall into the lipid family. They are made of 20 carbon fatty acids **(Figure 15.4)**, and are initially synthesized from arachidonic acid (a fatty acid) found in cellular membranes.

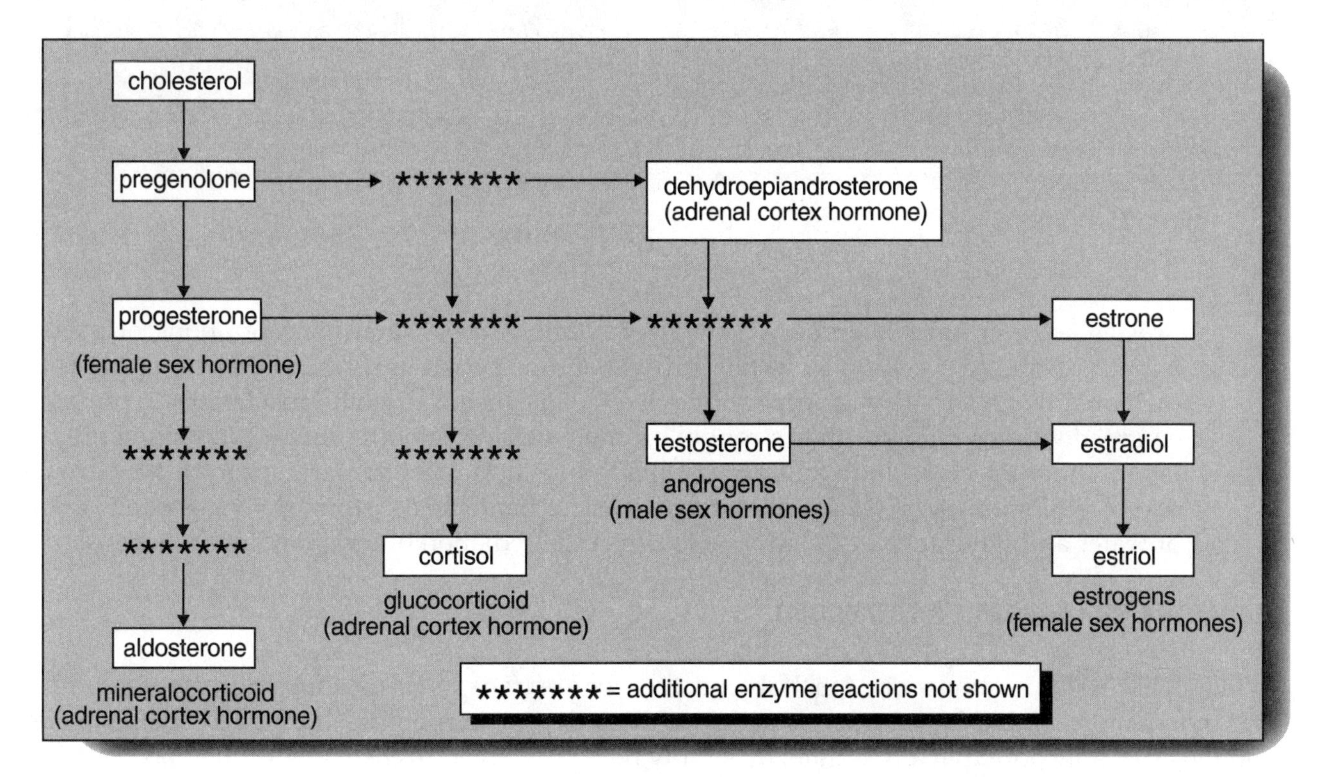

FIGURE 15.2

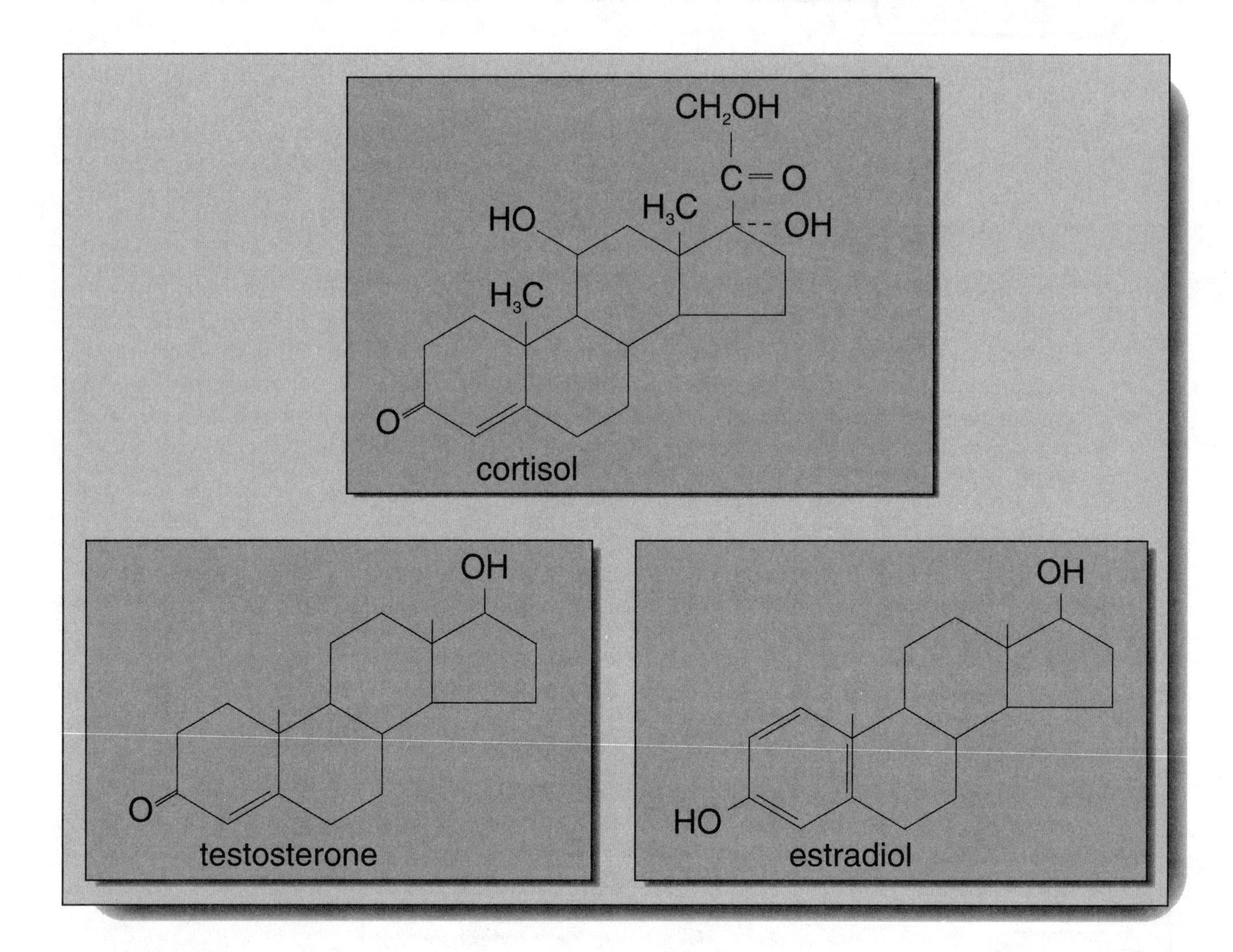

FIGURE 15.3

Prostaglandins primarily produce their effects locally, in very small concentrations, on a wide variety of different cell types. While prostaglandins promote many effects on target cells mediated by cAMP, they are especially effective in causing smooth muscle cells to undergo contractions. For example, they strongly promote uterine smooth muscle contraction during labor. Other effects include (but are not limited to) hormonal stimulation from the adrenal cortex, inhibition of the release of acid from gastric glands in the stomach mucosa, regulation of water and salt transport in kidney tubules, and mediation of inflammatory reactions.

HORMONE TRANSPORT IN THE BLOOD

While all hormones are transported in the blood, the mode of transport is not the same for water soluble and non-water soluble hormones. Catecholamine and protein/peptide hormones—being water soluble—are transported in a free state, by simply being dissolved in the water of the plasma.

On the other hand, steroid hormones and thyroid hormones dissolve only poorly in water, and are thus transported bound to carrier proteins in the plasma. While common plasma proteins (such as albumin) are certainly involved in this transport, additional specialized plasma proteins (such as sex steroid binding globulin) also play an important role in this process. While the majority of the water-insoluble hormones are transported bound to plasma carriers, small concentrations are present in the free form. Thus, an equilibrium relationship exists between the free form of the hormone and the bound form, as follows:

Free hormone + binding protein ∩ hormone-protein complex

In the above example, the total hormone concentration is calculated from the sum of the bound and free, but only the free hormone exerts biological activity. This is because only the free hormone is capable of crossing the capillary endothelium to reach the target cells. Therefore, any factors affecting the concentration of free hormone can alter the physiology of hormone action at the target tissues.

PLASMA CONCENTRATION OF HORMONES

The plasma concentration of a particular hormone is dependent on a number of interrelated processes **(Figure 15.5)**. Initially the production of a particular hormone by the endocrine cell requires an adequate supply of precursor substances. In the absence of such precursors the hormone may not be produced, or will be produced in decreased amounts.

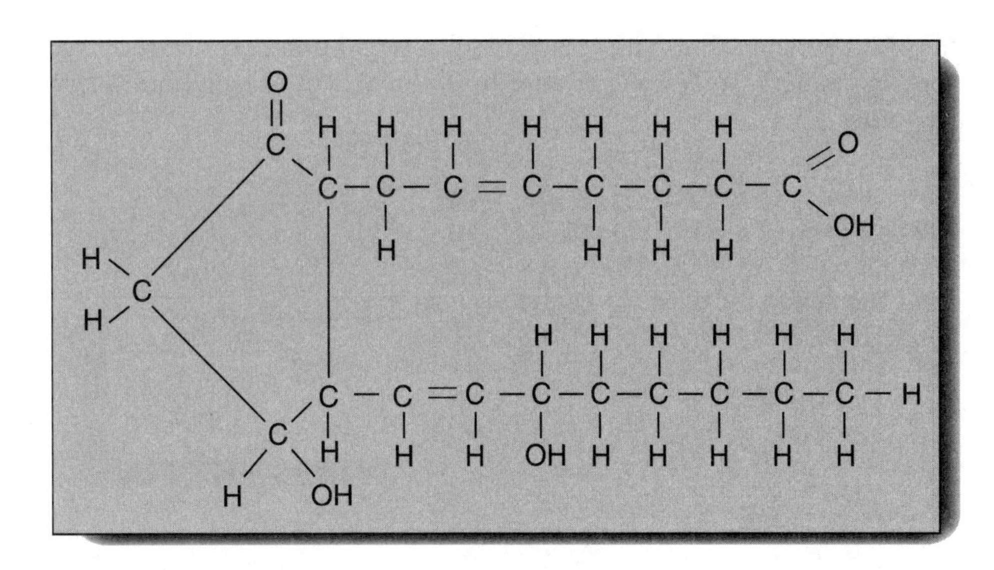

FIGURE 15.4

For example, the synthesis of thyroid hormones by the thyroid follicles depends on the presence of iodine. Thus, in the absence of iodine thyroid hormones will not be produced, and the thyroid, hormones will not be found in the circulation. A number of other factors can also affect the concentration of hormones in the plasma.

1) Clearance of circulating hormones, or their metabolic products, by the kidney will reduce the plasma concentration.

2) Inactivation of hormones by the liver or other tissues will also reduce the levels in the plasma.

3) Removal of hormones at the target tissues through receptor binding and endocytosis will reduce the levels in the plasma.

4) For steroid and thyroid hormones that are transported bound to protein carriers, the concentration of the carrier itself will affect the plasma concentration of the hormone. Since these carriers are synthesized by the liver, factors that alter liver function (for example, cirrhosis) may also produce endocrine disorders because the synthesis of these carrier molecules is affected.

INTERACTIONS OF HORMONE RECEPTORS

Hormone receptors present on target cells are usually highly specific for individual hormones. In addition, the more specific the binding site for the particular hormone, the greater will be the affinity of hormone binding. After being released by the endocrine cells, the circulating plasma hormone may directly bind to the target cell receptors. Alternatively, the circulating form may first be activated by metabolic events and the activated metabolites may then bind to the target cell receptors.

In addition, some hormones may catalyze the formation of active secondary hormones through conversion of plasma proteins, and these newly synthesized hormones may then bind to target cell receptors.

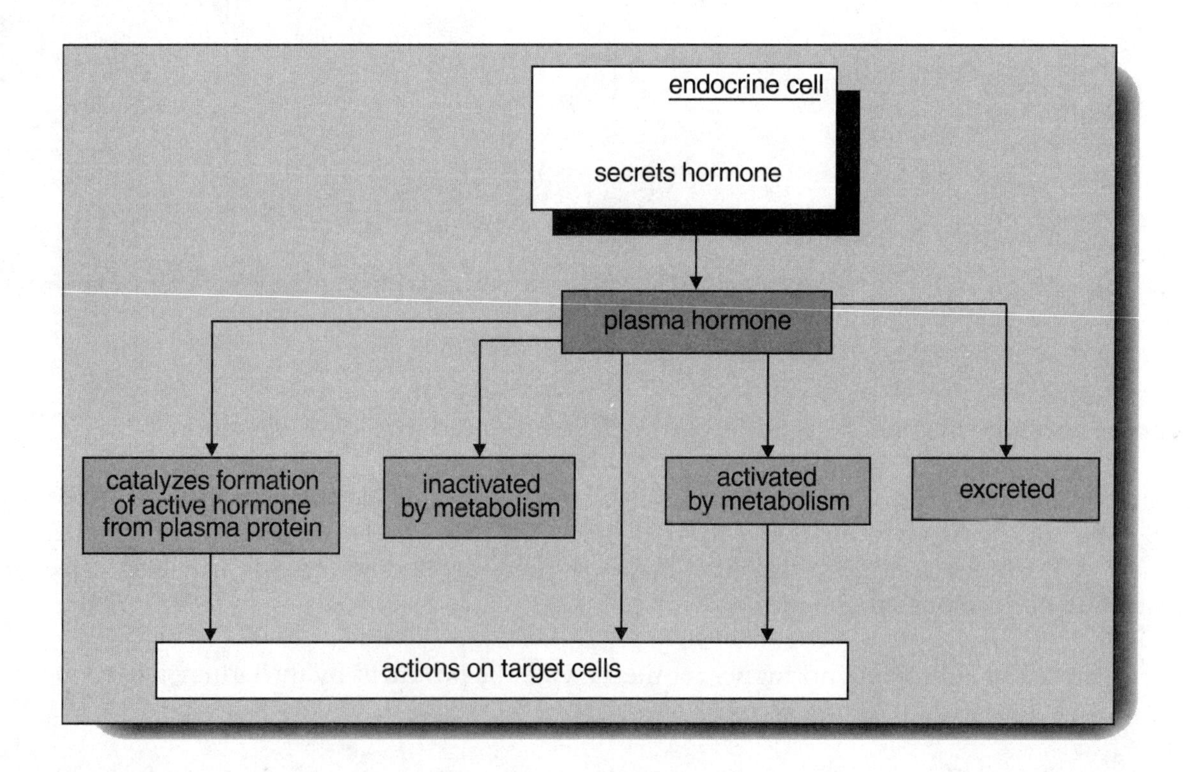

FIGURE 15.5

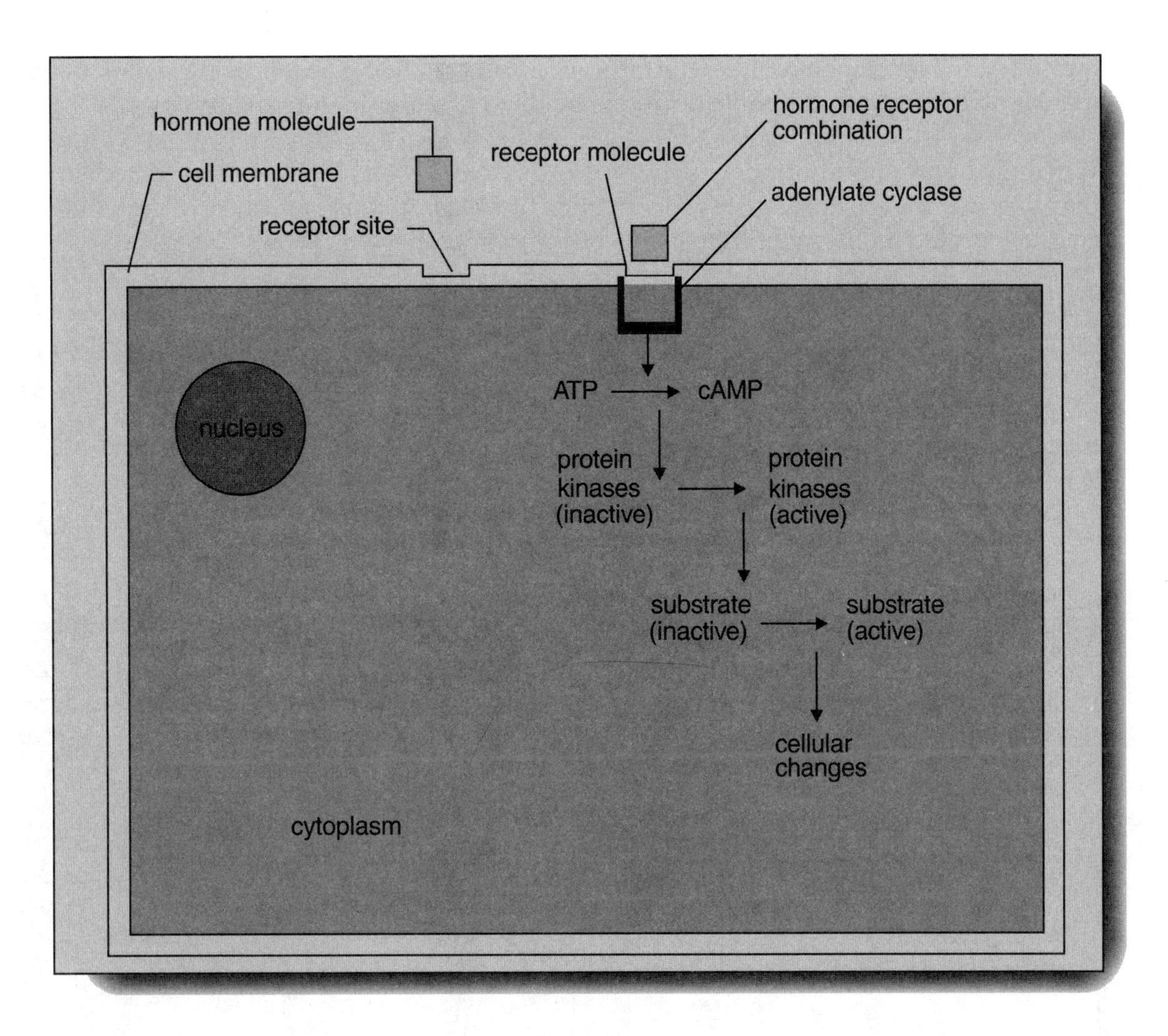

FIGURE 15.6

An example of this last mechanism can be found in the *renin-angiotensin* pathway, where the renin released from the juxtaglomerular apparatus catalyzes the conversion of the plasma precursor protein *angeotensinogen* (synthesized by the liver) to form angiotensin I.

CLASSIFICATION OF HORMONE RECEPTORS

Hormone receptors can be classified by their cellular location: namely those receptors that are located on the *outer membrane of target cells*, and those receptors that are located within the *intra-cellular compartment*. While both types promote cellular responses, the intermediate events leading to the final outcome are quite different.

Outer Membrane Hormone Receptors. *Hydrophilic hormones* cannot pass through the lipid bilayer of the plasma membrane because they are water soluble and are not soluble in lipid. Thus, in order for these hormones to promote intra-cellular changes, they must be able to bind to receptors

located on the outer membrane of the target cells **(Figure 15.6)**. The following events are known to take place after a hydrophilic hormone binds to its outer membrane receptor:

1) When the hydrophilic hormone (known as the *first messenger*) binds to its receptor located on the outside of the cell, this causes the activation of *adenylate cyclase* which is a membrane-bound enzyme located on the cytoplasmic side of the membrane.

2) The active adenylate cyclase converts ATP into *cyclic AMP* called *cAMP*. This cAMP acts as an intra-cellular messenger within the cytoplasmic compartment, and is thus described as the *second messenger*.

3) Inactive protein kinase present within the cytoplasmic compartment is activated by cAMP. The active protein kinase now phosphorylates other cytoplasmic enzymes.

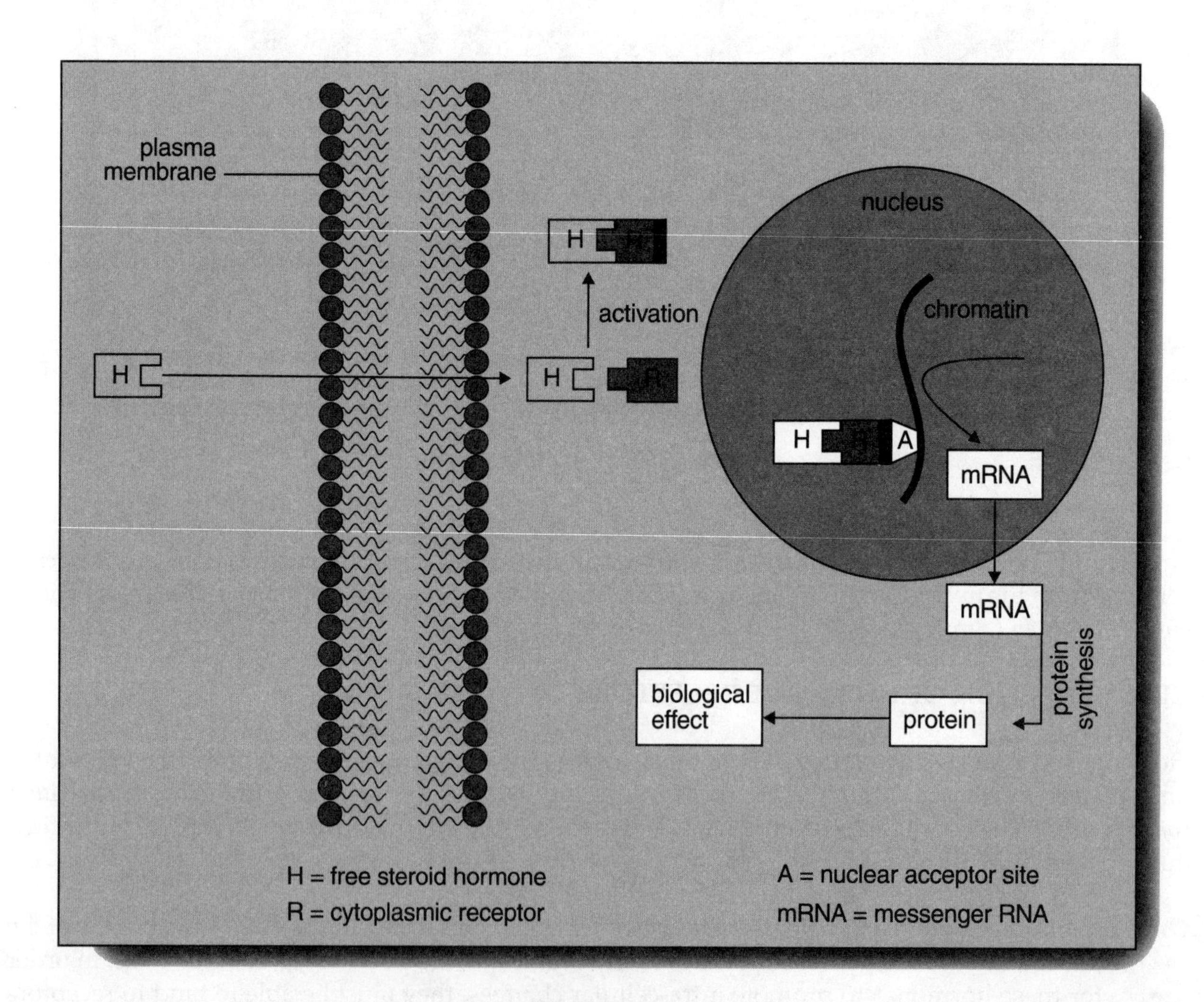

FIGURE 15.7

4) When cytoplasmic enzymes become phosphorylated, depending on the type of enzyme involved, the outcome may be either to activate enzymes that were previously inactive, or to inactive enzymes that were previously active. In either case, the effect is to alter cytoplasmic enzyme activity in the target cell.

5) Alteration of cytoplasmic enzyme activity will bring about metabolic changes in the target cells.

6) Upon removal of the hormone (first messenger), the cytoplasmic cAMP is degraded through the action of the enzyme phosphodiesterase. Once the cAMP is removed the cell reverts back to its original non-hormone stimulated state.

Transmembrane Transport. In addition to the *second messenger mechanism* outlined above, hydrophilic hormones, after binding to hormone receptors on the outer membrane of target cells, may also function by altering the transmembrane transport of substances into or out of the cell. Mechanisms involved in altering transmembrane transport include opening or closing of membrane channel proteins, or affecting facilitated diffusion by altering availability of membrane carriers. For example, binding of insulin to insulin receptors on target cells (muscle and fat cells) increases glucose transport into these target cells.

Cytoplasmic Hormone Receptors. Because steroid hormone structure is based on cholesterol, steroids are capable of passing through the lipid bilayer of the plasma membrane. Given this, the receptors to bind steroid hormones are located in the cytoplasmic compartment and not on the outer membrane of the target cell. Specifically steroid hormones promote effects **(Figure 15.7)** on target cells when:

1) The free steroid (having been released from the protein carrier in the blood), now diffuses into the cytoplasm of the target cell and binds specifically to cytoplasmic receptors.

2) The formation of the steroid hormone-receptor complex *activated* through a conformational change in the protein structure.

3) The activated hormone-receptor complex now translocates into the nuclear compartment where it binds to a specific *acceptor site* on the chromatin.

4) Binding of the hormone-receptor complex to the acceptor site alters the *transcription* of mRNA from one or more specific genes. Usually the effect is to increase mRNA transcription, but in some instances this process can reduce mRNA transcription.

5) If the mRNA transcription is increased, the mRNA passing back across the nuclear membrane into the cytoplasm will bind to ribosomes and increase the translation process leading to the formation of various new peptides or proteins in the target cell. On the other hand, if mRNA transcription is decreased, protein synthesis in the target cell would be reduced.

6) Thus the final biological effect of steroids on target cells is to alter protein synthesis in these cells. Since proteins can function not only as structural elements but also as enzymes, steroid

3,5,3',5', - tetraiodothyronine (thyroxine, T_4)

3,5,3', - triiodothyronine (T_3)

FIGURE 15.8

hormones can indirectly alter metabolism in target cells by altering the type and concentration of enzymes present here.

Thyroid Hormones. It should be noted that because *thyroid hormones* **(Figure 15.8)** are hydrophobic, they also function much like steroids by diffusing directly into the cytoplasmic compartment. However, thyroid hormones do not bind to their receptors in the cytoplasm, but instead pass directly into the nucleus in the free form where they then bind to nuclear receptors. By altering gene expression, thyroid hormones produce wide-ranging effects on the synthesis of enzymes and thus increase metabolic function in their target cells.

ENDOCRINE GLANDS

The following endocrine glands are present in the body: pineal, hypothalamus, pituitary, parathyroid, thyroid, thymus, heart, stomach, duodenum, adrenal, pancreas, kidney, skin, ovaries in the female, testes in the male, placenta in the pregnant female.

In addition, various cells of the immune system generate a variety of substances called *cytokines*. While many of these cytokines are involved in regulatory actions relating to immune response, it is now known that immune cells also synthesize and release various classic hormones (ACTH, TSH, GH). In addition, certain immune system derived hormones and cytokines can act on classic endocrine targets. For example, *Interleukin-1* functions at the level of the pituitary or adrenal to release cortisol, and thymic hormones can release GnRH from the hypothalamus.

INTERACTIONS BETWEEN THE NERVOUS AND ENDOCRINE SYSTEMS

The nervous system functions in close association with the endocrine system because nervous innervation can regulate the release of endocrine hormones **(Figure 15.9)**. Here are some cases in point.

1) The hypothalamus is responsible for synthesizing and directly regulating release of the hormones *oxytocin* and *vasopressin* from the *posterior pituitary.*

2) The hypothalamus produces both *release* and *release inhibiting* hormones that regulate the release of hormones from the *anterior pituitary.*

3) The autonomic nervous system (also regulated by hypothalamic inputs) controls the release of the hormones *epinephrine* and *norepinephrine* from the *adrenal medulla* (a sympathetic ganglia).

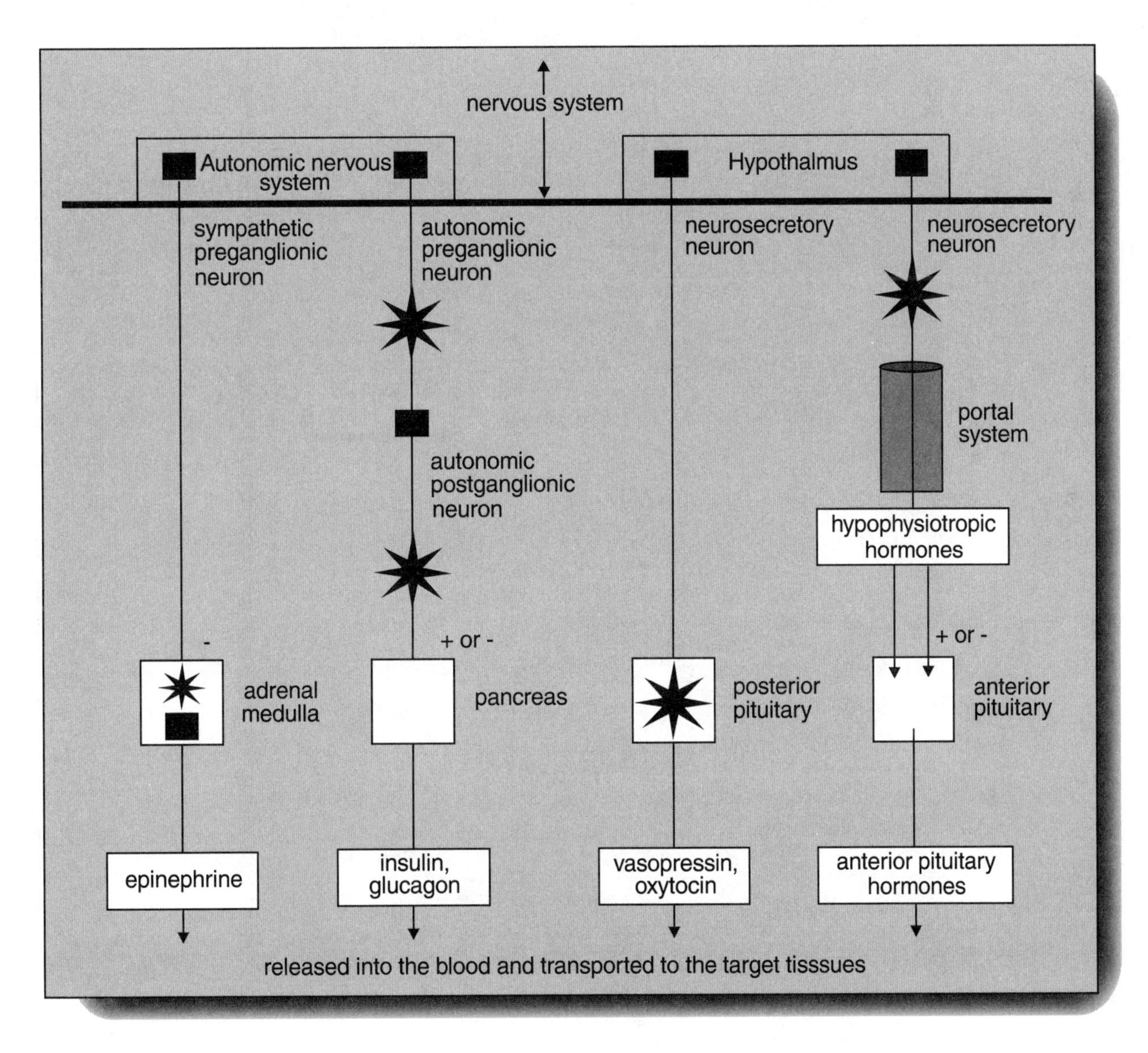

FIGURE 15.9

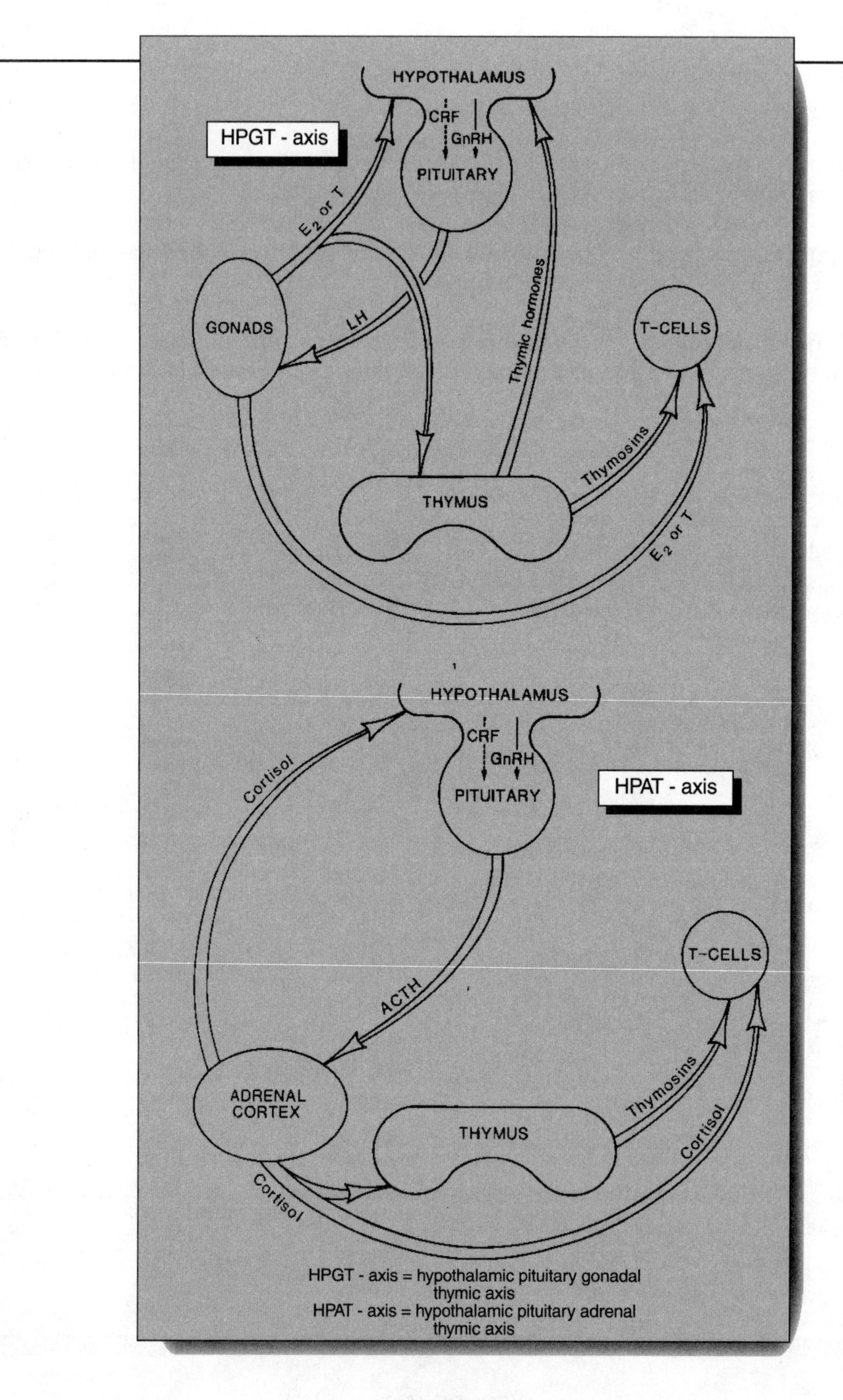
HPGT - axis
HYPOTHALAMUS
CRF
GnRH
PITUITARY
E_2 or T
GONADS
LH
Thymic hormones
T-CELLS
Thymosins
THYMUS
E_2 or T
HYPOTHALAMUS
CRF
GnRH
PITUITARY
HPAT - axis
Cortisol
T-CELLS
ACTH
ADRENAL CORTEX
THYMUS
Thymosins
Cortisol
Cortisol
HPGT - axis = hypothalamic pituitary gonadal thymic axis
HPAT - axis = hypothalamic pituitary adrenal thymic axis

FIGURE 15.10

4) The autonomic nervous system can stimulate or inhibit the release of hormones from various endocrine glands. For example it can regulate the release of *insulin* from the pancreas.

ADDITIONAL COMPLEX LEVELS OF ENDOCRINE INTERACTIONS

Endocrine-Immune Interactions. The endocrine system controls, and is controlled by, the immune system through a series of hormonal pathways (axes) involving the hypothalamus, pituitary, thymus, gonads, pineal and adrenals **(Figure 15.10)**. Hormones that play a central role in these regulatory axes include the gonadal steroids (estrogen, testosterone and progesterone), the adrenal steroid cortisol, growth hormone and prolactin from the anterior pituitary, and melatonin from the pineal.

Hormone Sources. Depending on the hormone source the same chemical substance may be classified as a hormone (norepinephrine from the adrenal medulla), or a neurotransmitter (norepinephrine from the post ganglionic sympathetic fiber).

Target Tissues and Multiple Hormones. The same target tissue can, and usually does, possess receptors for a variety of different hormones. Thus the same target tissue can be affected by many different hormones.

Hormones and Multiple Target Tissues. The same hormone may bind to receptors in many different target tissues. For example, growth hormones have wide-ranging effects on multiple tissues in the body. In addition a single hormone can promote different effects in different target tissues. Thus, vasopressin (ADH) causes water reabsorption in the kidney tubules, but causes vasoconstriction of the peripheral arteriolar smooth muscle and increased systemic blood pressure.

Hormonal Sources. Identical, or nearly identical hormones, can be secreted by different target tissues. For example, ACTH and GH are secreted by the anterior pituitary and also by lymphocytes; somatostatin is secreted by the hypothalamus and by the pancreas, and so on.

FACTORS THAT REGULATE ENDOCRINE CELLS

Multiple factors can regulate the release of hormones from endocrine cells **(Figure 15.11)**. Thus, in addition to the control exerted by neurons:

Other Hormones. They can target endocrine cells to release hormones; (ACTH causing the release of cortisol from the adrenal cortex; TSH causing the release of thyroid hormones from the thyroid gland).

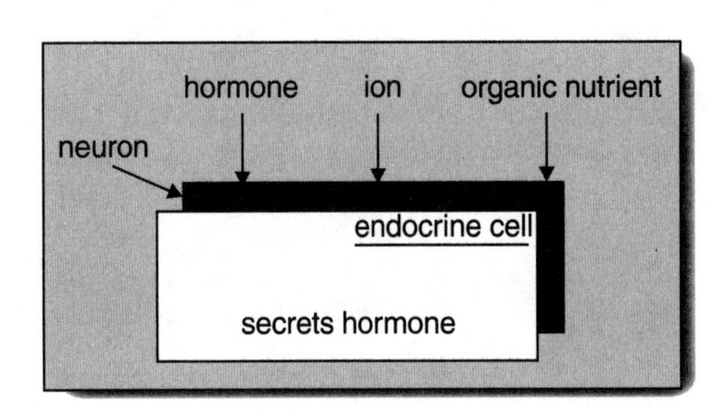

FIGURE 15.11

Various Ions. Various ions can act on target endocrine cells to release hormones; (elevated K^+ causing the release of aldosterone from the adrenal cortex; and low levels of Ca^{++} causing the release of parathyroid hormone from the parathyroids glands).

Organic Nutrients. They can act on target endocrine cells to release hormones; (elevated glucose causing the release of insulin from pancreatic islets; low levels of glucose causing the release of glucagon from pancreatic islets; low levels of glucose, acting on the hypothalamus and causing the release of GH from the pituitary).

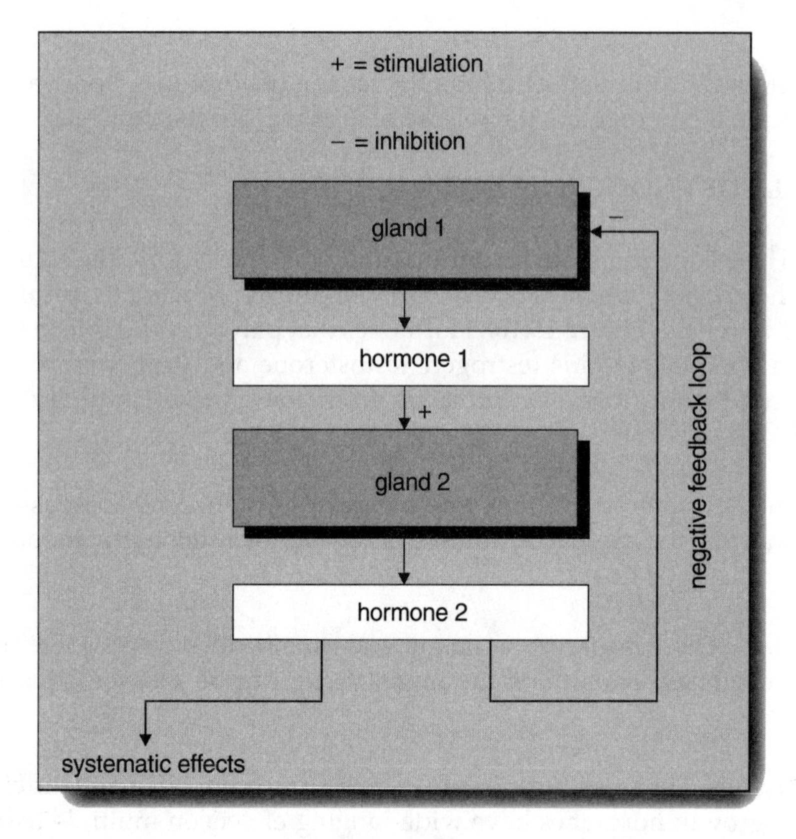

FIGURE 15.12

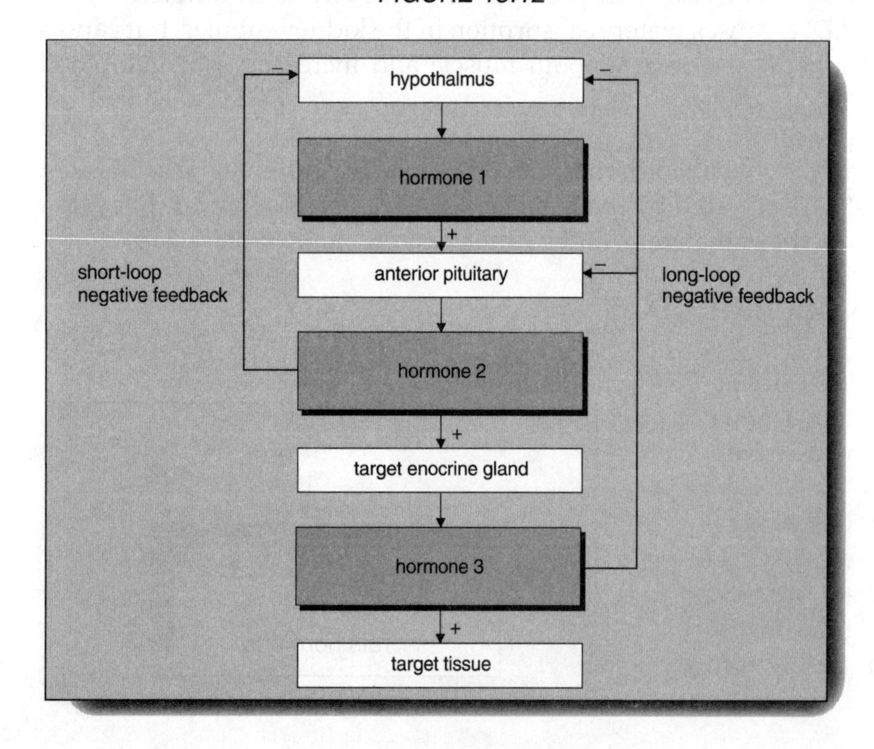

FIGURE 15.13

NEGATIVE FEEDBACK

To maintain homeostatic control of the hormone levels (and thus of function as well) negative feedback mechanisms are centrally involved in the endocrine system. In its simplest form, such negative feedback pathways involve either two endocrine glands, or one endocrine gland and an organic nutrient or ion.

1) Two endocrine glands: Here releases *hormone 1* which binds to *gland 2*. *Hormone 2* is released from *gland 2* when *hormone 1* binds here. *Hormone 2* circulates back to *gland 1* and *inhibits* the release of *hormone 1* **(Figure 15.12)**.

Short and long loop negative feedbacks are both involved in regulation of endocrine hormones **(Figure 15.13)**. The pathway between hormone 2 and the hypothalamus would be considered as an example of a short loop feedback, while the pathway between hormone 3 and the anterior pituitary, or hormone 3 and the hypothalamus, would constitute long loop negative feedbacks.

2) *One endocrine gland and an organic nutrient*: Here **(Figure 15.14)**, increased blood glucose results in the release of insulin from the pancreas. The insulin reduces the levels of circulating glucose by acting on target cells (muscle, or fat tissue), and the fall in this organic nutrient then turns off the insulin release from the pancreas.

DIURNAL RHYTHMS

Hormonal secretions can vary as a function of time of day. Termed *diurnal* or *circadian rhythm*, these variations are controlled by a complex sequence of events involving an internal neurological clock. In addition, the intrinsic rhythm of this clock can be influenced by light-dark stimuli transduced through the release of melatonin by the pineal gland. Accordingly, such hormones as cortisol from the adrenal, sex steroids from the gonads, and thymosin from the thymus gland express secretory patterns that are coupled with the light-dark cycle. Specifically:

1) The secretion of melatonin is increased in the dark and reduced in the light.

2) In the dark, melatonin released by the pineal, inhibits CRH released by the hypothalamus. This reduces ACTH released from the anterior pituitary and leads to a reduction of cortisol secretion from the adrenal gland during the dark phase.

3) In the dark, melatonin released by the pineal, inhibits GnRH released by the hypothalamus. This reduces LH and FSH released from the anterior pituitary and leads to a reduction in sex steroids from the gonads during the dark phase.

↑ plasma glucose concentration

insulin-secreting cells

↑ insulin secretion

↑ plasma insulin concentration

insulin's target cells

↑ actions of insulin

↓ plasma glucose concentration

FIGURE 15.14

NEUROENDOCRINE REFLEX PATHWAYS

Neuroendocrine reflexes are involved in over-riding existing negative feedback set points, under conditions where sudden increases in hormone secretions are warranted. For example, during stress

responses, a temporary, and transient, increase in glucocorticoid secretion takes place allowing the individual to overcome the stressor. Such a transient increase is mediated by autonomic influences acting on the hypothalamus. Note, however, that after the stressor is no longer present the original negative feedback regulation is reestablished and glucocorticoid levels return to pre-stress concentrations. Other examples of an overriding neuroendocrine reflex can be found in:

1. The transient increase in thyroid hormone secretion present during cold stress (in children).

2. The positive feedback involved in increased oxytocin secretion elicited during suckling and during labor.

3. The positive feedback involved in increased prolactin secretion elicited during suckling.

HORMONES OF THE ENDOCRINE SYSTEM

Hormones are named for their site of action and not by the endocrine gland that produced them. The following four tables should be used together. **Table 1** lists the names and target tissue actions of the hormones that are produced by the hypothalamus. **Table 2** lists the names and target tissue actions of the hormones produced by the anterior pituitary. **Table 3** lists the names and target tissue actions of the hormones produced by the posterior pituitary. **Table 4** lists the names and target tissue actions of hormones produced by other endocrine glands.

Hormones of the Hypothalamus (Table 15.1)

Name	Abbreviation	Action
Corticotropin releasing hormone	CRH	Releases ACTH from anterior pituitary.
Thyrotropin releasing hormone	TRH	Releases TSH from anterior pituitary.
Growth Hormone releasing hormone	GHRH	Releases GH from anterior pituitary.
Growth Hormone release inhibiting hormone; (Known previously as somatostatin)	GHIH (or SS)	Inhibits the release of GH from the anterior pituitary.
Prolactin releasing factor	PRF	Releases prolactin from the anterior pituitary.
Prolactin release inhibiting factor	PIF	Inhibits the release of prolactin from the anterior pituitary.
Gonadotropin releasing hormone	GnRH	Releases LH and FSH from the anterior pituitary.

Hormones of the Anterior Pituitary (Table 15.2)		
Name	**Abbreviation**	**Action**
Growth hormone, or Somatotropin; (release stimulated by GHRH; inhibited by GHIH)	GH	Plays a central role in the stimulation of growth of bones and soft tissue. Metabolically causes glucose conservation, elevation of blood glucose, fat mobilization and protein anabolism. In both sexes, enhances immune response. Targets the liver to cause release of somatomedin (Insulin-like Growth Factor-I; IGF-I).
Adrenocorticotropic hormone(release stimulated by CRH)	ACTH	Stimulates the release of Cortisol from the zona fasciculata of the adrenal cortex. Also promotes the growth of the zona fasciculata and zona reticularis of the adrenal cortex.
Thyroid stimulating hormone	TSH	Stimulates the release of thyroxin (T3) and [release stimulated triiodothyronine (T4) from by TSH] the thyroid.
Prolactin (release stimulated by PRF, inhibited by PIF])	Prl	In females stimulates development of breast tissue, and milk formation and secretion from the glandular tissue. In males may enhance action of LH on interstitial cells. In both males and females enhances immune response.

Hormones of the Anterior Pituitary (Table 15.2 cont.)		
Name	Abbreviation	Action
Follicle stimulating hormones (released by GnRH)	FSH	In females stimulates growth of ovarian follicles and estrogen secretion. In males acts on the seminiferous tubules to stimulate sperm production.
Luteinizing hormone (Originally called interstitial Ccll stimulating hormone in males) [released by GnRH]	LH (ICSH)	In females acts on the ovary to cause ovulation of the follicle. Stimulates secretion of estrogen and progesterone. In males acts on the Interstitial cells (Lydig cells) of the testes to stimulate testosterone secretion.

Hormones of the Posterior Pituitary (Table 15.3)		
Name	**Abbreviation**	**Action**
Oxytocin [released by neuroendocrine reflexes]		Acts on the smooth muscle of the uterus to cause contractions during labor. Released by neuroendocrine reflex initiated by cervical stimulation. Acts on the myoepithelial cells of the mammary glands to cause contraction and milk let-down. Releases by neuroendocrine reflex initiated by suckling.
Vasopressin, or Antidiuretic Hormone [released by neuroendocrine reflexes] o	ADH	Acts in kidney on distal convoluted tubule and collecting duct to increase water permeability and thus water reabsorption. Acts n smooth muscles of arterioles to cause constriction and increased systemic blood pressure.

Hormones of Other Endocrine Glands (Table 15.4)		
Gland	**Hormone/[Stimulus]**	**Action**
Adrenal medulla	Epinephrine (Epi), Norepinephrine (Norepi) [Release stimulated by stress when sympathetic nervous system is avtivated]	Sympathetic-type affects all over body. Epi also increases blood glucose.
Adrenal cortex	Corticosteroids	Three zones produce steroid hormones.
1. Zona glomerulosa	Aldosterone (Mineralocorticoid) [Release stimulated by Renin-angiotensin system, and elevated. K+ blood levels]	Increases Na+ reabsorption coupled with K+/H+ secretion in distal convoluted tubule of kidney
2. Zona fasciculata	Cortisol (Gucocorticoid) [Release stimulated by ACTH]	Increases blood glucose, increases utilization of protein and fat stores. Elevated levels are immuno-inhibitory. Assists in stress adaptation, blood pressure regulation, and energy mobilization and utilization. Target is most cells in body.
3. Zona feticularis	Androgens (dehydroepiandrosterone, DHEA) [Release stimulated by ACTH]	In females DHEA targets sites in bone and in brain. Responsible for Growth spurt at puberty, and labido. DHEA is also released in males but effects are over-shadowed by androgen from testes.
Thyroid follicles	Thyroxin (T4), Triiodothyronine (T3) [Release stimulated by TSH]	Target is most cells in the body. Increases overall cellular metabolism. Must be present for normal nerve growth and development to take place.
Thyroid C-cells	Calcitonin [Released when levels of plasma Ca++ are elevated]	Targets osteocytes in the bone matrix. Results in increased Ca++ uptake into bone, thus reducing blood Ca++ levels.

Hormones of Other Endocrine Glands (Table 15.4 cont.)		
Gland	**Hormone/[Stimulus]**	**Action**
Parathyroids	Parathyroid Hormone (PTH) [Release stimulate by reduced levels of with phosphate plasma Ca++]	Acts on kidney to increase Ca++ reabsorption coupled excretion. Acts on Intestine to increase Ca++ reabsorption as mediated by vitamin D activation. Acts on bone cells to increase removal of calcium and phosphate from the bone matrix. Overall result is to increase the low blood Ca++ levels while keeping the phosphate blood levels unchanged.
Endocrine Pancreas (islets of Langerhans)	Insulin (ß-islet cells) [Release stimulated by elevated levels of blood glucose]	Targets many cell types, including skeletal and cardiac muscle, and adipose. Acts through insulin receptors to facilitate glucose transport into target cells from the blood. Also effects storage and utilization of nutrients in target cells. Reduces blood glucose.
	Glucagon (á-islet cells) [Release stimulated by reduced levels of blood glucose]	Target is most cells. Promotes glycogeno-lysis and gluconeo-genesis. Elevates blood glucose.
	Somatostatin(D-islet cells) [Release stimulated by concentration of nutrients in the blood]	Target cells are present in the digestive system, and pancreatic islets. Exerts inhibitory influences on digestion and nutrient absorption.
Kidneys	Renin (juxtaglomerular cells) [Release stimulated by fall in blood Na+, fall in blood pressure in kidney]	Converts Angiotensingen into Angiotensin I. Angiotensin I is converted to Angeotensin II in lungs.
	Angiotensin II (Converted from Angiotensin I by converting enzyme in lungs)	Causes arteriolar smooth muscles to constrict. Increases blood pressure. Acts on the Zona Glomerulosa of the Adrenal Cortex to release aldosterone.

Hormones of Other Endocrine Glands (Table 15.4 cont.)		
Gland	**Hormone/[Stimulus]**	**Action**
Stomach and duodenum	Gastrin, Secretin Cholecystokinin, Gastric Inhibitory Peptide [Release stimulated by various nutrient feedbacks, parasympa-thetic nerves]	Target cells in the stomach, intestines, pancreas, gall bladder. Regulate various aspects of the digestive process.
Pineal gland	Melatonin [Release stimulated by sympathetic nerves and by light-dark cycle. Released in the dark, inhibited in the light]	Inhibits hypothalamic release of CRH, GnRH. Thus, inhibits the adrenal and gonadal axes functions in the dark. Involved in regulating overall circadian rhythm.
Gonads Ovaries in female	Estrogen [Release stimulated from the follicle fand corpus luteum by FSH and LH]	Required for development of ollicles, development of female secondary sexual characteristics. Maintains female reproductive organs in functional state. During peg-nancy necessary for growth of mammary tissues. Stimulates uterine contractions during parturition. Involved in Ca++ bone homeostasis. Regulates immune response.
	Progesterone[Release stimulated from corpus luteum by LH]	Prepares uterine endometrium for embryo implanttin. Inhibits uterine contractions. Regulates immune response.

Hormones of Other Endocrine Glands (Table 4 cont.)		
Gland	**Hormone/[Stimulus]**	**Action**
Testes in male	Testosterone [Release stimulated from the Interstitial (Lydig) Cells of the testes by LH]	Maintains male reproductive system in functional state. Maintains malesecondary sexual characteristics. Stimulates muscle growth, libdo, aggressive behavior.
Placenta	Estrogen, progesterone, Human Placental Lactogen (HPL), Human Chorionic Gonadotropin (HCG), other hormones. [Generally stimuli involved in release of these hormones remains unknown]	Involved in maintaining pregnancy to term by modifying uterine endometriumand myometrium. Effect overall body, develop mammary gland structure, depress maternal immune response.
Liver	Somatomedins (Insulin-like growth factor I; IGF-I) [Release stimulated by GH]	Causes growth of soft tissues and of bone.
Thymus	Thymosins (various thymic hormones, not all well defined.) [Release inhibited by sex steroids, cortisol, other stimuli]	Regulate immune response. Feedback of thymic hormones to the hypothalamus stimulates the release of GnRH, and LH.
Heart	Atrial Natriuretic Factor [Release stimulated by increased pressure in right atrium]	Acts on kidney tubules to inhibit reabsorption of Na+.Effect is to cause increased loss of Na+ and water from body reducing blood pressure.
Skin	Vitamin D (Generated in the presence of sunlight) Activation dependent on parathyroid hormone function in kidneys.	Acts in duodenum to increase Ca++ uptake.

CENTRAL ENDOCRINE ORGANS

Central endocrine organs, located in the brain, exert their regulatory functions on the peripheral endocrine organs and other target tissues in the body through the release of hormones into the systemic circulation.

The *hypothalamus-pituitary* unit (sometimes called *hypothalamus-pituitary* axis) is primarily responsible for central endocrine control of peripheral endocrine glands. Because the hypothalamus is part of the central nervous system, the over-riding influences on endocrine system regulation is provided by the nervous innervation; thus the concept of neuroendocrine involvement. Additionally, various short and long loop hormonal feedbacks to the hypothalamus and/or pituitary provide additional exquisite control of the entire network.

THE HYPOTHALAMUS AND THE PITUITARY GLAND

The *pituitary gland* (*hypophysis*) is located in a saddle-like indentation in the *Sphenoid* bone of the cranium called the *Sella turcica*. It is attached to the hypothalamus through a stalk called the *infundibulum* **(Figure 15.15)**. The pituitary gland is divided both anatomically and functionally into two lobes: the *anterior lobe* (adenohypophysis or anterior pituitary) and the *posterior lobe* (neurohypophysis or posterior pituitary). Hormones are released from both anterior and posterior lobes into the systemic blood, but the method of release is quite different for the anterior pituitary hormones and posterior pituitary hormones.

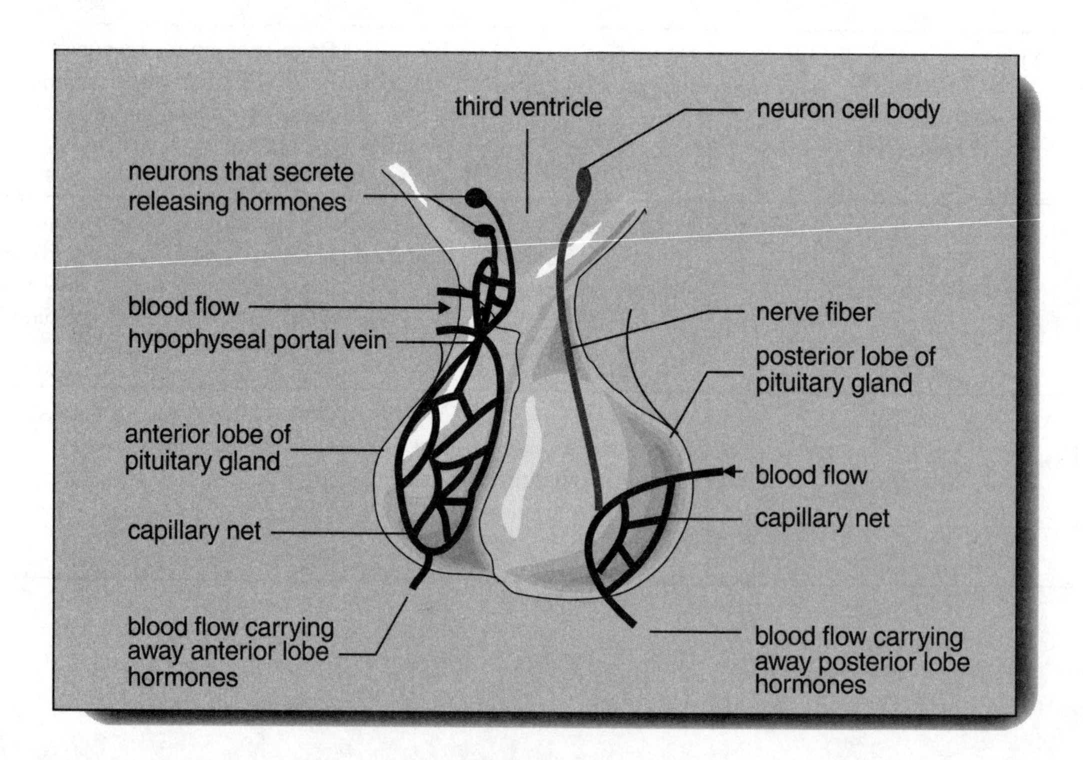

FIGURE 15.15

POSTERIOR PITUITARY

Although two hormones (called *Oxytocin* and *ADH*) are released from the posterior pituitary, neither is synthesized here. Oxytocin is synthesized in the *Paraventricular* nucleus of the hypothalamus, and ADH (Vasopressin) is synthesized in the *Supraoptic* nucleus of the hypothalamus **(see Figure 15.15)**. Nerve cell bodies located in these hypothalamic nuclei project axons down the pituitary stalk into the posterior lobe. These *hypothalmo-hypophyseal* tracts carry the synthesized ADH and oxytocin into the posterior lobe where they "synapse" on capillaries and release these hormones into the blood. Various stimuli act on these hypothalamic nuclei to release these hormones.

For example, oxytocin is released through neurological reflexes originating either in the uterus or in the nipple. Vasopressin is released upon sympathetic nervous system stimulation, (as ,for example, during the Baroreceptor reflex). In addition, inputs from the hypothalamic Osmoreceptors supply potent stimuli for increased ADH release.

Oxytocin. Uterine smooth muscle and myoepithelial cells in the mammary glands are targets for oxytocin. It has been clearly demonstrated that oxytocin is centrally involved in stimulating contractions of uterine smooth muscle during labor. Oxytocin also acts at the myoepithelial cells that line the lactiferous ducts in the mammary glands. Release of this hormone during suckling causes milk release (let-down). These effects will be considered in detail in the chapter on the reproductive system.

Vasopressin (Antidiuretic Hormone, ADH). The actions of this important hormone have been described fully in the chapter on the kidney. However, as the reader may recall, ADH functions in the nephrone to increase water reabsorption. ADH also increases blood pressure by acting on the precapillary sphincters to cause constriction, and thus to increase total peripheral resistance.

ANTERIOR PITUITARY

Unlike the posterior pituitary, hormones released from the anterior lobe are actually synthesized here in secretory cells. Five types of cells are found in the anterior pituitary, namely the *somatotropes* (that secrete GH), the *corticotropes* (that secrete ACTH), the *thyrotropes* (that secrete TSH), the *gonadotropes* (that secrete LH and FSH), and the *mammatropes* (that secrete PRL). In order for each to release its particular hormone into the blood a specific releasing factor synthesized in the hypothalamus must be present. In addition, hypothalamic inhibitory hormones regulate some of these anterior pituitary-derived hormones. To understand this process clearly, it is necessary to first consider the anatomical structures presented here.

The anterior lobe is connected to the hypothalamus through a *portal system* (*artery-capillary-vein-capillary-vein*) **(see Figure 15.15)**.

Beginning at the top of the pituitary stalk and ending in the anterior lobe, this vascular link transports the releasing and release-inhibiting hormones from their origins in hypothalamic nuclei to the anterior

lobe secretory cells. Specifically at the top of the pituitary stalk, an artery carrying systemic blood enters and supplies capillaries. Secretory nerve cell axons synapse here and release their hormones into the capillary blood. These capillaries attach to veins (hypothalmo-hypophyseal portal system) which then pass into the anterior pituitary where they supply blood to capillaries (sinusoids) in this region. Surrounding these sinusoids are the anterior pituitary secretory cells. As the blood infuses into these sinusoids, any releasing or release-inhibiting hormones in the blood can diffuse out of the capillaries and bind to receptors on the secretory cell membranes. In response, the secretory cell releases its particular hormone (GH, PRL, TSH, LH & FSH, or ACTH) which enters the blood and leaves the anterior pituitary via the vein. Thus, these pituitary hormones enter the systemic circulation where they promote a variety of endocrine effects.

Pineal Gland. It is deeply located between the cerebral hemispheres on the anterior surface of the brain. This gland secretes melatonin, which is synthesized from serotonin. Increased production of melatonin takes place in response to the absence of light stimuli from the eyes. This information reaches the pineal via nerve tracts involving the hypothalamus, reticular formation and spinal cord. Accordingly, the fluctuating concentration of melatonin is involved in the regulation of circadian rhythms.

In addition, since melatonin inhibits the production of GnRH and CRH from the hypothalamus, this, in turn, will affect the function of both the gonadal and adrenal axes. The circadian effect involving the gonadotropins can also play a role in the female menstrual cycle, although other factors are also involved.

Hormones of the Anterior Pituitary. Hormones produced by the anterior pituitary exert wide-ranging effects. They function either by targeting peripheral endocrine glands, or through direct effects on target tissues all over the body.

ANTERIOR PITUITARY HORMONES THAT EXERT DIRECT EFFECTS

Growth Hormone, or somatotropin, and prolactin exert direct effects. But the growth hormone functions *both* directly and indirectly to cause body growth. Initially, the hypothalamus releases growth hormone releasing hormone (GHRH) and growth hormone inhibiting hormone (GHIH) **(Figure 15.16)**. Thus, depending on the balance between these two hormones reaching the cells of the anterior pituitary, more or less GH will be released. GH affects multiple target tissues throughout the entire body to promote the growth of bones and soft tissue. In this capacity, GH functions as an insulin antagonist, stimulating glucose conservation leading to the elevation of blood glucose, fat mobilization, and protein anabolism. In addition to these classic, and well studied effects, GH also enhances immune function in both males and females. Some of the responses attributed to GH are now known to be mediated by somatomedins (Insulin-like Growth Factor-I; IGF-I) produced by the liver upon stimulation with GH.

GH secretion is under the control of various feedback pathways involving both hormones and organic nutrients.

1) Hypoglycemia increases the release of GH acting at the levels of the hypothalamus, either by stimulating the release of GHRH and/or inhibiting the release of GHIH (somatostatin). Since

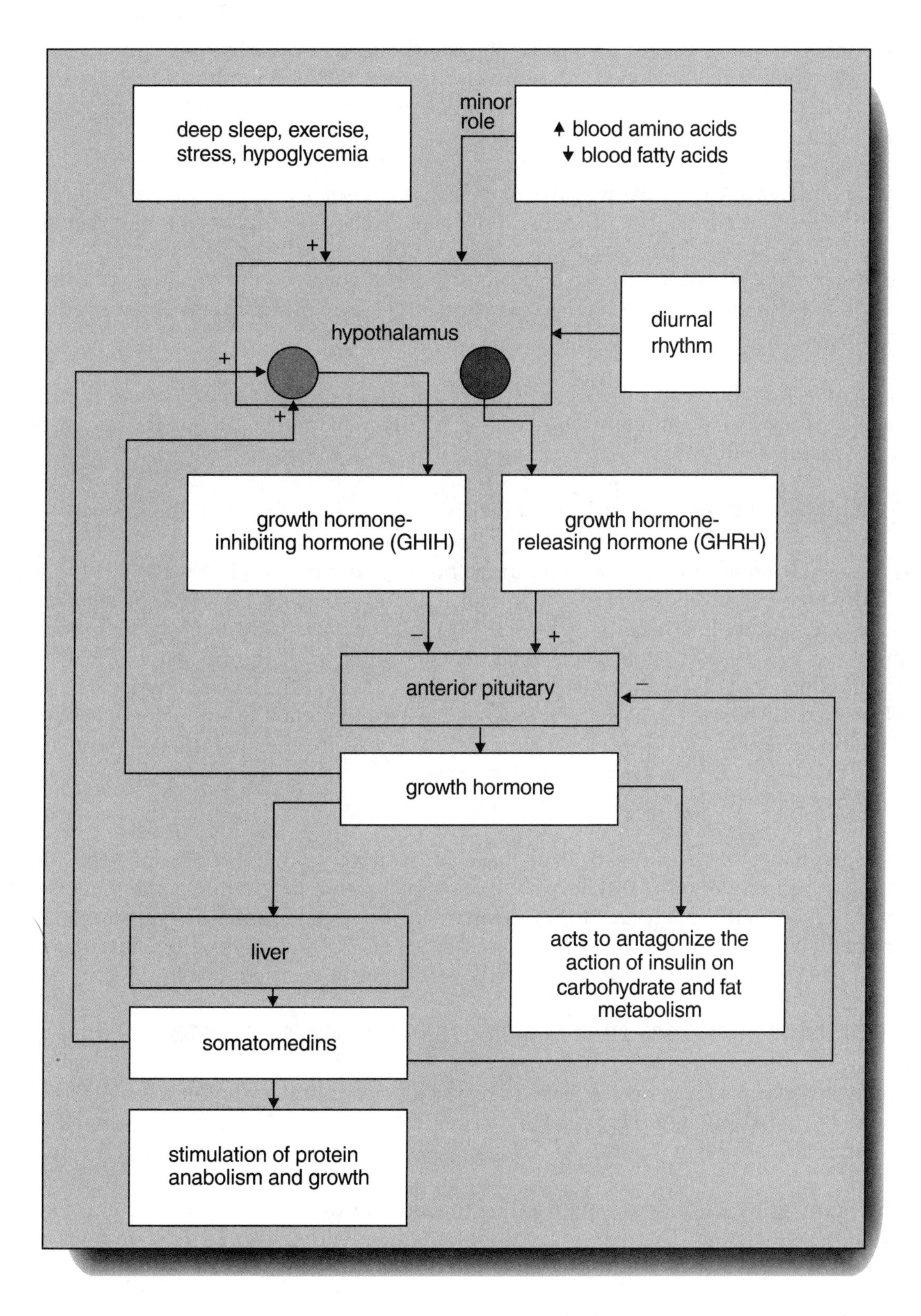
deep sleep, exercise, stress, hypoglycemia
minor role
↑ blood amino acids
↓ blood fatty acids
+
hypothalamus
diurnal rhythm
+
+
growth hormone-inhibiting hormone (GHIH)
growth hormone-releasing hormone (GHRH)
−
+
anterior pituitary
−
growth hormone
liver
acts to antagonize the action of insulin on carbohydrate and fat metabolism
somatomedins
stimulation of protein anabolism and growth

FIGURE 15.16

GH antagonizes the effects of insulin on the metabolism of carbohydrates and fats, it will produce *hyperglycemia*. As blood glucose levels increase, this will inhibit GH release through GHRH and/or GHIH regulation at the hypothalamus, thus a negative feedback (mediated through glucose) controls systemic GH levels.

2) A similar increase in GH release is also produced by exercise, stress and possibly by deep sleep, but again the exact mechanism of action at the level of the hypothalamus is not clarified.

3) Increased levels of blood amino acids and decreased levels of blood fatty acids also increase GH release, but again the exact mechanism of action at the level of the hypothalamus remains to be determined.

4) *Somatomedins* (*IGF-I*) are released from the liver upon GH stimulation. IGF-I acts at the levels of the hypothalamus to stimulate the release of GHIH, which then inhibits the release of GH from the anterior pituitary.

5) IGF-I may also act directly at the level of the anterior pituitary to inhibit the release of GH.

Prolactin is released from the anterior pituitary in the presence of PRF, and its release is inhibited by PIF. Normally PRL levels are very low in males due to increased PIF, while in females PRL levels are also quite low (due to increased PIF), but may fluctuate slightly with hormonal changes during the menstrual cycle. This effect results from the presence of fluctuating estrogen which stimulates PRL release. During pregnancy, PRL levels increase as estrogen levels rise because estrogen acts at the level of the hypothalamus to stimulate release of PRF and inhibit release of PIF. However, although PRL levels are elevated during pregnancy, no milk is produced by the mammary glands because the elevated estrogen at this time antagonizes the PRL effect at the glandular tissue.

At the end of pregnancy, after estrogen levels have fallen, PRL continues to be present and this accounts for the production of milk. PRL levels will, however, begin to fall unless maintained by the neuroendocrine reflex of suckling. If suckling does take place, afferent impulses are transmitted via the cord to the hypothalamus where they act to increase release of PRF and inhibit PIF. Thus, if suckling continues, PRL will continue to be released and milk will continue to be produced.

ANTERIOR PITUITARY HORMONES THAT ACT ON OTHER ENDOCRINE GLANDS

ACTH, Adrenocorticotropic Hormone, is released by the anterior pituitary under the control of CRH from the hypothalamus. ACTH primarily targets the Zona fasciculata of the Adrenal Cortex stimulating the release of cortisol. ACTH also causes the growth of the adrenal cortex.

TSH, Thyroid Stimulating Hormone, is released by the anterior pituitary under the control of TRH from the hypothalamus. TSH acts on the thyroid follicles within the thyroid gland which then release thyroid hormones.

LH and FSH, Luteinizing Hormone and Follicle Stimulating Hormone, are both released by the anterior pituitary under the control of GnRH from the hypothalamus. FSH and LH primarily target the gonadal tissues in both males and females causing development of gametes as well as release of sex steroid hormones.

PERIPHERAL ENDOCRINE ORGANS

Peripheral endocrine organs include those glands that are under the control of the hypothalamus-pituitary axis, as well as endocrine glands that respond to various stimuli, and are not under pituitary control.

Adrenal Gland. The adrenal glands (also called the *suprarenal glands*) are located on top of each kidney. These glands are subdivided into an outer region called the *adrenal cortex,* and an inner portion called the *adrenal medulla.* These two regions are derived in the embryo from different precursor tissues, and are not functionally associated later in life. However, both secrete hormones into the blood.

Adrenal Medulla. The Adrenal Medulla comprises about 20% of the adrenal tissue, and is actually a sympathetic ganglia of the autonomic nervous system. As such, it is activated by sympathetic preganglionic fibers arising from the central nervous system. Postganglionic chromaffin cell bodies in the adrenal medulla synthesize, store and release the two related amines—epinephrine and norepinephrine into the blood. These chromaffin cells are actually modified postganglionic cell bodies that have lost their axons and thus release their neurotransmitters directly into the blood. However, epinephrine is released in much larger quantities than is norepinephrine. Synthesis of these amines takes place as follows:

1. The amino acid tyrosine is converted (by tyrosine hydroxylase) into dopa.
2. Dopa is converted (by dopa decarboxylase) into dopamine.
3. Dopamine is converted (by dopamine betahyrdoxylase) into norepinephrine.
4. Some norepinephrine (about 10%) is stored and released in this form. However, the remaining 90% is converted (by phenylethanoloamine N-methyltransferase) into epinephrine, which is also stored and released. Thus, 80% to 90% of the catecholamine secretions produced by the adrenal medulla is epinephrine and only a small fraction (10% to 20%) is norepinephrine.

Functionally, the amines acting as hormones generally produce the same responses at target tissues, as does norepinephrine when it is released by adrenergic sympathetic fibers. For example, sympathetic stimulation causes an increase in heart rate, as does circulating epinephrine and norepinephrine. Sympathetic stimulation causes pupil dilation, as does circulating epinephrine and norepinephrine. Sympathetic stimulation causes dilation of bronchiolar smooth muscle, as does circulating epinephrine and norepinephrine, and so on. In all cases, the various effects produced by these hormones are mediated by alpha and beta adrenergic receptors, (described in an earlier chapter). Regarding these receptors, the following should be noted:

1) Epinephrine binds to alpha receptors, and to beta-1 receptors. It stimulates a tenfold greater effect at target tissues than does norepinephrine.

2) Epinephrine binds to alpha receptors, beta-1 receptors and beta-2 receptors.

3) The effects elicited on the target tissue depend on the type and number of receptor present. Alpha and beta-2 receptors are found in a wide variety of target tissues. Some target tissues may have only alpha or only beta-2 receptors, while others have both together. Generally, binding of hormone (epinephrine or norepinephrine) to alpha or beta-2 receptors produces an excitatory effect in the target tissues, but not in all cases. On the other hand, beta-1 receptors are found primarily on cardiac muscle tissue, and when epinephrine binds to them the effect is usually inhibitory.

4) Since epinephrine circulates in the blood, it can not only reach all alpha and beta-2 receptors, but can also reach all beta-1 receptors as well. Thus, the effects elicited by epinephrine are more extensive than those produced by direct sympathetic stimulation of postganglionic adrenergic fibers. In addition, the target tissue responses depend on the number and type of specific receptors in each target tissue.

5) In addition to the standard sympathetic-like effects elicited by epinephrine, this hormone also elicits metabolic actions at its target tissues. (For example, acting on the liver epinephrine stimulates conversion of stored glycogen into glucose and thus promotes hyperglycemia.) Norepinephrine has only limited metabolic action at target tissues, and no effect on blood glucose levels.

DISORDERS OF THE ADRENAL MEDULLA

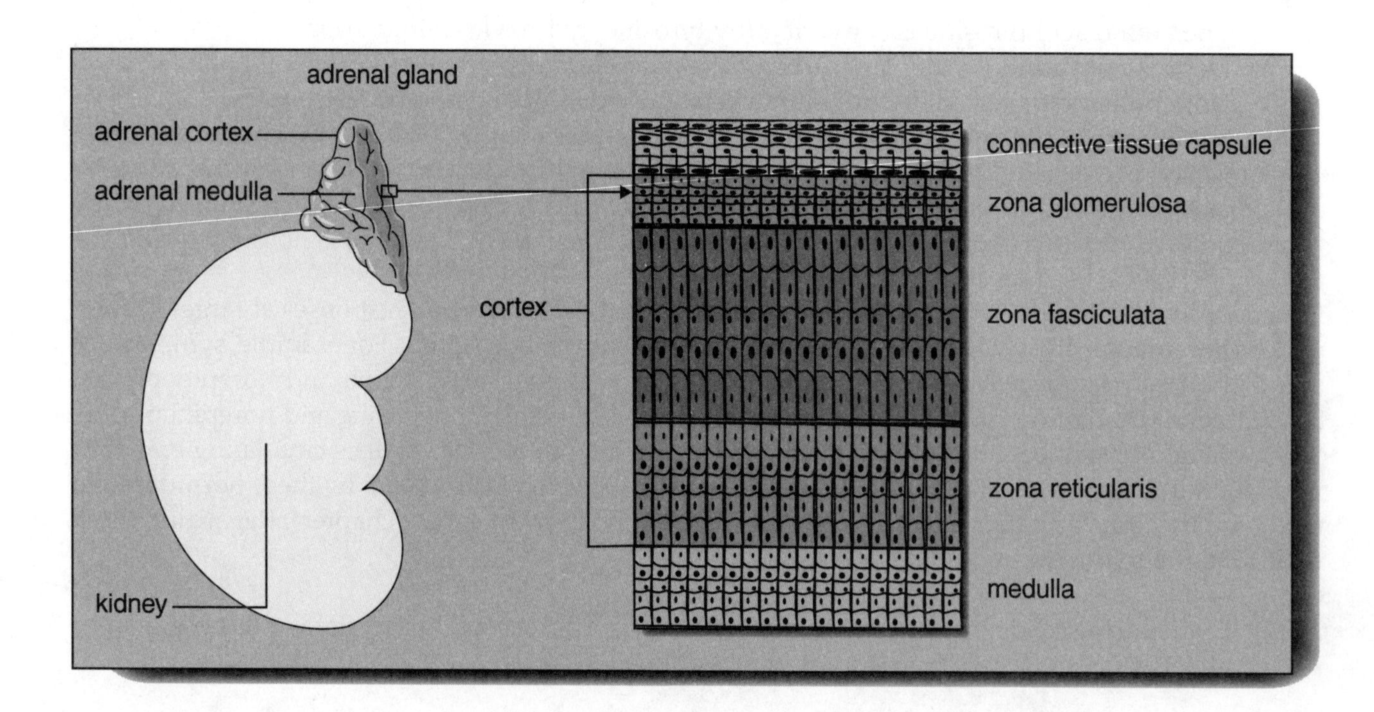

FIGURE 15.17

Hyposecretions of hormones from the adrenal medulla do not produce any apparent systemic effect. This is probably because stimulation by adrenergic sympathetic fibers allows the body to overcome stress in the absence of secretions by the adrenal medulla. In addition, other hormones can also promote changes much like epinephrine (i.e., cortisol, thyroid hormone, and glucagon).

Hypersecretion, predominantly of norepinephrine, but also sometimes of epinephrine, from adrenal medullary tumors (pheochromocytomas) may result in prolonged and excessive sympathetic-like responses (i.e., hypertension, increased heart rate, hyperventilation, pupil dilation, hyperglycemia). This condition is cured by surgical removal of the affected adrenal gland.

ADRENAL CORTEX

The adrenal cortex is divided into three zones: the zona glomerulosa, the zona fasciculata, and the zona reticularis **(Fig 15.17)**. Each zone synthesizes and releases steroid hormones (*corticosteroids*, or *adrenocortical hormones*).

Zona glomerulosa: This outermost zone of the cortex synthesizes steroids called *mineralocorticoids*. The mineralocorticoid of importance synthesized here is *aldosterone* and its function is to regulate the concentration of Na^+ and K^+ in the body. Aldosterone release takes place when:

1. Angiotensin II acts on the zona glomerulosa cells. Angiotensin II is synthesized through the Renin-Angiotensin pathway.
2. In addition, aldosterone is also released in the presence of elevated blood K^+. The complete pathways involved have been described previously in the chapter on fluid and electrolytes.
3. Functioning at the distal convoluted tubules of the nephrone in the kidney, aldosterone increases Na^+ reabsorption while concomitantly increasing the secretion of K^+ (and also under certain conditions H^+).

The secretion of Aldosterone is regulated by direct *negative feedback* of blood K^+ concentration, and also depends on the various factors that stimulate release of Renin from the kidney, namely, blood pressure, blood volume, blood Na^+/Cl^-, and stress.

Zona fasciculata: This middle zone of the cortex synthesizes *glucocorticoids*, primarily the substance called *cortisol*. Cortisol is a potent corticosteroid with wide-ranging effects on the total body. For example, cortisol:

1. Increases blood glucose through hepatic gluconeogenesis. In this role, it stimulates the liver to convert amino acids to glucose to maintain blood glucose levels, after glycogen stores in liver and muscle tissue have been depleted.

2. Increases the release of fatty acids into the blood by stimulating lipolysis of stored fat. The increased circulating fatty acids are then available for use as an alternate energy source by cells.

3. Increases the formation of amino acids by breaking down protein in muscle and other body tissues. These amino acids then enter the blood where they can be utilized in gluconeogenesis or other metabolic events.

4. Inhibits the uptake and utilization of glucose by many body tissues, but not by nerve tissue. This effect (coupled with the other effects promoted by cortisol), guarantees that adequate blood glucose will be available to the nervous system. This is of primary importance since the brain can only use glucose as a metabolic energy source, and neurons can only store limited amounts of glycogen. Thus, without blood glucose the nervous system will rapidly cease to function.

5. Assists in stress adaptation and blood pressure regulation through the mechanisms just listed. In addition, elevated levels of cortisol stabilize lysosomes, reducing autolysis of cells under stress conditions.

6. Inhibits immune response by down-regulating effector lymphocytes. Inhibits steps in lymphocyte synthesis in the bone marrow and thymus. Also modifies lymphocyte populations within the spleen. Reduces inflammation by limilting the inflammatory pathways that depend on lymphocyte activation. However, these immuno-inhibitory effects take place primarily at physiologically elevated levels. Low levels of cortisol have been reported to be immun-stimulatory.

Regarding regulation, the hormonal pathway involved is outlined in **Figure 15.18**. This pathway is described as the *hypothalamic-pituitary-adrenal axis*, or simply as the *HPA-axis*.

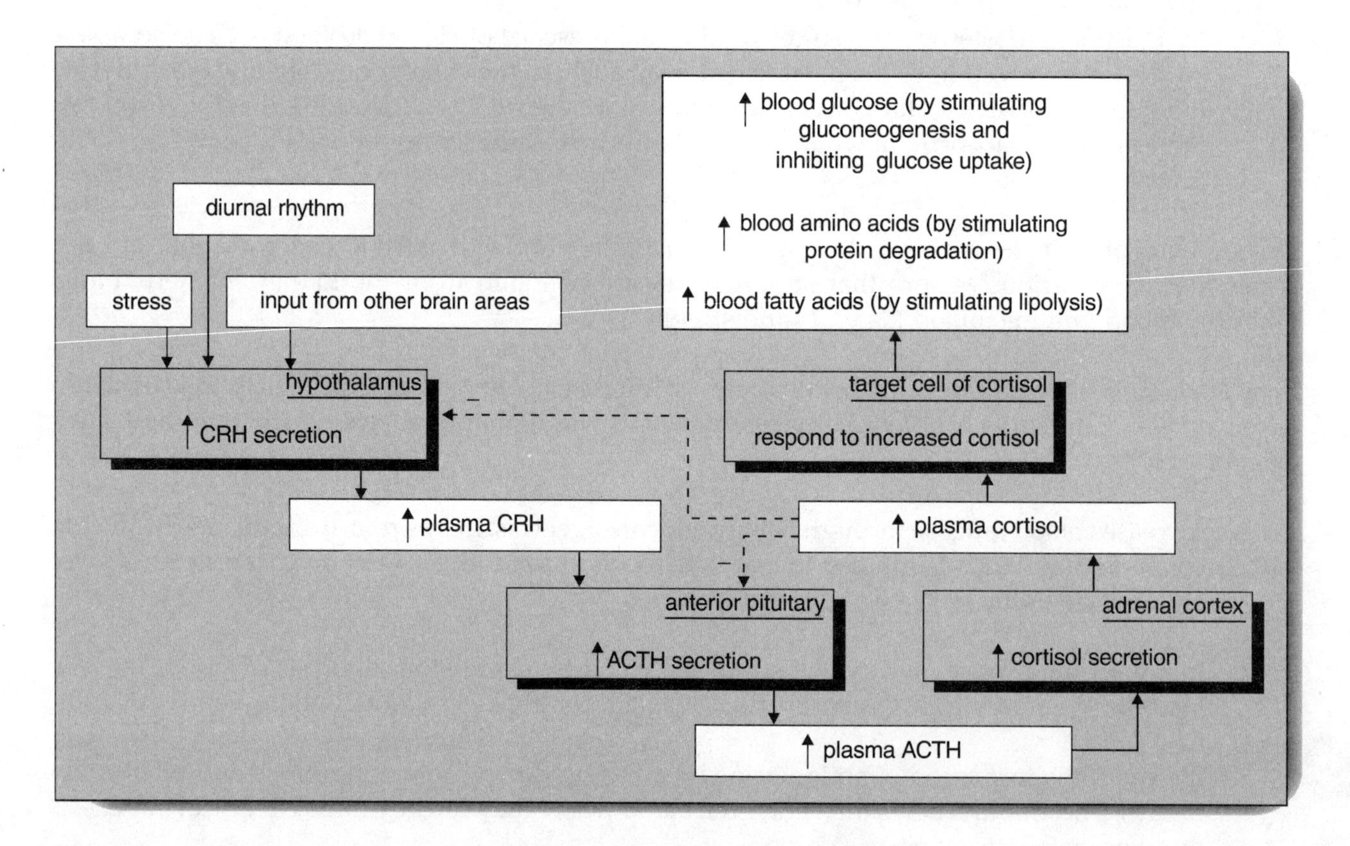

FIGURE 15.18

1) Cortisol secretion is initiated by the release of CRH from the hypothalamus, stimulating the release of ACTH from the anterior pituitary.

2) ACTH in turn, then stimulates release of cortisol from the zona fasciculata of the adrenal cortex. This causes cortisol blood levels to rise. Since ACTH acts as a tropic hormone for the zona fasciculata (and zona reticularis), it also causes the growth of these regions of the adrenal cortex.

3) The circulating cortisol promotes the various effects on the body as described. In addition ,it also gives negative feedback at the levels of both the hypothalamus and anterior pituitary.

4) This negative feedback reduces secretion of both CRH and ACTH. The final outcome is a reduction in cortisol secretion from the adrenal cortex, causing blood levels to fall.

5) Negative feedback of cortisol not only regulates its own blood levels, but inputs from higher brain areas can also act upon the hypothalamus to increase cortisol release. Stimuli that initiate such inputs from higher brain areas include stress and diurnal (circadian) rhythm.

   a) *Stress* inputs are initiated as a result of exposure of the individual to various physical or emotional stimuli. Increased secretion of cortisol acts to combat such stressors because it facilitates metabolic processes that maintains available glucose, amino acids and fatty acids to support brain functions. The presence of these circulating nutrients also assist in tissue repair processes. Elevated levels of cortisol also inhibit excessive, or destructive, immunological responses such as excessive inflammation and hypersensitivity.

   b) *Diurnal Rhythms*: Over a 24-hour period, plasma cortisol levels rise and fall in a specifically repetitive pattern coupled to the day-night cycle (termed *diurnal rhythm*). Generally levels are elevated during the day and fall significantly at night. The mechanism to account for this *circadian rhythm* is complex, involving pineal gland function, and the intrinsic clock that maintains day/night activity patterns. Researchers studying this diurnal pattern of cortisol secretion were quick to point out that this might be important in surgical procedures, or drug therapy. Perhaps a patient would respond better to surgery or drug therapy if these procedures were performed at a particular time of day or night. Work continues in this interesting and important area.

Zona reticularis: This inner zone of the adrenal cortex also secretes some cortisol, and like the zona fasciculata, is also under the control of ACTH. In addition, the zona reticularis also secretes adrenal androgens and estrogen which are similar to the sex steroids produced by the gonads in both sexes. However, normally the secreted amount of adrenal sex steroids is small in comparison to the production of sex steroids by the gonads.

The most important adrenal sex hormone is the weak androgen Dehydroepiandrosterone (DHEA). Although DHEA is produced by the adrenal in both males and females, it is only of significance in females. This is because in males the high levels of testosterone produced by the testes supersedes any effects produced by DHEA. However, in females where only small amounts of androgen are produced

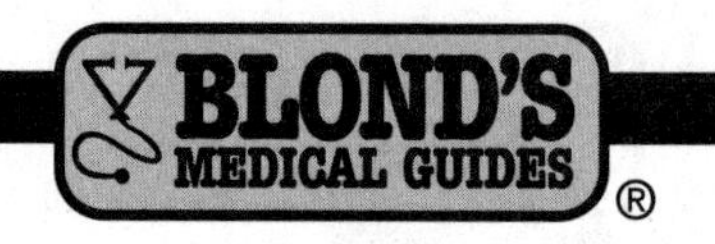

by the ovaries (but much estrogen), the weak androgen action of adrenal DHEA is of importance. In females, adrenal DHEA:

1. Enhances the pubertal growth spurt
2. Stimulates growth of axillary and pubic hair
3. Stimulates libido (sex drive)

Regulation of DHEA Secretion. ACTH is responsible for stimulating secretion of DHEA from the zona reticularis. Thus, normally the production of cortisol and of DHEA rise and fall together. (However, at puberty adrenal androgen secretion rises above that of cortisol.) Note also that unlike cortisol, DHEA does not inhibit, by negative feedback, the secretion of CRH and ACTH. Instead, it functions at the hypothalamus to inhibit GRH, and at the pituitary to inhibit GnRH. Thus, DHEA acts much like testosterone to down-regulate the Hypothalamic-Pituitary-Gonadal axis.

DISORDERS OF THE ADRENAL CORTEX

Disorders of adrenal hormone secretion fall into two major categories: *hyposecretion* disorders (not enough hormone is produced), and *hypersecretion* disorders (excessive amounts of hormone are produced).

Hyposecretion Disorders. They can be subdivided into *primary adrenocortical insufficiency* (*Addison's disease*), and *secondary adrenocortical insufficiency*.

1) **Addison's Disease** results from an inability of the adrenal cortex to secrete cortisol and aldosterone. In addition, secretion of sex hormones is frequently also reduced. While a reduction in adrenal sex hormones is not life-threatening, in severe Addison's disease, the reduced levels of cortisol and aldosterone can lead to rapid death. Addison's disease results from destruction of the adrenal cortical tissues for reasons that are frequently not clear (called *idiopathic disease*). However, it is very likely that an autoimmune mechanism is involved. The symptoms of Addison's disease are:

a) Decreased circulating glucose (hypoglycemic), as well as other nutrients due to the absence of cortisol.
b) Inability to effectively respond to stress stimuli due to the absence of cortisol.
c) Loss of Na^+ in the urine leading to decreased Na^+ blood levels (hyponatremia) due to the absence of aldosterone.
d) Decreased K^+ secretion into the urine in the kidney leading to increased K^+ blood levels (hyperkalemia) due to absence of aldosterone.
e) Coupled with the increased Na^+ excretion in the urine, there will also be an increased loss of water. This, associated with the hyperglycemia, will lead to dehydration and a dangerous fall in blood pressure (hypotension).
f) The hyponatremia, hyperkalemia and decrease in extra-cellular fluid volume will seriously alter the osmolarity and electrolyte concentrations in both the extra-cellular and intra-cellular compartments. This may affect the ability of nerve and muscle cells to conduct action potentials and lead to problems in muscle function and in brain/CNS activity (like disorientation and

disfunction).

g) Increased skin pigmentation due to increased levels of ACTH. This is an interesting observation and comes about because reduced levels of cortisol do not feedback to the hypothalamus and pituitary. In the absence of negative feedback, ACTH levels rise, but the increased ACTH has nothing to stimulate since the tissue of the adrenal cortex is destroyed. Elevated ACTH stimulates melanocytes in the skin because its structure is similar to that of melanocyte stimulating hormone. Stimulation of melanocytes leads to increased skin pigmentation in many Addison's patients.

2) **Secondary Adrenocortical Insufficiency** results in an inability of the anterior pituitary-hypothalamic axis to secrete ACTH. In the absence of ACTH, the cortex cannot secrete adequate amounts of cortisol and thus these patients would suffer from hypoglycemia; reduced levels of other circulating nutrients (amino acids, lipids); and a reduced ability to withstand stress. However, there would probably be no electrolyte imbalances since aldosterone is not affected, and no skin pigmentation since ACTH is reduced, not elevated.

TREATMENT OF ADRENOCORTICAL HYPOSECRETIONS

Treatment of Addison's disease requires the use of replacement mineralocorticoids and glucocorticoids. Since the cortical tissue, which was destroyed, does not regenerate, such patients must receive replacement hormone therapy for the rest of their lives.

Treatment of Secondary Adrenocortical Insufficiency would depend on the cause of the hypothalamic-pituitary malfunction, but could require replacement with glucocorticoids.

HYPERSECRETION DISORDERS

They can be subdivided into **excess cortisol secretion** (**Cushing's syndrome**), **excess aldosterone secretion**, and **adrenogenital syndrome**).

Cushing's Syndrome. This results from excess secretion of cortisol which primarily promotes excessive gluconeogenesis and resultant hyperglycemia. The outcome of this elevated conversion of amino acids to glucose creates many problems due to the systemic protein shortage including muscle and bone weakness, and fatigue. The elevated blood glucose causes glucosuria (loss of glucose in the urine) and thus the patient appears to be diabetic (diabetes mellitus). However, since the cause of the glucosuria is elevated cortisol and not hypoinsulinism, this condition is termed *adrenal diabetes*. In addition to the hyperglycemia, patients also experience abnormalities in body fat deposition, as a result of glucose being converted to excess fat. This fat is commonly deposited in the face (described as moon face), and in the abdomen.

Excess Aldosterone Secretion (or Aldosterone Hypersecretion). In the presence of elevated levels of this potent mineralocorticoid, Na^+ is retained and K^+ is lost in the urine. The hypernatremia and hypokalemia can lead to problems in muscle and nerve function. Hypernatremia causes water retention and swelling of the extra-cellular compartment, overall body edema, and hypertension. Aldosterone Hypersecretion results from:

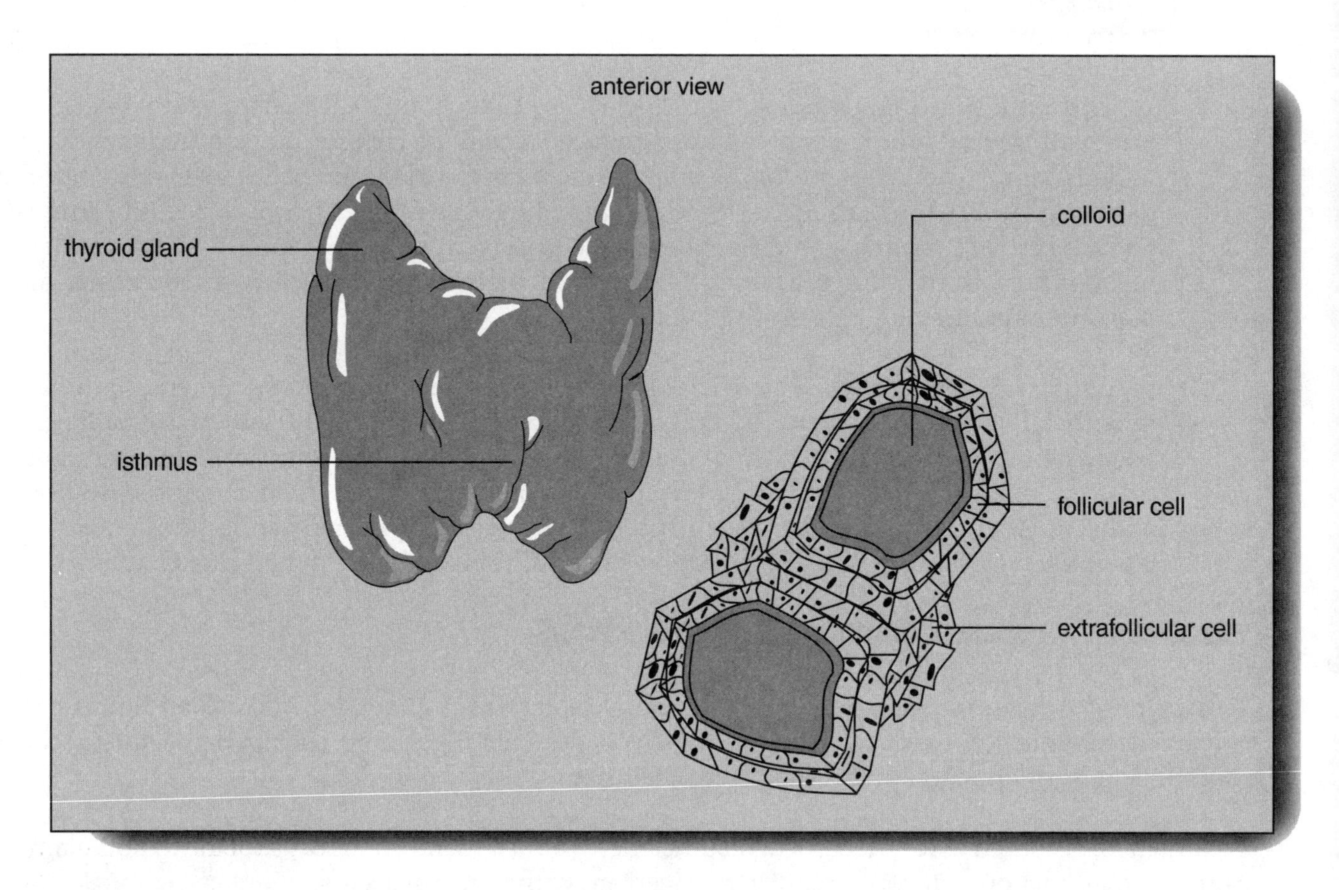

FIGURE 15.19

1) *Primary Hyperaldosteronism,* also called *Conn's Syndrome,* caused by a tumor in the zona glomerulosa of the adrenal cortex that secretes aldosterone.

2) *Secondary Hyperaldosteronism* caused by elevated renin-angiotensin activity.

Adrenogenital Syndrome. It is a disorder of the adrenal cortex (possibly zona reticularis) resulting in either *Excess Adrenal Androgen Secretion* (sometimes called *Adrenal Androgen Hypersecretion*), or *Excess Adrenal Estrogen Secretion*. Excess secretion of androgen from the adrenals is more commonly reported, and in females can cause significant masculinization. Excess estrogen secretion from the adrenals is a very uncommon condition, but, if present, can cause feminization in males. Effects vary depending on the age and sex of the afflicted person.

Causes of Hypo and Hyper Secretion Disorders of the Adrenal Cortex. As mentioned, hyposecretion disorders can result from abnormalities of the adrenal cortical tissue itself (possibly resulting from an autoimmune disease), or can be due to the inability of the pituitary to secrete ACTH. In the case of hypersecretions, tumors of the cortical tissue can secrete excess hormones without support of any ACTH from the pituitary. On the other hand, excess ACTH from the pituitary or another source can also cause hypersecretion disease. If excess ACTH is secreted by the pituitary it could be due to a tumor of the pituitary, or absence of negative feedback regulation. ACTH not of pituitary origin (ectopic

ACTH) can also be secreted by tumors in other locations (such as the lungs). In the case of adrenogenital syndrome, the most common cause is a deficiency in an enzyme pathway necessary for cortisol synthesis. Under these conditions, the reduction in cortisol production increases ACTH release from the pituitary and the adrenal cortex (which cannot manufacture cortisol) responds by increasing androgen production instead.

THYROID GLAND

This gland is located below the larynx in the neck and also covers the trachea in this region. It is formed from connective tissue and *thyroid follicles*. These follicles have a cuboidal epithelial cell layer. Within the center of each follicle is a space filled with thyroid colloid primarily comprised of the glycoprotein, called *thyroglobin* **(Figure 15.19)**.

MANUFACTURE AND TRANSPORT OF THYROID HORMONES

Thyroid hormones are manufactured by the epithelial cells of the follicles, and either released into the blood, or stored in the colloid for later release. Two hormones are manufactured by the follicle cells, and the synthesis of both requires the amino acid tyrosine and also iodine. These thyroid hormones are *thyroxine* (*tetraiodothyronine*) which contains four molecules of iodine (and thus is termed T_4), and *triiodothyronine* which contains three molecules of iodine (and is thus called T_3) **(Figure 15.20)**.

While T_4 is secreted in much larger quantities than is T_3 (90% T_4:10% T_3), T_3 is many times more active at the target tissues than is T_4. Circulating T_4 is converted into T_3 through peripheral removal of one iodine molecule. This activation step takes place not only in the target tissues, but also in many other peripheral tissues such as liver and kidneys.

Thyroid hormones are transported in the blood bound to the plasma proteins as follows:

Name	Percent Binding of	
	T4	**T3**
Thyroxine-binding globulin	55%	65%
Thyroxine-binding prealbumin	35%	—
Albumin	10%	35%

BIOLOGICAL ACTION OF THYROID HORMONES

Probably all body tissues are targets for thyroid hormones.

Thyroid hormones function primarily to elevate the rate of metabolic reactions within cells, thus increasing heat generation. The action of thyroid hormones is therefore considered to be *calorigenic* to body tissues. Additionally, thyroid hormones also promote intermediary metabolism, acting to regulate both anabolism and catabolism of carbohydrates, proteins and fats. With regard to the cardiovascular system, thyroid hormones indirectly increase heart rate because they increase the sensitivity of cardiac muscle to circulating epinephrine. Finally, thyroid hormones are also required for normal growth and development of the central nervous system in children, and normal function of the brain and other CNS structures in adults.

REGULATION OF THYROID HORMONE SECRETION

The secretion of hormones from the thyroid is regulated by a standard negative feedback pathway involving the hypothalamic-pituitary axis. Accordingly, elevation of TRH from the hypothalamus stimulates increased release of TSH from the anterior pituitary and this, in turn, stimulates increased release of T_3 and T_4 from the thyroid. Negative feedback of T_3 and T_4 at the level of the anterior pituitary reduces release of TSH, thereby decreasing levels of circulating thyroid hormones **(Figure 15.21)**.

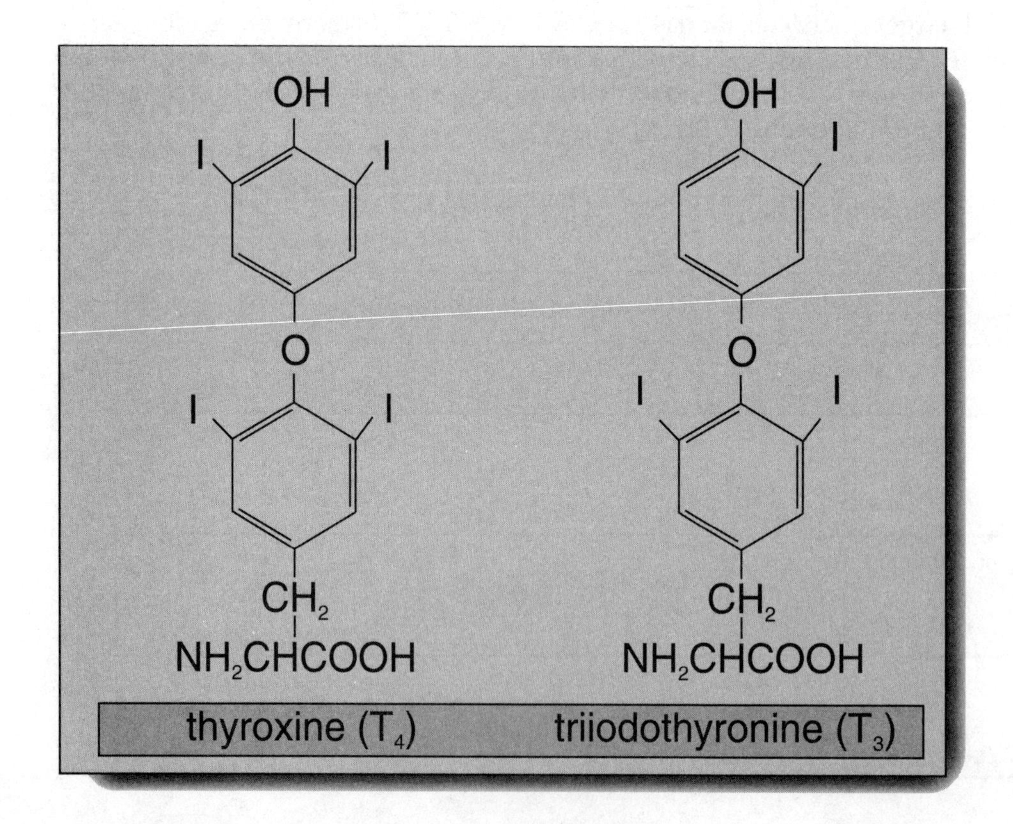

FIGURE 15.20

In addition to this standard negative feedback that constantly regulates blood levels of thyroid hormones, stress stimuli and cold stimuli (especially in infants) also act at the level of the hypothalamus to up-regulate the hypothalamic-pituitary-thyroid axis and increase levels of circulating thyroid hormones. Elevated levels of thyroid hormones assist in combating the stress in two ways. They:

1) Stimulate an overall increase in metabolic activity.

2) Stimulate an increased responsiveness of catecholamine (epinephrine, norepinephrine) sensitive target tissues. Thus elevated thyroid hormone synergistically acts in conjunction with the sympathetic nervous system, and adrenal medullary hormones, to assist the body in coping with stress conditions.

DISORDERS OF THE THYROID GLAND

Problems of thyroid hormone secretion fall into two major categories: *hyposecretion* disorders (or *hypothyroidism*), and *hypersecretion* disorders (or *hyperthyroidism*).

Hypothyroidism (or ***Myxedema***). is manifested by reduced levels of thyroid hormone secretion. Based on the cause of the hormonal decrease, this condition can be further subdivided into **primary hypothyroidism**, *secondary hypothyroidism*, and *iodine deficiency*.

1) In *primary hypothyroidism*, the secretory tissue of the thyroid gland itself is unable to manufacture and release thyroid hormones. One probable cause is destruction of the follicular cells, or thyroglobin, through autoimmune-like interactions (*Hashimoto's disease*). In the absence of circulating blood levels of thyroid hormones, there is no negative feedback at the level of the pituitary and this causes elevated blood levels of TSH to be present.

2) In *secondary hypothyroidism*, the thyroid gland is unable to secrete thyroid hormones because:

a. The cells of the anterior pituitary cannot manufacture and release TSH, or

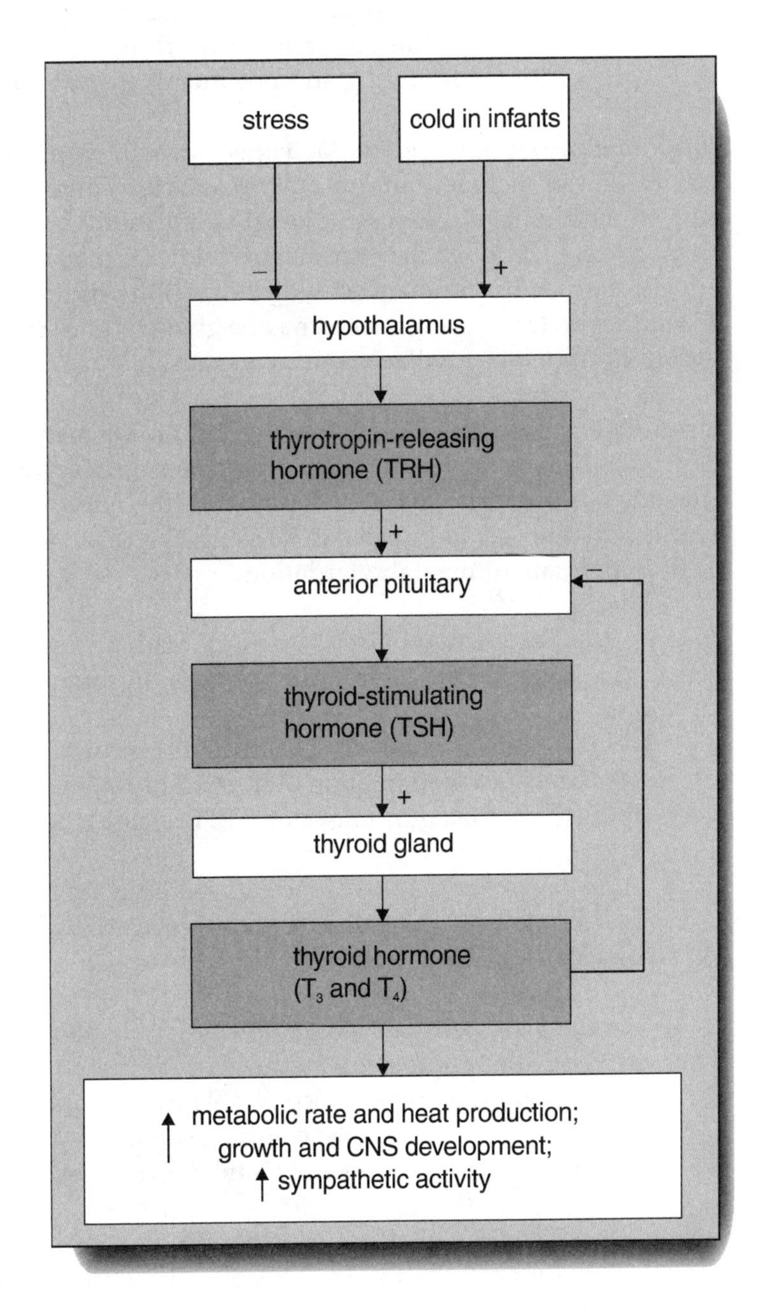

FIGURE 15.21

b. TRH is not released from the hypothalamus, which results in no TSH secretion from the pituitary.

Thus, in secondary hypothyroidism circulating levels of TSH are reduced or absent.

3) *Iodine deficiency* can produce a form of hypothyroidism because the thyroid follicle cells cannot manufacture thyroid hormones without iodine. As in primary hypothyroidism, elevated levels of TSH will be present due to the absence of negative feedback regulation of thyroid hormones at the pituitary. Note also that when elevated TSH is present, for any reason, this may stimulate growth of the thyroid gland producing an *endemic goiter* (see following material).

Symptoms of hypothyroidism, as expected, will primarily be related to a reduction in metabolic activity. These include, but are not necessarily limited to, increased sensitivity to cold exposure, reduced metabolic rate, excessive weight gain and reduced mental activity. In addition, such patients are more easily fatigued than normal individuals, may complain of feeling cold all the time, will have reduced cardiac function and may have swelling in the extremities (correctly known as *myxedema*). Depending on the cause, patients may be given iodine supplements in their diets, or may need to take thyroid hormone (T_4) replacement by mouth.

Cretinism is a disorder resulting from hypothyroidism in babies. Since thyroid hormones are necessary for proper development and function of the nervous system, the absence of thyroid hormones at birth will seriously affect normal development of the nervous system in these children. While it is easily curable if diagnosed and treated in its early stages, if it is allowed to continue for a few months, it will result in permanent mental retardation.

Hyperthyroidism. A number of causes all lead to elevated levels of thyroid hormones. These are increased release of TSH, thyroid tumor, and thyroid stimulating immunoglobin.

1) *An increase in the release of TSH* from the pituitary may result from an insensitivity in negative feedback, down regulation of thyroid hormone at the pituitary, or it may be due to problems in the hypothalamus (elevated TRH release), or problems at the levels of the pituitary itself.

2) *A thyroid tumor* may be capable of secreting large amounts of thyroid hormone in the absence of TSH support. Note that here the elevated levels of thyroid hormone will inhibit the release of TSH from the pituitary.

3) *An autoimmune disease* (*Graves disease, or thyrotoxicosis*) can produce elevated levels of an auto-antibody that is capable of acting like TSH. Such *Thyroid Stimulating Immunoglobin* (*TSI*) can act at the thyroid gland in place of TSH to cause release of elevated levels of thyroid hormones. Graves disease is the most common cause of hyperthyroidism and here the elevated levels of circulating thyroid hormones inhibit TSH release from the pituitary.

Symptoms of hyperthyroidism, as expected, will primarily be related to an increase in metabolic activity. These include, but are not necessarily limited to, increased heat production and sweating, increased appetite and weight loss, loss of muscle and fat stores, muscle weakness, emotional instability, sleeplessness, and an increase in heart rate, pulse and possible arrhythmias. In addition patients may also have *exophthalmos* (pop-eyes) due to localized edema behind the eyeballs.

Goiters result from increased secretion of TSH that stimulates thyroid tissue. If the increased TSH is due to the absence of thyroid hormone, resulting from iodine deficiency, the resulting goiter is said to be *endemic* or *simple*. If the elevated TSH release results from some primary problem at the levels of the hypothalamus-pituitary axis, the goiter is described as a *toxic goiter*.

Treatment of Hyperthyroidism necessitates reduction in the levels of circulating thyroid hormones. This can be accomplished either with surgical removal of portions of thyroid tissue; with use of radioactive iodine with a very short half-life, which concentrates in the thyroid tissue and destroys it; or with certain drugs (*propylthiouracil or methimazole*) that block thyroid hormone synthesis/release.

Thyroid C Cells. The *thyroid gland* not only contains thyroid follicles but also contains *Calcitonin secreting cells* (or *C cells*). Calcitonin is released from these cells when blood Ca^{++} levels are elevated. This hormone then causes uptake of Ca^{++} into the bone matrix (probably by inhibiting osteoclast activity), and in so doing causes blood levels of Ca^{++} to return to the normal range.

Parathyroids. Four parathyroid glands are imbedded in the thyroid tissue. These bean-sized glands are not actually part of the thyroid, but are only anatomically located within the thyroid tissue. They secrete *Parathyroid Hormone* (or *Parathormone*) when blood levels of Ca^{++} fall below the normal range and the effect is to elevate blood Ca^{++} (as described previously). This is accomplished because parathormone:

1. Increases absorption of Ca^{++} from the GI tract. This process also involves vitamin D.
2. Increases Ca^{++} (and phosphate) removal from the bone matrix.
3. Decreases Ca^{++} loss (and increases phosphate loss) in the kidney tubules.

Pancreatic Islets. The pancreas contains endocrine tissue called I*slet of Langerhans*. There are four distinct types of cells in these islets: *alpha cells*, *beta cells*, *delta cells*, and *PP cells*, and each secretes a different hormone. However, the alpha and beta cells are the most important, and their functions will be described below. The delta cells secrete *somatomedin*, and the PP cells secrete *pancreatic polypeptide*.

1) *Alpha cells* secrete *glucagon*. This hormone, also called the *hyperglycemic factor*, is released from the alpha cells when blood glucose levels are low. It functions, as its name implies, to elevate blood glucose levels. Glucagon elevates blood glucose by acting at the liver to convert stored glycogen into glucose (*glycogenolysis*). In addition, it also stimulates catabolism of fats and fatty acids into glycerol, as well as the conversion of noncarbohydrates into glucose (*gluconeogenesis*).

2) *Beta cells* secrete *insulin*. This hormone is released by the beta cells when the levels of blood glucose are elevated. The insulin functions to reduce blood glucose by facilitating the uptake of glucose into most tissues of the body, and is especially effective in promoting glucose uptake into skeletal muscle, cardiac muscle and adipose tissue. However, working muscle, brain and liver are not dependent on the presence of insulin to promote glucose uptake. In addition, insulin also mediates other effects on cellular metabolism as follows:

a) It stimulates the formation of glycogen from glucose (*glycogenesis*) in skeletal muscles and in the liver.
b) It inhibits the catabolism of glycogen into glucose and therefore increases the amount of stored carbohydrate in cells.
c) It inhibits the conversion of amino acids into glucose (glyconeogenesis).
d) It stimulates the formation of glycerol, fatty acids and triglycerides from glucose, and inhibits the conversion of triglycerides into fatty acids. It also facilitates the movement of fatty acids from the circulation into adipose tissue.
e) It decreases the concentration of circulating amino acids by facilitating the movement of amino acids from the blood into muscle tissues.
f) It increases protein synthesis in target cells.

Regulation of Insulin Release. As described, the primary stimulus mediating insulin release from the beta cells of the islets is elevated levels of blood glucose. However, these other factors also play a role in this process.

1. Insulin release is stimulated by elevated levels of amino acids in the blood.
2. Insulin release is stimulated by parasympathetic nerves acting directly on the beta cells of the islets.
3. Insulin release is inhibited by sympathetic stimulation acting directly on the beta cells of the islets.
4. Insulin release is stimulated by hormones (gastric inhibitory peptide) released from the walls of the gastrointestinal tract.

Diabetes Mellitus. This is an endocrine disorder leading to extreme metabolic problems coupled with serious imbalances in fluids and electrolytes. The term *diabetes* refers to the production of large volumes of urine, while *mellitus* indicates that the urine has a sweet taste. Individuals with this disorder thus not only produce larger volumes of urine (*polyuria*) containing large amounts of glucose (*glucosuria*), but they also must continue to drink great quantities of water (*polydipsia*) to prevent dehydration. The presence of glucose in the urine is abnormal and results from elevated levels of glucose in the blood (hyperglycemia).

<u>Note the difference between diabetes mellitus and Diabetes insipidus.</u> Patients suffering from diabetes mellitus generate large volumes of urine containing glucose. Patients with diabetes insipidus also generate large volumes of urine which are however, very dilute and does not contain any glucose. (Since "insipidus" refers to tasteless, these terms—*mellitus* and *insipidus*—would suggest that physicians in the "old days" would taste the urine to determine a diagnosis).

<u>The causes of diabetes mellitus and diabetes insipidus are quite different.</u> Diabetes mellitus is caused by dysfunctions in the insulin-glucose regulatory pathways. The resulting large volumes of urine generated are due to the elevated glucose in the urine that acts as an osmotic diuretic. On the other hand, diabetes insipidus is due to the absence of ADH (Vasopressin) secretion from the pituitary, leading to increased loss of water (diuresis) in the kidneys.

Diabetes Mellitus is Actually Two Different Diseases. It can be subdivided into *Insulin-Dependent Diabetes Mellitus* also known as *Juvenile Onset Diabetes Mellitus*, or *Type I Diabetes Mellitus*; and

Noninsulin-Dependent Diabetes Mellitus, also known as *Maturity Onset Diabetes Mellitus,* or *Type II Diabetes Mellitus.*

1. **Type I Diabetes Mellitus (Juvenile Onset)**: Type I diabetes is present in 10% to 20% of all diabetic patients. As the name implies, this disease usually appears in childhood and the underlying cause is almost certainly due to an autoimmune disease, where the patient's own immune system attacks and destroys the beta cells of the islets. Thus, the immediate cause of the symptoms of Type I is the absence of insulin leading to the inability of many target tissues (especially skeletal muscle, cardiac muscle and adipose tissues) to transport glucose across the cell membranes by facilitated diffusion.

It has been suggested that the trigger for this autoimmune response may be infection with a virus that in some unclear way primes the patient's immune system to then attack the beta cells. Note also that this disease appears to be associated with certain genetic factors which explains why Type I seems to run in families. Thus, both genetic and environmental switches appear to be a prerequisite to develop Type I Diabetes Mellitus.

Symptoms of Severe (Type I) Diabetes Mellitus. As mentioned, the symptoms include hyperglycemia, diuresis or polyuria, glucosuria, polydipsia and severe dehydration.

In addition, patients with this disease also do not synthesize adequate amounts of proteins and fats. In the absence of insulin secretion, their insulin-dependent target tissues cannot utilize glucose. These tissues must use protein and fat in place of the glucose as an energy source. This leads to protein wasting and rapid weight loss. In addition, the increased utilization of fat as an energy source, coupled with the decreased synthesis and storage of fat, leads to abnormally elevated levels of circulating fatty acids, and ketone bodies (a metabolic byproduct of fatty acid catabolism). Elevated levels of circulating fatty acids may damage the walls of capillaries and seriously affect circulatory system integrity. Elevated levels of ketone bodies will cause increased osmotic diuresis and dehydration, coupled with loss of Na^+ in the urine. The outcome is acidosis, electrolyte imbalances, and dehydration that can lead to severe CNS neurologic disfunction, diabetic coma and death.

It is interesting to also note that high levels of ketone bodies in the blood will not only be excreted in the urine, but will also be present in the exhaled air from the lungs. The odor of these ketone bodies is much like that of acetone and is a strong indication that the patient is very severely ill, and if not quickly treated, may rapidly lapse into coma and death.

Treatment of Type I Diabetes Mellitus. To treat this disease, insulin injections are given. However, it is almost impossible to maintain stable levels of blood glucose to the extent present in normal people. Thus, even under the best of conditions blood glucose levels in a diabetic patient will be more volatile. In addition, dietary controls are also utilized to maintain a modified intake of carbohydrate in an attempt to maintain more regulated levels of blood glucose. Future treatments will undoubtedly include:

1) Insulin pumps that will deliver a more regulated level of the hormone. Such pumps may even contain a glucose sensor that monitors blood glucose levels and releases exactly the correct amount of insulin to maintain constant levels of blood glucose.

2) Transplantation of beta cells back into the islets of the patients to repair the damage.
3) Controling the underlying autoimmune reactions in children potentially at risk, thus preventing initial development of disease.

2. **Type II Diabetes Mellitus (Maturity Onset)**: Type II diabetes is present in 80% to 90% of all patients with diabetes.

This form of diabetes is not due to a lack of insulin. Indeed, in many patients with this disease the levels of insulin are elevated above the normal range. Commonly, the patients with Type II are obese adults, and the onset may take many years to develop. The underlying cause is at the cells which lose their insulin receptors over time. Obviously, as insulin receptors are lost, sensitivity of the target tissue for insulin is reduced. Eventually, the target tissues no longer responds to circulating insulin, facilitated diffusion of glucose into the target cells ceases and blood levels of glucose increase.

Patients with Type II will therefore also have glucosuria and some of the other symptoms found in Type I, but usually in a far less severe form. Additionally, Type II patients will also have elevated levels of circulating insulin because they continue to eat large amounts of carbohydrate. The digested carbohydrate will elevate their blood glucose and stimulate increased release of insulin from the islet cells.

Underlying Causes of Type II Diabetes Mellitus. While there appears to be a genetic predisposition, the obvious environmental cause is over-eating which results in continuously elevated levels of insulin in the circulation. This *hyperinsulinism* reduces insulin-receptor re-expression on target cells because it blocks recycling and replacement on the outer cell membranes. With the loss of insulin-receptors, target cells become refractory to insulin stimulation.

Treatment of Type II Diabetes Mellitus. To treat this disease the patient is placed on a strict diet, and the intake of all food and especially of carbohydrates is carefully controlled. The patient loses weight and blood glucose levels fall. As blood glucose levels are reduced, insulin secretion from the pancreas also is reduced. As insulin levels in the circulation fall, insulin-receptors will be resynthesized and expressed on target cells. The cells are no longer refractory to insulin and the Type II diabetes is cured, so long as the patient continues to control his/her eating behavior. In some cases, to assist in curing Type II diabetes, patients may also be treated with oral hypoglycemic drugs in an attempt to further reduce the levels of circulating insulin.

THYMUS

It is found in the mediastinum and plays an important roll in immune system development (as discussed previously in the chapter on immunology). However, the thymus also functions in the capacity of an endocrine gland because it secretes a variety of hormones (e.g., thymosins, thymulins, thymopoietin) that are involved in immune system regulation. The production and release of at least some of these thymic hormones are under the control of both sex steroids (estrogen, androgen) and glucocorticoids (cortisol).

1. *The hypothalamic-pituitary-gonadal-thymic axis* functions as an additional negative feedback pathway that involves gonadal release of sex steroids. Here Thymosin B_4 feeds back to the hypothalamus and stimulates the release of GnRH. This, then increases the release of LH which then up-regulates release of sex steroids (estrogen, androgen) from the gonads. The increased levels of sex steroids then turn off the release of thymosin B^4 from the thymus. This reduces release of GnRH from the

hypothalamus, LH levels fall and sex steroids from the gonads are reduced **(see Figure 15.10)**.

2. *The hypothalamic-pituitary-adrenal-thymic axis* also regulates release of thymic hormones through the action of cortisol at the levels of the thymus.

In this pathway, cytokines are released from activated cells of the immune system (lymphocytes and macrophages) during an on-going immune response. These cytokines then stimulate release of CRH from the hypothalamus, and ACTH from the pituitary. The ACTH increases release of cortisol from the adrenal gland. Elevated levels of cortisol feedback to the thymus and inhibit the release of thymic hormones. As thymic hormones fall, the on-going immune response is turned off. Cytokine levels fall and this reduces release of cortisol. This pathway thus down-regulates immune responses that might otherwise get out of hand and possibly produce autoimmune damage **(see Figure 15.10)**.

HEART

The right atrium of the heart releases Atrial Natriuretic Factor (ANF) upon stimulation by increased blood pressure. ANF then acts on the kidney to inhibit reabsorption of Na_+. The effect is to increase Na^+ loss in the urine which in turn will then cause water loss, reducing blood volume and thus reducing blood pressure.

KIDNEYS

They produce the hormone *renin* which is involved in the *renin-angiotensin-aldosterone pathway* described previously. This pathway not only regulates blood Na^+, K^+ and H^+, but also is involved in regulation of blood pressure, and blood volume. Additionally, the kidneys produce the hormone *erythropoietin* when O_2 transport by the blood falls below the normal range. This hormone then acts at the bone marrow to stimulate RBC production. Over a period of time additional RBCs are released into the blood and the O^2 carrying capacity of the blood increases. When this happens, erythropoietin release by the kidney is turned off.

GONADS

The gonads (ovaries in the female, testes in the male) produce sex steroids (estrogen, progesterone in the female, testosterone in the male) under the control of the gonadotropins (LH, FSH) from the anterior pituitary. The sex steroids are necessary for gamete production (ovum in the female, sperm in the male), and also must be present to maintain the functions of the reproductive structures as well as secondary sex characteristics

PLACENTA

The placenta functions in the pregnant female during the second and third trimesters. It is necessary to maintain the pregnant state and additionally produces a number of important hormones, including estrogen, progesterone and human placental lactogen. A complete discussion of the gonads and placenta will be presented in the chapter on the reproductive system.

Chapter 16
THE REPRODUCTIVE SYSTEM

DEVELOPMENT OF THE REPRODUCTIVE SYSTEM

The processes that determine the sex of an individual are complex and begin even before fertilization takes place. Major events involved in this process are:

1. Formation of gametes by meiosis.
2. Fertilization and genetic sex.
3. Embryonic development.

FORMATION OF GAMETES

Human cells contain 46 chromosomes (which is the *diploid number*), comprised of 22 pairs of autosomes and 1 pair of sex chromosomes. However, the genetic information encoded on individual pairs is not identical because one of the pair was originally derived from the father and the other from the mother. When regular tissue cells divide by the process of mitosis, identical copies of these 46 chromosomes end up in the two new daughter cells. However, in the process of fertilization, a cell from the father must combine with a cell from the mother to form the new individual (called the *zygote*).

If a male cell containing 46 chromosomes were to fertilize a female cell containing 46 chromosomes, the resulting zygote would now contain 92 chromosomes. Because this would be double the number of chromosomes found in normal cells, the zygote would not be viable.

To get around this problem, the reproductive system generates *gametes* (or *sex cells*) that contain only half (or 23 chromosomes, which is the *haploid number*) of chromosomes found in tissue cells. The process that leads to the formation of mature sperm and ovum containing the haploid number of chromosomes is called *meiosis* (or *reduction division*). The steps involved in meiosis will be outlined in detail later in this chapter. Suffice it to say, at this point, that mature sperm and ovum will each contain 22 autosomes and 1 sex chromosome.

Thus, when a sperm fertilizes an ovum, the zygote will possess the appropriate number of chromosomes (23 + 23 = 46 chromosomes) and development can now proceed correctly.

GENETIC SEX

In males, the process of meiosis leads to two types of sperm, those that contain either 22 autosomes + a Y chromosome, or those that contain 22 autosomes + an X chromosome. On the other hand, in females, meiosis produces only one type of ovum containing 22 autosomes + an X chromosome.

1. If an X containing sperm fertilizes the ovum, the zygote will have an XX genotype (a genetic female).
2. If a Y containing sperm fertilizes the ovum, the zygote will have an XY genotype (a genetic male).

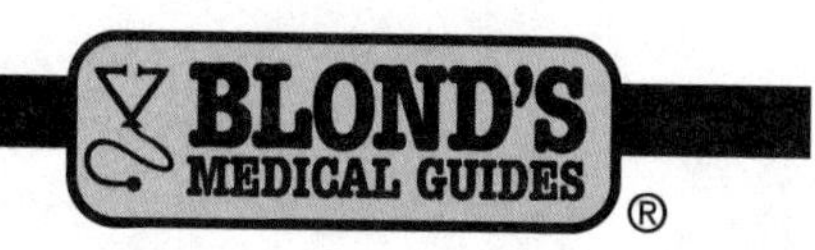

Thus, it is the sperm that determines the genetic sex of the embryo. Once the *genetic sex* is determined, this will lead to the gonadal sex, and this will result in the *phenotypic sex*. To understand the stages in this complex process, consider what takes place during embryonic development.

THE EMBRYO V. THE FETUS

The term *embryo* refers to the dividing cell mass formed after fertilization. Embryo is used to describe growing cell mass for the first 2 months (8 weeks) after fertilization has taken place. From 9 weeks to parturition, the term *fetus* is used.

SEXUAL DIFFERENTIATION DURING EMBRYONIC/FETAL DEVELOPMENT

Prior to about 42 days of gestation, and regardless of the genetic sex, the embryonic gonads of males and females are identical.

Differentiation into defined testes or ovaries takes place after this period and depends on the processes outlined later. However, formation of a functional male or female reproductive system also requires the differentiation of primordial genital ducts into defined structures. The primordial male ducts (*the Wolffian ducts*), and the primordial female ducts (*the Mullerian ducts*) are both present in the embryo/ fetus by the seventh week of gestation. In the presence or absence of substances produced by the male gonads, one of these ducts develops, while the other regresses. The final outcome, barring any misprogramming, will be an individual equipped as either a functionally phenotypic male or functionally phenotypic female.

STEPS IN THE FORMATION OF A PHENOTYPIC MALE

1) In genetic (XY) male embryos, the genetic information (*testicular determining factor gene, TDF gene*) present in the Y chromosome codes for the protein, *H-Y antigen*. The H-Y antigen is inserted into the plasma membrane of the cells that make up the undifferentiated primordial gonads.

2) In the presence of the H-Y antigen, the primordial gonads differentiate into functional testes. Thus, through this process, the *gonadal sex* of the embryo has been irreversibly made male.

3) Within the functional testes, *Leydig cells* (*Interstitial Cells*) secrete the male steroid *testosterone*. *Sertoli cells* within the developing *seminiferous tubules* also manufacture and secrete the substance *Mullerian Inhibitory Factor* (*MIF*).

4) The testosterone acts on the Wolffian ducts to form the epididymis, vas deferens (or ductus deferens), ejaculatory ducts and seminal vesicles.

5) The testosterone is also metabolically converted into *Dihydrotestosterone* (*DHT*), which acts on the undifferentiated external genitalia to form the penis and scrotum.

6) The Mullerian Inhibitory Factor acts on the primordial Mullerian ducts causing them to degenerate.

7) The outcome of the above processes is to create a *phenotypic male.*

STEPS IN THE FORMATION OF A PHENOTYPIC FEMALE

1) In genetic (XX) female embryos, there is no Y chromosome and, thus, no *TDF gene*. Therefore no *H-Y antigen* is manufactured.

2) In the absence of any H-Y antigen, the undifferentiated primordial gonads develop into ovaries. Thus, through this process, the *gonadal sex* of the embryo has been irreversibly made female.

3) The ovaries in the female do not secrete any testosterone. They also do not manufacture any *Mullerian Inhibitory Factor*.

4) In the absence of testosterone, the Wolffian ducts regress.

5) In the absence of testosterone the undifferentiated external genitalia develop into the clitoris and labia

6) In the absence of Mullerian Inhibitory Factor the Mullerian ducts develop into the oviducts and uterus.

7) The outcome of the above process will produce a *phenotypic female.*

ABNORMALITIES IN SEXUAL DIFFERENTIATION

It is interesting to note that sexual development within the embryo will proceed towards a female phenotype, unless it is specifically shunted towards the male phenotype through the application of testosterone and MIF. However, for testosterone and MIF to produce a phenotypic male, they must be supplied only at specific stages in the process of development. For example, if testosterone is supplied to a phenotypic female, it will not cause the reproductive structures to convert into those of a male. It will, however, cause certain defined masculinizing changes in secondary sexual characteristics (such as increased growth of facial and body hair, increased libido, clitoral enlargement, and so on).

A summary of stages in the developmental process is as follows:

	Gestational Age	Developmental Stage
Male	45 days	Bipotential gonads are seeded with primordial germ cells.
	60 days	Testes form containing seminiferous tubules and Leydig cells.
	60 days	Mullerian ducts regress.
	70 days	Penis and prostate formed.
Female	45 days	Bipotential gonads are seeded with primordial germ cells.
	45 days	Mullerian ducts begin to develop.
	80 days	Meiosis begins in the ovaries. Vaginal structures begin to develop.
	25 weeks	Primary follicles present in ovaries. Internal female reproductive structures developed. Female external genitalia developed.

Misprogramming at any stage can derail the process leading to individuals who may be genetically one sex and phenotypically another, or who may possess characteristics of both sexes. For example:

1. A true hermaphrodite possesses both ovarian and testicular structures.
2. A male pseudohermaphrodite possesses testes but the genital ducts or external genitalia are partially or fully female.
3. A female pseudohermaphrodite possesses ovaries but the genital ducts or external genitalia are partially or fully male.

One example of male pseudohermaphrodism is the disorder called *testicular feminization syndrome*. Here a genetically XY male with abdominal testes develops phenotypically as a female because this individual does not possess androgen receptors in the target tissues.

During early development, MIF inhibits the formation of the internal female reproductive ducts, but the testosterone cannot masculinize the external genitalia due to the absence of testosterone receptors. As a result, the external genitalia are female, but the vagina does not lead anywhere since there is no uterus, ovaries, oviducts, and so on. However, the high levels of circulating testosterone are converted peripherally into estrogen and this hormone stimulates development of female secondary sexual characteristics (breast development, female fat deposition, and so forth). Such individuals are therefore phenotypically female, and although they cannot have children, they can otherwise fulfill all female roles in society.

NEUROLOGICAL SEXUAL DEVELOPMENT

There is much controversy with regard to this area. The central question relates to whether gender orientation is a learned response or is programmed during early development.

While it has been hypothesized that early exposure to sex hormones can permanently fix the nervous system into a male or female mode, the studies to data remain equivocal. Certainly studies in primates and some other animals, such as the rat, would seem to suggest that early exposure of the brain to androgens leads to certain types of male behavior (such as aggression) in the adult. Studies have also suggested that sex steroid receptors are present in various brain areas. However, in human studies the data is unfortunately too limited to lead to any concrete conclusions.

It has been reported that women, whose mothers were treated with an androgen analogue during pregnancy, demonstrated an increased tomboy-like behavior as children, but as adults this group did not appear to be significantly different when compared to controls.

Tentatively these studies would support the view that while sex hormones may influence certain forms of behavior, in humans the major determining factor for gender orientation is a learned response, imposed by society and by early experiences. However, further studies are certainly warranted.

PRIMARY V. ACCESSORY REPRODUCTIVE ORGANS

In both the male and female reproductive systems, the gonads (testes in the male and ovaries in the female) are the *primary reproductive organs*. The function of the primary reproductive organs is to produce germ cells (sperm in the male, ovum in the female), and to manufacture and release sex steroids (testosterone in the male, estrogen and progesterone in the female). On the other hand, *accessory reproductive organs* in the male and female include all of those ducts, tubes, glands and other structures required to maintain a functional reproductive system, and guarantee that the germ cells will be released to form the zygote. These accessory reproductive organs are dependent on sex hormones for maintenance of their structural/functional integrity. Removal of sex hormones by castration/ovariectomy will cause these organs to regress and lose their function. Accessory reproduc-

tive organs can be further subdivided into internal accessory reproductive organs and external secondary reproductive organs.

THE MALE REPRODUCTIVE SYSTEM

In the male, the primary reproductive organs are the testes. The secondary reproductive organs include the epididymis, vas deferens, seminal vesicle, prostate, bulbourethral glands and penis.

TESTES

The testes are the primary reproductive organs in the male and produce both germ cells (sperm) and the male hormone testosterone. Each testis is covered by a fibrous capsule (*tunica albuginea*). Internally, there is a thickened area of connective tissue (*mediastinum testis*) that form into the *septa* which subdivide each testis into about 250 lobules. Each lobule contains coiled seminiferous tubules; there is at least one, but there can be up to four **(Figure 16.2)**. The seminiferous tubules unite in the region of the mediastinum testis to form a series of interconnected tubules called the *rete testis*. The rete testis connects to a number of *efferent ductules*, which in turn connect to the *epididymis*.

Descent of the Testes. In the male fetus, the testes descend in the *inguinal canal*, and pass through the abdominal wall and into the scrotal sac, at about 1 or 2 months prior to birth.

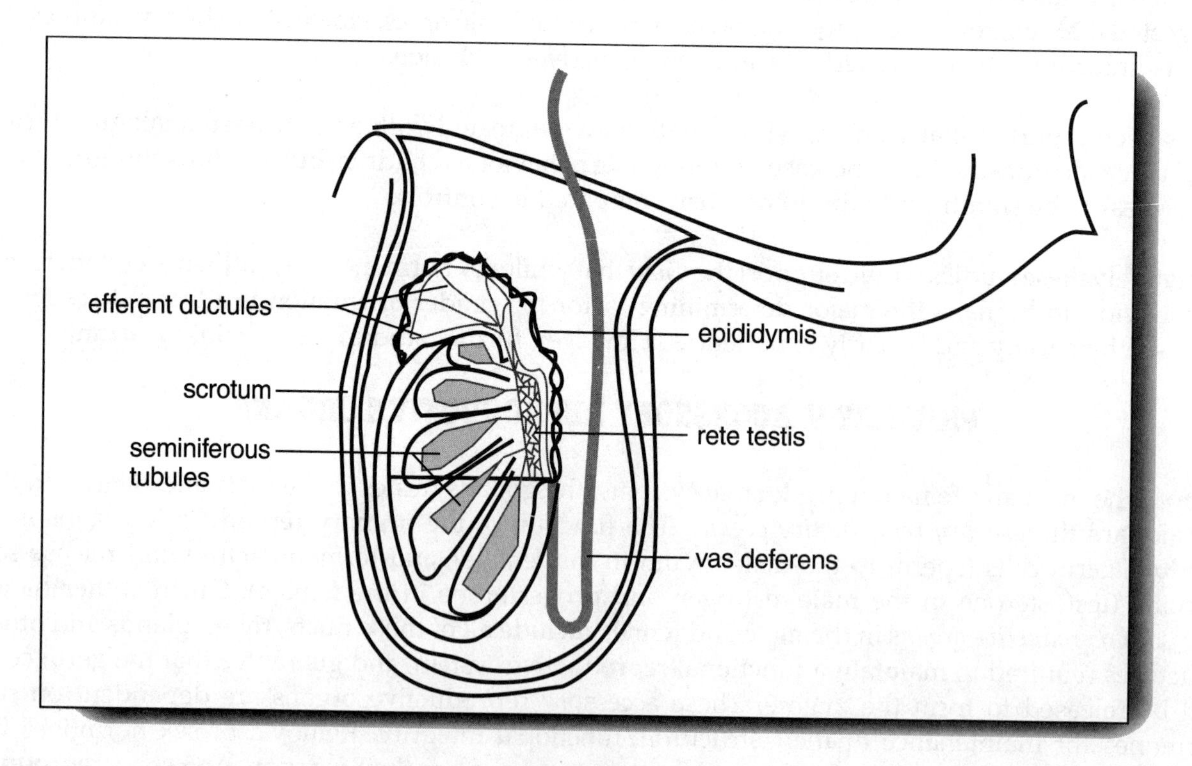

FIGURE 16.1

The process of descent is stimulated by rising levels of testosterone and is accomplished by the contraction of the gubernaculum (a fibro-muscular chord). In humans, once the testes are in the scrotum, they remain there. However, in certain animals like the rat, the testes can be withdrawn back up into the abdominal cavity in times of stress, and descend again when the animal is sexually stimulated.

The testes must be in the scrotal sac to allow spermatogenesis to take place because one of the steps in this process leading to sperm production is temperature sensitive. If the testes remain in the body cavity, no sperm production will take place, although testosterone production would continue since this is not affected by body temperature. Thus, to promote sperm production, the testes must be in the scrotum, which is a few degrees cooler than in the abdomen. In some children, the testes fail to descend at the proper time. This condition is called *cryptorchidism* and must be corrected before puberty (by testosterone treatment or surgery). Otherwise cells in the seminiferous tubules will degenerate and the individual will be sterile as an adult.

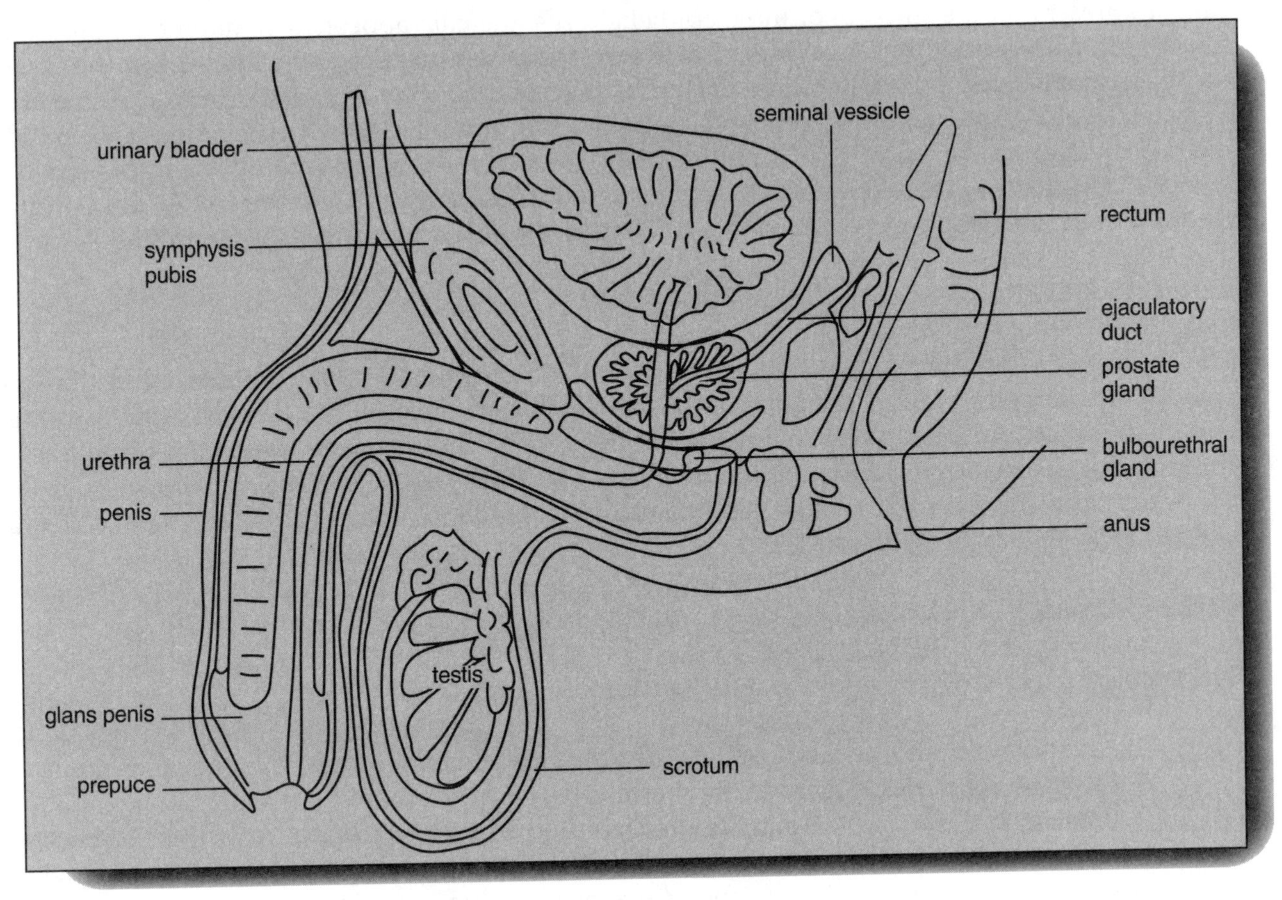

FIGURE 16.2

THE SPERMATIC CORD

The spermatic cord is the structure that is comprised of the vas deferens, the testicular artery and vein, and the nerve which supplies the testes, all enclosed in a sheath that passes from the inguinal canal into the scrotal sac.

MALE INTERNAL ACCESSORY ORGANS

The secondary reproductive organs in the male are dependent on the presence of testosterone to maintain their structural and functional integrity. Testosterone is also necessary in the male to promote the development of the male secondary sexual characteristics. In addition, testosterone is also metabolically converted into the substance dihydrotestosterone which acts as a potent androgen in certain target tissues.

THE EPIDIDYMIS

The epididymis is a thin tube, about six meters in length, which is tightly coiled on the top of the testis. After passing downward along the surface of the testes the epididymis connects to the *vas deferens*. Within the epididymis, the lining is comprised of pseudostratified columnar epithelium. Although cilia are present on the surfaces of these epithelial cells, the cilia appear to be nonmotile (and their function remains under investigation). However, it is known that the epithelial cells in this region secrete glycogen, and probably other substances that maintain the stored sperm in a viable state. Elevated levels of testosterone present here are believed to stimulate sperm maturation. The epididymis also reabsorbs the majority of the fluid that enters from the seminiferous tubules, and thus increases the sperm concentration by a factor of 100 (for example, 50 sperm/ml is concentrated, through fluid absorption, to become 50 sperm/0.01 ml, which is actually 5000 sperm/ml).

THE VAS DEFERENS

The vas deferens (or ductus deferens) is a tube about 45 cm in length that contains a thick layer of smooth muscle in the walls. The inside of the vas deferens is lined with a layer of pseudostratified columnar epithelium. The vas deferens begins at the epididymis and passes out of the scrotal sac, up through the spermatic cord, into the inguinal canal, to finally enter the abdominal cavity. It then continues upward, curving around behind the urinary bladder to end at the enlargement called the *ampulla of the vas deferens* **(Figure 16.2)**.

THE SEMINAL VESICLES

The *seminal vesicles* are two sac-like glands that secrete:

1. An alkaline fluid to neutralize the acid pH of the semen arriving from the epididymis.
2. Fructose to provide energy to the sperm cells.
3. Prostaglandins that act within the female reproductive tract to cause muscular contractions. Such contractions can assist the sperm in moving up the oviducts and fertilizing the ovum.
4. Large volumes of fluid that expands the semen. Seminal secretions probably account for 50% of the final ejaculatory volume.

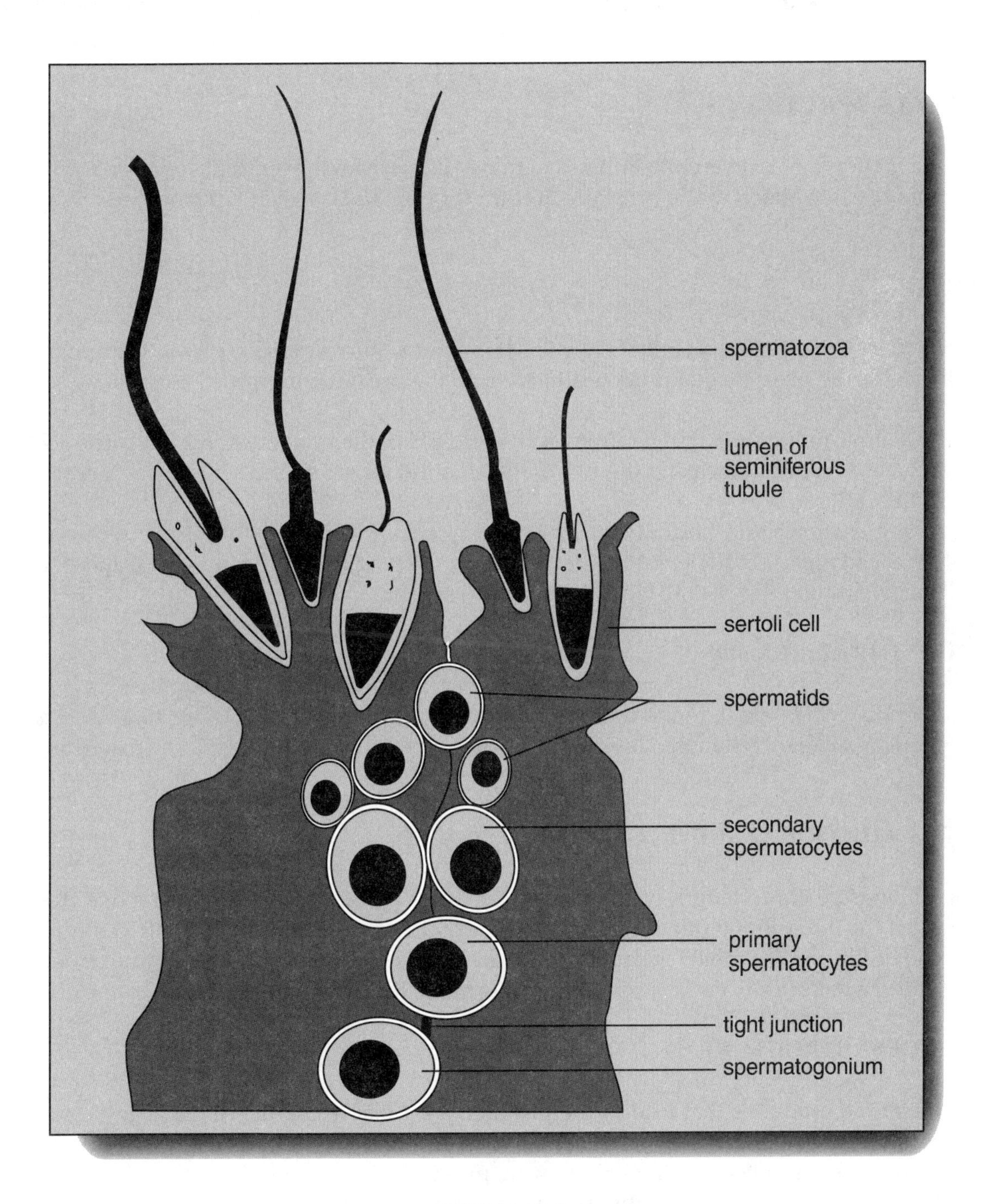

FIGURE 16.3

Secretion of fluid by the seminal vesicle is under the control of the sympathetic nervous system. Such stimulation takes place during emission and ejaculation. Fluid leaves the seminal vesicle through a short connecting duct which combines with the ampulla of the vas deferens to become the ejaculatory duct **(Figure 16.2)**.

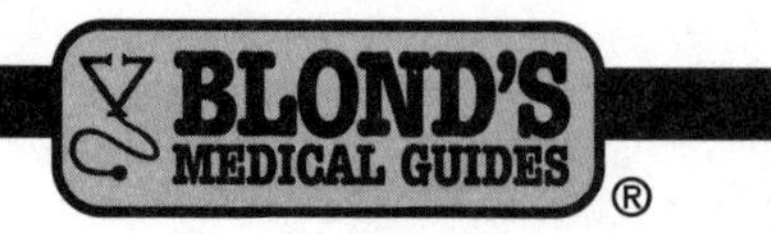

THE EJACULATORY DUCTS

The ampulla of the two vas deferens combines with two ducts from the seminal vesicle to form the two *ejaculatory ducts*. They attach to the prostatic urethra **(Figure 16.2)**, which is surrounded by prostate tissue.

THE PROSTATE

The *prostate gland* is positioned directly below the bladder and surrounds the prostatic urethra **(Figure 16.2)**. Ducts from the prostate gland open directly into the prostatic urethra. The prostate secretes:

1) An alkaline, milky fluid that neutralizes the acid pH of the semen and probably the acid in the vagina as well. By increasing the pH to the alkaline range, sperm viability is increased as is sperm motility.
2) Clotting enzymes and fibrinolysin. The clotting enzymes initially cause the semen to "clot" after ejaculation. This keeps the sperm within the vagina after removal of the penis. The fibrinolysin then digests the clot, releasing the sperm.

BULBOURETHRAL GLANDS

Two *bulbourethral glands* attach to the urethra directly below the prostate **(Figure 16.2)**. These glands produce a mucous-like secretion that lubricates the glans (head) of the penis prior to insertion into the vagina.

URETHRAL MUCOUS-SECRETING GLANDS

Positioned along the entire length of the urethra are *mucous-secreting glands*. Secretions from these glands protect the inner lining of the urethra and prevent bacterial invasion up into the vagina. The length of the urethra in the male (as compared to the female), coupled with these mucus secretions, explains why males suffer from far less bladder infections than do females.

MICROSCOPIC STRUCTURE OF THE TESTIS

To understand the processes that lead to the production of viable sperm we must first consider the microscopic structure of the testis itself. As mentioned above each testis is subdivided into lobules that contain seminiferous tubules. In cross section many seminiferous tubules are seen to be packed together. Between the seminiferous tubules are areas of testicular *interstitium* contain *Interstitial Cells* (or *Leydig cells*). These Leydig cells are the source of the male hormone, *testosterone*.

Within the seminiferous tubules are two major types of cells; the *germ cells* and the supporting cells (or *Sertoli cells*) **(Figure 16.3)**. Starting at the outer wall of the seminiferous tubule, the germ cells are arranged in order of their progressing stage of differentiation, with the least differentiated cells at the margin of the wall and the most differentiated at the luminal surface.

THE FUNCTION OF THE SERTOLI CELLS

The Sertoli cells are tall cells, and extend from the margins of the tube up to the luminal region in the center of the tube. They are intimately involved in all stage of the meiotic processes of germ cells development.

1) The Sertoli cells are responsible for creating the *blood-testes barrier*. The barrier results from the tight junctions formed between adjacent Sertoli cells **(Figure 16.3)**, and separates the *Spermatogonium* from the next stage in germ cell development—the *Primary Spermatocyte*. Accordingly, this blood-testes barrier isolates the developing germ cells from outer chemical influences It is responsible for maintaining the correct internal chemical environment necessary for sperm cell development. In addition, it prevents cells of the immune system from reacting against the developing germ cells. Thus auto-antibodies are not generated, and viability of the sperm is maintained.

2) The Sertoli cells phagocytize defective sperm cells, and cytoplasmic components released from sperm cells during their remodeling phase (when *Spermatids* remodel to form *Mature Spermatozoa*).

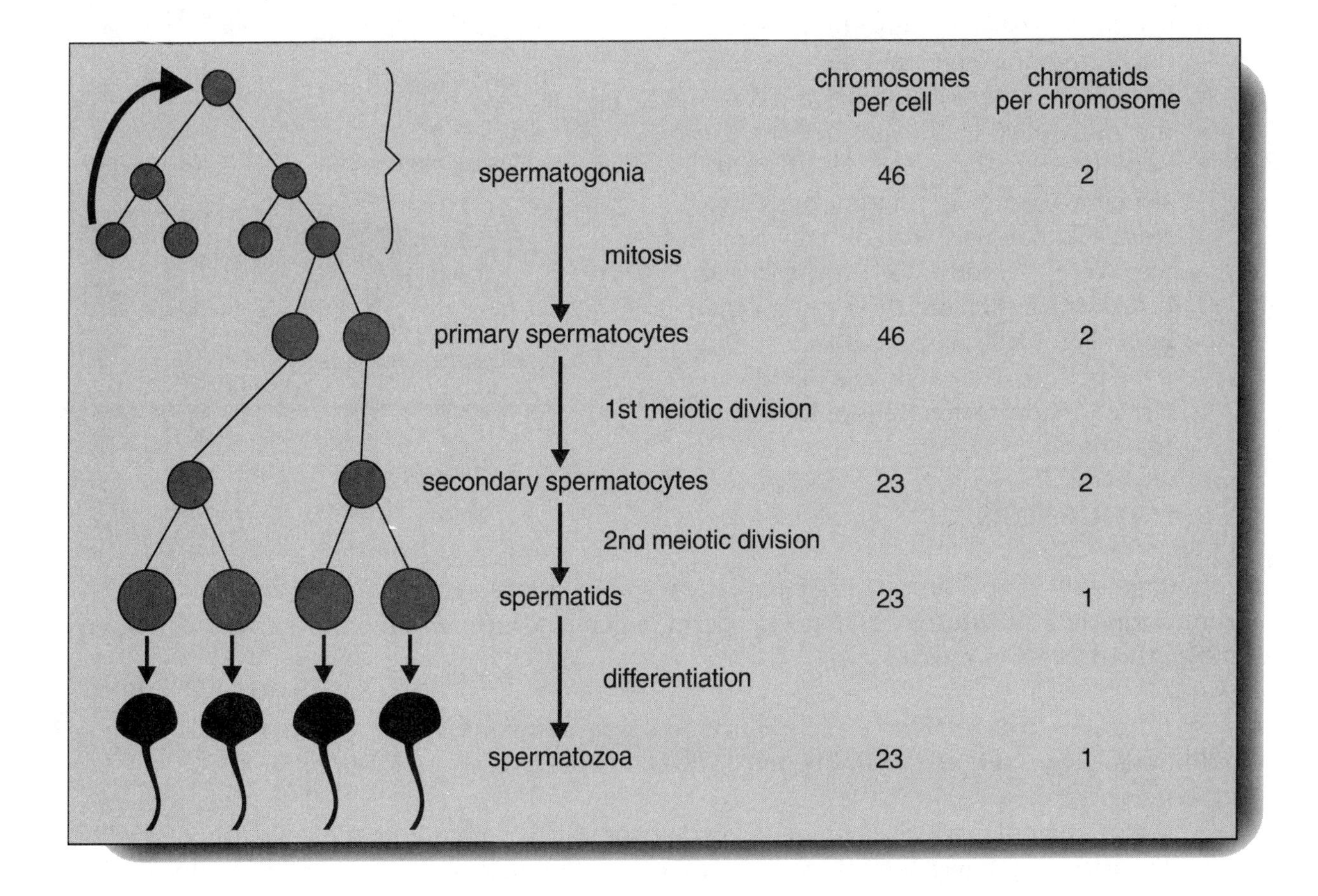

FIGURE 16.4

3) The Sertoli cells provide nutrition to the sperm cells throughout all stages of their development.

4) The Sertoli cells guide the germ cells through the various stages in the meiotic process. They may accomplish this through direct contact with the developing sperm cells. In addition they secrete various hormones and other substances that are necessary for sperm cell development.

5) The Sertoli cells generate large volumes of *seminiferous tubular fluid* which washes the mature spermatozoa, present in the lumen of the seminiferous tubules, into the epididymis. (As mentioned, the majority of this fluid is then reabsorped in the epididymis to concentrate the sperm stored here a hundred fold.)

6) The Sertoli cells manufacture and secrete *androgen-binding protein* which binds testosterone. Because of the presence of this androgen-binding protein (coupled with the presence of the blood-testes barrier), the local testosterone concentration, bathing the germ cells, is very high. This high concentration is necessary to allow the sperm production, and development to proceed correctly.

7) The Sertoli cells control spermatogenesis because they release the hormone *inhibin*. Inhibin acts at the level of the anterior pituitary to inhibit the release of FSH. FSH acts at the Sertoli cells to promote spermatogenesis. This negative feedback, coupled with an additional negative feedback involving testosterone, LH and the Leydig cells, not only controls the processes of sperm production, but also the levels of circulating testosterone (see following material).

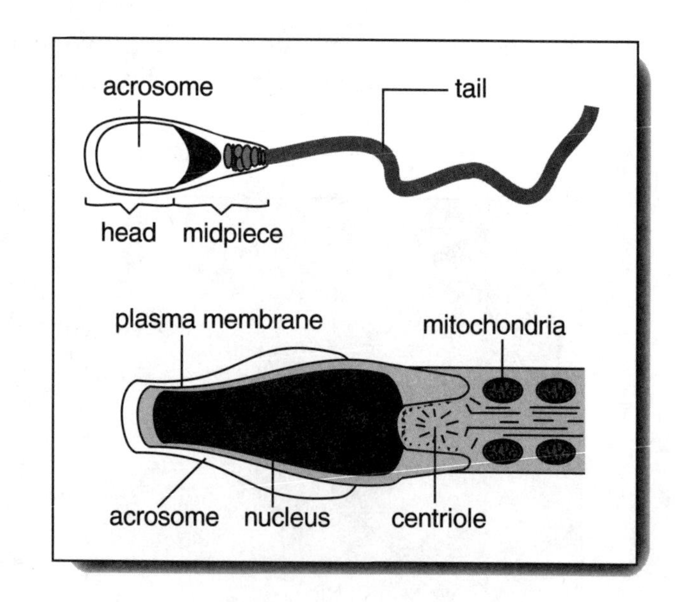

FIGURE 16.5

SPERMATOGENESIS

Spermatogenesis is the process that produces viable sperm containing 23 chromosomes (1 sex chromosome and 22 autosomes). During spermatogenesis, differentiation of germ cells takes place through the process of meiosis.

Meiosis (also known as *reduction division*) is in some ways similar to the process of mitosis, but consists of additional stages **(Figure 16.4)**. The major events in meiosis are as follows:

1) Spermatogonia, which contain 46 chromosomes like other somatic cells, divide by *mitosis*. Production of new spermatogonia by mitosis begins at puberty and continues throughout the reproductive life of the male. In males, sperm may continue to be produced into old age. This is unlike the female (as described below). Dividing spermatogonia produce daughter cells

which also continue to divide to form new clones. During each subsequent division the cells differentiate slightly, and eventually some of these clone cells move up, through the tight junction between the Sertoli cells, differentiating in the process into the *primary spermatocytes.*

2) *Primary spermatocytes,* which also contain 46 doubled chromosomes (2 sex chromosomes, each composed of two chromatid; and 44 autosomes each composed of two chromatids), now undergo the *first meiotic division* (*Meiosis I*). This process results in the formation of two *secondary spermatocytes,* each of which contain 23 doubled chromosomes (1 sex chromosome, comprised of two chromatids; and 22 autosomes, each comprised of two chromatids).

3) *Secondary spermatocytes* now undergo the *second meiotic division* (*Meiosis II*) where the chromatid pairs split and 1 copy (1 sex chromosome and 22 autosomes) end up in each of two spermatids. Thus, from 1 secondary spermatocyte 2 spermatids are produced.

4) The spermatids do not continue dividing, but now undergo *reorganization* losing cytoplasm and developing a flagella tail. The final result of this process is the formation of *spermatozoa.* Thus from 1 primary spermatocyte, 4 mature sperm are produced.

5) These sperm contain either an X or a Y sex chromosome plus 22 autosomes. Mature spermatozoa collect in the lumen of the seminiferous tubules and then pass into the epididymis where they continue the process of maturation, as described.

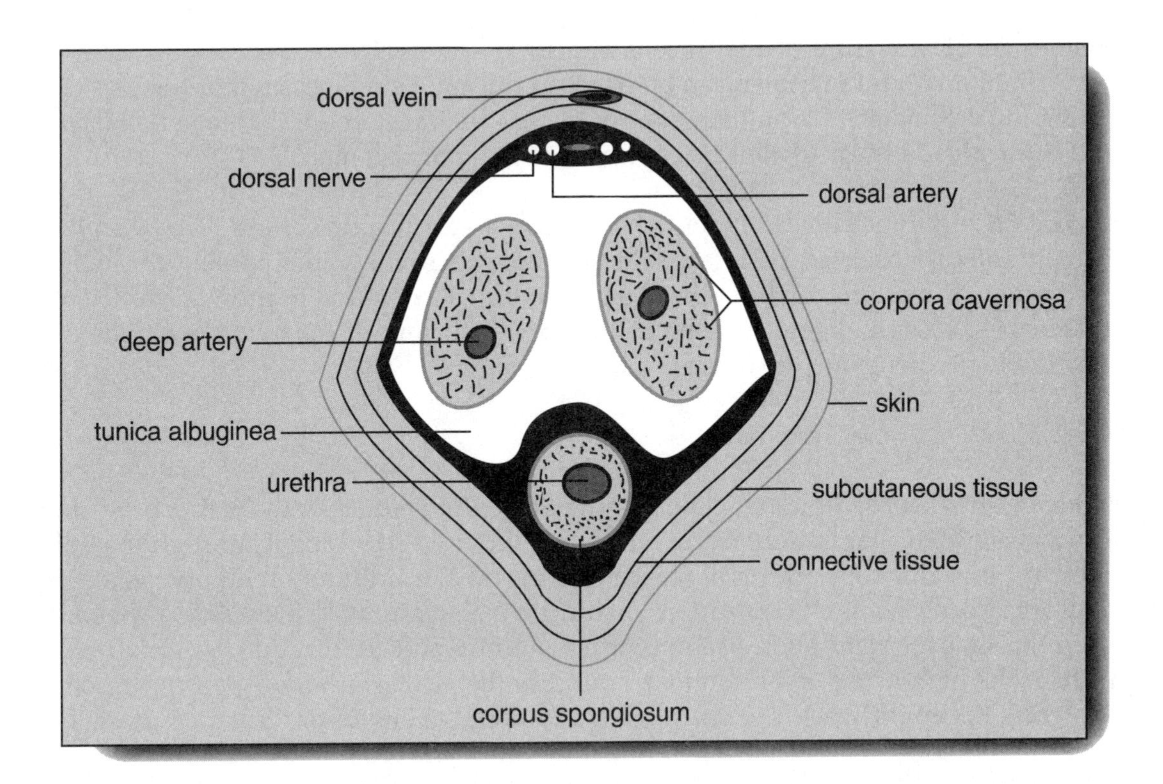

FIGURE 16.6

STRUCTURE OF THE SPERM

Mature sperm can be divided into the head, midpiece (or body) and tail regions **(Figure 16.5)**.

1) The *head* contains the nuclear material and practically no cytoplasm. In addition, covering the head is the *acrosome* (or *acrosomal cap*). The acrosome is a vesicle-like structure filled with enzymes that can penetrate the ovum during fertilization. The acrosome must first undergo capacitation (see below) in order for the enzymes to be released when the sperm head contacts the ovum membrane.

2) The *midpiece* (or *body*) contains mitochondria used to supply ATP to power the flagellum tail. However, since there is little, or no, cytoplasm in the sperm, carbon containing compounds used by the mitochondria to generate the ATP must be supplied from outside of the sperm cell. These energy-containing substances are present in the fluid of the epididymis, in the ejaculatory fluids generated by the seminal vesicle and prostate, and also in secretions present in the female reproductive tract.

3 The *tail* (or *flagellum*) is organized by the *centriole* located in the midpiece. The movement is generated by the sliding of the *microtubules* that comprise the flagellum, and is powered by ATP.

SEMEN

Semen consists of a mixture of sperm cells and fluid secreted from the various accessory glands. Between 2.5 ml and 6 ml of semen can be released during a single ejaculatory event, with each ml containing approximately 120 million sperm. On the average, about *300 to 400 million* sperm are released during a single ejaculation in a normal (sexually fertile) male.

A male considered to be clinically infertile will have a sperm count of *less than 20 million/ml.* He is functionally infertile because, first, large numbers of sperm die while passing through the female reproductive tract, and second, large numbers of sperm are required to provide sufficient quantities of acrosomal enzymes to digest the zona pellucida that surrounds the ovum (even though only one sperm can actually penetrate the ovum).

MOTILITY AND CAPACITATION OF SPERM

Sperm are incapable of swimming while passing through the ducts of the testes, but gain the ability to swim during their maturation process in the epididymis. Movement through the epididymis is provided by peristaltic contractions of the walls of the tube. In addition, sperm present in the ducts of the testes are also incapable of fertilization, but during their maturation process in the epididymis, they also gain this ability. Maturation of the sperm's motility and fertilization capacity occurs through exposure to the high levels of testosterone bound to the *androgen-binding-protein* present in tubular fluid entering the epididymis. The sperm gain additional ability to fertilize an ovum when they are exposed to secretions in the female reproductive tract, and this is known as *capacitation.* During capacitation the membrane of *acrosomal cap* is altered. Release of these acrosomal enzymes can now take place if the sperm contacts the ovum.

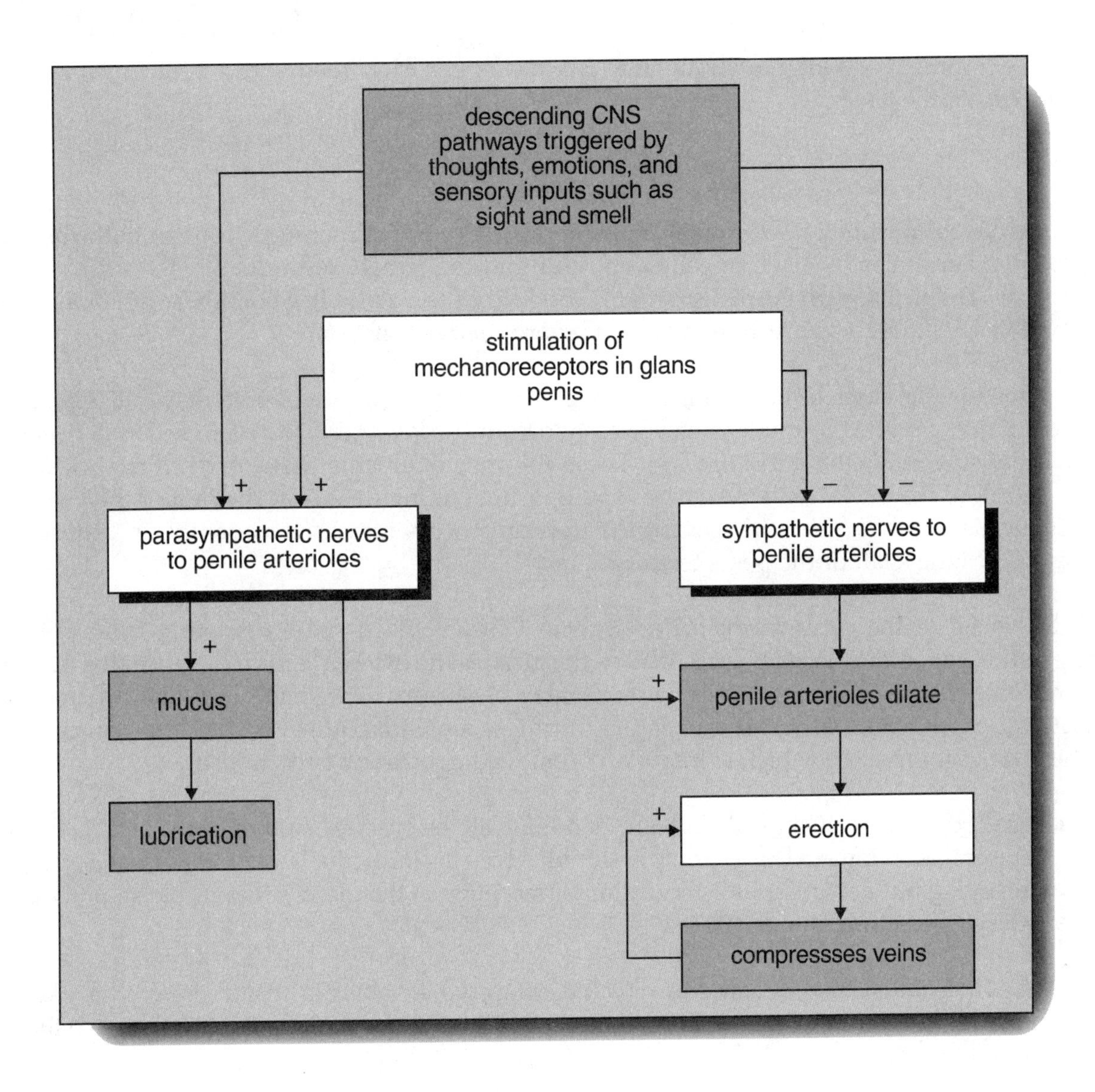

FIGURE 16.7

MALE EXTERNAL ACCESSORY ORGANS

Both *scrotum* and *penis* are classified as male external reproductive organs. Early development of these structures depends on the presence of dihydrotestosterone which is synthesized from testosterone.

SCROTUM

A scrotum holds the testes outside the body proper, and is comprised of skin and subcutaneous tissue. The wall of the scrotum is lined with *dartos muscle,* (a smooth muscle layer). Contraction of this muscle layer results in wrinkling and shortening of the scrotal skin, and tightening the scrotum against the

testes. The scrotum is divided internally into two chambers, each lined with serous membrane. Each contains one of the testes.

PENIS

The penis has a dual function within the male reproductive tract. It conveys urine from the bladder to the outside, and it conveys sperm cells and fluid into the female reproductive tract during sexual intercourse. The structure of the penis is highly specialized to accomplish both these functions because it is capable of undergoing erection to fulfil its reproductive function.

In cross section, the shaft, or body of the penis, contains three columns of erectile tissue **(Figure 16.6)**; an upper pair of *Corpora Cavernosa*, each of which surround a *deep artery*, and a single *Corpus Spongiosum* below (that surrounds the *penile urethra*). These columns of erectile tissue extend backward for the entire length of the penis, with the corpus spongiosum ending deeply at the root of the penis in the structure called the *bulb of the penis*. During the erection process, increased blood flow into these erectile tissues causes the shaft of the penis to stiffen.

The distal end of the penis forms an enlargement of erectile tissue, originating from the corpus spongiosum, called the the *glans penis*. Within the glans is the external urethral orifice (the opening of the urethra). The glans also contains a large number of sensory nerve endings which are involved in sensations leading to orgasm and ejaculation during sexual intercourse. Finally, the glans is covered by the *prepuce* (or *foreskin*) which is frequently removed at birth by *circumcision*.

Erection. In order for erection to take place, blood must fill the vascular sinuses of the erectile tissues.

1) The process is initiated by a variety of stimuli which the individual interprets as sexually arousing, including but not limited to stimulation of receptors at the glans, other tactile stimuli, olfactory stimuli and visual stimuli **(Figure 16.7)**.

2) This then initiates a reflex, mediated by the sacral cord, resulting in an increase in parasympathetic stimulation reaching the penile arterioles, while inhibiting sympathetic stimulation to the penile arterioles.

3) Parasympathetic stimulation also increases mucus release by the bulbourethral glands. This leads to lubrication of the glans in preparation for intercourse.

4) In the presence of increased parasympathetic stimulation, and decreased sympathetic stimulation, the penile arterioles dilate.

5) Increased blood enters the vascular spaces of the erectile tissues and erection takes place.

6) As the erectile tissue begins to fill with blood, it swells and compresses the veins which drain the erectile tissues. This further increased erection.

Orgasm. Orgasm is the culmination of sexual stimulation. It is accompanied by a pleasurable feeling of both physiological and physical release as described in the sexual response cycle (see following material). During orgasm in the male, ejaculation also takes place. In the female, orgasm alone takes place.

Ejaculation. Ejaculation results from an increased sympathetic stimulation acting through a spinal reflex mediated by the sacral cord. Initially sympathetic stimulation causes sequential smooth muscle contractions to take place in the accessory reproductive organs in the following order:

1. In the prostate (releasing prostatic secretions).
2. In the epididymis (releasing sperm and fluid).
3. In the vas deferens (promoting pumping action).
4. In the ejaculatory duct (promoting pumping action).
5. In the seminal vesicle (releasing more secretions).

The effect is to deliver *semen* (sperm + various fluids) into the urethra. As the semen moves into the urethra, this is termed the *emission phase of the ejaculatory reflex.*

As the urethra fills with semen, this triggers a reflex causing skeletal muscle at the base of the penis to contract rhythmically. Semen is now forced out of the penis, and this is termed the *expulsion phase of the ejaculatory reflex.*

MALE AND FEMALE SEXUAL RESPONSE CYCLE IS SIMILAR

The direct responses of erection, orgasm, and ejaculation in the male; and erection and orgasm in the female (to be discussed more fully later) describe only a limited view of the processes actually taking place during sexual intercourse. In order to more fully encompass the plethora of physiological effects prior to, during, and after sexual intercourse, we must consider the four phases of the sexual response cycle.

1. *Excitement phase*: erection can take place in both males and females. Physical sensitivity increases, accompanied by an increase in sexual awareness.
2. *Plateau phase*: heart rate, blood pressure, respiration, and muscle tension increase. Sexual awareness intensifies.
3. *Orgasmic phase*: physical pleasure is perceived accompanied by ejaculation and release of semen in the male.
4. *Resolution phase*: during which the overall body relaxes and returns to a normal, unstimulated state.

HORMONAL REGULATION OF MALE REPRODUCTION

Hormonal regulation in the male is under the control of gonadotropins from the anterior pituitary, and depends on two negative feedback pathways **(Figure 16.8)**.

1) The Hypothalamus releases *GnRH* which stimulates the release of FSH and LH from the anterior pituitary.
2) *Follicle Stimulating Hormone* (*FSH*) stimulates the Sertoli cell which then promotes spermatogenesis. In addition the Sertoli cell also responds to FSH by releasing the hormone Inhibin.
3) *Luteinizing Hormone* (*LH*) (previously named *Interstitial Cell Stimulating Hormone* (*ICSH*) in the male), acts on the *Leydig cell* (or *interstitial cell*) to stimulate the production of testosterone (T).

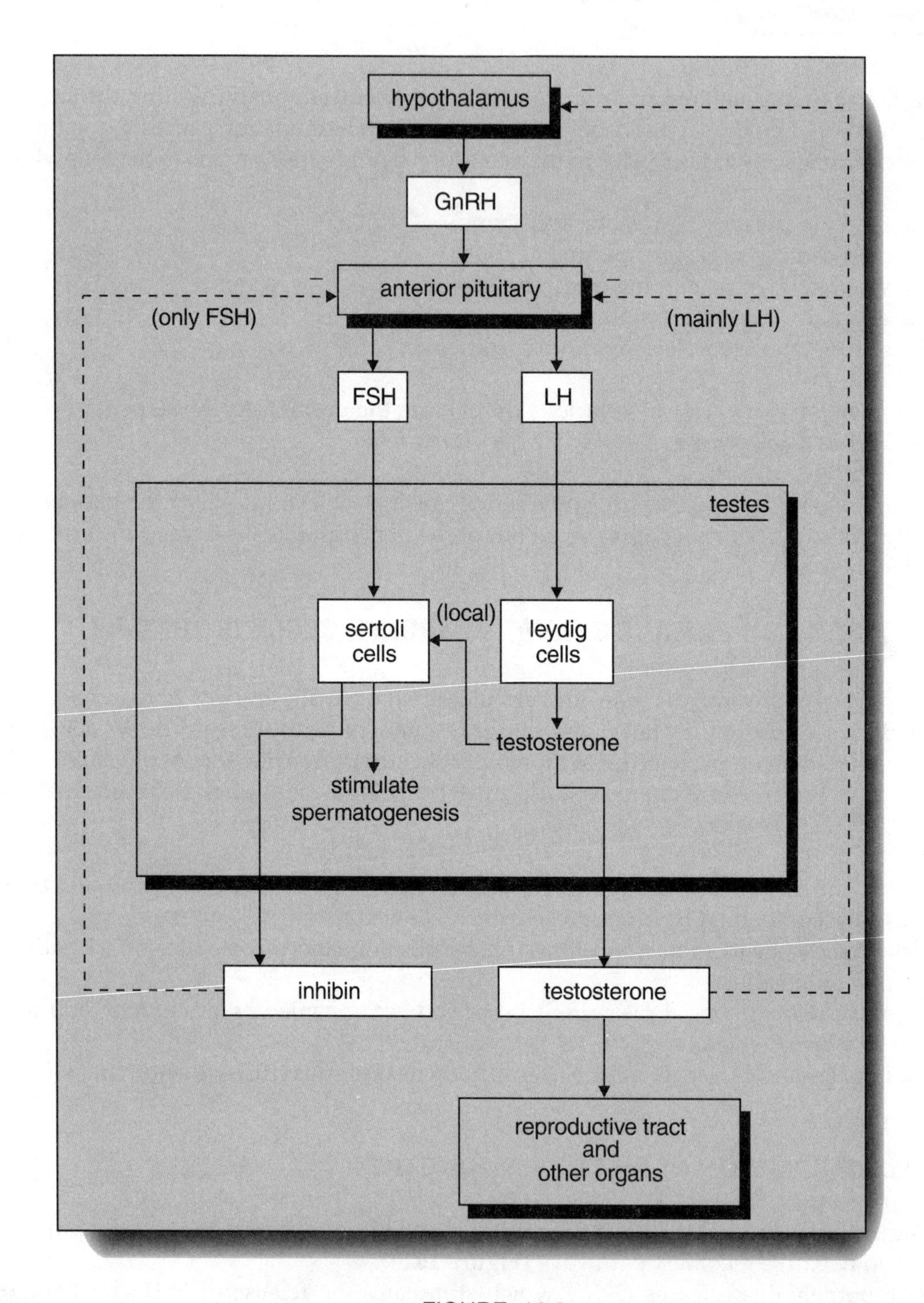
hypothalamus
GnRH
anterior pituitary
(only FSH)
(mainly LH)
FSH
LH
testes
sertoli cells
(local)
leydig cells
testosterone
stimulate spermatogenesis
inhibin
testosterone
reproductive tract and other organs

FIGURE 16.8

4) Testosterone binds to androgen-binding protein (synthesized by the Sertoli Cell). This results in locally elevated levels of testosterone that facilitate spermatogenesis.
5) Testosterone enters the blood and promotes many systemic androgen dependent changes in the male (see following material). In addition, the circulating testosterone feeds back to the anterior pituitary and the hypothalamus.
6) Testosterone acting at the hypothalamus inhibits the episodic release of GnRH. This indirectly reduces the release of LH at the anterior pituitary.
7) Testosterone acting at the anterior pituitary inhibits the release of all of the LH, and some of the FSH, by reducing the response of the pituitary secretory cells to GnRH.
8) Inhibin enters the blood and feeds back to the anterior pituitary where it inhibits the release of any FSH not already inhibited by the presence of testosterone-induced negative feedback.

SYSTEMIC EFFECTS OF TESTOSTERONE

Testosterone (or its metabolite dihydrotestosterone) are responsible for development and maintenance of male secondary sexual characteristics. For example, testosterone is required for:

1 Beard growth.
2. Axial and pubic hair growth in both sexes.
3. Male pattern baldness.
4. Enlargement of the larynx—deepening of the voice.
5. Thickening of the bones and bone growth. However, while elevated levels of testosterone initially cause the growth spurt, the effect is also to cause the epiphysis to become ossified and close. Thus, linear growth ceases around this time.
6. Increased hypertrophy of skeletal muscle tissue (This is why athletes use anabolic steroids.)
7. Increased sebaceous secretions from the skin. This is why acne develops during puberty in both males and females (androgen levels also rise in the female at puberty).
8. Male pattern of fat deposition.
9. Aggression.
10. Libido (sex drive) in both males and females.

FEMALE REPRODUCTIVE SYSTEM

In the female the primary reproductive organs are the *ovaries*. The internal accessory reproductive organs consist of the uterus, uterine tubes and cervix; while the vagina and its associated structures are considered to be the external organs.

OVARIES

Ovaries generate the female germ cells called *ovum* and also secrete the female sex steroids: *estrogen* and *progesterone*. The ovaries are suspended from the pelvic brim through the *broad ligaments*, and the *suspensory ligaments* and to the uterus through the *ovarian ligaments*. The ovaries are comprised of an *internal medulla* and an *outer cortex*. The medulla, which maintains the structure of the ovary and supports the cortex, consists of loose connective tissues, blood vessels, nerves and lymphatics. By contrast, the ovarian cortex has compact tissue and is responsible for generating and expelling the mature ovum. Cells in the cortex also manufacture and secrete the female sex steroids.

The Microscopic Structure of the Cortex. The ovarian cortex is covered by a thin layer of germinal epithelium of cuboidal cells. Within the cortex itself are many *primary follicles* (each about 40 um in diameter). Under microscopic examination, these primary follicles can be identified because they contain a large central primary oocyte surrounded by a single layer of flattened *follicular cells* (*granulosa cells*). Additionally, the cortex may also contain other follicles in later stages of development, as well as structures called *corpra lutea* and *corpra albicans* (see following material).

1) The process that leads to the formation of a mature oocyte begins during early embryonic development. At this time oogonia divide by *mitosis*, forming several million *primary oocytes*. However, many of these oocytes undergo *atresia* and degenerate. Thus, by the time of birth, there are about one million primary oocytes remaining in the ovaries.

2) After birth, follicular atresia continues and this means that by the onset of puberty, there are only about 400,000 primary oocytes present in the ovaries.

3) From puberty through menopause, some of these primary oocytes begin to undergo *meiosis* at each menstrual cycle.

4) However, during each menstrual cycle only one (or sometimes two) reach the stage where they ovulate releasing an ovum into the oviduct. The majority of oocytes never reach maturity but continue to be lost by atresia at each menstrual cycle.

5) Thus, during the reproductive life of the female, it is estimated that only about 400 of these primary follicles actually ovulate; the remainder (400,000 - 400 = 399,600, or about 99.99%) disappear by atresia.

6) When all of the primary follicles have been used up, this signals the onset of menopause (as described later in this chapter).

STAGES IN FOLLICULAR MATURATION

Within the cortex of the ovary, follicular development progresses through a series of defined stages. As stated, at the beginning of each new menstrual cycle a group of primary follicle begin the process of meiosis. However, only one usually reaches the stage where it ovulates. The others all undergo atresia, but the mechanisms remain under investigation. For example, it has been suggested that the estrogen/androgen (E/A) ratio plays a role in this process, and that follicles containing more estrogen in their follicular fluid survive, while those with more androgen regress. The sensitivity of the growing follicles to FSH may also be involved in determining which undergo atrisia and which survive to ovulate. Further studies are clearly warranted.

The stages of follicular maturation are as follows:

1) At the beginning of a new menstrual cycle, a *primary follicle* (actually more than one) begins to mature. This primary follicle contains a *primary oocyte* in the center, and is surrounded by a single layer of *granulosa cells*. Under the stimulus of rising FSH from the pituitary, the granulosa

cells proliferate, and the oocyte in the center enlarges.

2) A layer of *theca cells* develop outside of the granulosa cell layer as the entire follicle continues to enlarge.

3) A space filled with *follicular fluid*, called the *antrum*, appears within the layer of proliferating granulosa cells.

4) As the follicle continues to enlarge, a clear zone surrounds the oocyte in the center. This clear region is called the *zona pellucida*. The follicle continues to grow larger and the antrum also enlarges.

5) Within the follicle the primary oocyte develops into the secondary oocyte by meiosis. This takes place close to the time of ovulation.

6) Estrogen is being secreted by the granulosa/theca cell layers of the follicle. As the follicle

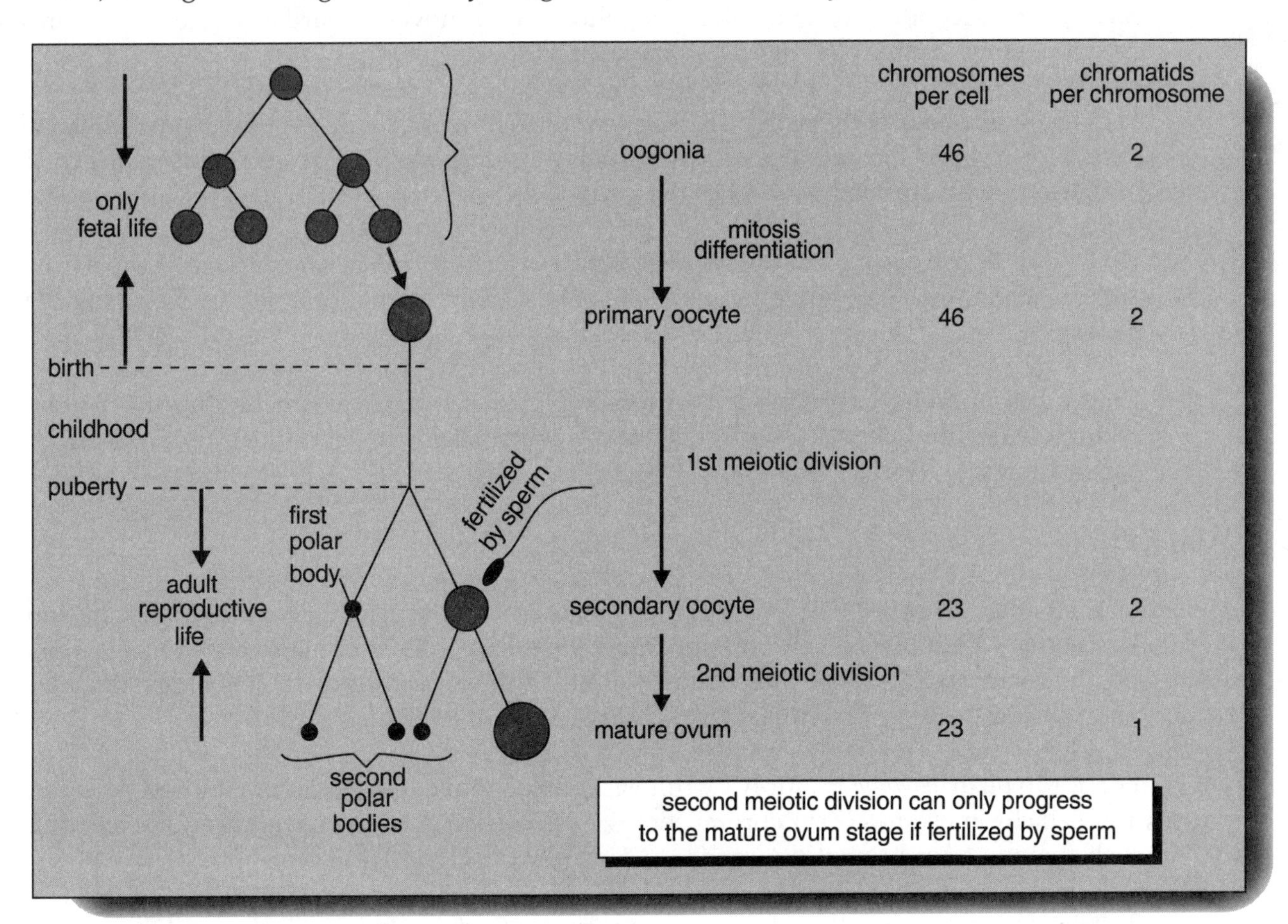

FIGURE 16.9

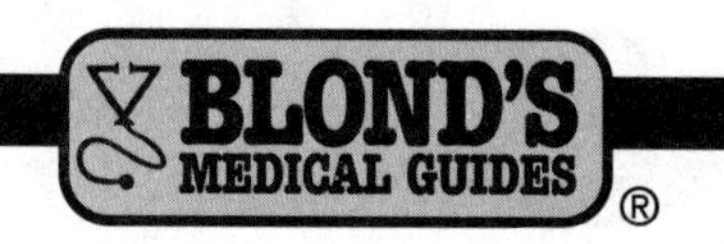

matures, estrogen secretion also increases.

7) Around day 14 of the menstrual cycle the follicle is ready to ovulate. It is now quite large, contains a big fluid filled antrum, and can be called a *Graafian follicle* (or a *preovulatory*, or *mature follicle*). It also protrudes slightly above the surface of the ovary. Within this Graafian follicle, the secondary oocyte is located at one side, and is surrounded by a mass of granulosa cells.

8) LH levels now rapidly rise and stimulate enzyme action in the bulging follicular wall at the ovarian surface. With increased enzyme action, the connective tissue layer is digested, and wall weakens and ruptures.

9) Ovulation now occurs and the secondary oocyte is ejected from the follicle. It is surrounded by a clearly defined Zona pellucida, outside of which is an adhering mass of granulosa cells called the *Corona radiata*.

10) The secondary oocyte, and its mass of granulosa cells, is now picked up by the *fimbria* of the *oviduct*. The secondary oocyte enters the oviduct and moves towards the uterus. If it meets sperm passing up from the uterus, it may be fertilized.

11) Meanwhile back at the ovary, the empty follicle now fills with a yellowish material and is called the *Corpus luteum*. The Corpus luteum is responsible for secreting estrogen and progesterone for the next 14 days of the cycle.

12) By day 28, the corpus luteum regresses (unless fertilization has taken place), and estrogen and progesterone levels fall. This signals the onset of menstruation and the beginning of the next cycle.

13) If fertilization has taken place, the corpus luteum continues to function for the next three months. After this period the placenta takes control and this continues until the onset of parturition.

MEIOSIS

Meiosis in the female **(Figure 16.9)** results in the formation of an ovum that contain 23 chromosomes (1 X chromosome + 22 autosomes). Comparison between this process in males and females demonstrates many similarities. However, there are also some significant differences. (It is suggested that the reader compare **Figure 16.4** of the male, with **Figure 16.9** of the female.)

1) Prior to birth, *oogonia* multiply by mitosis. The oogonia, like other somatic cells, contain 46 doubled chromosomes (2 sex chromosomes, each consist of two chromatid; and 44 autosomes each chaving two chromatids).

2) These oogonia eventually differentiate into *primary oocytes* containing 46 doubled chromosomes (2 sex chromosomes, each having two chromatid; and 44 autosomes each two chromatids). Each primary oocyte is packaged in the center of a layer of *granulosa cells*, and the entire structure is called a *primary follicle*.

3) When a primary oocyte is chosen to begin the process of differentiation, it enlarges and proceeds through the *first meiotic division* (*Meiosis I*). During this process the primary oocyte divides into two cells; a *secondary oocyte*, and the *first polar body*. The secondary oocyte contains 23 doubled chromosomes (1 sex chromosome composed of 2 chromatids, and 22 autosomes, each of 2 chromatids), and also most of the cytoplasm from original primary oocyte. The first polar body, contains the other 23 doubled chromosomes, but little or no cytoplasm.

4) The secondary oocyte will now proceed through the *second stage of meiosis* (*Meiosis II*) to form the mature ovum. However, this will only take place *if the secondary oocyte is fertilized by a sperm*. Thus, it is the secondary oocyte that is actually ovulated on day 14 of the cycle, and it is the secondary oocyte that enters the oviduct. If the secondary oocyte is fertilized, it will then divide into two cells: the *mature ovum* and the *second polar body*. The mature ovum contains 23 single chromosomes (1 single X chromosome—of only 1 chromatid—and 22 single autosomes, with 1 chromatid each), and all of the cytoplasm. The second polar body is comprised of the other 23 chromosomes and little or no cytoplasm. Note also the first polar body may divide into two, additional second polar bodies. None of these polar bodies are viable, and all will disappear.

5) Thus, from 1 primary oocyte, only 1 mature ovum is produced. This is different from the

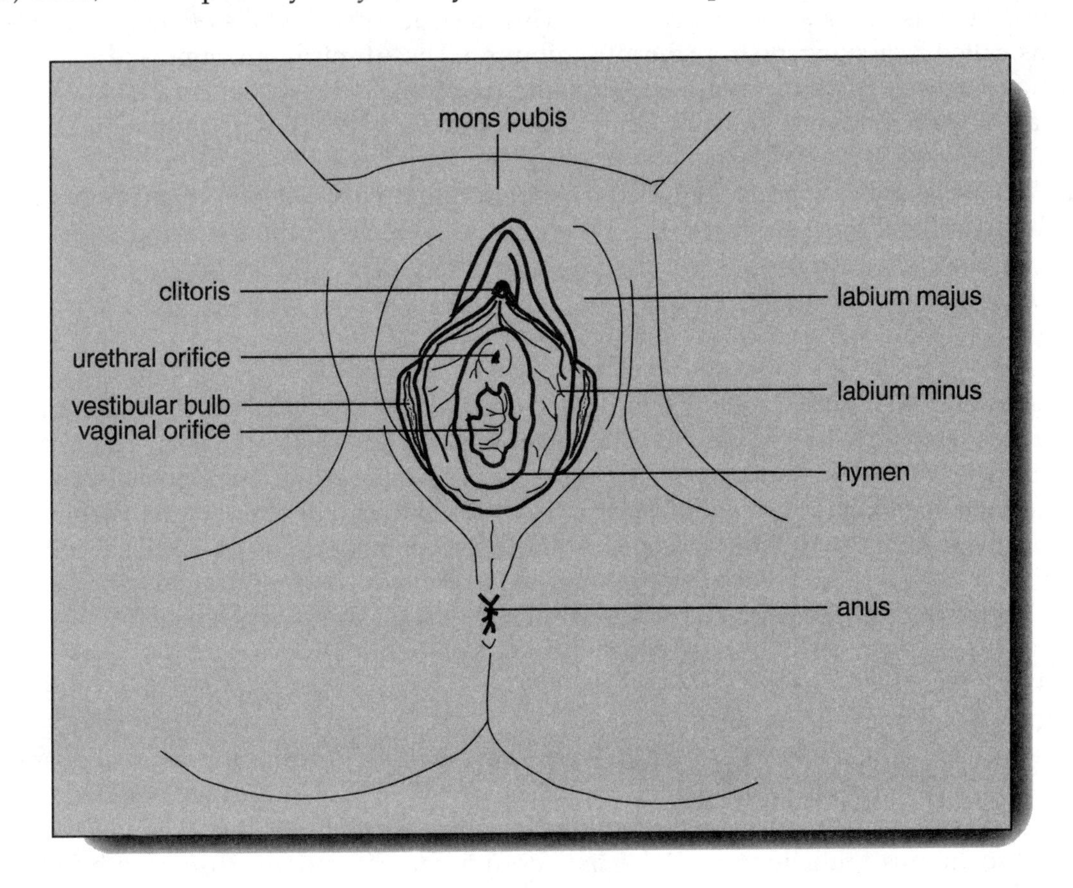

FIGURE 16.10

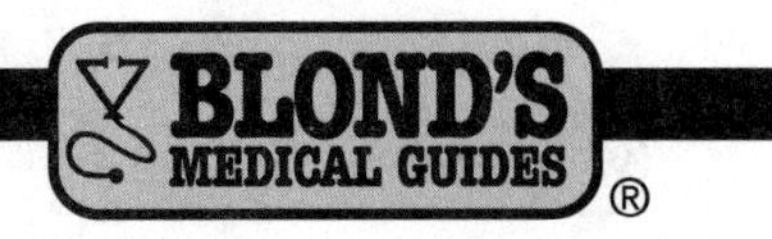

process in the male where from 1 primary spermatocyte, 4 viable mature sperm are generated.

6) In the oviduct, the sperm fertilizes the secondary oocyte by penetrating the zona pellucida and the plasma membrane. The sperm head then enters the cytoplasm of the secondary oocyte. This stimulates the secondary oocyte nucleus to undergo Meiosis II and throw off the second polar body. The mature ovum nucleus is then formed, and while this is all going on, the sperm head continues to remain in the cytoplasm of the ovum. Once the mature ovum nucleus is formed, the sperm nucleus and the ovum nucleus immediately combine. This forms the zygote containing 46 chromosomes. The zygote begins to divide by mitosis as it moves down the oviduct.

7) If the secondary oocyte is not fertilized, it will degenerate in the oviduct within about 24 hours.

OVIDUCTS

The *oviducts* (*uterine tubes, fallopian tubes*) are responsible for trapping the secondary oocyte after it is released from the follicle during ovulation. In addition, the oviducts transport the ovum down towards the uterus, and the sperm up from the uterus, such that fertilization takes place in the oviduct. The end of the oviduct that is closest to the ovary enlarges into a structure called the *infundibulum*. Finger-like projections, called *fimbria*, extend from the infundibulum. These fimbria surround and cover the ovary, and during ovulation guide the ovum into the infundibulum. This process is assisted by movement of the cilia lining the fimbrae, which produce a suction-like current pulling the ovum along.

What are the walls of the oviduct like? There is an inner lining of ciliated epithelium, a middle layer of smooth muscle and an outer layer of peritoneum. Movement of the ovum down the oviduct is accomplished by ciliary action.

UTERUS

The uterus, in the nonpregnant state, is about 7 cm long, 5 cm wide and 2.5 cm in diameter. The upper portion of the uterus is called the *body*, and it is here that the oviducts attach. The uterus narrows down towards the open end called the *cervix*, where the cervical opening is located. Mucous-secreting glands are present in the cervical region, and they supply mucus to the inner lining of the vagina (which has no mucous glands of its own). The walls of the uterus are: an inner lining of epithelial cells called the *endometrium*; a thick middle layer of smooth muscle (composed of longitudinal, circular and spiral fibers) called the *myometrium*; and an outer serosal layer called the *perimetrium*.

VAGINA

Another external organ, the vagina, is about 9 cm long and extends from the cervical opening to the *vestibule* (the space enclosed by the *labia minor*). The cervix projects into the vagina and forms the regions called the *fornix*. The walls of the vagina are composed of three layers: an *inner mucosal layer* of stratified squamous epithelium; a *middle muscular layer*, mostly smooth muscle (with a little skeletal muscle at the lower open end); and an *outer fibrous layer*.

EXTERNAL ACCESSORY ORGANS

The external reproductive organs in the female form the structure called the *vulva*. They consist of *labia major, labia minor, clitoris,* and *vestibular glands* **(Figure 16.10)**.

VESTIBULE

The vestibule is the space surrounded by the labia minor. The separate openings of the vagina and the urethra (*urethral orifice*) are found within this space. Within the vestibule, on either side of the vaginal orifice, are the openings of the ducts from a pair of *vestibular glands*. The secretions from these glands supply lubrication during intercourse. Thus, the vestibular glands correspond to the bulbourethral glands found in the male. In addition, lying to either side, within the mucosa of the vestibule, are regions of vascular tissue called the *vestibular bulbs*. During sexual arousal these regions fill with blood, causing the area to swell.

LABIA MAJOR

This is the female counterpart of the male scrotum. It is composed of folds of adipose tissue, and contains hair follicles and sweat glands on its outer side.

LABIA MINOR

The labia minor surround the vestibule and are located between the labia major. The labia minor attach to the labia major at their posterior margin, while at their anterior margin, they form the covering of the *clitoris*.

CLITORIS

The clitoris is located at the top of the vulva where the labia minor converge. It is the counterpart of the penis in the male. Within are found two columns of vascular tissue called the *corpora cavernosa*, which fill with blood and cause the clitoris to become erect during sexual arousal. The head of the clitoris is called the *glans*, and it too is composed of erectile tissue. The glans is also well supplied with sensory nerve endings.

FEMALE SEXUAL RESPONSE

During sexual arousal, there is increased stimulation of the parasympathetic nervous system, originating from the sacral cord, and causing dilation of the arteries supplying the erectile tissues of the clitoris, and labia minor.

Increased filtration of fluid from the vaginal capillaries, coupled with mucus secretions from the vestibular glands (and from the male bulbourethral glands), supplies lubrication during sexual intercourse. Orgasm in the female, in response to continued sexual stimulation, is triggered through the sympathetic nervous system. This can cause contractions of the muscles in the pelvis and in the

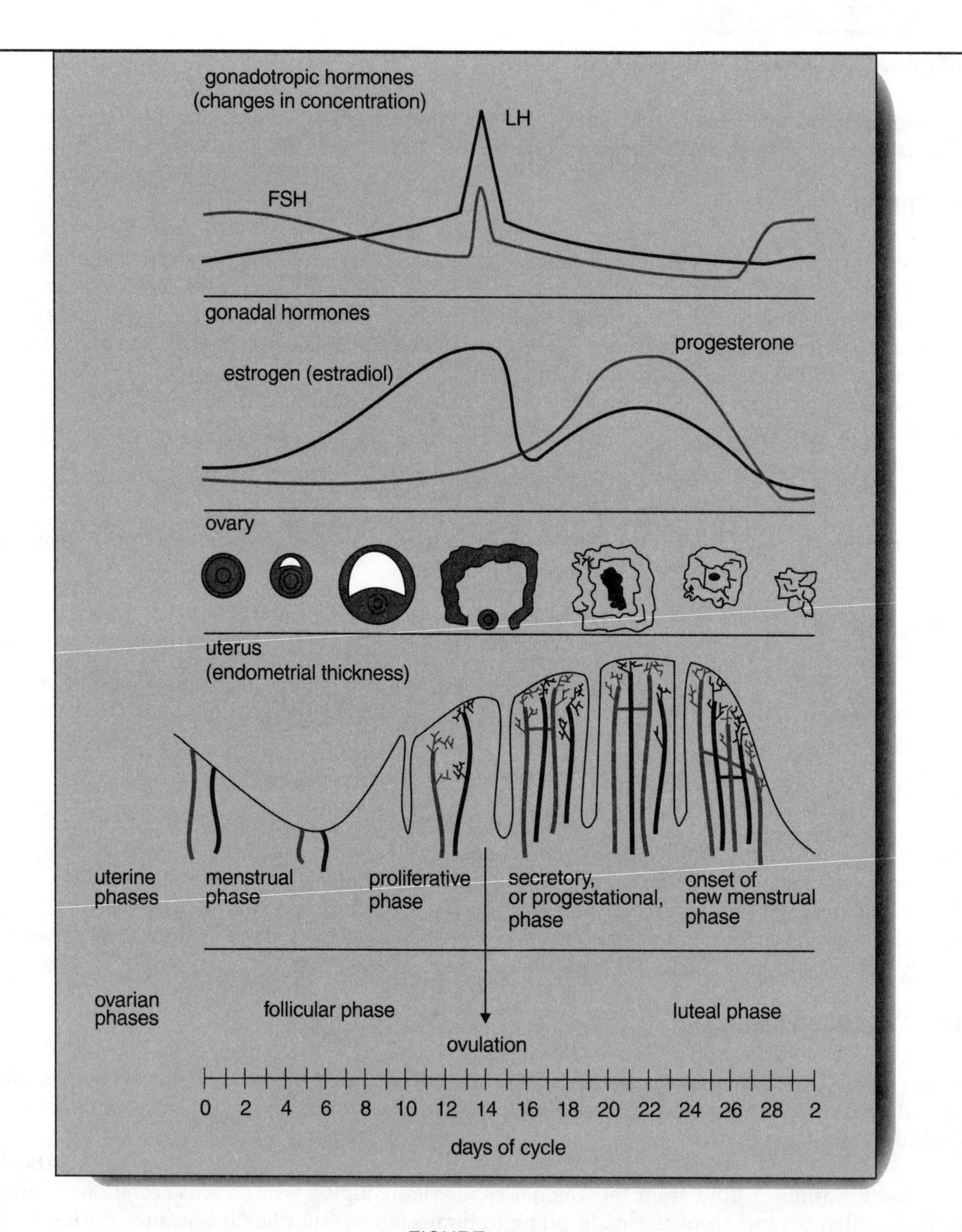
gonadotropic hormones
(changes in concentration)
LH
FSH
gonadal hormones
progesterone
estrogen (estradiol)
ovary
uterus
(endometrial thickness)
uterine
phases
menstrual
phase
proliferative
phase
secretory,
or progestational,
phase
onset of
new menstrual
phase
ovarian
phases
follicular phase
luteal phase
ovulation
0
2
4
6
8
10
12
14
16
18
20
22
24
26
28
2
days of cycle

FIGURE 16.11

outer two-thirds of the vaginal canal.

While the systemic effects during orgasm are basically the same in both males and females (increased heart rate, increased respiration, and so on), in the female there is no ejaculatory response. Furthermore, after an orgasm, males become refractory and cannot immediately experience additional orgasms. However, since females do not become refractory, they can continue to experience multiple orgasms upon continued sexual stimulation.

MENSTRUAL CYCLE

The *menstrual cycle* is a term used to describe the recurring changes that take place in the uterus, as the uterine endometrium first grows and then sloughs. A cycle is associated with monthly menstrual bleeding called *menstruation*.

On the average, the menstrual cycle is about 28 days long, but individual variations are certainly not uncommon. During these monthly changes in the female reproductive tract, other associated changes also take place in the general bodily physiology. The first menstruation, called *menarche*, begins at puberty and the cycle then continues uninterrupted (except by possible pregnancies) until it finally ends at *menopause*. The purpose of the menstrual cycle is to grow and then ovulate a mature ovum, and also to prepare the uterine endometrium for the possible implantation of the embryo.

To understand how the menstrual cycle function, consider the following:

1. The changes that take place in the follicle.
2. The changes that take place in the uterine endometrium.
3. The effect of FSH and LH on the ovary.
4. The production of estrogen and progesterone by the ovary.
5. The hormonal regulatory centers in the hypothalamus and pituitary.

CHANGES IN THE FOLLICLE DURING THE CYCLE

When considering the changes that take place in the follicle during the menstrual cycle **(Figure 16.11)**, day 1 is identified as the first day of the menstrual flow. To follow these cyclic changes, you can use the *follicle* as a guide, thus:

Follicular Phase. From day 1 until day 13 the follicle is growing. Early growth is not dependent on FSH stimulation, but FSH, LH and estrogen are required for later growth of the follicle. Estrogen is released from the theca/granulosa cells of the growing follicle. During the follicular phase, estrogen levels rise significantly, LH levels rise slightly, but FSH, which on day 1 is elevated, falls by day 10 and remains low (except for a slight peak on day 14) until the next cycle begins.

Ovulatory Phase. On day 14 a surge of LH from the pituitary causes the Graafian follicle to ovulate, releasing the secondary oocyte. A slight peak in FSH secretion also takes place on day 14. Estrogen levels, which had peaked just prior to day 14, now fall sharply.

Luteal Phase. From day 15 until 28 the corpus luteum is operating. It originates from the empty follicle,

synthesizes large amounts of progesterone, and synthesizes lesser amounts of estrogen. These hormones rise to a peak around day 21 to 23 and then fall, as the corpus luteum begins to degenerate from day 25 to 28. Degeneration is complete by day 28, and the uterine endometrium begins to slough. Both FSH and LH levels fall slowly from their high on day 14 to day 28. They rise with onset of the next cycle. (Note that if pregnancy takes place, the corpus luteum will not regress on day 28, but will continue to operate for the next 3 months.)

CHANGES IN THE UTERUS DURING THE CYCLE

When considering the changes that take place in the uterus during the menstrual cycle **(Figure 16.11)**, day 1 is determined as the first day of the menstrual flow. To follow these cyclic changes, you can use the *uterine endometrium* as a guide, thus:

Menstrual Phase. From day 1 until about day 4 the uterine endometrium is sloughing, with associated bleeding from the coiled arteries. Sloughing takes place because the levels of estrogen and progesterone are falling. Without estrogen and progesterone support, the endometrium regresses. In addition, the fall in these steroids also stimulates the uterine tissue to release prostaglandins. The prostaglandins constrict the vessels supplying the endometrium, and this is also responsible for sloughing.

Proliferative Phase. From about day 5 until day 14 the endometrium, as well as the myometrium, is growing because the levels of estrogen are rising at this time. The estrogen acts through

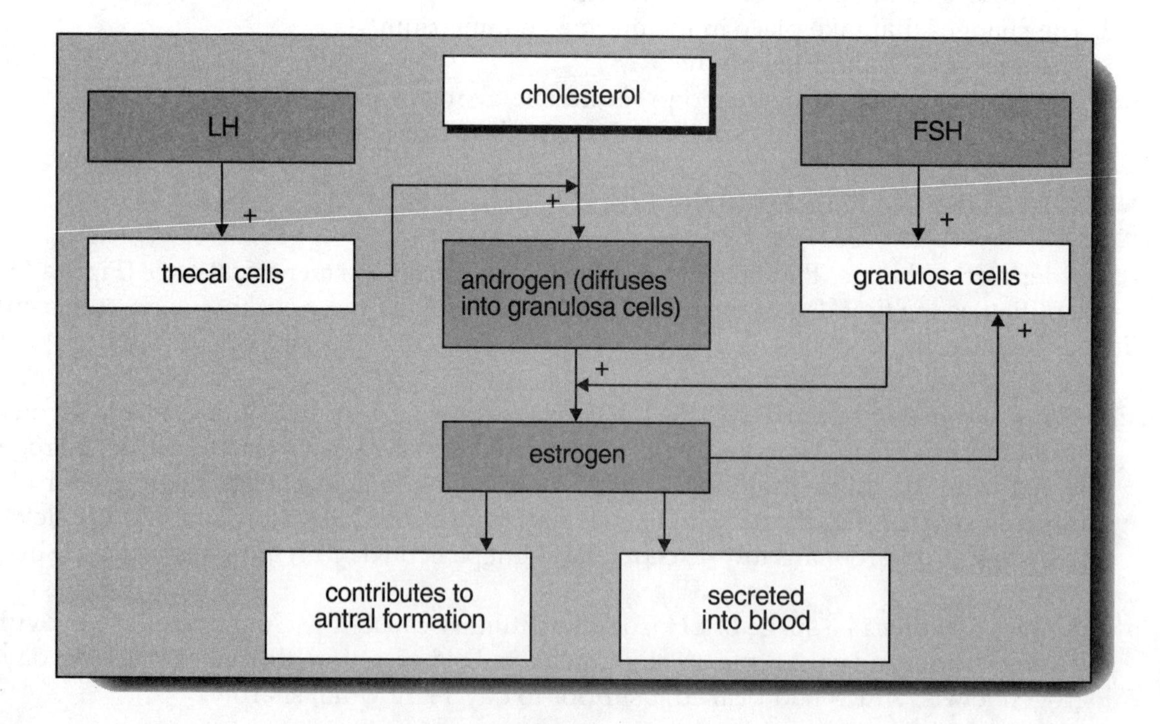

FIGURE 16.12

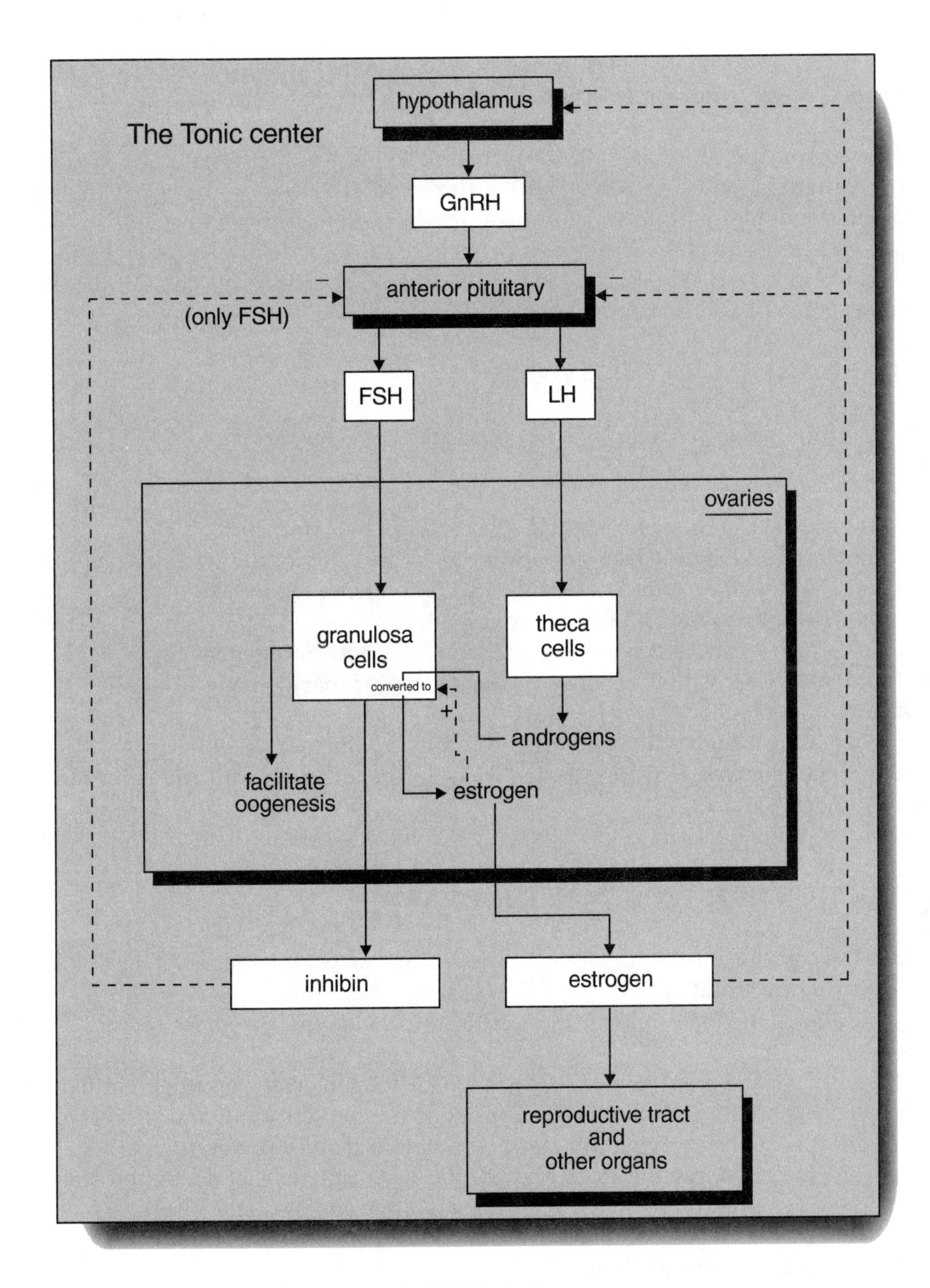
The Tonic center
hypothalamus
GnRH
anterior pituitary
(only FSH)
–
–
–
FSH
LH
ovaries
granulosa cells
theca cells
converted to
+
androgens
facilitate oogenesis
estrogen
inhibin
estrogen
reproductive tract and other organs

FIGURE 16.13

estrogen receptors in the endometrium and myometrium. Estrogen also stimulates the formation of progesterone receptors in the cells of the endometrium.

Secretory Phase. From day 15 until day 28 the increased levels of both progesterone and estrogen cause endometrial glands to develop, and then to secrete and store glycogen. Progesterone also promotes endometrial edema by causing fluids and electrolytes to collect in the connective tissue spaces. The endometrium is now ready for a possible implantation by the embryo. (If pregnancy does not take place, the corpus luteum regresses by day 28, the endometrium sloughs and a new cycle begins.)

EFFECT OF FSH AND LH ON THE OVARY

Early growth of the preantral follicle is not dependent on the presence of FSH, LH or estrogen. However:

1. FSH is necessary in order for the follicular antrum to form.
2. A combination of FSH and estrogen are necessary in order for the granulosa cells to proliferate.
3. LH and FSH together stimulate estrogen secretion from the follicle.
4. The relatively sharp rise in LH prior to day 14:
   a. Blocks further estrogen synthesis by the follicle, and estrogen levels fall.
   b. Stimulates reactivation of meiosis causing the primary oocyte to become the secondary oocyte.
   c. Stimulates prostaglandin synthesis, resulting in follicular swelling.
   d. Stimulates enzyme activation which weakens the follicular wall and results in ovulation of the secondary oocyte.
   e. Stimulates the follicular cells in the empty follicle to differentiate into luteal cells.

THE PRODUCTION OF ESTROGEN BY THE OVARY

Production of estrogen by the growing follicle is dependent on the interaction of the theca cells and the granulosa cells **(Figure 16.12)**.

1. LH stimulates the *theca* cells of the follicle to convert the precursor steroid *cholesterol* into *androgen.*
2. FSH stimulates the granulosa cells of the follicle to convert the androgen into estrogen.
3. Estrogen acts on the granulosa cells to increase the conversion of androgen into estrogen.
4. Estrogen acts on the growing follicle to stimulate antrum formation.
5. Estrogen is secreted into the blood and regulates the functions of the hypothalamus-pituitary unit, as well as promoting many other reproductive and systemic effects.

HORMONAL REGULATION OF REPRODUCTION

Gonadotropin secretion is regulated by three different mechanisms during the follicular, ovulatory and luteal phases of the cycle.

FOLLICULAR PHASE REGULATION OF LH AND FSH

The control of LH and FSH secretion during the follicular phase depends on the negative feedback of inhibin and of low levels of estrogen **(Figure 16.13)**. The hypothalamus-pituitary unit that is involved in this negative feedback is sometimes called the *Tonic Center*.

1. GnRH release from the hypothalamus stimulates the release of FSH and LH from the anterior pituitary.
2. LH stimulates the theca cells to convert cholesterol to androgen.
3. FSH stimulates the granulosa cells to convert the androgen to estrogen.
4. FSH plus estrogen acting through the granulosa cells facilitates oogenesis, and the release of inhibin by the granulosa cells.
5. Low rising levels of estrogen act on the reproductive tract and other organs of the female. Low rising levels of estrogen also cause negative feedback to the hypothalamus and anterior pituitary to reduce GnRH and FSH.

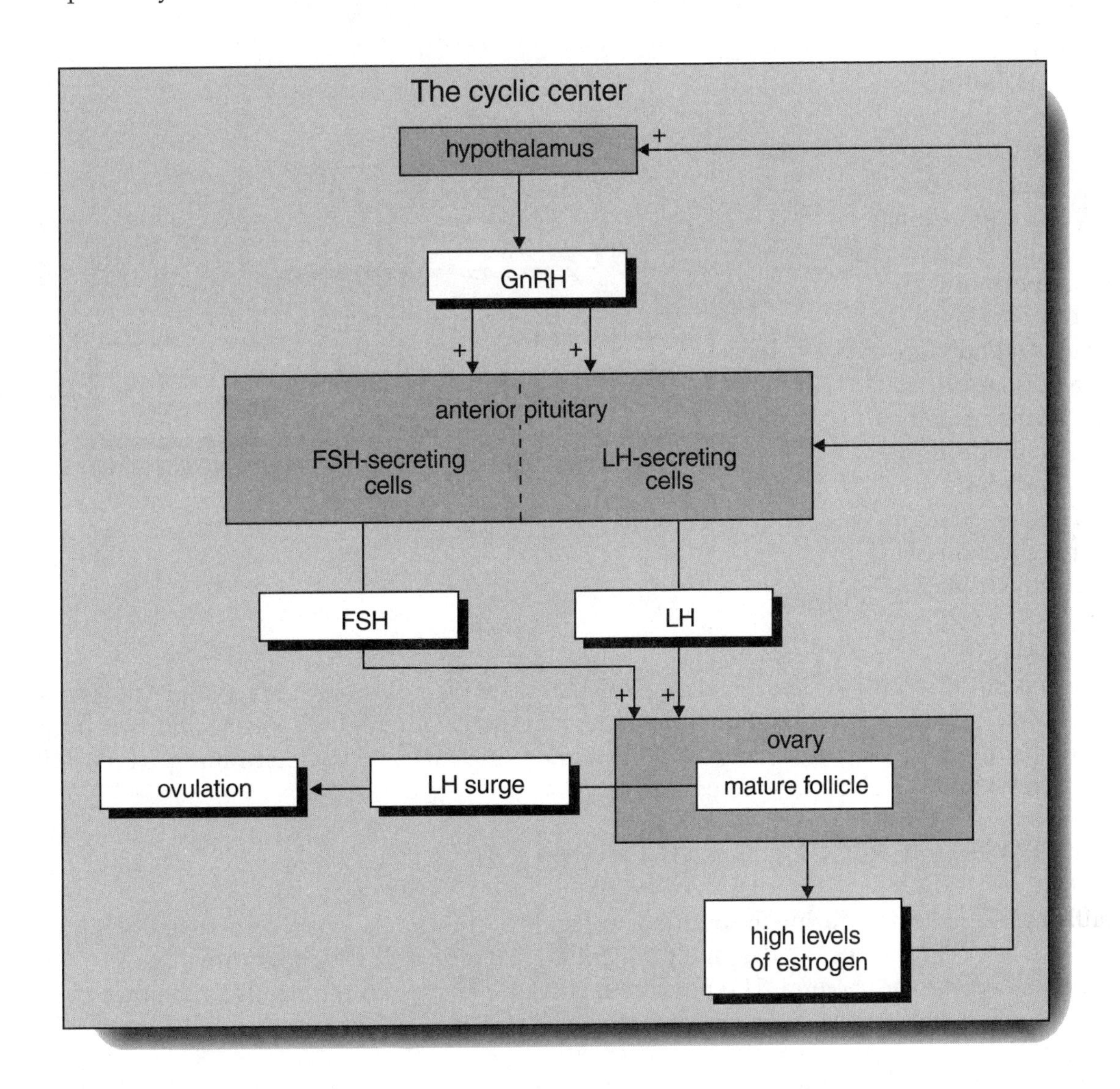

FIGURE 16.145

6. Inhibin causes negative feedback to the anterior pituitary to reduce FSH.

This pathway is responsible for the fall of FSH by day 10, through the negative feedback of low but rising levels of estrogen. Estrogen acts at the hypothalamus to inhibit GnRH. It also acts at the anterior pituitary to reduce the GnRH sensitivity of FSH producing cells. Since its action at the anterior pituitary is more important than at the hypothalamus, FSH levels will fall during the follicular phase, but LH levels actually rise slightly. In addition, inhibin feedback also reduces FSH secretion, but has no effect on LH secretion (much like in the male).

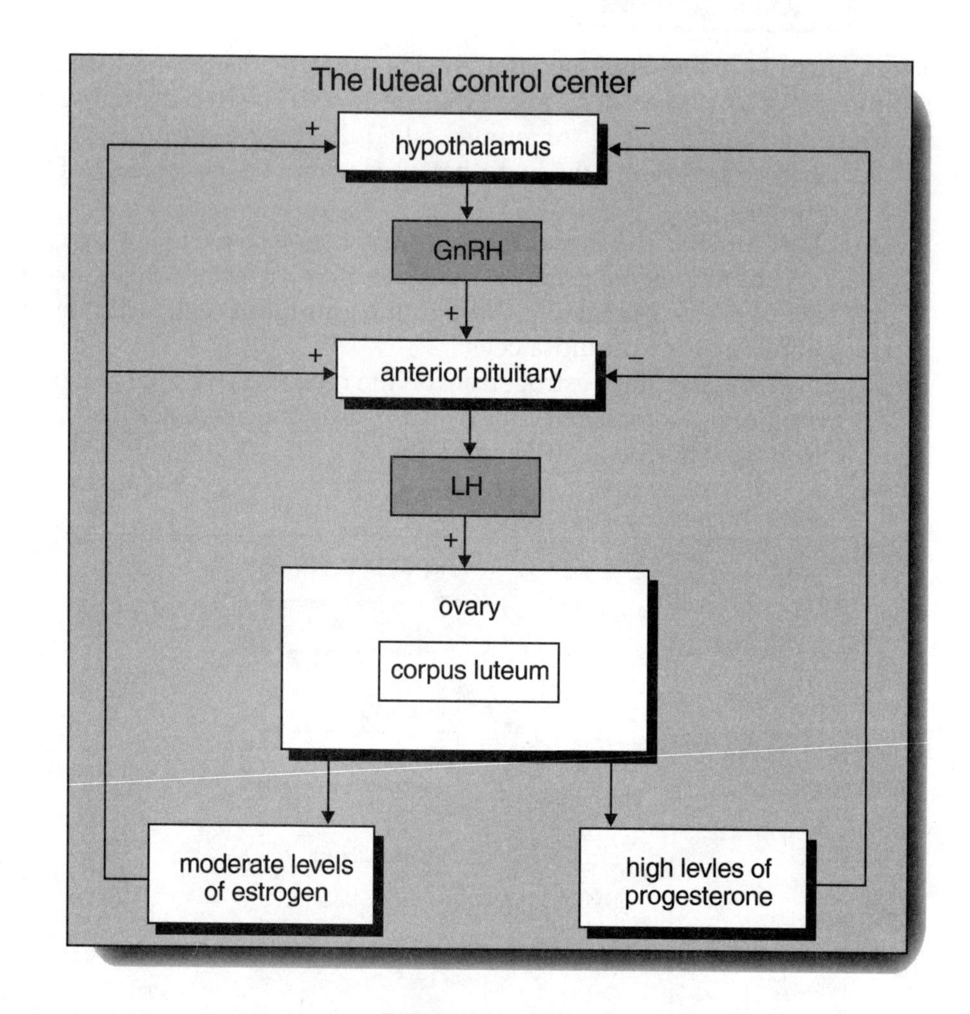

FIGURE 16.15

As FSH levels fall, atresia of the less developed follicles takes place. Usually only a single, more mature, follicle is able to continue to the Graafian stage of development. Differential concentrations of androgen v. estrogen in the follicular fluid may also play some unclear role, in determining which follicles undergo atresia and which survive to ovulate.

OVULATORY PHASE REGULATION OF LH AND FSH

The control of the LH surge during the ovulatory phase depends on the positive feedback of high rising levels of estrogen **(Figure 16.14)**. The hypothalamus-pituitary unit which controls this LH surge is sometimes called the *cyclic center*. The positive feedback of estrogen that leads to the mid-cycle LH surge is described as *"priming of the cyclic center."*

1. Rising levels of LH late in the follicular phase stimulate high and rising levels of estrogen. The estrogen stimulates increased tonic LH secretion (but inhibits an LH surge). The overall effect

is to cause estrogen levels to peak at the end of the follicular phase (on day 13).
2. The very high levels of estrogen stimulate increased release of GnRH from the hypothalamus.
3. The very high levels of estrogen act on the LH secreting cells in the anterior pituitary, making them more sensitive to GnRH stimulation.
4. On day 14 the increased release of GnRH, coupled with the increased sensitivity of the LH secreting cells in the anterior pituitary promote the LH surge. A smaller FSH surge also takes place at this time because the FSH secreting cells in the anterior pituitary, are not as sensitive to GnRH as are the LH secreting cells.

Prior to ovulation the rising level of LH induces the primary oocyte to enter meiosis I, leading to the formation of the secondary oocyte. The mid-cycle (day 14) LH surge triggers ovulation of the follicle and this is accompanied by a significant reduction in the secretion of estrogen by the granulosa cells.

LUTEAL PHASE REGULATION OF LH

The control of tonic LH secretion during the luteal phase depends on the negative feedback of progesterone coupled with the positive feedback of estrogen **(Figure 16.15)**.

1. The corpus luteum forms from the cells of the empty follicle under the influence of LH. LH also promotes the secretion of high levels of progesterone and lower levels of estrogen by the corpus luteum. (FSH is not involved in regulating sex steroid production from the corpus luteum.)
2. The moderately high levels of estrogen secreted by the corpus luteum is positive feedback to the hypothalamus and anterior pituitary. This estrogen is responsible for promoting LH secretion.
3. Feedback of progesterone limits the amount of LH secreted (although some is released by estrogen feedback). Feedback of progesterone also almost completely inhibits FSH secretion. Thus no new follicle can form during the luteal phase.
4. Feedback of high levels of progesterone to both the hypothalamus and anterior pituitary inhibits additional estrogen priming of the cyclic center, and prevents another LH surge from taking place. Thus, even if a follicle were able to develop, it would be unable to ovulate during the luteal phase.
5. The corpus luteum functions from day 15 through about day 25 and then begins to degenerate. By day 28 it ceases to function. While the underlying cause for this decline is not fully clarified, it has been suggested that:
   a. As progesterone levels rise, LH falls. Eventually the concentration of LH falls to a point where the corpus luteum is no longer being supported.
   b. Prostaglandin released by luteal cells may constrict arteriolar smooth muscle, reducing blood flow into the corpus luteum and causing its regression. Estrogen synthesized by the corpus luteum may be associated in some way with the mechanism leading to prostaglandin release. However, further studies are required to clarify the mechanisms involved.
6. After the corpus luteum ceases to function, the fall in both estrogen and progesterone signal the onset of the next cycle.

EFFECTS OF ESTROGEN AND PROGESTERONE

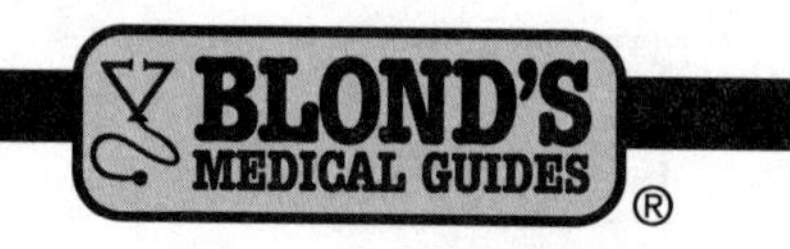

ESTROGEN

1. Acts on the ovary and follicle to stimulate growth and development.
2. Generally acts on the epithelial cell lining, and smooth muscle in the reproductive tack, causing growth and development of these structures. For example, it stimulates the growth of the uterine endometrium.
3. Acts on the hypothalamus-pituitary unit through both negative and positive feedback pathways.
4. Acts on the uterus to increase the contractile activity of the myometrium.
5. Acts on the uterus/cervix to stimulate large amounts of clear watery mucus. This type of mucus is easily penetrated by sperm as they move into the female reproductive tract.
6. Stimulates the formation of progesterone receptors in the endometrial cells, and thus prepares the endometrium to respond to progesterone, which increases after day 15.
7. Acts on the vaginal lining to stimulate epithelial cornification.
8. Acts on the oviducts to stimulate increased smooth muscle peristaltic contractions and increased ciliary action. Thus, it increases ovum transport down the oviduct.
9. Acts on the external genitalia to stimulate growth and development.
10. Acts on the mammary glands to promote growth, and to develop milk ducts. It also stimulates fat deposition in mammary glands.
11. Generally acts systemically to stimulate female body configuration, including fat deposition in thighs and buttocks, and hip development.
12. Stimulates fluid retention in tissues.
13. Inhibits Ca^{++} loss from bone matrix and thus prevents osteoporosis.
14. Stimulates epiphyseal disc closure limiting linear growth.
15. Acts on the skin to increase vascularization.
16. Acts on the sebaceous glands to stimulate a less viscous, more fluid-like secretion, as compared to that promoted by testosterone in males.
17. Develops female pubic hair pattern, but the actual growth of pubic and axillary hair is dependent on androgen.
18. Is involved in the regulation of vascular blood flow; at menopause the absence of estrogen, coupled with elevated levels of gonadotropins produce "hot flashes."
19. Acts on the hypothalamus to stimulate prolactin secretion.
20. Blocks the effect of prolactin on milk production in the mammary glands.
21. May play some unspecified role in protecting the female against atherosclerosis.
22. Regulates immune response, and may assist in preventing fetal rejection.

PROGESTERONE

1. Acts on the hypothalamus-pituitary unit through a negative feedback pathway.
2. Acts on the cervix to promote the production of thick, sticky mucus secretions. This plugs the cervix and prevents sperm penetration from the vagina.
3. Acts on the endometrial glands to stimulate glycogen secretions.
4. Acts on the uterine and oviductular smooth muscle to decrease contractile activity.
5. Causes thinning of the vaginal lining, thus it reduces cornification.
6. Acts at the mammary gland to stimulate growth of the glandular tissue.
7. Acts at the mammary gland to inhibit prolactin induced milk production.
8. Regulates immune response, and may assist in preventing fetal rejection.

EFFECTS OF ANDROGEN IN THE FEMALE

1. Stimulates the growth of axillary and pubic hair.
2. Stimulates libido.
3. Regulates the immune response.

PUBERTY

Puberty describes the sexual changes that take place in both males and females at the age of 10 to 14 years, approximately. At this time:

1. The gonads and accessory organs complete the maturation process, and sexual reproduction is now possible.
2. Secondary sexual characteristics develop.
3. The growth spurt takes place.
4. In females the first menstruation, called *menarche*, takes place.
5. Libido increases in both sexes.
6. In some males, aggressive behavior may increase due to the increasing levels of testosterone.
7. In both males and females, changes in social behavior take place; (although it is not clear as to how much of this is hormonal, and how much is a learned response).
8. Acne appears transiently due to the increase in sebaceous gland secretion, primary under the control of testosterone.
9. All of these changes are stimulated by an increase in the levels of circulating sex steroids. This programmed rise of sex steroids at puberty results from increased GnRH release from the pituitary, because at this time:
   a) Direct neural inhibitory influences from the higher centers acting at the hypothalamus are removed.
   b) There is a decreased feedback inhibition at the hypothalamus to sex steroids. Thus, it takes higher levels of circulating sex steroids to inhibit GnRH.
   c) It is possible that the anterior pituitary cells which release FSH and LH become more sensitive to GnRH stimulation.
10. It has been suggested that removal of these various inhibition influences takes place:
    a) Through preprogramming in the central nervous system related to age.
    b) At a critical threshold level in body weight/mass.
    c) At a critical threshold level in the percentage of body fat.
    d) When melatonin secretion from the pineal gland is reduced during the dark phase.

MENOPAUSE

Menopause is a term used to describe the loss of reproductive function in the female. It usually takes place from 45 to 55 years of age and is characterized by the cessation of the menstrual cycle accompanied by various other anatomical and physiological changes in the reproductive organs. Menopause does not take place instantly, but is proceeded by increasingly irregular cycles, accompanied by loss of estrogen, and physical and emotional changes.

In the male, spermatogenesis continues throughout life. However, in the female, oogenesis ceases before birth and thus the initial number of primary oocytes (primary follicles) is not replaced as they are used up. When the supply of primary follicles is finally exhausted, estrogen levels fall and remain permanently low. In the absence of estrogenic support, structural and functional changes now take place in the target tissues previously dependent on estrogen. Additionally, in the absence of negative feedback to the hypothalamus and pituitary, circulating gonadotropin levels rise significantly.

Maintenance of male and female secondary sexual characteristics is not dependent on androgen alone in the male, or estrogen alone in the female. Rather it is driven by the estrogen/androgen (E/A) ratio. Thus, in the female prior to menopause, estrogen dominates the E/A ratio, while during and after menopause, androgen dominates the E/A ratio. This accounts for the bodily and emotional changes that take place at this time. For example:

1) Breasts and genital organs gradually undergo a certain amount of atrophy.
2) Loss of Ca^{++} from the bone matrix can reduce the bone mass and strength. This is termed *osteoporosis*.
3) Libido (sexual drive) does not usually decrease, and may actually increase.
4) Dilation of skin arterioles, due to estrogen deficiency, increases blood flow to the skin causand can cause *hot flashes*. However, how estrogen mediates this effect remains unclear.
5) In the overall population, the incidence of cardiovascular diseases (hypertension, atherosclerosis) in women is less than in men. However, after menopause the incidence of cardiovascular disease becomes equal in both sexes. Thus, estrogen, or estrogen coupled with progesterone, appears to play a protective role in prevention of cardiovascular disease.
6) Although estrogen levels fall significantly in the absence of follicular function, small amounts of estrogen are still manufactured through peripheral conversion of adrenal androgen to estrogen.
7) Estrogen, or estrogen/progesterone replacement therapy, is frequently prescribed to counteract the more severe changes that may accompany menopause in some women.

In males, changes in reproductive function with aging are much less pronounced. Testosterone levels begin to decline after age 40, but sperm production can continue even through age 80. Aging in males may cause some to experience emotional problems or depression, but the underlying causes are not clarified. Such effects are termed *male menopause*.

LIFESTYLE OF THE SPERM AND OVUM

VIABILITY

After ejaculation, sperm can live within the female reproductive tract for approximately 48 to 72 hours. After the secondary oocyte has been ovulated, it can survive in the oviduct for approximately 10 to 24 hours. Thus, for fertilization of the oocyte to take place, intercourse must occur from 72 hours before ovulation to 24 hours after ovulation (approximately 3 days before to 1 day after).

In the real population, germ cell survival times will vary from these average figures. Also, it is very difficult to accurately predict the exact time of ovulation. Although the *rhythm method* of contraception

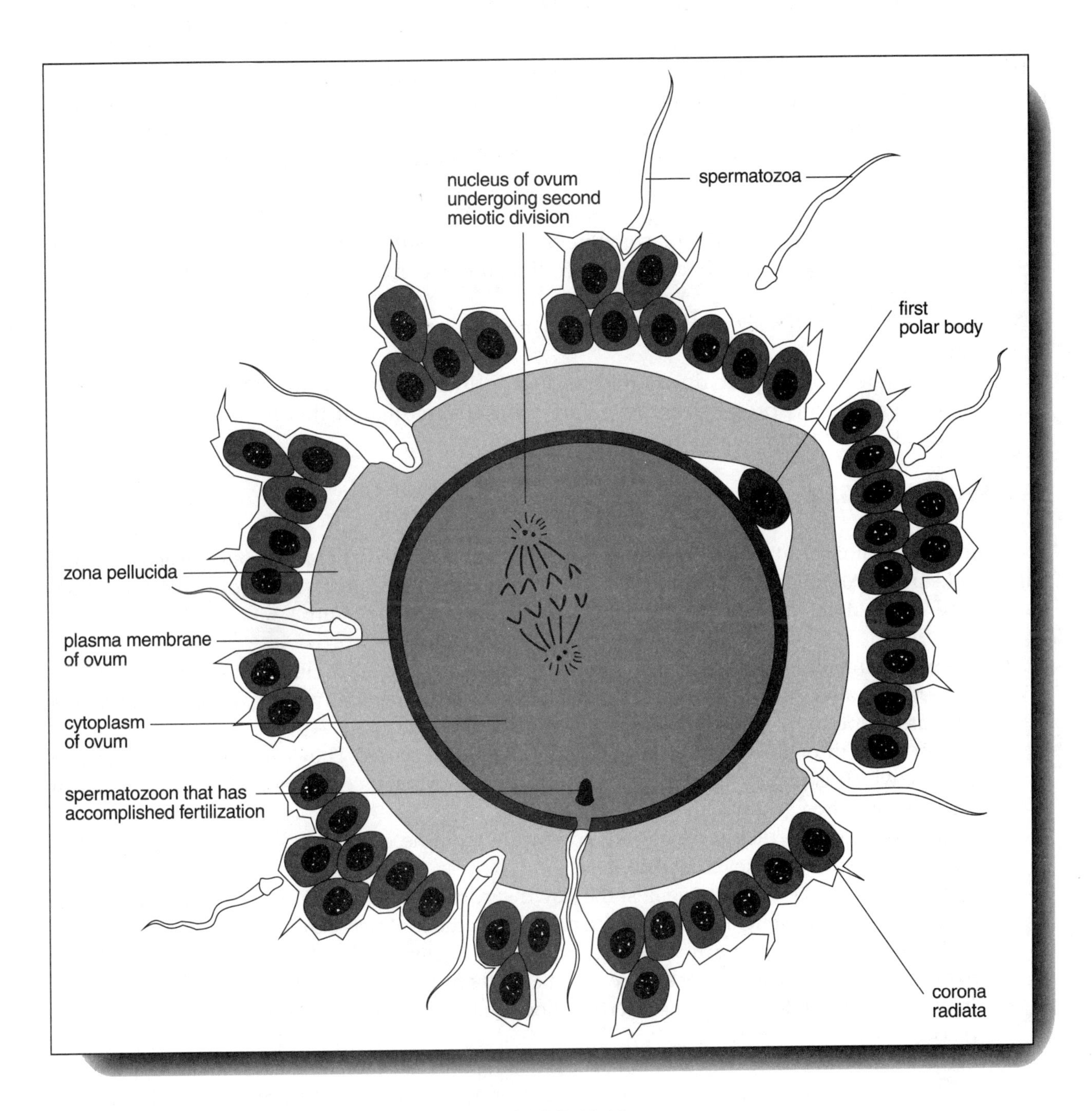

FIGURE 16.16

(see following material) attempts to use this sort of information to limit pregnancies, the effectiveness is low, given the sources of variation intrinsically present.

TRANSPORT OF OVUM

After ovulation the ovum is swept into the mouth of the uterine tube through the action of cilia located on the fimbriae. Initially, the movement of the ovum in the uterine tube is accomplished through ciliary

action coupled with contractions of the smooth muscles in the walls of the tube. This rapid rate of transport propels the ovum into the ampulla of the oviduct. Ciliary action alone then takes over and the movement slows significantly. Thus, it takes four days for the ovum to reach the uterus. During this four-day period, while the ovum is in the oviduct, fertilization must take place due to the short life-span of the secondary oocyte.

1) Fertilization usually takes place while the ovum is in the ampulla.
2) If the ovum (secondary oocyte) is not fertilized while in the uterine tube, it will degenerate and disappear.
3) If the ovum is fertilized it will complete meiosis, become a zygote, and begin dividing by mitosis.
4) The rate of transport down the oviduct is increased by estrogen and decreased by progesterone. Since progesterone levels are rising after ovulation, this may help to maintain the slow rate of transport down the oviduct.

TRANSPORT OF SPERM

Sperm in the ejaculate are capable of swimming. But they must also receive help from outside or fertilization will not take place. Although a few hundred million sperm enter the vagina, the majority die. Only a few hundred penetrate all the way up the uterine tube to reach the waiting ovum.

1) Sperm are initially pumped into the vagina by ejaculatory pressure during intercourse.
2) If the female also experiences an orgasm, this will causes contractions within vagina and uterus that may assist sperm movement towards the uterus and oviducts.
3) Sperm swim through the cervical canal under their own power. They can penetrate the cervical mucus, which is watery during the ovulatory period. Mucous cells generate watery mucus in response to estrogen stimulation.
4) Contractions of the uterus propel the sperm towards the oviducts. Such contractions can be stimulated by elevated levels of estrogen.
5) Sperm swim up the oviducts, against the downwardly directed ciliary beating. They are assisted in their upward movement by reverse (or anti) peristaltic contractions of the smooth muscle in the wall of the oviduct. Uterine tube contractions are stimulated by elevated estrogen levels just prior to ovulation. Contractions may also be stimulated by prostaglandins present in the seminal fluid.
6) In order for sperm to be capable of fertilization, first capacitation, and then activation must take place.
   a) *Capacitation*: For sperm capacitation to occur, the sperm must remain in the female reproductive tract for six to seven hours. During this period of maturation, the surface layer of glycoprotein molecules, derived from epididymal secretions, are removed.
   b) *Activation*: As the sperm moves into the vicinity of the ovum, they are activated. During activation, enzymes in the acrosome are exposed, the wave-like beating of the flagella changes into a more rapid whip-like movement, and the sperm plasma membrane is altered to allow it to fuse with surface membrane of the ovum.

FERTILIZATION

The acrosomal enzymes allow the sperm to penetrates through the mass of *corona radiata* cells surrounding the ovum **(Figure 16.16)**. Once the sperm have passes through this barrier, they must then specifically bind to receptor sites located on the surface of the *zona pellucida* layer. The first sperm to bind here will penetrate the zona pellucida, the sperm plasma membrane will then fuse with the ovum (secondary oocyte) membrane, and the sperm head will enter the ovum cytoplasm. (Note that sperm and ovum must be of the same species. For example, a canine sperm cannot bind to the receptors on a feline ovum. This prevents cross species fertilization.)

1) After the sperm binds to the oocyte plasma membrane, it is drawn through the oocyte membrane into the oocyte cytoplasm.
2) The oocyte releases enzymes present within secretory vesicles located under the membrane. These enzymes pass out of the oocyte and enter the space between the oocyte membrane and the zona pellucida. This alters the zona preventing further sperm penetration, and inactivates any sperm that have already penetrated. Thus, only one sperm is allowed to fertilize the ovum. This process is called *block to polyspermy*.
3) Entrance of the sperm head into the ovum cytoplasm also activates additional enzyme pathways that are necessary for early embryonic development (*embryogenesis*) of the zygote.
4) Although the sperm head (containing the genetic information) enters the ovum cytoplasm, the tail remains outside, and is lost.
5) Entrance of the sperm head, stimulates the secondary oocyte to undergo Meiosis II, forming the mature ovum nucleus.
6) The sperm nucleus, containing 23 chromosomes, now fuses with the ovum nucleus containing 23 chromosomes, forming the 46 chromosome zygote. Fusion occurs within an hour of sperm head penetration into the ovum.

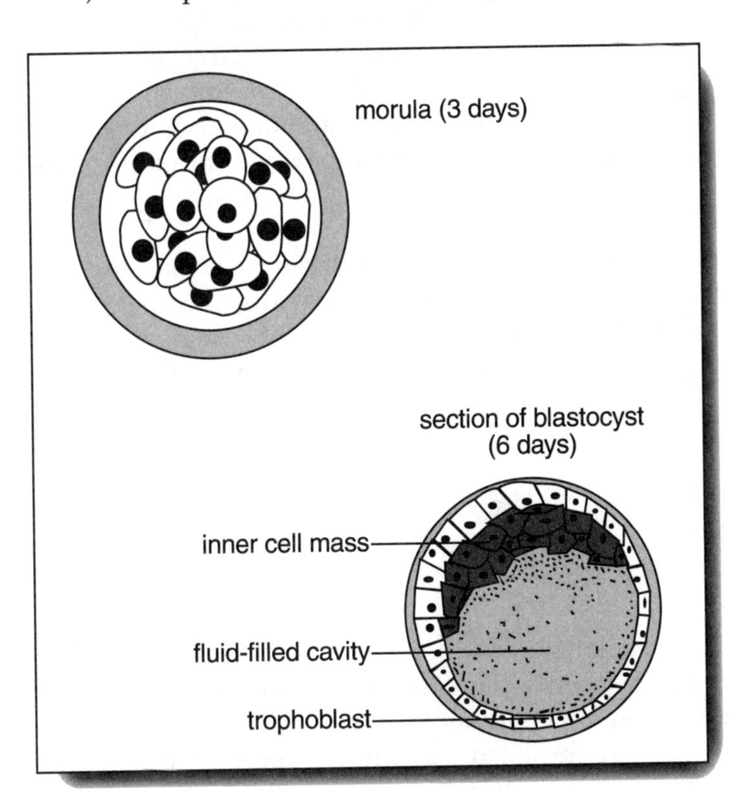

FIGURE 16.17

EARLY DEVELOPMENT

After formation, the zygote undergoes *cleavage*, dividing repeatedly by mitosis. As division continues the mass of cells continues to move down the oviduct towards the uterus. This movement is facilitated by ciliary action. After 3 days, the dividing mass of about 16 cells, called the *morula*, enters the uterus **(Figure 16.17)**.

1) The mass of cells now floats freely in the fluid of the uterine cavity for an additional 3 days. During this 3-day interval, the zona pellucida degenerates and division continues. At the end of this period the cell mass has developed into a hollow ball, called the *blastocyst*.
2) The blastocyst now implants in

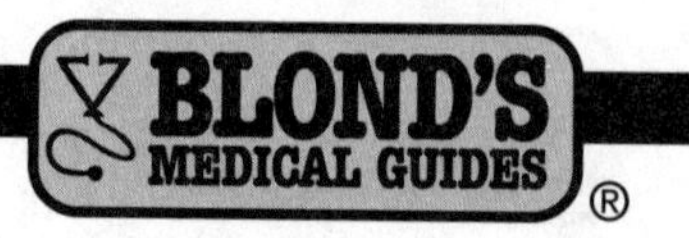

the endometrial layer, which being in the secretory phase, is ready for implantation of the embryo. Thus, 6 days after ovulation, implantation takes place. If ovulation occurs on day 14, this means that implantation takes place on about day 20 of the menstrual cycle.

3) At the blastocyst stage, the embryo **(Figure 16.17)** appears as a hollow ball of cells. This hollow ball contains an *inner cell mass*, a fluid-filled cavity, and outer *trophoblast cells.*
   a) The inner cell mass is destined to form the fetus.
   b) The trophoblast cells, in conjunction with cells of the maternal endometrium, will form the placenta.
   c) The fluid-filled cavity, called the *blastocoele*, will become the fluid-fiilled *amniotic sac*, surrounding and protecting the fetus.
4) The developing cell mass is called the *embryo* until the end of the week 8. After this time, it is called a *fetus*.
5) The blastocyst implants in the endometrium with the inner cell mass side towards the wall. The trophoblast cells on the side closest to the inner cell mass release *proteolytic enzymes*, which make this side "sticky." The proteolytic enzymes digest away the endometrium and the embryo now sinks completely into the uterine lining.
6) Endometrial tissue at the site of implantation becomes the *decidua*, and secretes prostaglandins that increase local blood flow, promote local edema and increase the availability of nutrients.

ECTOPIC PREGNANCY

When implantation of the embryo takes place outside of the uterus, this is termed an *ectopic pregnancy*. This happens very infrequently, and usually the embryo implants in the oviduct. Thus, this kind of ectopic pregnancy is also called a *tubal pregnancy*. However, in very rare instances the implantation site may be at some other location such as the cervix, ovary, or an organ in the abdominal cavity. In any event, all ectopic pregnancies need to be terminated since they have the potential of harming the mother.

Tubal Pregnancy. If progress down the oviduct is too slow, the morula develops into the blastocyst in the oviduct. In this case, the blastocyst will implant in the wall of the oviduct. As the embryo enlarges, it stretches the wall of the oviduct, causing pain. If such a tubal pregnancy is not terminated, it will rupture the oviduct, potentially causing severe hemorrhage and possible death.

TWINS

When two embryos share the same uterus during the same pregnancy, and are then born during the same labor, they are called twins.

Fraternal Twins. They develop when two follicles ovulate during the same cycle. These two separate oocytes are then fertilized by two separate sperm. Since the two zygotes produced are genetically different, they will form two different people, and may be of different sexes. Note that if more than two follicles ovulate and are fertilized by different sperm, then this will produce multiple births. Women who have such multiple births are almost always taking fertility drugs.

Identical Twins. They develop from a single fertilized oocyte, that forms a single zygote. During the very early process of division, the single embryo divides completely into two genetically identical embryos. Thus, the two people created from this process will each have exactly the same genotype, and will be of the same sex.

PREGNANCY

The prenatal period is 40 weeks (or 10 lunar months) long.

To Calculate the Expected Time of Birth:

1. Fertilization age = date of onset of last menstruation + 14 days.
2. Expected time of birth = Fertilization age + 266 days.
3. This is a relatively accurate method since most fetuses are born between 10 to 15 days of this calculated date.

After implantation, the trophoblast cells rapidly digest the endometrium and by the 12th day of pregnancy, the embryo is completely embedded in the uterine wall. The trophoblast cell layer thickens, and forms into the *chorion*, which continues to digest the *decidua*.

PLACENTA

Blood-filled cavities appear in the decidua, and the chorionic tissue of the developing embryo extends into these pools. Capillaries from the embryo/fetus pass into these chorionic projections and form *placental villi*. The capillaries in the placental villi are separated from the maternal blood supply by a thin membrane which allows exchange of nutrients, gases, waste product, salts, water, and so on—but not blood cells. This complex arrangement of fetal/chorionic tissue-maternal/decidual tissue forms the organ called the *placenta*.

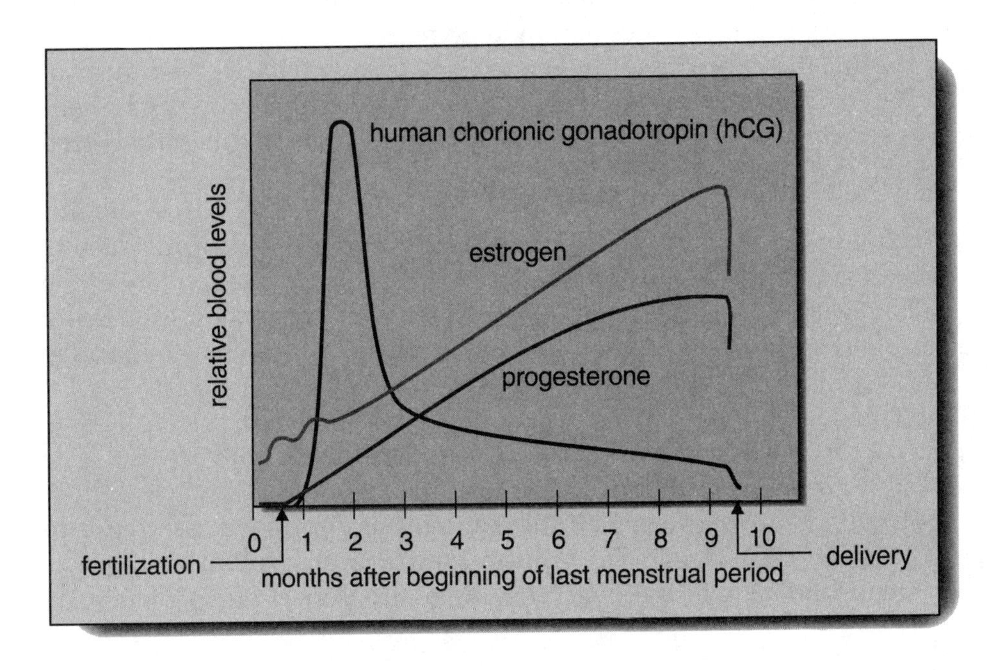

FIGURE 16.18

1) Depending on the type of substance involved, movement across the placental barrier may be by passive diffusion, or by special carrier mediated transport.
2) The placental barrier does not protect the fetus from a variety of harmful agents, (such as drugs, alcohol, certain microorganism). Thus, during pregnancy, the mother should be especially careful about taking into her body substances that might affect the development of her baby. If in doubt, the best choice is to forgo the questionable substance or drug. Note that maternal antibodies can cross the placental barrier. This means that the fetus is protected from many potential infections, because its mother has already developed immunization against these diseases.
3) The placenta manufactures various hormones that are essential for maintenance of the pregnant state. Certain steroid hormones produced by the placenta must be manufactured in conjunction with the fetus itself (this is called the *fetal-placental unit*).
4) The placenta is also believed to play a central role in stimulating the onset of parturition.

HORMONES OF PREGNANCY

The trophoblast cells manufacture the hormone Human Chorionic Gonadotropin. The placenta manufactures the hormones Estrogen, Progesterone, Human Chorionic Somatomammotropin, and Relaxin. All of these hormones are absolutely necessary if a viable fetus is to be delivered at term. The placenta may also manufacture other less important, or less well defined substances, which will not be considered here.

HUMAN CHORIONIC GONADOTROPIN

To maintain the pregnant state, the secretory endometrium cannot slough on day 28. If it were to slough, the imbedded embryo would be lost, and pregnancy would be terminated. To prevent the endometrium from sloughing, the corpus luteum does not undergo regression on day 28. Instead, it continues to function through the first three months of pregnancy (*first trimester*) under the control of *Human Chorionic Gonadotropin* (*hCG*). During the first trimester, hCG also stimulates the Leydig cells in the embryonic male testes to secrete testosterone. Testosterone secretion during this early period stimulates the development of the male reproductive tract.

1. Functionally, hCG is similar to LH, and can thus take its place to support the corpus luteum. Levels of hCG rise early in gestation, peak by 60 days after the last menstruation (about the second month), and fall by week 10 of gestation. For the remainder of pregnancy, hCG levels remain low, but detectable **(Figure 16.18)**. (Note that hCG is the substance measured in the pregnancy test.)
2. Immediately upon implantation, the trophoblast cells in the embryo begin to secrete *hCG*.
3. Supported by hCG, the corpus luteum continues to secrete rising levels of both progesterone and estrogen for the next th ree months **(Figure 16.18)**.
4. By the beginning of the fourth month of pregnancy, the placenta has developed sufficiently to secrete increasing concentrations of estrogen and progesterone (as well as other hormones).
5. Although the corpus luteum remains in the ovary for the entire period of pregnancy, its function is no longer necessary.

ESTROGEN AND PROGESTERONE

Estrogen and progesterone levels rise during the luteal phase and remain elevated to the end of gestation. The elevated levels of these hormones suppress the release of LH and FSH from the pituitary. This prevents maturation and ovulation of any new follicles during the entire period of gestation. Although the systemic and reproductive actions of estrogen and progesterone have already been described in detail in this chapter, here is some additional information to consider:

1. Estrogen (E), acting with progesterone (P), is at least partly responsible for preventing the immune system of the mother from rejecting the fetus, which is actually a foreign "tissue transplant" because it expresses paternal antigens.

2. Progesterone acts on the myometrium to suppress contractions. Since progesterone and estrogen have an opposite effect on myometrial excitability, the P/E ratio plays a significant role in the onset of contractions at parturition (see below). Progesterone also stimulates the formation of the cervical mucus plug. This plug closes off the uterus during pregnancy and prevents bacterial contamination and infection.

Human Chorionic Somatomammotropin (HCS). HCS is also called *Human Placental Lactogen* (*HPL*). This placental hormone assists in preparation of the mammary glands for lactation. It also assists in maintaining a positive protein balance, mobilizes fat to generate energy, and elevates

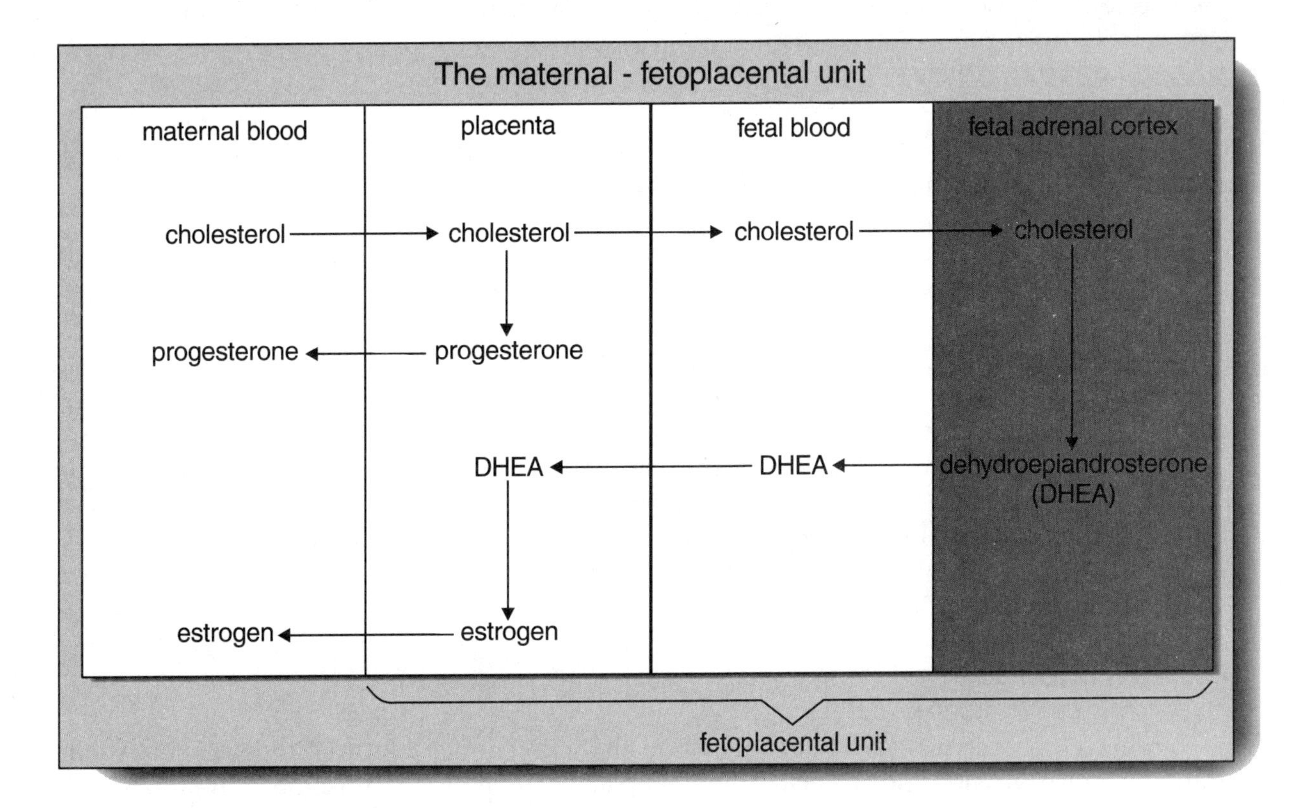

FIGURE 16.19

and stabilizes blood glucose levels to supply the needs of the fetus. The hyperglycemic (insulin-antagonistic) effects of HCS on maternal plasma glucose explains why some women experience gestational diabetes.

Relaxin. It prepares the cervix and pelvis for labor, and softens the cervix making it able to dilate; it also softens the connections between the bones of the pelvis, making these joints more pliable. This allows the pelvic bones to stretch apart during labor, allowing the fetal head to pass through. Relaxin may also act on other joints of the body making them feel more elastic. Relaxin is secreted by the corpus luteum.

Prolactin. This secretion from the anterior pituitary increases during pregnancy, due to elevated levels of estrogen. Prolactin assists in preparation of the mammary glands for lactation. However, during pregnancy, no milk is produced by the glands because estrogen blocks the effect of prolactin on mammary tissue.

EFFECTS OF PREGNANCY ON OTHER HORMONES

1. Vasopressin levels increase
2. Secretion of aldosterone and cortisol increases
3. ACTH secretion increases
4. Secretion of Parathyroid Hormone increases
5. Secretion of increased levels of renin, erythropoietin and 1,25-dihydroxyvitamin D_3

FETAL PLACENTAL UNIT (FPU)

While the placenta can directly synthesize progesterone, synthesis of estrogen by the placenta requires the assistance of the fetal adrenal gland **(Figure 16.19)**.

1) Cholesterol passes from the maternal blood into the placenta where it is converted by a placental enzyme pathway into progesterone. The progesterone passes back into the maternal blood to mediate various systemic and reproductive effects.
2) Cholesterol also passes from the maternal blood, to the placenta, to the fetal circulation, and finally into the fetal adrenal cortex. Here it is converted into the androgen, Dehydroepiandrosterone (DHEA).
3) DHEA passes into the fetal circulation, and from there back into the placenta. In the placenta DHEA is converted into estrogen, which enters the maternal blood to mediate various systemic and reproductive effects. Since the primary estrogenic compound synthesized by the FPU is estriol (not estradiol), measurement of estriol in the urine of the pregnant women can be used to clinically ascertain the viability of the fetus.

MORNING SICKNESS

This common clinical sign of early pregnancy, is usually first noticeable after implantation. While the underlying causes of morning sickness are incompletely understood, it has been hypothesized that hCG may play a role. Possibly, hCG acts on the chemoreceptors in the hypothalamic vomit center to trigger this response. Another suggestion is that this effect is related to fluctuating blood glucose levels in early pregnancy, which stabilize after HCS levels become elevated.

SYSTEMIC CHANGES DURING PREGNANCY

The pregnant physiological state is quit different from the nonpregnant physiological state. During pregnancy a new homeostatic equilibrium must be established because the fetus acts as a sort of "parasite," and must be supported by the mother.

1. Salts and water are retained, edema may develop. This results from elevated levels of estrogen, vasopressin (ADH) and aldosterone.
2. Enlargement of mammary glands takes place transiently during pregnancy. This results from elevated levels of estrogen, progesterone, prolactin and HCS.
3. Blood volume increases as a result of the action of hormones that increase salt (aldosterone) and water (ADH), and erythrocytes (erythropoietin).
4. Blood pressure normally stays constant, although cardiac output increases. This results from a drop in total peripheral resistance, through vasodilation in the uterus, skin, mammary glands, GI tract and kidneys. The overall effect is to increase circulation (but not change blood pressure).
5. Respiration increases and pCO_2 decreases.
6. Metabolic rate increases. Plasma glucose increases due to increased levels of HCS and cortisol which act as insulin antagonists. These hormones also cause an increase in gluconeogenesis and fatty acid mobilization.
7. Calcium balance becomes positive. This results from increased levels of parathyroid hormone and 1,25-dihydroxyvitamin D_3.

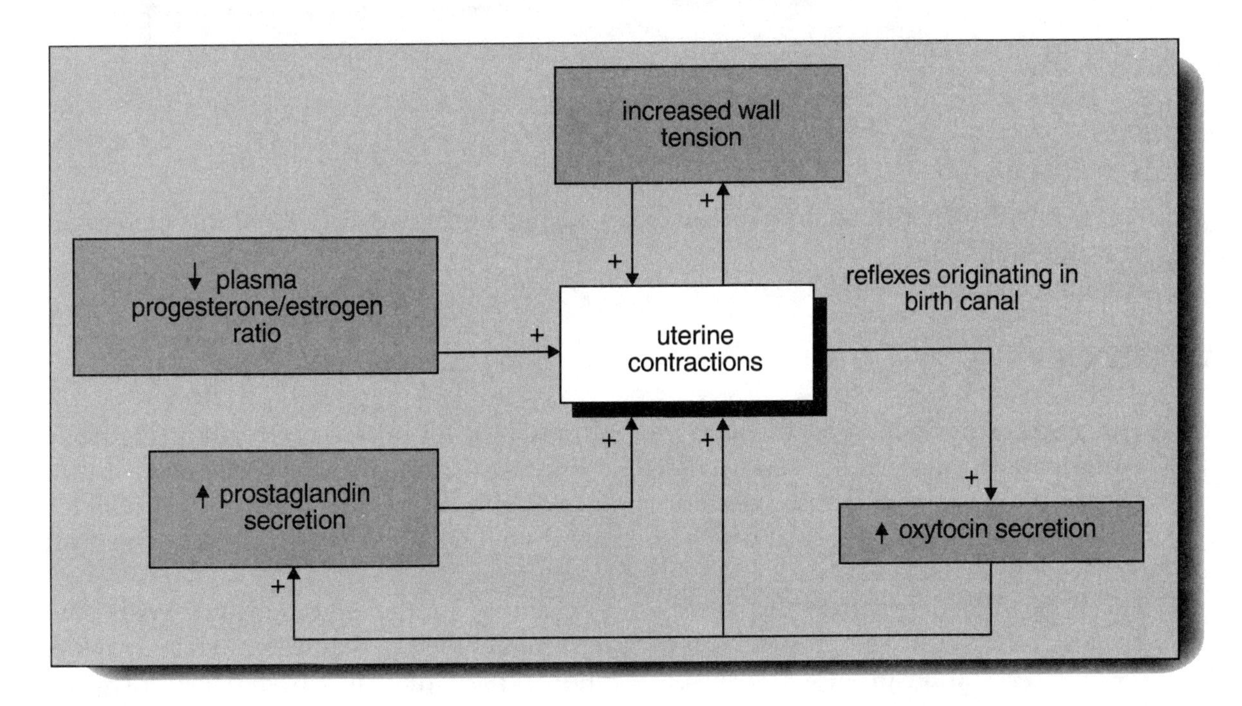

FIGURE 16.20

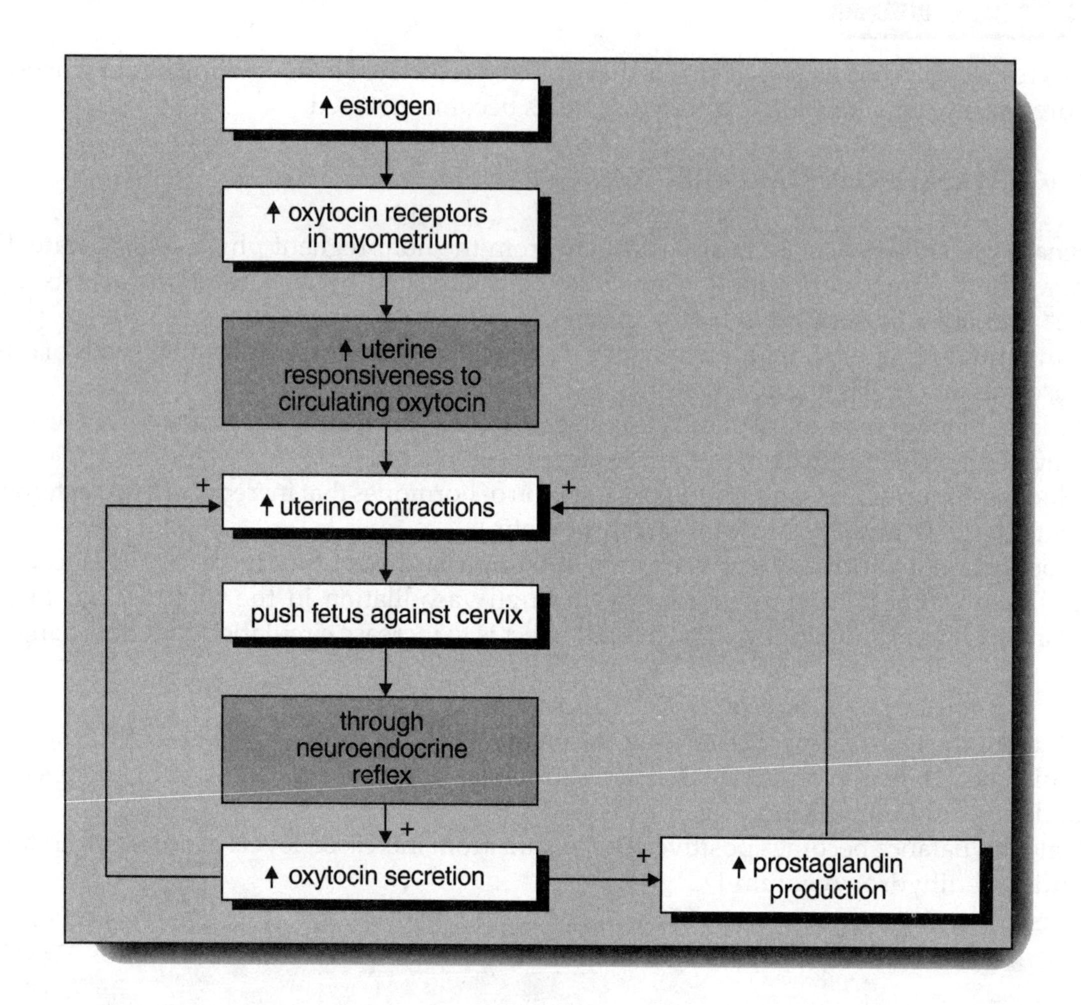

FIGURE 16.21

8. Body weight increases on the average by about 12.5 kg. However, 60% of this increase is due to water retention.
9. Appetite and thirst both increase.

PARTURITION

Various physical and hormonal factors play a role in stimulating the onset of parturition **(Figure 16.20)**.

1) When smooth muscle cells are stretched their intrinsic contractility increases. Since the uterine myometrium is also composed of smooth muscle, with enlargement of the fetus, wall tension increases. By the end of gestation, the increased wall tension reaches a point where uterine contractions are stimulated.
2) The P/E ratio determines the intrinsic contractility of the myometrium. With elevated progesterone v. estrogen, the smooth muscle is maintained in a quiescent state. However, at the end of gestation, progesterone levels fall more rapidly than do estrogen levels. This primes the myometrium to become more intrinsically contractile.
3) Oxytocin and Prostaglandin play a central role in labor **(Figure 16.21)**. Elevated levels of

estrogen stimulate the formation oxytocin receptors in the uterine myometrium. Thus, the myometrium becomes increasingly responsive to the presence of oxytocin.
a) As uterine contractions begin, the fetal head engages with, and presses into the cervix. This activates a neuroendocrine reflex that releases oxytocin from the posterior pituitary.
b) As oxytocin levels rise, uterine contractions increase. This promotes further positive feedback via a neuroendocrine reflex. Thus myometrial contraction will increase significantly in frequency, strength and duration, leading to expulsion of the fetus.
c) Oxytocin can act directly on the uterine myometrium to stimulate contractions. However, it can also release prostaglandins from the uterine tissues that will, in turn, also act on the uterine myometrium to increase contractions.

INTERACTIONS OF THE FETAL-PLACENTAL UNIT

In sheep, increasing stress of the fetus, four weeks prior to parturition, results in the release of increasing levels of cortisol from the fetal adrenal. The increased cortisol acts on enzymes in the placenta to promote an increase in synthesis of estrogen and a decrease in synthesis of progesterone. As estrogen dominates the P/E ratio, contractions are initiated. While this pathway has not as yet been directly demonstrated in humans, it makes biological sense because it allows the fetus to signal the mother when it has reached the appropriate stage of maturity, and is ready to be born.

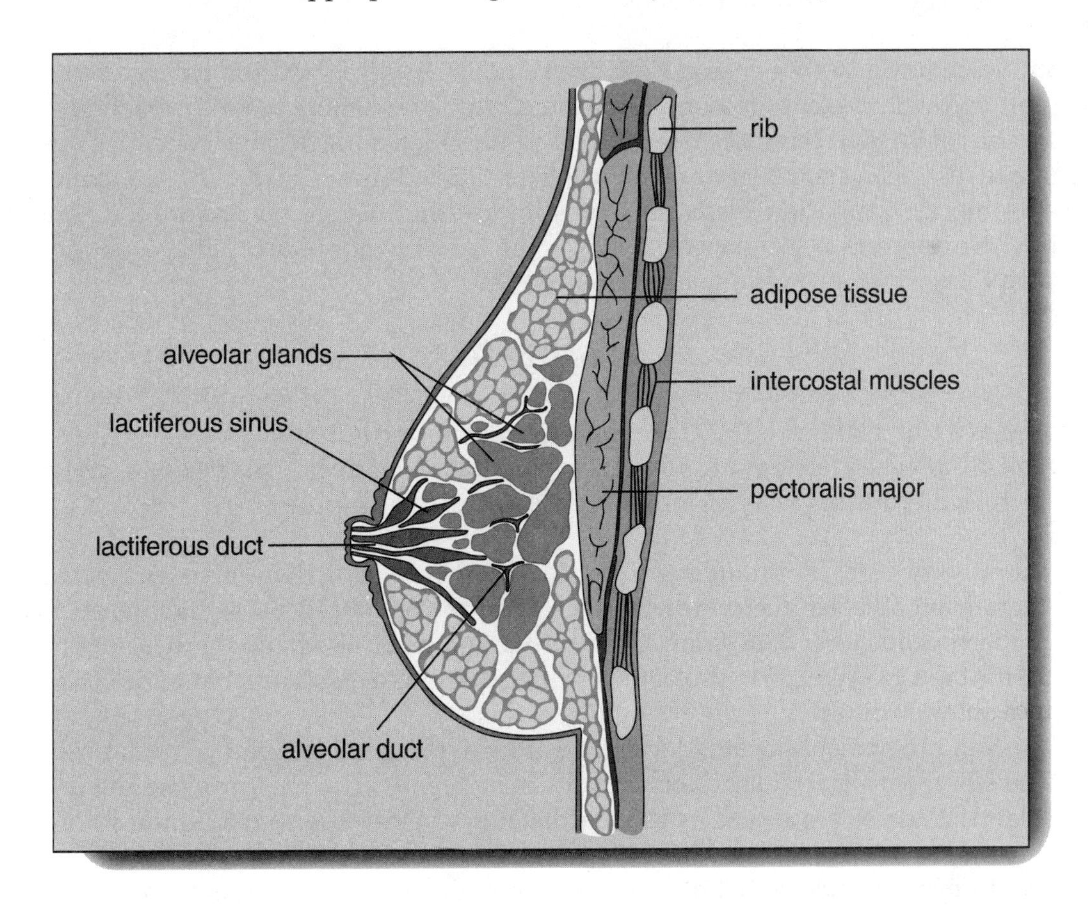

FIGURE 16.22

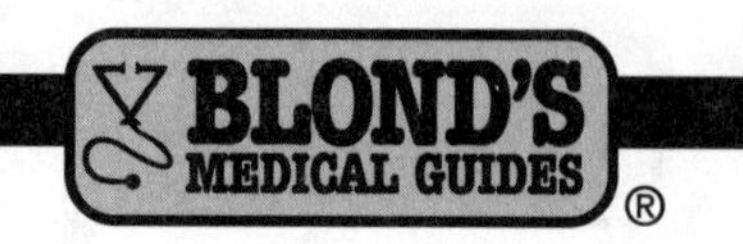

STAGES OF LABOR

There are three stages in labor:

1. **Stage I — which ends with complete cervical dilation.** Maximum dilation is about 10 cm which allows the baby's head to pass through. Stage I is also the longest of the three stages. It may last for as long as 24 hours, if this is the first pregnancy.
2. **Stage II — during which the baby is delivered.** During this stage, the mother can utilize her abdominal muscles to help push the baby out of the birth canal. This stage usually lasts about 30 to 90 minutes.
3. **Stage III — during which the placenta is delivered.**

After the baby is born, the placenta separates from the wall. Additional uterine contractions now cause the placenta (*afterbirth*) to be expelled.

POSTPARTUM PERIOD

The postpartum period lasts about six weeks. During this time all of the body systems are returning to the nonpregnant state. At the time of parturition, the placental hormones are rapidly withdrawn and many tissues experience severe reactions.

For example, removal of progesterone, estrogen, and HCS will affect fluid balance between compartments, salt and water retention, nutrient balance, cardiac function, blood pressure, respiration and appetite. The central nervous system is especially affected, having become dependent on the elevated levels of both estrogen and progesterone during gestation. It is not surprising, therefore, that women experience mood swings and bouts of depression during the postpartum period. One wonders if women who are prone to premenstrual syndrome may be more susceptible to severe postpartum depression then women who do not experience PMS.

LACTATION

The mammary glands **(Figure 16.22)** are comprised of many individual sac-like glands called *alveoli*. The alveoli are attached to many alveolar ducts. These alveolar ducts converge to a series of lactiferous ducts. Each lactiferous duct has an individual opening at the nipple.

1) At puberty, estrogen stimulates the growth and branching of the alveolar ducts. However, no significant alveolar gland growth takes place at this time. Most of the breast enlargement at puberty is due to fat deposition. Cyclic progesterone levels after puberty also stimulate growth of the breasts. At puberty prolactin levels also rise, as a result of direct estrogen stimulation on the anterior pituitary.
2) During pregnancy, elevated levels of estrogen, progesterone and prolactin cause significant development of alveolar ducts and alveolar glands. During early pregnancy the alveolar glands begin to grow, and by the middle of pregnancy they have significantly enlarged and lumen have formed.
3) Additional hormones that also play a role in development of the mammary glands during pregnancy are: *Human Chorionic Somatomammotropin, Growth hormone, and Thyroid hormone.*

4) Prolactin levels are elevated during pregnancy, beginning by the eighth week, due to the elevated levels of estrogen acting directly at the anterior pituitary to release prolactin. However, although the mammary glands are fully developed by the end of pregnancy, no milk is produced because estrogen (and progesterone) inhibit prolactin action on the mammary tissue.
5) At the termination of pregnancy, estrogen levels fall. In the absence of the blocking effect of estrogen, prolactin now stimulates milk formation by the glandular tissue.
6) Although prolactin levels fall after parturition, nipple stimulation during suckling releases prolactin from the anterior pituitary.
7) For the first five days postpartum, the substance secreted by the mammary glands is *colostrum*. In comparison to milk, colostrum contains lower concentrations of lactose and fat, but higher concentrations of the proteins *lactoferren and immunoglobulins*.

8) Both lactoferren and immunoglobulins protect the newborn against many bacterial diseases because:

a) Lactoferren is a bactericidal protein.

b) Immunoglobulins are antibodies. IgG and IgA are the classes of antibodies found in these secretions.

c) In addition to the above, breast milk may also actively stimulate the newborn's immune system, helping it to mature.

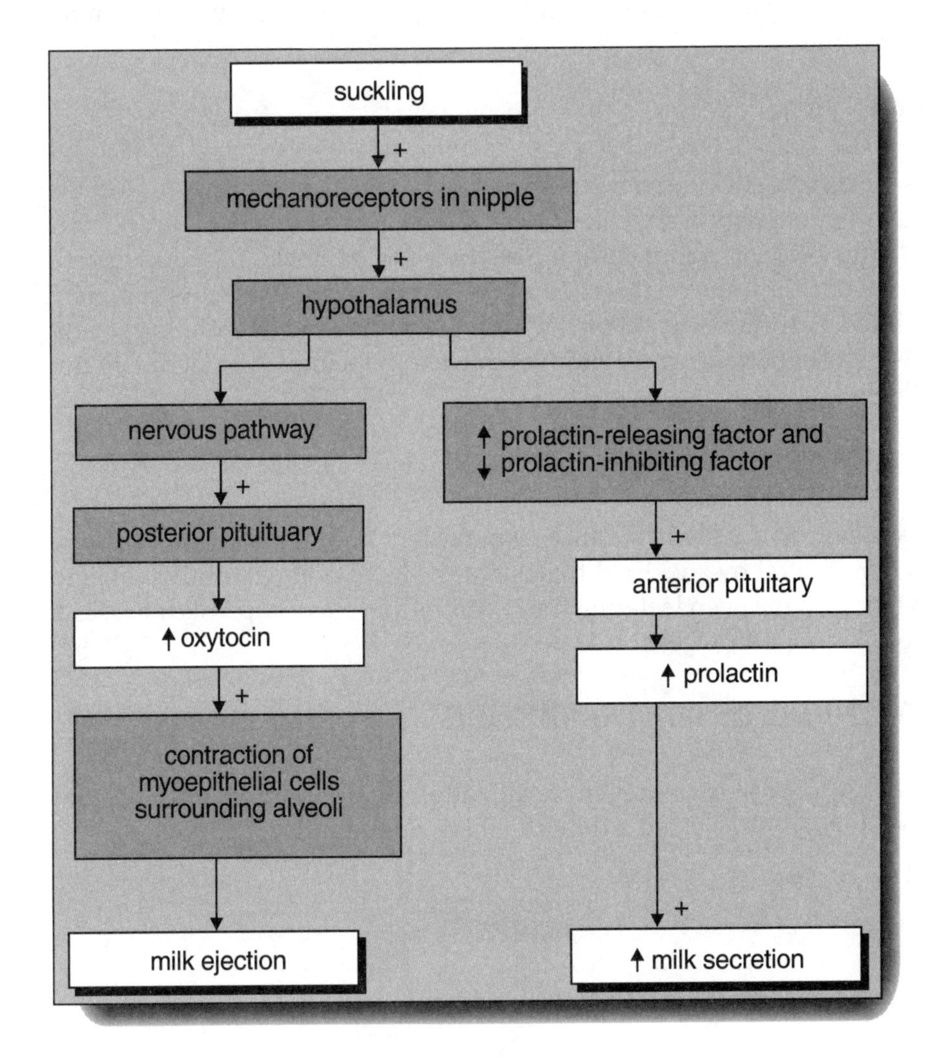

FIGURE 16.23

PROLACTIN-RELEASING REFLEX

This reflex is initiated by suckling **(Figure 16.23)** which stimulates nipple mechanoreceptors. Nerve pathways are activated and this stimulates hypothalamic nuclei. PRF (prolactin releasing factor) is stimulated, and possibly PIF (prolactin inhibitory factor; actually dopamine) may be inhibited. The anterior pituitary releases prolactin, which returns to the mammary glands and stimulates milk secretion.

OXYTOCIN-RELEASING REFLEX

This reflex is initiated by suckling **(Figure 16.23)**, which stimulates nipple mechanoreceptors. Nerve pathways are activated and this stimulates hypothalamic nuclei. The hypothalmo-hypophyseal nerve tracts are stimulated and this causes the release of oxytocin from the posterior pituitary.

Oxytocin returns to the mammary glands and causes the contraction of the myoepithelial cells that line the lactiferous sinuses and ducts. The sinuses and ducts contract and milk is forced out of the nipple. This is called *milk let-down*.

During suckling, oxytocin also circulates to the uterus causing strong myometrial contractions. These contractions are helpful in causing the uterus to return to its nonpregnant size. These strong uterine contractions are also initially accompanied by menstrual-like bleeding. Eventually, the oxytocin receptors present in the myometrium disappear (without estrogen support) and the myometrium no longer contracts in response to suckling.

FACTORS THAT AFFECT LACTATION

Milk production will continue as long as the suckling reflex is present. However, once the woman stops breast feeding, milk production will cease within a week or two, as PIF is reestablished and (PRF is inhibited). Because the hypothalamus controls the release of prolactin and oxytocin, nervous stimuli originating from higher centers can also impinge on this regulation. For example, extreme stress may cause milk production to be inhibited. Additionally, women who are lactating and hear the cry of any baby—not necessarily their own—may experience milk letdown due to oxytocin release mediated not by suckling, but by cerebral influence.

LACTATION AND CONTRACEPTION

In lactating women, prolactin usually, but not always, causes a persistent anovulation. This effect is mediated by surges of prolactin (stimulated by suckling) that act at the level of the hypothalamus to inhibit GnRH secretion. In women who have been lactating for 3 months and then stop breast feeding, the average time to reestablish ovulation is 17 weeks.

METHODS OF CONTRACEPTION

Contraception is any process whose purpose is to prevent fertilization of an ovum, and thus avoid pregnancy. A variety of contraceptive methods exist, although all are not equally effective.

Effectiveness of Various Contraceptive Methods		
Method	**Failure Rate**	**Comments**
Chance (no protection)	90.0%	Dumb, unless you are trying to get pregnant!
Abstinence	0	Commendable, but difficult for most people.
Tubal ligation	Less than 0.5%	Effective, but difficult or impossible to reverse.
Vasectomy	Less than 0.5%	Same as above.
Combined birth control pills	.5% — 3.0%	Minor side effects, but some serious health problems in a small percentage of users.
Contraceptive implants	0.5% — 2.5%	Same as above. Is effective for up to 5 years.
IUD	4.5% — 6.0%	May cause increased bleeding. Effective for 1 year.
Condom	9.5% — 12.0%	Provides increased protection against sexually transmitted diseases, including AIDS.
Diaphragm + spermicide	14.5% — 19.0%	Can be inserted up to 2 hours before intercourse.
Creams, foams, jellies and suppositories	12.0% — 21.0%	Must be applied no more than 1 hour before intercourse.

Effectiveness of Various Contraceptive Methods (cont.)

Method	Failure Rate	Comments
Vaginal sponge	10.0% — 15.0%	Easy to insert, but may be difficult to remove, and may fragment.
Coitus interruptus (penis withdrawal)	6.5% — 23.0%	Definitely not the method of choice. (Also see the first comment in this chart.)
Rhythm method (natural family planning)	15.5% — 24.0%	Difficult to apply effectively because of problems in determining the exact time of ovulation.

TUBAL LIGATION

In the female, the uterine tubes are cut and tied. This prevents the sperm from fertilizing the oocyte. The oocyte degenerates naturally. A tubal ligations does not affect female hormones or female sex drive. This is a relatively permanent method and should only be considered if the individual is certain that she wants no more children. While some tubal ligations can be reversed (a *reanastomosis* can be preformed), the potential for becoming pregnant is significantly reduced.

VASECTOMY

With the male, the vas deferens are cut and tied. This prevents the sperm from exiting the male reproductive tract during ejaculation. A vasectomy does not effect male hormones or male sex drive. This is a relatively permanent method and should only be considered if the individual is certain that he wants no more children.

In an individual with a vasectomy, sperm cells continue to be manufactured by spermatogenesis, but cannot get out of the epididymis. These sperm apparently degenerate, and are removed by immunological processes. Some researchers have suggested that this presents excessive autoantigen to the individuals immune system, and could result in autoimmune disease. However, to date, there is no convincing evidence to support this conjecture.

SEQUENTIAL AND COMBINED BIRTH CONTROL PILLS

Birth control pills function by convincing the follicle not to ovulate, while convincing the uterine endometrium that hormonally, all is normal.

These pills are formulated with low levels of a synthetic estrogenic analogue, accompanied by an additional synthetic progestational analogue. They can be taken for three weeks either sequentially (estrogen, then progesterone) or in some combination (various concentrations of estrogen plus progesterone). The pills are then withdrawn for an additional week (or sugar pills are substituted), and then the procedure is repeated.

Most women who use birth control pills experience only minor side effects, similar to those of early pregnancy, during the first three months of use. However, a very limited number may experience major complications, such as blood clotting disorders and hypertension. Women who take birth control pills should be seen regularly by a physician.

MECHANISM OF ACTION

1. Menstruation proceeds for the first 7 days. Estrogen (or combination) containing pills are then taken for the next 7 days. The levels of the estrogen in these pills are relatively low (especially in the combined version) but remain constant (unlike the natural state where estrogen would slowly be increasing.
2. After day 14, the progesterone (or combination) containing pills are taken through day 28.
3. The sugar pills are taken for the next week.

As a result of the steroid hormone regime outlined above, the individual: first, cannot grow a follicle; second, cannot ovulate a follicle; and third, the uterine endometrium cycles normally. This is because:

1) The constant level of estrogen replacement acting at the hypothalamus and anterior pituitary (tonic center) inhibits the release of FSH. Therefore a follicle cannot grow to maturity.
2) The constant level of estrogen replacement does not prime the cyclic center (hypothalamus and anterior pituitary). This means that there will not be an LH surge on day 14. Thus, even if a follicle were to grow, it could not ovulate.
3) Estrogen replacement for the first half of the cycle, followed by progesterone (+ estrogen) for the second half of the cycle, stimulates the uterine endometrium to first proliferate and then enter the secretory phase.
4) As in the normal condition, withdrawal of the steroid support on day 28 causes the endometrium to slough, with associated menstrual flow.

PROGESTATIONAL PILLS

Cervical mucus secretions are affected by estrogen and progesterone. If a progesterone-containing pill (*minipill*) is given, it will increase the viscosity of the cervical mucus and prevent sperm penetration. Progestational analogues may also reduce the contractions of the oviducts which will reduce sperm transport. Progestational pills also inhibit the priming of the cyclic center, and thus no LH surge takes place on day 14. Minipills may produce nausa and irregular bleeding in some women. However, given time these side effects usually disappear.

CONTRACEPTIVE IMPLANTS AND LONG-ACTING INJECTABLES

Silastic implants containing a synthetic progestational compound can be inserted into the muscle of the upper arm. These implants release the steroid for up to 5 years. The effect is to inhibit ovulation by mechanisms similar to that described above for the minipill. Side effects include nausea and irregular bleeding.

However in the majority of women, these side effects will disappear in a few months. Implants can be removed before the end the of 5-year period if the woman plans to become pregnant. Synthetic progestational steroids can also be injected and function as describe above, but only for a period of about 3 to 6 months.

IUD

The Intrauterine Device is inserted into the uterus and remains in place for about 1 year. Its appears to act by induction of inflammation in the endometrial lining. This prevents implantation of the embryo, and thus pregnancy is terminated. The IUD should only be used in a women who has already been previously pregnant. In some women, it may cause increased bleeding during menstruation. IUDs may also increase the risk of pelvic inflammatory disease. Furthermore, if a women with an IUD becomes pregnant, there may be a slightly increased risk of the pregnancy being ectopic.

CONDOM

This very old form of contraception also protects against the spread of sexually transmitted diseases including AIDS. Rubber condoms must be placed on the penis immediately before intercourse. One complaint associated with this form of contraception is that it blunts the sexual sensation. This device can break if not used correctly and should be removed immediately after intercourse is completed to prevent leakage of sperm. Condoms are more effective if the woman also uses spermicide.

A new version, the *female condom*, is actually a large size condom that is inserted into the vagina. It is used not only to prevent pregnancy, but also to protect against the spread of sexually transmitted diseases.

DIAPHRAGM PLUS SPERMICIDE

The diaphragm is a barrier placed across the mouth of the cervix. A similar device called a *cervical cap* can be placed over the mouth of the cervix. With either type of device, spermicide must also be used to assure a sperm-free closure. Diaphragms come in different sizes and must be fitted to each women's cervix. These barrier devices can be inserted up to 2 hours before intercourse. Use of the diaphragm may increase the risk of urinary tract infections.

CREAMS, FOAMS, JELLIES AND SUPPOSITORIES

These preparations contain spermicides that are applied to the interior of the vagina to block sperm entrance into the cervix. To be functional, they must be applied just prior to intercourse if possible, and certainly within one hour of intercourse. Some men and women become allergic to these preparations. They work best if used in conjunction with a barrier device, such as a diaphragm or condom.

VAGINAL SPONGE

This is a relatively new device consisting of a sponge-like vaginal insert containing spermicide. It can be inserted up to 24 hour before intercourse. It does not need to be fitted like a diaphragm (one size fits all), and it is disposable.

However, in some women it may be hard to remove, fragmenting in the process. It may also irritate the vaginal mucosa in some women. In a small number of users it has been implicated in cases of toxic shock. It is interesting to note that this device is reported to have a higher failure rate in women who have given birth.

COITUS INTERRUPTUS (PENIS WITHDRAWAL)

Coitus interruptus (as the name implies) involves removing the penis from the vagina, just prior to ejaculation. Although a very old form of contraception, it is not reliable because the male partner may be unable to control his ejaculation. In addition, some sperm may be released prior to the main ejaculatory event.

RHYTHM METHOD (NATURAL FAMILY PLANNING)

This method depends on being able to ascertain exactly when ovulation takes place. With this information the couple abstains during the women's fertile period (at least three days before ovulation to one day after ovulation).

Prediction of the ovulation is dependent on keeping careful records of the cycle, and recording body temperature each morning. A slight surge in body temperature takes place at the time of the LH surge (which is day 14). The temperature-rhythm method can only be used to determine when sex can resume *after* ovulation. It does not tell the couple when to stop having sex *before* ovulation. For this, the women has to depend on records from previous cycles. Since cycles vary, this is not always an accurate method, and the failure rate is 20 to 30%.